Advanced Studies in Urban and Regional Planning

Advanced Studies in Urban and Regional Planning

Guest Editors

Shisong Cao
Yang Shi
Shuo Li
Meizi Yang

Basel • Beijing • Wuhan • Barcelona • Belgrade • Novi Sad • Cluj • Manchester

Guest Editors

Shisong Cao
School of Geomatics and Urban Spatial Informatics
Beijing University of Civil Engineering and Architecture
Beijing
China

Yang Shi
School of Architecture and Urban Planning
Beijing University of Civil Engineering and Architecture
Beijing
China

Shuo Li
School of Architecture and Urban Planning
Beijing University of Civil Engineering and Architecture
Beijing
China

Meizi Yang
School of Architecture and Urban Planning
Beijing University of Civil Engineering and Architecture
Beijing
China

Editorial Office
MDPI AG
Grosspeteranlage 5
4052 Basel, Switzerland

This is a reprint of the Special Issue, published open access by the journal *Buildings* (ISSN 2075-5309), freely accessible at: https://www.mdpi.com/journal/buildings/special_issues/YQ9JQ46GG8.

For citation purposes, cite each article independently as indicated on the article page online and as indicated below:

Lastname, A.A.; Lastname, B.B. Article Title. *Journal Name* **Year**, *Volume Number*, Page Range.

ISBN 978-3-7258-5483-7 (Hbk)
ISBN 978-3-7258-5484-4 (PDF)
https://doi.org/10.3390/books978-3-7258-5484-4

Contents

buildings

MDPI

Article

Study on Community Detection Method for Morning and Evening Peak Shared Bicycle Trips in Urban Areas: A Case Study of Six Districts in Beijing

Yao Sun [1], Zheng Wen [1,2], Dongwei Tian [1,3,*], Man Zhang [1] and Yue Hou [4]

1 School of Architecture and Urban Planning, Beijing University of Civil Engineering and Architecture, Beijing 100044, China; sunyao@bucea.edu.cn (Y.S.); wenzheng_bucea@163.com (Z.W.); zhangman@bucea.edu.cn (M.Z.)
2 School of Geomatics and Urban Spatial informatics, Beijing University of Civil Engineering and Architecture, Beijing 100044, China
3 Research Center for Urban Big Data Applications, Beijing University of Civil Engineering and Architecture, Beijing 100044, China
4 Beijing Tsinghua Tongheng Urban Planning & Design Institute, Beijing 100085, China; hy840329@163.com
* Correspondence: tiandongwei77@163.com

Abstract: Examining the clustering characteristics and fluctuations within urban areas during peak hours through the lens of bike-sharing is of utmost importance in the optimization of bike-sharing systems and urban transportation planning. This investigation adopts the principles of urban spatial interaction network construction and employs streets as the fundamental units of analysis to model bike-sharing activities during morning and evening peak hours within Beijing's six central districts. Subsequent to this, a comprehensive analysis of the network's structural attributes was carried out. A Walktrap method, rooted in modularity analysis, was introduced to discern and scrutinize the clustering patterns and characteristics of communities within the network across different temporal intervals. Empirical findings reveal a predominant usage pattern of shared bicycles for short-distance travel during both morning and evening peak hours. Notably, distinctive community structures manifest during these periods, characterized by two large communities and multiple smaller ones during the morning peak, while the evening peak showcases a single large community alongside several medium-sized and smaller ones. Moreover, the extended interaction radius points to an expanded geographic range of interactions among streets. These findings bear significant implications for the management of urban transportation, bike-sharing enterprises, and urban residents, proffering valuable insights for the optimization of bike-sharing schemes and transportation strategies. These research findings not only contribute to enhancing urban transportation planning and bike-sharing systems but also provide robust guidance for advancing more efficient and sustainable urban transportation solutions, thereby fostering the sustainable development of cities.

Keywords: spatial interaction network; community detection; bike-sharing; urban mobility; Walktrap method

1. Introduction

In recent years, urban bike-sharing has undergone significant development and has emerged as a prominent sustainable and convenient mode of transportation in cities globally [1]. The rapid expansion of urban bike-sharing programs has resulted in a substantial increase in the proliferation of bike-sharing stations and users [2]. This expansion has yielded positive effects, encompassing the reduction of traffic congestion, curbing carbon emissions, fostering healthier lifestyles, and affording urban residents access to affordable and adaptable transportation alternatives [3]. Nonetheless, concomitant with the widespread adoption of bike-sharing systems, an array of challenges and concerns have surfaced, impinging on their operational efficiency and overall efficacy. These challenges encompass imbalanced bike distribution among stations, the overutilization of popular stations leading to bike shortages, incomplete coverage within specific communities, as well as considerations concerning bike maintenance and parking [4,5]. As a nascent manifestation of internet-based transportation, bike-sharing systems are conventionally furnished with positioning devices capable of real-time location recording, thereby furnishing an abundance of invaluable data amenable to analysis [6]. To surmount these challenges and optimize bike-sharing systems, a multitude of investigations have concentrated on comprehending the behavior and patterns characterizing bike-sharing travel. Within this context, the present study aims to comprehensively investigate the clustering characteristics and patterns of bike-sharing travel within the urban core sub-regions, utilizing an established foundation of existing research. Employing spatial interaction network analysis techniques, this study delves into the travel patterns of bike-sharing during the morning and evening peak hours. The ultimate objective is to provide recommendations for urban transportation planning, bike-sharing operations, and residents' travel preferences.

The utilization of bike-sharing among urban residents has resulted in diverse levels of spatial interactions spanning various areas within the city, constituting what is termed as spatial interactions [7]. The scrutiny of this phenomenon is approached by constructing a spatial interaction network encompassing urban zones, grounded in the principles of complex network theory [8–10]. The realm of network science has undergone rapid advancements since the 1970s [11]. As posited by the complex network theory, individuals' travel behaviors between regions can be harnessed to establish a prototypical two-dimensional weighted spatial network, recognized as the spatial interaction network. In this network framework, regions are represented as nodes, travel activities as edges, and edge weights denote the magnitude of travel between points, abstracting the overarching travel interactions among all regions. Once the network is meticulously constructed, structural network metrics and statistical models can be applied to qualitatively and quantitatively assess the interplay between travel behavior and regions, thereby evaluating the spatial interconnections between these regions. Numerous researchers have harnessed the tenets of complex network theory for relevant inquiries. To illustrate, Beck et al. devised a bicycle network to probe the interrelationship between environmental characteristics and cycling behavior [12]. Liu et al. evaluated the accessibility of dockless bike-sharing from a network vantage point, proffering decision-making support for urban planners, policymakers, and bike-sharing providers to fine-tune bike-sharing systems [13]. Hu et al. delved into the travel patterns of bike-sharing by constructing a spatial interaction network and juxtaposed the disparities between weekdays and weekends [14]. While prior investigations have predominantly centered around comprehensive analyses and examinations of bike-sharing travel patterns, scant research has ventured into the clustering characteristics and systematic attributes of bike-sharing travel regions. To bridge this research lacuna, this study leverages community detection methodologies rooted in complex network theory to scrutinize the clustering attributes of bike-sharing travel regions, subsequently investigating clustering patterns grounded in modularity analysis.

Community structure is an essential and distinctive feature within the realm of complex networks [15]. Investigating and discerning community structures from a network-oriented perspective facilitates an in-depth comprehension of the clustering patterns inherent in the network. This process, in turn, unveils the intricate interconnections interwoven among distinct sub-regions within urban environments, thereby unraveling the intricate tapestry of travel patterns. Such an investigation holds profound ramifications for the judicious deployment of bike-sharing initiatives, the efficacy of urban transportation management, and the formulation of strategic travel plans for urban inhabitants. Within the intricate fabric of complex network theory, the subject of community detection has enjoyed a longstanding position of prominence [16]. In the context of urban settings, a multitude of scholars have embarked on probing the spatiotemporal clustering phenomena through the prism of community identification methodologies. For instance, the work of Chen W and

Wei C leverages the Infomap algorithm to delineate urban agglomerations and economic regions within city networks [17,18]. Fangyu and colleagues advance the field by harnessing an improved BBO algorithm, specifically tailored to unearth community structures inherent in intelligent city network systems [19]. Similarly, Gao P et al. harness the efficacy of the fast greedy algorithm to partition functional domains within the urban agglomeration of the middle reaches of the Yangtze River [20]. Parallel to these methodologies, a cadre of classic community detection algorithms, such as the GN algorithm [21], the Label Propagation Algorithm (LPA) [22], the Louvain algorithm [23,24], and the Walktrap algorithm [25,26], stand as cornerstone tools for such research endeavors. Nonetheless, despite these scholarly advancements, the existing landscape of community detection research predominantly gravitates towards urban agglomerations, leaving a perceptible gap in our understanding of community clustering dynamics within the nucleus of cities' core functional regions. The core areas within urban centers constitute vital components that prominently reflect urban characteristics. Presently, numerous urban planning and transportation studies have designated the urban core regions as focal points of their investigations [27–30]. This underlines the pressing need for studies that uncover and elucidate community structures within these essential urban zones.

Furthermore, the classification of urban traffic based on its temporal distribution yields distinct categories: daily traffic (weekday traffic), weekend traffic, and other types of traffic [31]. Among these categories, weekdays stand out as a prominent temporal segment characterized by an extensive duration, a significant volume of travel, and pronounced travel attributes, thus warranting the highest degree of attention and research scrutiny [32]. Investigating weekday traffic bears the potential to yield a profound comprehension of urban residents' habitual travel behaviors and travel patterns. The morning and evening peak hours on weekdays exhibit the most concentrated travel flow within a day. This heightened concentration of travel during peak periods imparts distinct characteristics to travel patterns, rendering them more reflective of actual travel behaviors. Consequently, these patterns hold significant research implications and value. Within this particular traffic category, individuals often confront a myriad of tasks, including work, education, and shopping, resulting in a complex and dynamic interplay of demands and usage patterns within the transportation system [33]. A substantial body of research has already unveiled the peak travel times [34], primary travel attributes [35], and popular travel destinations [36] associated with weekday traffic. These research findings form pivotal cornerstones for urban transportation planning and management, offering crucial underpinnings to alleviate congestion, enhance transportation efficiency, and ameliorate the overall travel experience of urban inhabitants [37].

Currently, urban vitality research extensively employs multi-source urban data, yet the analysis units often revolve around blocks or parcels. However, residents' economic, social, and cultural activities primarily congregate around streets and their proximate areas [38]. By evaluating the urban vitality of streets, the quality of urban streets can be enhanced, thereby infusing urban areas with vibrancy and livability, fostering organic urban renewal and driving sustainable development [39]. In this study, streets are adopted as the fundamental analytical units, and the travel OD flow of shared bicycles between streets during morning and evening peaks is extracted and distributed. The Walktrap community detection algorithm is employed to explore street clustering characteristics, and an analysis and summary of clustering patterns are conducted based on modularity changes. Furthermore, a comparative analysis of the differences between the morning and evening peak periods is undertaken. By employing streets as the analytical units, the study aims to reveal the intricate spatial patterns of shared bicycle travel aggregation within streets while identifying correlations between streets and shared bicycle travel. This approach provides recommendations and references for urban transportation planning, shared bicycle operations, and residents' travel choices [39,40].

The remainder of this paper is organized as follows. Section 2 describes our study area and data. Section 3 introduces the Walktrap community detection algorithm and the steps for modularity analysis. Section 4 explores the regional clustering characteristics and patterns from the perspective of urban bike-sharing travel during the morning and evening peak hours. Section 5 discusses the paper's findings. Section 6 provides a concise summary of the paper.

2. Study Area and Data

2.1. Research Data Study Area

This study centers on exploring the central area of Beijing, encompassing Dongcheng District, Xicheng District, Haidian District, Fengtai District, Shijingshan District, and Chaoyang District. Known as the heart of the city, this region boasts a dense population, accounting for approximately 60% of Beijing's total inhabitants, and hosts nearly 70% of its industrial establishments. Its strategic location and economic significance make it a hub of activity, resulting in high volumes of pedestrian traffic, especially during peak hours. In addition to the bustling foot traffic, shared bicycles play a crucial role in the transportation landscape within this area.

The focus of our investigation lies in the intersection of the operational area of shared bicycles with Beijing's central region. By narrowing our scope to this specific area, we aim to uncover valuable insights into the spatiotemporal dynamics of shared bike usage in an environment characterized by heavy pedestrian movement and economic activity. Analyzing the patterns and trends of shared bike utilization within this context will offer significant implications for urban transportation planning and management. For clarity and reference, Figure 1 provides a visual representation of the study area's geographical extent within Beijing. This visualization aids in understanding the spatial context and serves as a vital reference point for subsequent analyses.

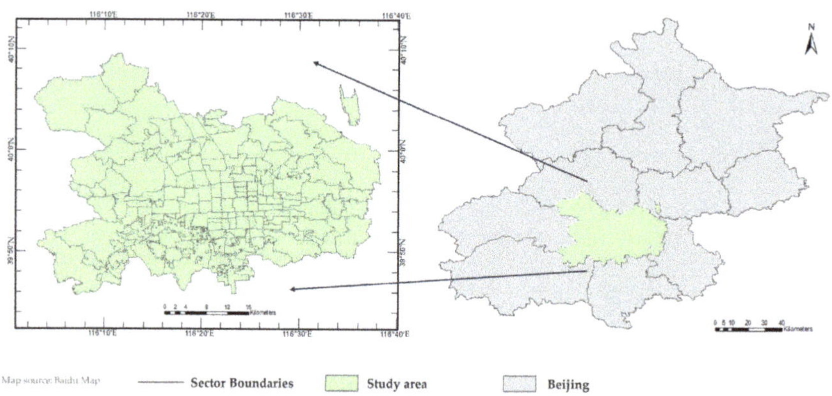

Figure 1. The study area in Beijing.

2.2. Data Sources and Preprocessing

This study utilizes shared bicycle data from the central area of Beijing, collected during a continuous and complete period of five consecutive working days, specifically from 26 July to 30 July 2021. The data comprise real-time location point data for all shared bicycles at the current moment. The data include the timestamp of data collection, an ID number, and location information. It is important to note that different bicycles do not share the same ID number, as illustrated in Table 1. The data collection intervals correspond to the morning and evening peak hours on weekdays, which are defined as 6:00 to 9:00 and 17:00 to 20:00, respectively. Defining the data intervals in this manner allows for the inclusion of a broader range of trips, preventing potential extensions of peak periods due to exceptional circumstances. It also ensures consistency in the duration of both morning and evening peak periods. Therefore, data are collected within two specific time intervals each day: 6:00 to 9:00 and 17:00 to 20:00.

Table 1. Table of data elements.

Data Name	Data Size	Data Field Content	Data Field Name
Bicycle Sharing Data Points	370,000 entries	Data Collection Time Bicycle ID number Latitude/° Longitude/°	TIME BICYCLE_ID LATITUDE LONGITUDE

After data collection, it is essential to proceed with data cleaning and preprocessing. Initially, data anomalies, such as duplicates or other irregularities arising from data acquisition, should be removed. Subsequently, the data in CSV format will be converted into the Shapefile format, consistent with zoning data, to facilitate spatial intersection with the zoning data.

In this investigation, the time frame from 6:00 to 9:00 was established as the commencement and conclusion of the morning peak period, whereas the interval spanning 17:00 to 20:00 was designated as the initiation and culmination of the evening peak period. The spatial alignment of bicycle location data with street zoning data was performed for every timestamp, facilitating the identification of the specific streets harboring bicycles during both the morning and evening periods. Subsequently, a linkage between data points for each timestamp was established through the unique bicycle ID number, culminating in the generation of an origin-destination (OD) matrix delineating bicycle trips amidst streets.

The OD matrix represents the flow of bicycle trips between streets, where the rows and columns correspond to the origin and destination street indices, respectively. The values in the matrix represent the number of bicycle trips between each origin and destination pair. The analysis of the OD matrix allows us to understand the spatial distribution of bicycle usage during the specified time periods and identify key routes and patterns of travel within the city. By associating the bicycle location data with the street zoning information, we can gain valuable insights into the utilization patterns of the bicycle-sharing system and its impact on urban transportation dynamics.

3. Methodology

3.1. *Community Detection in Complex Networks*

Complex networks provide a powerful abstraction for understanding real-world systems, encompassing a diverse range of actors or entities and their intricate relationships [14]. In these networks, individual entities are represented as nodes, while the connections or interactions between them are captured by edges. Owing to the varying degrees of interconnectivity among nodes, complex networks often exhibit a phenomenon known as community structure or clustering. Communities refer to cohesive subgraphs of nodes within the network, wherein nodes within the same subgraph are tightly interconnected, while nodes in different subgraphs have weaker connections [41]. Figure 2 presents an illustrative example of a simple network with a visible community structure.

Complex networks serve as a robust conceptual framework for comprehending intricate real-world systems, encompassing an array of actors or entities and their intricate relationships [14]. Within these networks, individual entities are symbolized as nodes, while the connections or interactions between them are encapsulated by edges. Due to the varying degrees of interconnectivity among nodes, complex networks frequently manifest a phenomenon termed as community structure or clustering. Communities denote cohesive subsets of nodes within the network, wherein nodes in the same subset are closely interconnected, while nodes in distinct subsets possess weaker connections [42]. Figure 2 provides an illustrative instance of a basic network displaying a discernible community structure.

This concept of community structure holds significant relevance across various fields, including social networks, biological systems, and technological infrastructures. It aids in uncovering latent patterns, functional modules, and inherent organizations within complex systems. Researchers utilize various algorithms and methods to identify and analyze community structures, contributing to a deeper understanding of system dynamics and behaviors. This concept's applications extend to enhancing information dissemination strategies, identifying key nodes for targeted interventions, and optimizing network design for improved efficiency and resilience. Consequently, the investigation of community structures within complex networks remains a crucial pursuit with broad interdisciplinary implications.

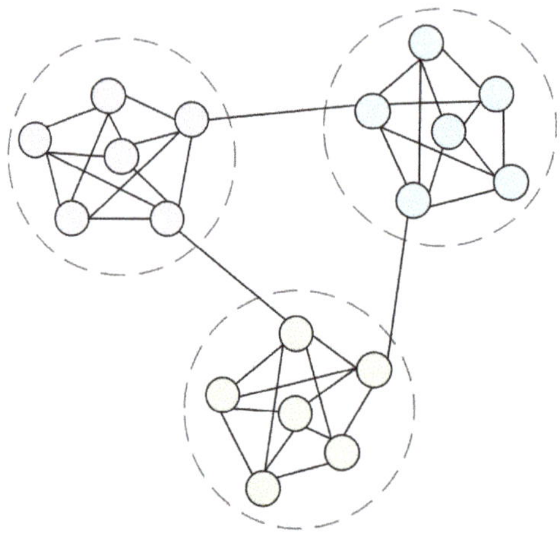

Figure 2. Community structure of a simple network.

3.2. Walktrap Community Detection Algorithm

With the development of complex network theory, researchers have proposed various methods for community detection, including the Walktrap algorithm introduced by Pons and Latapy [29]. This algorithm is based on the concept of random walks and utilizes the analysis of transition probabilities of node-to-node jumps within the network to achieve community detection. Random walks tend to become trapped within subgraphs formed by densely connected nodes, which correspond to communities, and this entire process forms a Markov chain. The algorithm establishes a distance metric to quantify the structural similarity between nodes or communities. This metric aids in evaluating the relationships between different elements within the network. The distance from node i to node j is computed using Equation (1):

$$r_{ij} = \sqrt{\sum\nolimits_{k=1}^{n} \frac{\left(P_{ik}^{t} - P_{jk}^{t}\right)^2}{d(k)}} \tag{1}$$

In the equation, r_{ij} represents the distance from node i to node j, t is a given time, $d(k)$ denotes the degree of node k, and P_{ik}^{t} represents the probability of reaching node k from node i within t steps. The calculation formula for P_{ik}^{t} is as follows:

$$P_{ij} = \frac{A_{ij}}{d(k)} \tag{2}$$

In this context, A_{ij} represents the value in the adjacency matrix A. Equation (3) calculates the distance from community C to node j. $|C|$ represents the number of nodes in association C. The calculation in Equation (3) is as follows:

$$P_{Cj}^{t} = \frac{1}{|C|} \sum_{i \in C} P_{ij}^{t} \tag{3}$$

Equation (4) calculates the distance from community C_1 to community C_2. The total community distance is obtained by calculating the square sum of the square roots between nodes from different communities. The calculation in Equation (4) is as follows:

$$r_{C_1C_2} = \sqrt{\sum_{k=1}^{n} \frac{\left(P^t_{C_1k} - P^t_{C_2k}\right)^2}{d(k)}} \quad (4)$$

After the initiation of the random walk, each node is initially considered as a separate community. The distances between adjacent nodes (communities) are computed, and two communities C_1 and C_2, are selected for merging into a single community based on minimizing the value of $\Delta\sigma(C_1, C_2)$. The calculation formula for $\Delta\sigma(C_1, C_2)$ is given by Equation (5).

$$\Delta\sigma(C_1, C_2) = \frac{1}{n} \frac{|C_1||C_2|}{|C_1| + |C_2|} r^2_{c_1c_2} \quad (5)$$

This process iterates continuously until all nodes are merged into a single community. The optimal community partitioning involves minimizing the distance within communities while maximizing the distance between communities.

3.3. Evaluation Methods for Community Detection

Modularity (Q) is an evaluation metric proposed by Girvan and Newman [42] for measuring the quality of community structures. It serves as a measure to assess the effectiveness of community detection in a network. A higher value of modularity indicates a better community structure, where nodes within communities are more densely connected. Conversely, a lower value of modularity suggests a weaker community structure with fewer internal edges. The calculation formula for modularity Q is given as follows:

$$Q = \frac{1}{2m} \sum_{ij} \left(A_{ij} - \frac{d_i d_j}{2m} \right) \delta(C_i, C_j) \quad (6)$$

In the context of directed networks, the modularity calculation takes into account both the out-degree and in-degree of nodes. Let A be the adjacency matrix, d_i and d_j represent the degree of nodes i and j, respectively, and let m be the total number of edges in the network. To assess whether nodes i and j belong to the same community, the Kronecker delta function $\delta(C_i, C_j)$ is utilized. When nodes i and j belong to the same community, $\delta(C_i, C_j)$ equals 1; otherwise, $\delta(C_i, C_j)$ is 0. The calculation formula for modularity in directed networks is as follows:

$$Q = \frac{1}{m} \sum_{ij} \left(A_{ij} - \frac{d_i^{In} d_j^{Out}}{m} \right) \delta(C_i, C_j) \quad (7)$$

In this context, d_j^{In} and d_i^{Out} represent the in-degree of node j and the out-degree of node i, respectively. The maximum value of modularity usually falls within the range of 0.3 to 0.7. A modularity value approaching 1 signifies a more favorable outcome in community detection.

It is important to note that the modularity value can be influenced by the overall scale of the network and the number of communities detected. For smaller networks or networks with only a few communities, the modularity value may be lower due to the limited structural complexity. Conversely, in larger networks with a higher number of communities, the modularity value might be higher due to the increased likelihood of finding more distinct communities. In summary, modularity is a valuable tool for assessing the effectiveness of community detection algorithms and provides insights into the underlying organizational principles of complex networks. Researchers aim to achieve high modularity values by optimizing the community detection process, as this corresponds to a more accurate and informative representation of the network's community structure.

The Walktrap algorithm possesses unique characteristics in community detection distinct from other methods: the number of communities progressively decreases as the algorithm proceeds. Consequently, the results of each community detection can be subjected to modularity calculations, utilizing the numerical variations in modularity to reflect the evolving patterns of community changes throughout the algorithmic process.

The algorithm starts by constructing the network structure using nodes and edges. Initially, each node is considered as a separate community based on the Walktrap algorithm. Then, the algorithm proceeds by iteratively merging the closest communities into larger ones while calculating the modularity score at each step. This process continues until all nodes are merged into a single community, marking the end of the algorithm. The flowchart of the algorithm is depicted in Figure 3.

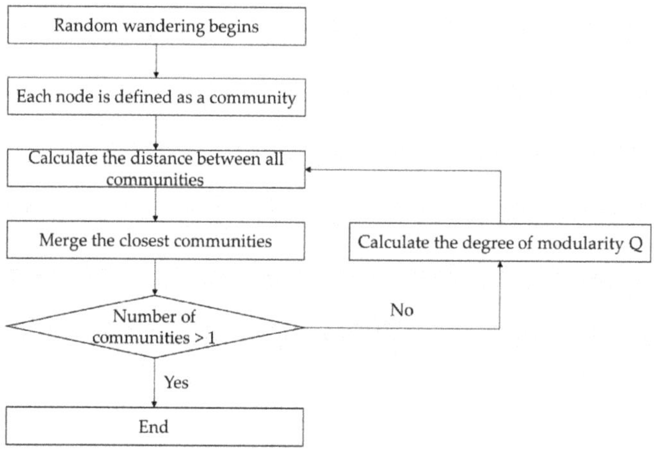

Figure 3. The flowchart of the Walktrap community detection algorithm based on modularity analysis.

4. Results

4.1. Construction and Analysis of Shared Bicycle Spatial Interaction Networks

The shared bicycle network is represented as a directed weighted network, consisting of three fundamental elements: "nodes," "directed edges," and "weights of directed edges". In this study, the "nodes" are represented by streets within the core area of the capital city. The "directed edges" of the network are defined by travel behavior between streets, indicating the direction of bicycle movement. The "weights of directed edges" are determined by the volume of bicycle usage, providing information on the intensity of travel between streets. Firstly, shared bicycle data for departure and arrival times were initially acquired and subsequently converted into SHP format. Following this, the streets where the bicycles were located were determined by spatially intersecting the data with administrative district data. Subsequently, by utilizing unique ID numbers, the shared bicycle data for departure and arrival times were linked, thereby providing the starting and ending street information for each bicycle. Lastly, trips with identical starting and ending points were aggregated, resulting in a collection of travel information organized by street as travel nodes.

According to the principle of network construction, the shared bicycle travel network is built based on data from two different periods: the morning peak and the evening peak. To construct the spatial interaction network for the core area of the capital, we employed ArcGIS 10.8 software for network visualization (Figure 4). The network consists of 135 nodes, with each node representing a street block. Figure 4 illustrates the differences in network structures between the morning peak and the evening peak, where flow strength is represented by the thickness of the edges: the wider the blue lines, the higher the flow strength, while narrower blue lines indicate weaker flow strength. This representation method provides an intuitive depiction of traffic volume, facilitating the analysis of travel characteristics and patterns. Table 2 provides information on the origin and destination of the top travel activities with high flow volume during the morning and evening peaks.

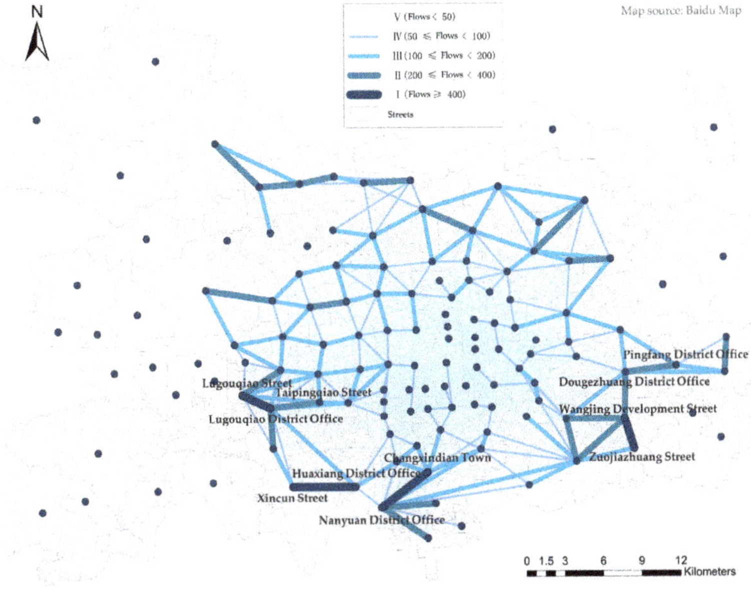

(**a**)

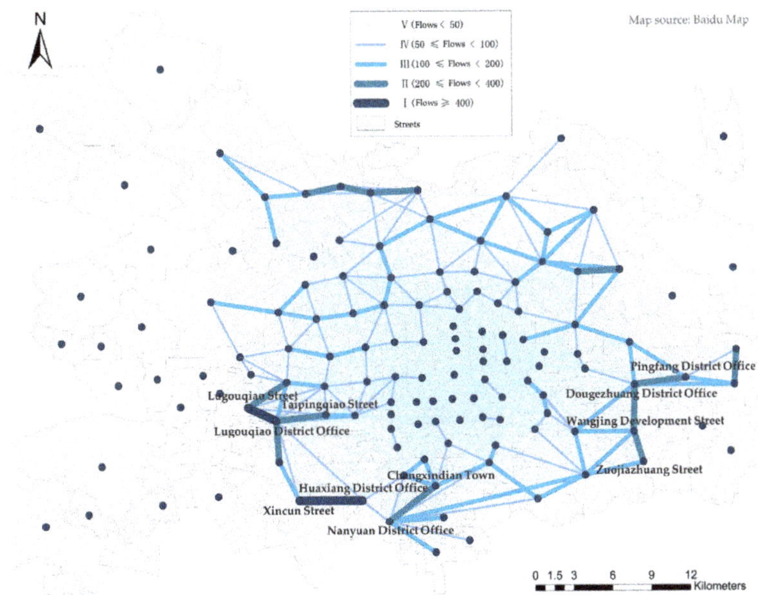

(**b**)

Figure 4. Spatial interaction network visualization: (**a**) Morning period; (**b**) Evening period.

Table 2. Traffic details and levels.

Periods	Origin	Direction	Destination	Flows Level
Morning period	Xincun Street	←→	Huaxiang District Office	I
	Nanyuan District Office	←→	Changxindian Town	I
	Lugouqiao Street	←→	Lugouqiao District Office	I
	Wangjing Development Street	←→	Zuojiazhuang Street	I
	Taipingqiao Street	←→	Lugouqiao District Office	II
	Dougezhuang District Office	←→	Pingfang District Office	II
Evening period	Xincun Street	←→	Huaxiang District Office	I
	Lugouqiao Street	←→	Lugouqiao District Office	I
	Nanyuan District Office	←→	Changxindian Town	II
	Wangjing Development Street	←→	Zuojiazhuang Street	II
	Taipingqiao Street	←→	Lugouqiao District Office	II
	Dougezhuang District Office	←→	Pingfang District Office	II

Based on the information obtained from Figure 4 and Table 2, several characteristics of the travel network can be summarized. Firstly, during both morning and evening peak hours, a significant portion of the shared bicycle trips takes place between adjacent street blocks, suggesting that shared bicycles are predominantly utilized for short-distance travel within the study area. Secondly, the flow of trips shows a concentration in specific areas, with notable travel activities observed between certain street blocks. For instance, considerable flow occurs between Huaxiang District Office and Xincun Sub-district Office, as well as between Lugouqiao Sub-district and Lugouqiao District Office, during both morning and evening peak periods. Among these, Huaxiang Street Office is a street known for its concentrated residential areas, while Xincun Street accommodates some enterprises and companies. Lugouqiao Sub-district and Lugouqiao District Office, similarly, are two adjacent streets that, respectively, concentrate residential and office areas. These regions appear to be popular destinations or origins for shared bicycle users, leading to intensified bicycle movements between these locations.

There are noticeable differences in the network structure between the morning and evening peak hours. The morning peak exhibits a higher level of flow concentration, with large-flow trips being more tightly clustered. In contrast, the evening peak shows a relatively lower level of flow concentration, and there are only two locations with significant flow. This analysis indicates that the morning peak trips are characterized by a higher level of purposefulness and concentration. The areas with high-flow edges in the network structure during weekdays are often associated with workplaces, indicating that the majority of trips during weekdays are related to commuting. Therefore, the street blocks with high-flow edges tend to be locations with more businesses or employment units, making them prominent destinations for commuters.

On the other hand, during the evening peak, people's trip destinations exhibit a higher degree of diversity, leading to a more scattered flow pattern. This suggests that people engage in more leisure activities after work, and their travel destinations are more influenced by personal preferences and varied interests. Furthermore, the distribution of workplaces within the study area appears to be more concentrated, while the distribution of residential areas is relatively more dispersed. This spatial pattern may be a reflection of the urban development and land-use characteristics, where workplaces tend to cluster in specific areas, whereas residential areas are more widely distributed.

4.2. Analysis of Modularity Calculation Results

This study investigates the community detection of the spatial interaction network using the Walktrap method based on the concept of random walks. This method effectively groups closely connected nodes into communities, enabling the analysis of inherent relationships between streets. Since modularity can be employed to reflect the quality of the current community detection results, calculating modularity continuously throughout the process of changing the number of communities can intuitively illustrate community aggregation patterns. The evolution of modularity values during the random walk process is computed in this research, as illustrated in Figure 5. According to the principles of the Walktrap community detection algorithm, the number of communities decreases gradually during the random walk process, eventually resulting in a single community that includes all nodes. The modularity values gradually increase as the number of communities decreases, with the morning period showing a faster increase in modularity compared to the evening period. Specifically, when there are 14 communities, the modularity value for the morning period reaches 0.3, and, when there are 7 communities, the modularity value for the evening period reaches 0.3.

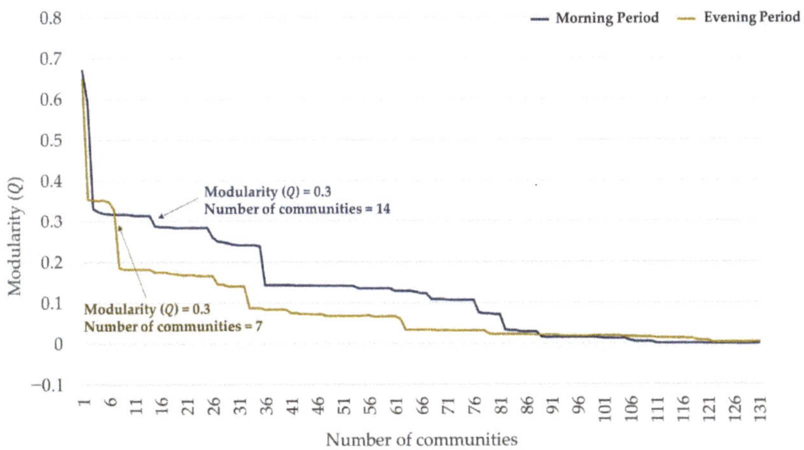

Figure 5. Modularity calculation results.

The analysis reveals that the spatial interaction network during the morning period exhibits stronger community structures, indicating a higher level of aggregation between streets. In other words, during the morning period, people tend to travel with more specific purposes between streets within the same community, resulting in a higher volume of intra-community travel and a lower volume of inter-community travel. This finding highlights the significant community structure in the morning period's spatial interaction network. The Walktrap community detection approach used in this study proves effective in uncovering underlying community structures within the spatial interaction network, shedding light on the characteristics of travel patterns and street connections during different time periods. The results contribute to a better understanding of urban mobility dynamics and can inform targeted urban planning and transportation management strategies.

4.3. *Analysis of Community Detection Results*

The visualization of the community detection results for 7 and 14 communities is presented in this study. Table 3 provides information about the number of communities and the corresponding number of streets in each community. The spatial visualization of the results of the community testing is shown in Figure 6. Within Figure 6, streets are represented as nodes located at the centers of street clusters, and each node is labeled with a unique street identifier, totaling 135 nodes. Different colors are employed to indicate their respective community affiliations.

Table 3. Community detection results.

Type of Community/Unit: Number	Morning Period			Evening Period
	14 COMM		7 COMM	7 COMM
①	68	1	68	91
②	49	1	52	23
③	3	1	4	14
④	2	1	3	2
⑤	2	1	3	1
⑥	1	1	2	1
⑦	1	1	1	1

Notably, there are significant differences in the community structures between the morning and evening periods. During the morning peak, the spatial interaction network exhibits the formation of two large communities along with multiple small communities (Figure 6a,b). In contrast, the evening peak reveals the identification of one large community, two medium-sized communities, and several small communities, with the majority of streets belonging to the large community (Figure 6c).

The results indicate distinct patterns in the spatial distribution of community formations during different time periods. The morning peak shows a more fragmented community structure, with several smaller communities interacting with each other, while the evening peak presents a more integrated community structure, with fewer but larger communities encompassing multiple streets. This finding suggests that travel behaviors and patterns during the morning and evening periods differ significantly, reflecting the varying travel purposes and destinations at different times of the day.

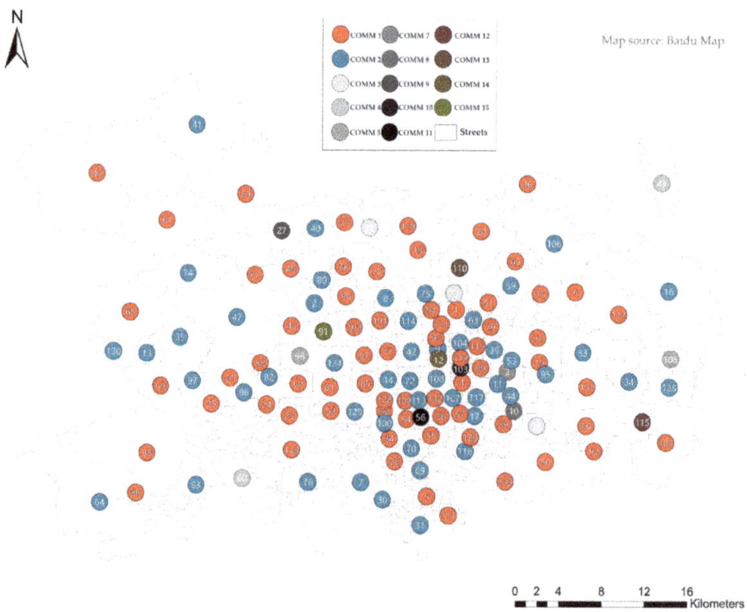

(a)

Figure 6. *Cont.*

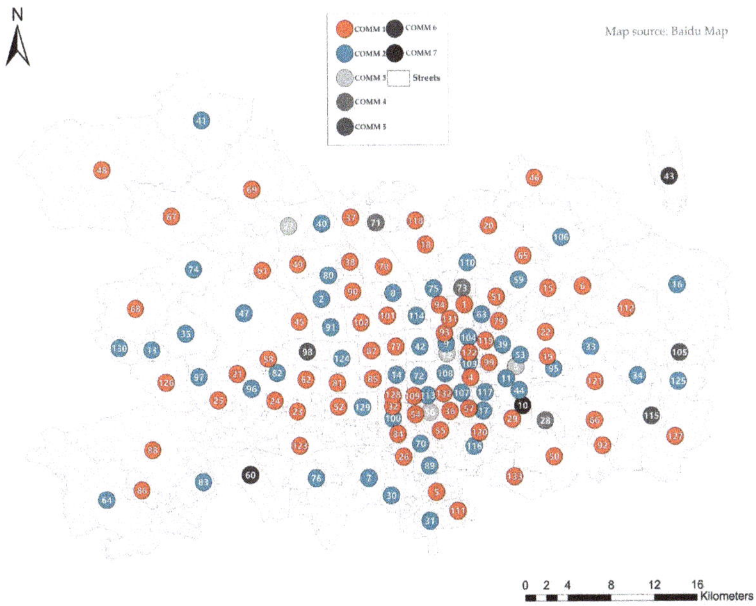

(**b**)

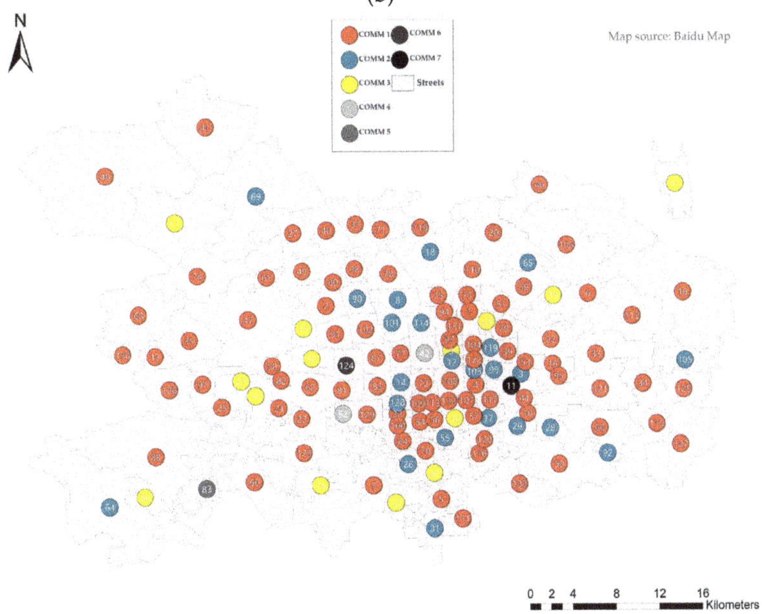

(**c**)

Figure 6. Spatial visualization of community detection: (**a**) 14 communities during the morning period; (**b**) 7 communities during the morning period; (**c**) 7 communities during the evening period.

The findings from the community detection analysis reveal characteristics of the spatial interaction network during different time periods. Specifically, during the morning period, the presence of two large communities and several small communities indicates that certain streets have stronger connections and are closely associated with each other. However, the lack of distinct geographical boundaries within these communities suggests that the interactions between streets are not solely determined by their physical proximity. Instead, it is likely influenced by the travel behavior of residents who commute to work areas. The high flow volume between certain street blocks can be attributed to the concentrated movement of commuters between residential and work areas, leading to the formation of cohesive communities.

In contrast, the evening period exhibits a different pattern, with a single large community and a few medium-sized and small communities. Although the communities lack geographical boundaries, there is a higher degree of interconnectivity among the streets compared to the morning period. This suggests that, during the evening, more streets are engaged in reciprocal interactions with each other, leading to a broader range of travel purposes. Leisure and social venues within the study area are characterized by a wide and scattered distribution. The increased diversity in destinations during the evening period indicates that people are engaged in various activities, such as leisure and social engagements, which may result in a more scattered flow of shared bicycle trips.

The longer interaction radius observed in both the morning and evening periods suggests that the influence of certain streets extends over a wider geographical scope. This implies that the spatial interactions between streets are not confined to a limited area but rather encompass larger regions within the study area. The broader scope of interactions has implications for urban planning and transportation management, as it indicates the importance of considering the connectivity between different areas when designing bike-sharing systems and transportation infrastructure.

The Walktrap community detection method proves to be effective in revealing the underlying structure of the spatial interaction network and provides valuable insights into the travel patterns and street connections during different time periods. The observed differences in community detection shed light on the dynamics of urban mobility and the factors influencing travel behavior. The research contributes to a better understanding of urban transportation systems and offers valuable information for optimizing bike-sharing schemes and urban planning strategies.

5. Discussion

Urban streets, when viewed through the lens of bike-sharing, exhibit distinct clustering tendencies and adhere to clustering patterns. By constructing a street-level spatial interaction network from the perspective of bike-sharing, it has been observed that shared bicycles are primarily utilized for short-distance trips within the study area. The flow of trips is concentrated among specific streets, with a higher degree of concentration during the morning peak period compared to the evening peak. This phenomenon may be attributed to the fact that the primary purpose during the morning peak is work commuting, and the streets with concentrated flow are often located near areas with a higher concentration of businesses or employment centers. In contrast, the evening peak exhibits a more diversified set of destinations, as workplaces are more centrally distributed within the urban region while residential areas tend to be more dispersed. Through community detection analysis within the network, it has been revealed that the spatial interaction network exhibits stronger community structure during the morning period, with more dispersed community configurations. Several community clusters interact with each other. Conversely, the evening period demonstrates more cohesive community structures, characterized by fewer but larger communities. Based on the above conclusions, recommendations are provided for urban transportation authorities, bike-sharing companies, and city residents.

For urban transportation authorities: Strengthen traffic flow monitoring and management during morning and evening peak hours: Given the observed differences in community structures during these periods, transportation authorities should focus on monitoring and managing traffic flow, especially in areas with concentrated commuting trips, to allocate traffic resources more efficiently and alleviate congestion. Optimize transportation planning and bike-sharing deployment strategies: Utilizing the outcomes of community detection, transportation authorities can strategically adjust the deployment and planning of bike-sharing services. In contrast to conventional dispatching strategies based solely on the usage volume at distinct locations, our research primarily emphasizes the provision of a universal framework for the analysis and scheduling of bike-sharing trips. This approach offers macro-level guidance for bike-sharing dispatching, which can be applicable to other urban core areas.

For bike-sharing companies: Increase the availability of bikes during morning and evening peak hours: In response to the unique travel patterns observed during these periods, bike-sharing companies can increase bike availability to cater to the commuting demand, thereby improving bike utilization rates. Enhance services in social and leisure areas during evening hours: As the research indicates a diversified travel pattern during evenings, bike-sharing companies can augment bike services, particularly in social and leisure areas, to fulfill the varied travel demands of residents during nighttime.

For city residents: Make informed choices for travel modes: Considering the suitability of bike-sharing for short-distance travel during morning and evening peak hours, city residents can prioritize bike-sharing for short trips to avoid traffic congestion. Explore diverse travel destinations: With a diverse range of travel destinations observed during evening hours, residents can choose travel destinations based on their preferences and needs, enjoying the rich nightlife and recreational activities in the city.

Overall recommendations: By formulating reasonable transportation planning and bike-sharing deployment strategies based on research results, urban transportation authorities can optimize traffic flow and resource utilization. Meanwhile, bike-sharing companies can offer more flexible and customized services according to different time periods and community characteristics, catering to the diverse travel needs of residents. City residents should make informed choices for travel modes and actively participate in and enjoy the diverse transportation and leisure activities in the city.

6. Conclusions

This study adopts streets as the research unit and builds a spatial interaction network for shared bicycles in the core area of the capital city. The research employs a random walk community detection method based on weighted modularity to explore and analyze the community structures within the urban area. Modularity values are utilized to assess the quality of community detection, thereby evaluating the regional community characteristics. The community detection results are further employed for street clustering analysis, facilitating the exploration of travel patterns. The experimental results indicate that the morning period's spatial interaction network exhibits stronger community structures, consisting of two large communities with higher street aggregation. Residents' travel during this period is predominantly for commuting purposes, showing a higher level of purposefulness. In contrast, the evening period comprises a single large community, with most streets interacting with each other in pairs, suggesting a broader range of travel purposes for residents compared to the morning period. Both time periods demonstrate characteristics of long interaction radii and wide interaction scopes.

The findings of this study provide decision-making assistance to relevant authorities. During the morning period, shared bicycle operators can strategically deploy bicycles to streets near locations where interactions between streets within the same community are likely to occur. In the evening period, due to the extensive usage of shared bicycles, transportation management departments should focus on regulating their usage to avoid traffic issues caused by overcrowding.

For future work, several research directions are suggested: expanding the representation of urban network nodes by including more representative urban areas, such as urban transportation hubs; exploring various community detection methods and investigating their impacts on community detection results; and enriching the types of travel data to enhance the universality of research on urban population travel patterns. Furthermore, in the theoretical methods section, further research and expansion are needed. More community detection methods can be integrated with evaluation methods to adapt to a wider range of application scenarios.

Author Contributions: Conceptualization, Y.S., D.T., M.Z. and Y.H.; data curation, Z.W. and D.T.; formal analysis, D.T. and Y.S.; funding acquisition, Y.H.; investigation, Z.W.; methodology, M.Z. and Y.H.; project administration, D.T.; resources, Y.S. and Y.H.; software, D.T.; supervision, M.Z.; writing—original draft, Z.W. All authors have read and agreed to the published version of the manuscript.

Funding: This work was supported by the Beijing Key Laboratory of Urban Spatial Information Engineering (20230110), the BUCEA Doctor Graduate Scientific Research Ability Improvement Project (DG2023001), and the Open Fund of Key Laboratory of Urban Spatial Information, Ministry of Natural Resources (Grant No. 2023PT002).

Informed Consent Statement: Not applicable.

Data Availability Statement: Not applicable.

Acknowledgments: The authors would like to thank all of the reviewers for their valuable contributions to this work.

Conflicts of Interest: The authors declare no conflict of interest.

References

1. Jensen, P.; Rouquier, J.B.; Ovtracht, N.; Robardet, C. Characterizing the speed and paths of shared bicycles in Lyon. *Transp. Res. Part D Transp. Environ.* **2010**, *15*, 522–524. [CrossRef]
2. Politis, I.; Fyrogenis, I.; Papadopoulos, E.; Nikolaidou, A.; Verani, E. Shifting to shared wheels: Factors affecting dockless bike-sharing choice for short and long trips. *Sustainability* **2020**, *12*, 8205. [CrossRef]
3. Li, X.; Zhang, Y.; Du, M.; Yang, J. Social Factors Influencing the Choice of Bicycle: Difference Analysis among Private Bike, Public Bike Sharing and Free-Floating Bike Sharing in Kunming, China. *KSCE J. Civ. Eng.* **2019**, *23*, 2339–2348. [CrossRef]
4. Dell'Amico, M.; Iori, M.; Novellani, S.; Subramanian, A. The bike sharing rebalancing problem with stochastic demands. *Transp. Res. Part B Methodol.* **2018**, *118*, 362–380. [CrossRef]
5. Chen, J.; Yang, Z.; Shu, Y.; Cheng, P. Rebalance bike-sharing system with deep sequential learning. *IEEE Intell. Trans-Portation Syst. Mag.* **2020**, *13*, 92–98. [CrossRef]
6. Federico, C.; Chiara, P.; Andrea, Z.; Michele, Z. A dynamic approach to rebalancing bike-sharing systems. *Sensors* **2018**, *18*, 512. [CrossRef]
7. Oshan, T.M. The spatial structure debate in spatial interaction modeling: 50 years on. *Prog. Hum. Geogr.* **2021**, *45*, 925–950. [CrossRef]
8. Wang, X.; Ding, S.; Cao, W.; Fan, D.; Tang, B. Research on network patterns and influencing factors of population flow and migration in the yangtze river delta urban agglomeration, china. *Sustainability* **2020**, *12*, 6803. [CrossRef]
9. Zhang, W.; Chong, Z.; Li, X.; Nie, G. Spatial patterns and determinant factors of population flow networks in china: Analysis on tencent location big data. *Cities* **2020**, *99*, 102640. [CrossRef]
10. Chong, Z.; Pan, S. Understanding the structure and determinants of city network through intra-firm service relationships: The case of guangdong-hong kong-macao greater bay area. *Cities* **2020**, *103*, 102738. [CrossRef]
11. Harris, J.K. *Introduction to Exponential Random Graph Models*; Shanghai People's Publishing House and Gezhi Publishing House: Shanghai, China, 2016.
12. Beck, B.; Winters, M.; Nelson, T.; Pettit, C.; Leao, S.Z.; Saberi, M.; Thompson, J.; Seneviratne, S.; Nice, K.; Stevenson, M. Developing urban biking typologies: Quantifying the complex interactions of bicycle ridership, bicycle network and built environment characteristics. *Environ. Plan. B Urban Anal. City Sci.* **2023**, *50*, 7–23. [CrossRef]
13. Liu, P.; Chen, J.; Sun, H.; Guo, X.; Wang, Y.; Zhu, Z. Assessing Accessibility of Dockless Sharing-Bike Networks by the Social Network Analysis Method. *J. Adv. Transp.* **2021**, *2021*, 5584008. [CrossRef]
14. Hu, L.; Wen, Z.; Wang, J.; Hu, J. Spatial Interaction Analysis of Shared Bicycles Mobility Regularity and Determinants: A Case Study of Six Main Districts, Beijing. *ISPRS Int. J. Geo-Inf.* **2022**, *11*, 477. [CrossRef]
15. Cherifi, H.; Palla, G.; Szymanski, B.K.; Lu, X. On Community Structure in Complex Networks: Challenges and opportunities. *Appl. Netw. Sci.* **2019**, *4*, 117. [CrossRef]
16. Li, X.; Wu, X.; Xu, S.; Qing, S.; Chang, P. A novel complex network community detection approach using discrete particle swarm optimization with particle diversity and mutation. *Appl. Soft Comput.* **2019**, *81*, 105476. [CrossRef]
17. Chen, W. Delineating the Spatial Boundaries of Megaregions in China: A City Network Perspective. *Complexity* **2021**, *2021*, 2574025. [CrossRef]

18. Hui, E.C.M.; Li, X.; Chen, T.; Lang, W. Deciphering the spatial structure of China's megacity region: A new bay area—The Guangdong-Hong Kong-Macao Greater Bay Area in the making. *Cities* **2018**, *105*, 102168. [CrossRef]
19. Liu, F.; Gang, X. A Fast Algorithm for Community Detection of Network Systems in Smart City. *IEEE Access* **2019**, *7*, 51856–51865. [CrossRef]
20. Gao, P.; He, D.; Sun, Z.; Ning, Y. Characterizing functionally integrated regions in the Central Yangtze River Megaregion from a city-network perspective. *Growth Chang.* **2020**, *51*, 1357–1379. [CrossRef]
21. Newman, M.; Girvan, M. Finding and Evaluating Community Structure in Networks. *Phys. Rev. E* **2004**, *69 Pt 2*, 026113. [CrossRef]
22. Kouni, I.; Karoui, W.; Romdhane, L.B. Node Importance based Label Propagation Algorithm for overlapping community detection in networks. *Expert Syst. Appl.* **2019**, *162*, 113020. [CrossRef]
23. Blondel, V.D.; Guillaume, J.L.; Lambiotte, R.; Lefebvre, E. Fast unfolding of communities in large networks. *J. Stat. Mech.* **2008**, *2008*, 10008. [CrossRef]
24. Austwick, M.Z.; O'Brien, O.; Strano, E.; Viana, M. The Structure of Spatial Networks and Communities in Bicycle Sharing Systems. *PLoS ONE* **2013**, *8*, e74685. [CrossRef]
25. Pons, P.; Latapy, M. Computing communities in large networks using random walks. *J. Graph Algorithms Appl.* **2016**, *10*, 191–218. [CrossRef]
26. Ke, W.; Wei, C.; Yu, Z. Uncovering Spatial Structures of Regional City Networks from Expressway Traffic Flow Data: A Case Study from Jiangsu Province, China. *Sustainability* **2017**, *9*, 1541. [CrossRef]
27. Tiznado-Aitken, I.; Vecchio, G.; Guzman, L.A.; Arellana, J.; Humberto, M.; Vasconcellos, E.; Muñoz, J.C. Unequal Periurban Mobility: Travel Patterns, Modal Choices and Urban Core Dependence in Latin America. *Habitat Int.* **2023**, *133*, 102752. [CrossRef]
28. Zhang, R.; Yao, E.; Liu, Z. School Travel Mode Choice in Beijing, China. *J. Transp. Geogr.* **2017**, *62*, 98–110. [CrossRef]
29. Zhang, Y.; Yao, E.; Zhang, R.; Xu, H. Analysis of Elderly People's Travel Behaviours during the Morning Peak Hours in the Context of the Free Bus Programme in Beijing, China. *J. Transp. Geogr.* **2019**, *76*, 191–199. [CrossRef]
30. Xie, P.; Li, T.; Liu, J.; Du, S.; Yang, X.; Zhang, J. Urban flow prediction from spatiotemporal data using machine learning: A survey. *Inf. Fusion* **2020**, *59*, 1–12. [CrossRef]
31. Zhang, Y.; Mi, Z. Environmental benefits of bike sharing: A big data-based analysis. *Appl. Energy* **2018**, *220*, 296–301. [CrossRef]
32. Gu, T.; Kim, I.; Currie, G. Measuring immediate impacts of a new mass transit system on an existing bike-share system in China. *Transp. Res. Part A: Policy Pract.* **2019**, *124*, 20–39. [CrossRef]
33. Mateo-Babiano, I.; Bean, R.; Corcoran, J.; Pojani, D. How does our natural and built environment affect the use of bicycle sharing? *Transp. Res. Part A: Policy Pract.* **2016**, *94*, 295–307. [CrossRef]
34. Du, M.; Cheng, L. Better Understanding the Characteristics and Influential Factors of Different Travel Patterns in Free-Floating Bike Sharing: Evidence from Nanjing, China. *Sustainability* **2018**, *10*, 1244. [CrossRef]
35. Yu, S.; Liu, G.; Yin, C. Understanding spatial-temporal travel demand of free-floating bike sharing connecting with metro stations. *Sustain. Cities Soc.* **2021**, *74*, 103162. [CrossRef]
36. Guo, Y.; Yang, L.; Lu, Y.; Zhao, R. Dockless bike-sharing as a feeder mode of metro commute? The role of the feeder-related built environment: Analytical framework and empirical evidence. *Sustain. Cities Soc.* **2020**, *65*, 102594. [CrossRef]
37. Li, Q.; Cui, C.; Liu, F.; Wu, Q.; Run, Y.; Han, Z. Multidimensional Urban Vitality on Streets: Spatial Patterns and Influence Factor Identification Using Multisource Urban Data. *ISPRS Int. J. Geo-Inf.* **2022**, *11*, 2. [CrossRef]
38. Liu, M.; Jiang, Y.; He, J.J.S. Quantitative Evaluation on Street Vitality: A Case Study of Zhoujiadu Community in Shanghai. *Sustainability* **2021**, *13*, 3027. [CrossRef]
39. Yao, Y.; Zhang, Y.; Tian, L.; Zhou, N.; Li, Z.; Wang, M. Analysis of Network Structure of Urban Bike-Sharing System: A Case Study Based on Real-Time Data of a Public Bicycle System. *Sustainability* **2019**, *11*, 5425. [CrossRef]
40. Zhang, H.; Zhuge, C.; Jia, J.; Shi, B.; Wang, W. Green travel mobility of dockless bike-sharing based on trip data in big cities: A spatial network analysis. *J. Clean. Prod.* **2021**, *313*, 127930. [CrossRef]
41. Newman, M. Analysis of weighted networks. *Phys. Rev. E* **2004**, *70*, 056131. [CrossRef]
42. Zuo, T.; Wei, H. Bikeway prioritization to increase bicycle network connectivity and bicycle-transit connection: A multi-criteria decision analysis approach. *Transp. Res. Part A Policy Pract.* **2019**, *129*, 52–71. [CrossRef]

buildings

MDPI

Article

Analyzing the Spatial Interaction Characteristics of Urban Area Shared Bicycle Systems: A Case Study of Beijing's Central Area

Dongwei Tian [1,2], Zheng Wen [3,*] and Yao Sun [1]

1 School of Architecture and Urban Planning, Beijing University of Civil Engineering and Architecture, Beijing 100044, China; tiandongwei77@163.com (D.T.); sunyao@bucea.edu.cn (Y.S.)
2 Research Center for Urban Big Data Applications, Beijing University of Civil Engineering and Architecture, Beijing 100044, China
3 School of Geomatics and Urban Spatial Informatics, Beijing University of Civil Engineering and Architecture, Beijing 100044, China
* Correspondence: wenzheng_bucea@163.com

Abstract: Shared bicycle systems play a crucial role in promoting sustainable urban transportation, addressing challenges such as traffic congestion and air pollution. Understanding the spatiotemporal patterns of shared bike usage is essential for optimizing bike-sharing infrastructure and improving transportation planning. In this study, we analyzed 2.4 million records of shared bicycle data to explore the spatial distribution, interaction patterns, and flow dynamics within Beijing's urban central area. We found that bike distribution peaks during commuting hours, particularly in central regions with employment centers. Complex networks are an important method for studying travel flows. Through a spatial interaction network, we identified key streets with high node strength and popularity, often concentrated in central areas. They experience heavy shared bicycle use during peak hours due to their employment-centric location. Conversely, peripheral areas see increased usage in the evenings, reflecting distinct commuting patterns. The morning exhibits higher positive central values compared to the evening, while negative values show the opposite trend. Based on these findings, we recommend enhancing bike infrastructure in high-density areas with bike lanes and ample shared bikes during peak hours. Implementing mixed-use zoning policies in the central region can reduce traffic congestion. Expanding shared bike services to peripheral regions can promote equitable access. This research underscores the importance of considering spatial and temporal factors in urban transportation planning. Future work should incorporate additional data sources, explore environmental impacts, and analyze usage in different seasons and special events, further contributing to sustainable urban mobility development.

Keywords: shared bikes; spatiotemporal dynamics; spatial distribution; spatial interaction network; net flow ratio

1. Introduction

Urban transportation systems play a crucial role in shaping the livability and sustainability of cities, and these roles are even more pronounced in larger cities with high levels of economic development [1,2]. With the increasing challenges posed by rapid urbanization and population growth, cities worldwide are seeking efficient and sustainable transportation alternatives, primarily due to the growing concerns of urban environment pollution [3,4]. Shared bicycles have emerged as a popular mode of transportation, offering a flexible and eco-friendly solution for short-distance trips, contributing to the development of urban transportation systems in more environmentally conscious ways [5,6]. Shared bicycles play a crucial role in integrating into the overall urban transportation system. To begin, they serve as a complement to urban public transportation, offering a solution for the last mile, thus alleviating the pressure on public transit during peak hours. These systems have the potential to alleviate traffic congestion, reduce carbon emissions, and improve urban mobility [7,8]. Shared bicycles can integrate with other transportation modes such as subways, buses, and electric scooters, enabling multimodal travel. This enhances the convenience of urban residents' journeys, reduces private car usage, and mitigates issues related to traffic congestion and environmental pollution. Beijing, as the capital of China, faces unique transportation challenges due to its immense population and complex urban structure [9]. The city's transportation infrastructure must accommodate the needs of millions of residents and ensure smooth mobility within its vast territory [10,11]. In recent years, shared bicycle systems have gained significant popularity in Beijing, providing an affordable and convenient transportation option for both daily commuters and occasional users. The extensive adoption of shared bicycles has raised intriguing questions regarding their usage patterns, spatial distribution, and interactions with other transportation modes [12,13].

Numerous studies conducted worldwide have extensively explored various aspects of shared bicycle systems, shedding light on user behavior, system performance, and spatial–temporal usage patterns [14,15]. These investigations have significantly contributed to understanding the factors that influence shared bicycle usage. Factors such as trip distance, weather conditions, availability of docking stations, and demographic characteristics of users have been identified as key determinants affecting the adoption and utilization of shared bicycles [16]. For instance, studies have revealed that shorter trip distances are more likely to be covered by shared bicycles, as they provide a convenient and time-efficient mode of transportation for short-distance travel. Similarly, the presence of docking stations in close proximity to popular destinations significantly influences the usage patterns of shared bicycles [17]. Furthermore, researchers have focused on analyzing the spatial distribution of shared bicycles and identifying hotspots of high demand. Spatial analysis plays a crucial role in understanding the spatial patterns and dynamics of shared bicycle usage [18,19]. By employing techniques such as Geographic Information Systems (GIS) and spatial clustering algorithms, researchers have successfully identified areas with high demand and concentrated shared bicycle activity [20,21]. These hotspots are typically characterized by their proximity to transportation hubs, commercial centers, educational institutions, and residential areas, reflecting the diverse needs and preferences of users in different locations [22,23].

However, despite the existing body of research, there remains a significant knowledge gap in comprehensively examining the spatiotemporal dynamics of shared bicycle usage within the central area of Beijing, particularly during peak hours of pedestrian activity [24]. The central area of Beijing encompasses the economic and administrative center of the city, hosting a substantial population and a multitude of industrial establishments [25]. As a result, this area experiences high volumes of pedestrian traffic during peak hours, presenting unique challenges for transportation planning and management [26,27]. In summertime, shared bicycles serve as a pivotal urban transportation option. High temperatures can cause discomfort for pedestrians or public transit commuters, but shared bicycles swiftly move people between points, mitigating heat-related inconveniences. Shared bicycles' eco-friendly attributes are accentuated in summer, reducing urban air pollution in contrast to temperature-induced pollution. Their cost-effectiveness is advantageous, especially for summer travelers. Thus, shared bicycles offer both a convenient mode of travel in extreme summer conditions and contribute significantly to advancing sustainable urban development.

Shared bicycles have distinct peak usage times within Beijing's transportation network on a daily, weekly, monthly, and yearly basis. Typically, daily peaks occur during the morning and evening rush hours, catering to residents' commuting needs. Weekly trends predominantly focus on workdays, with reduced usage on weekends, although there may be a slight increase during leisure or special event periods. Monthly usage patterns are influenced by seasons and weather, with more activity during the summer compared to winter. Yearly peak usage times are often linked to seasons and special events, such as increased usage during the Spring Festival. Furthermore, shared bicycles are closely associated with commuting time and travel duration. They reduce commuting time, making them especially suitable for individuals working close to their destination, as cycling is faster than walking. Simultaneously, the quick accessibility of shared bicycles reduces waiting time and addresses the issue of parking difficulties.

From a methodological point of view, complex network analysis is a valuable approach for studying urban transportation systems [28–30]. It provides a holistic understanding of the dynamics and interconnections within these complex systems by representing transportation infrastructure, flows, and interactions as nodes and edges. In urban transportation research, complex network analysis has been applied to areas such as traffic flow, public transportation systems, and shared mobility services [31,32]. It allows researchers to examine network properties and understand the resilience and robustness of urban mobility networks. By identifying critical nodes or links, infrastructure planning can be guided to enhance overall network performance [33]. Complex network analysis has also been used to study traffic congestion patterns and identify influential nodes or bottlenecks. Centrality measures help pinpoint locations where congestion accumulates and spreads, informing targeted interventions to alleviate congestion and improve traffic flow efficiency [34]. In public transportation, complex network analysis explores the structure and dynamics of transit systems. By analyzing passenger flows and interactions between different lines or modes, researchers gain insights into system performance and identify areas for improvement in terms of connectivity and service quality [35,36]. Additionally, complex network analysis has been applied to shared mobility services like bike-sharing and ride-sharing systems. It helps understand usage patterns, spatial distributions, and the impact of factors such as network connectivity and user behavior on system performance [28]. This knowledge supports optimizing station placement, rebalancing strategies, and demand forecasting for more efficient and sustainable shared mobility services. Overall, complex network analysis offers an integrative approach to studying urban transportation systems. It captures the complex interactions and dependencies between transportation infrastructure, flows, and users, supporting evidence-based decision-making, policy formulation, and urban planning efforts to create efficient, sustainable, and resilient urban transportation systems [11,37]. Previous studies have primarily utilized complex network methods to investigate spatial interactions between different regions, with limited focus on researching internal population flow patterns within a specific area. We employed this approach to analyze the overall spatial flow patterns within the region.

Three problems are proposed and resolved in this paper. What are the network characteristics and interaction patterns of shared bicycles in urban areas? What are the travel patterns of shared bicycles during peak periods in the city? How do the travel patterns during morning and evening peak periods differ? To address these issues, this study attempts to (i) construct a spatial interaction network of shared bicycles in urban areas and investigate the interaction patterns using various network statistical indicators; (ii) develop an interaction model based on shared bicycle data during peak periods in the city to examine the travel characteristics; (iii) compare and analyze the travel patterns during morning and evening peak periods.

This paper is structured as follows: The second part presents the research area and data sources, as well as the principles employed for network construction and the selection of indicators for the network model. In the third part, the network structure and the analysis of network indicators are discussed. The fourth part comprises the conclusions drawn from the study and a detailed discussion of the findings.

2. Materials and Methods

2.1. Research Data

The central area of Beijing, comprising Dongcheng District, Xicheng District, Haidian District, Fengtai District, Shijingshan District, and Chaoyang District, is situated in the central part of the city. This area is densely populated, accounting for approximately 60% of the total population of Beijing, and accommodates nearly 70% of its industrial establishments. Due to its central location and economic significance, the central area experiences high volumes of pedestrian traffic, particularly during peak hours. Furthermore, shared bicycles are extensively utilized within this region, contributing to the overall transportation landscape. For the purpose of this study, the research scope focuses on the intersection of the operational area of shared bicycles with the central area of the capital. By examining this specific region, we can gain valuable insights into the dynamics of shared bike usage in an area of high pedestrian activity and economic concentration. The selected study area encompasses over 90% of the central area of the capital, ensuring comprehensive coverage for the analysis.

To illustrate the location of the study area within Beijing, Figure 1 presents a visual representation. It highlights the spatial extent of the central area of the capital and its intersection with the operational area of shared bicycles. This depiction aids in understanding the geographical context of the study and provides a clear reference for subsequent analyses. In summary, the central area of the capital, situated in the central part of Beijing, serves as the focal point of this study. Its significance lies in its high population density, concentration of industrial establishments, and the substantial usage of shared bicycles. By investigating the region where the central area intersects with the operational area of shared bicycles, we aim to uncover valuable insights into the spatiotemporal dynamics of shared bike usage in an area characterized by heavy pedestrian traffic and economic activity. The location of the study area within Beijing is visually represented in Figure 1.

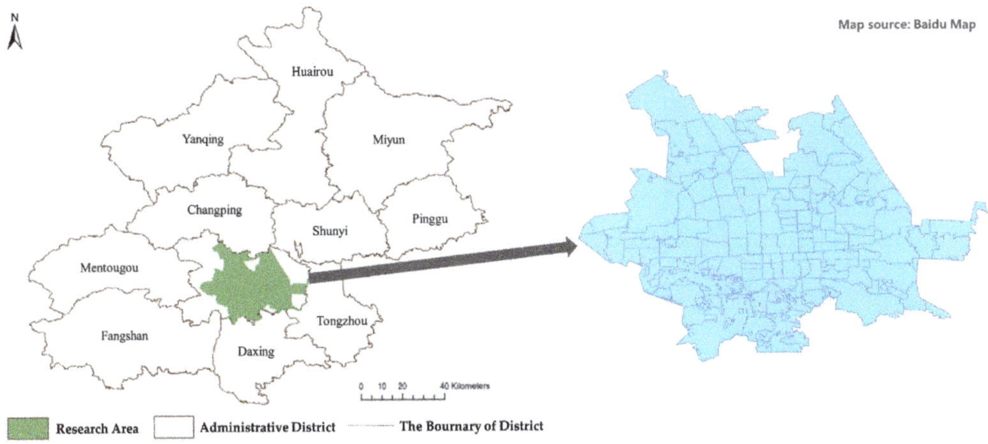

Figure 1. Research scope map.

The shared bicycle data utilized in this research are derived from collaborative data with enterprises. This dataset comprises shared bicycle location data within the central area of the capital. We continuously gathered location data for all shared bicycles within the study area throughout the data collection period. Each data entry includes attributes such as bicycle ID, time, and latitude–longitude coordinates, totaling 2.4 million records. Taking into account the impact of the COVID-19 pandemic, data collection was conducted during the morning peak hours (6:00–9:00) and evening peak hours (17:00–20:00) in July 2021. During this period, the pandemic situation in Beijing was relatively optimistic, with minimal impact on travel. Consequently, residents were more inclined to use shared bicycles for commuting.

The collected data underwent preprocessing, beginning with data screening. Records associated with abnormal weather conditions, such as precipitation or high temperatures, were excluded. Subsequently, the data format was transformed from CSV to Shapefile for spatial intersection processing with administrative division data. This conversion facilitated spatial analysis and integration with administrative boundary information.

For the analysis of the spatial distribution and changes in shared bicycles before and after the entire commuting period, the study focused on examining the spatial distribution changes of shared bicycles during the morning and evening peak hours. The starting times of 6:00 and 17:00 were defined as the beginning of the morning and evening peak hours, respectively, while the ending times of 9:00 and 20:00 were defined as the end of the morning and evening peak hours, respectively. The data corresponding to the starting and ending times were associated with bicycle IDs and spatially intersected with the street administrative divisions, allowing the identification of the streets where each bicycle started and ended, thus generating street-level origin–destination (OD) trajectories.

After filtering out abnormal data, a challenge emerged: the remaining data often had discontinuous dates. To facilitate rigorous experimental validation, our study required a continuous dataset spanning several consecutive working days. Therefore, data from 6 July, 7 July, and 8 July 2021 were meticulously chosen due to being representative of typical working days. These specific dates provided the necessary continuous data over a three-day period, free from irregularities, and reflective of regular daily resident mobility patterns. The list of experimental data used in this study is presented in Table 1. The administrative division data for the central area of the capital can be downloaded from the OpenStreetMap website (https://www.openstreetmap.org/, (accessed on 1 August 2021)). Please note that Table 1 and the provided website link are for illustrative purposes and should be replaced with the actual references and data sources used in the research.

Table 1. List of experimental data.

Data Name	Data Format	Data Collection Time	Data Size	Field Name	Field Explanation
Bicycle sharing data	CSV	6 July, 7 July and 8 July, 2021	2.4 million records	TIME BICYCLE_ID LATITUDE LONGITUDE	Data collection time Bicycle ID number Latitude/° Longitude/°
Administrative district data	Shapefile	April 2021	122	FID SHAPE NAME	Street serial number Element type Street name

2.2. Research Methods

2.2.1. Construction Method of Spatial Interaction Network for Shared Bicycle System

In the realm of complex systems, various interaction relationships found in the real world can be effectively organized and analyzed using complex network theory [34]. Complex networks serve as highly abstract representations of these intricate systems, where individual entities are abstracted as nodes, and the relationships between these entities are depicted as edges in the network [35]. When a network consists of a large number of nodes and edges, it is referred to as a complex network, exhibiting two fundamental characteristics: small-worldness and scale-free property [38].

In the context of urban areas, streets play a crucial role as an important spatial scale for analysis, especially when considering administrative divisions. Exploring the characteristics of streets within urban settings directly supports the decision-making processes at the neighborhood level [39,40]. Consequently, when constructing a spatial interaction network for shared bicycles in urban areas, the first step involves abstracting streets as network nodes based on the principles of complex network theory. The travel behavior between streets is then captured as directed edges in the network, with each edge's weight representing the volume of bicycle usage. Through the establishment of a travel relationship matrix among streets within the central area of the city, the construction of the network is successfully accomplished. By leveraging the framework of complex network theory, this approach enables a comprehensive understanding of the interplay between streets and the dynamics of shared bicycle usage. Relevant metrics from complex network theory can be utilized to quantify the flow characteristics between streets. Quantifying and visualizing spatial interactions form the basis of a spatial interaction network. The resulting spatial interaction network provides a powerful tool for examining the patterns and characteristics of shared bicycle systems within urban areas.

2.2.2. Network Analysis Methods

In the field of complex network theory, there exists a range of metrics and methods concerning nodes, edges, and network construction. Given that the network constructed in this paper is a directed weighted network, the network's edges possess two distinct attributes: direction and weight. Although complex network theory offers a range of network statistical metrics, the number of metrics suitable for assessing directed and weighted spatial interaction networks is limited. These metrics should be capable of considering the network's overall characteristics while allowing for a qualitative or quantitative evaluation and analysis of edge direction and magnitude. Consequently, this study opted to employ three metrics, namely network density, node strength, and degree centrality. These three metrics enable a comprehensive analysis of the network from various perspectives. Network density assesses the connectivity density within the network, while node strength evaluates the capacity of nodes to absorb or emit traffic. Degree centrality, on the other hand, provides insights into the flow of traffic into and out of nodes. Utilizing the three mentioned indicators allows for research from different perspectives, including the overall network, node mobility, and attraction of traffic. This enables a comprehensive analysis of spatial interaction networks' interaction patterns.

Network density is used to reflect the degree of connectivity between nodes, where nodes in this study represent streets [11,30]. A higher value of node strength indicates a stronger relationship and a greater level of mutual influence between nodes. The calculation formula is shown in Equation (1), where D represents the network density, k denotes the number of nodes (i.e., streets), and d represents the number of edges in the network.

$$D = \frac{2d}{k(k-1)} \tag{1}$$

Node strength is defined as the sum of the weights of all edges connected to that node, which reflects the magnitude of travel associated with the street. The calculation formula is shown in Equation (2), where S_i represents the node strength of node i, N_i denotes the set of nodes connected to node i, and W_{ij} represents the weight of the edge connecting nodes i and j.

$$S_i = \sum_{j \in N_i} W_{ij} \tag{2}$$

In a directed network, node strength can be divided into in-strength and out-strength based on the direction of edges, allowing the calculation of net flow ratio (NFR). The calculation is described in Equations (3)–(5), where $S_{in}(i)$ and $S_{out}(i)$ represent the in-strength and out-strength of node i, respectively, v_{in} represents the set of nodes that flow towards node i, W_{ij} represents the edge weight, v_{out} represents the set of nodes to which node i flows, and NFR represents the net flow ratio. When the NFR is greater than zero, it indicates that the in-strength of the node is greater than the out-strength, indicating an increase in prominence, and vice versa [29].

$$S_{in}(i) = \sum\nolimits_{j \in v_{in}} W_{ij} \tag{3}$$

$$S_{out}(i) = \sum\nolimits_{j \in v_{out}} W_{ij} \tag{4}$$

$$NFR = \frac{S_{in}(i) - S_{out}(i)}{S_{in}(i) + S_{out}(i)} \tag{5}$$

Centrality is one of the indicators used to study the state and function of networks, reflecting the status of nodes in the network. The higher the centrality of a node, the greater its influence on the network. Degree centrality normalizes the node degree and is used for comparisons between networks at different scales. It represents the proportion of other nodes connected to a specific node and is an important indicator for reflecting network centrality. The calculation formula for degree centrality is shown in Equation (6), where $C_D(v)$ represents the degree centrality of node v, $deg(v)$ represents the degree of the node, and N represents the total number of nodes.

$$C_D(v) = \frac{deg(v)}{N-1} \tag{6}$$

In directed networks, degree centrality can be categorized based on the direction of edges. Using out-degree centrality and in-degree centrality, the capabilities of nodes to send and receive traffic can be represented. The calculations are shown in Equations (7)–(10), where $C_{O,i}$ and $C_{I,i}$ represent the out-degree centrality and in-degree centrality of a node, respectively. W_{ij} represents the flow from node i to j, and N is the total number of nodes. The $Insurplus$ and $Outsurplus$ metrics are defined to measure the ability of nodes to attract and export traffic in a directed network, with traffic tending to leave nodes with larger $Outsurplus$ values and flow towards nodes with larger $Insurplus$ values.

$$C_{O,i} = \sum_{j=1, j\neq 1}^{N} \frac{W_{ij}}{N-1} \tag{7}$$

$$C_{I,i} = \sum_{j=1, j\neq 1}^{N} \frac{W_{ij}}{N-1} \tag{8}$$

$$Insurplus = C_{I,i} - C_{O,i} \tag{9}$$

$$Outsurplus = C_{O,i} - C_{I,i} \tag{10}$$

2.2.3. Technical Approach

The technical approach of this study is illustrated in Figure 2. In the proposed methodology, the first step involves extracting the operational range of shared bikes from the shared bike data. This is achieved by identifying the boundaries within which the bikes operate. The operational scope is then overlapped with the urban administrative boundary data, enabling the identification of the specific area of interest for this study. Next, street units are selected as the network nodes. These street units are extracted from the study area and serve as the fundamental components of the spatial interaction network. By representing each street unit as a node, the network can capture the spatial relationships and interactions between different streets.

To construct the spatial interaction network, the shared bike trajectory data are subjected to two essential preprocessing steps: outlier data cleansing and format conversion. Outliers, which may arise due to various factors, are removed from the trajectory data to ensure the accuracy and reliability of subsequent analyses. Additionally, the data are converted into a suitable format that facilitates the construction of the network. The OD trajectories are then extracted from the preprocessed shared bike data. Each OD trajectory represents the movement between two locations within the study area. These trajectories are used as the edges of the spatial interaction network, connecting the corresponding street units.

With the network structure established, a range of network statistical indicators are applied to analyze the spatial interaction patterns within the urban street network. These indicators provide insights into various aspects of the network, such as node connectivity, traffic flow, and centrality measures. By examining these network characteristics, a comprehensive understanding of the spatial interactions facilitated by shared bikes in the urban environment can be obtained.

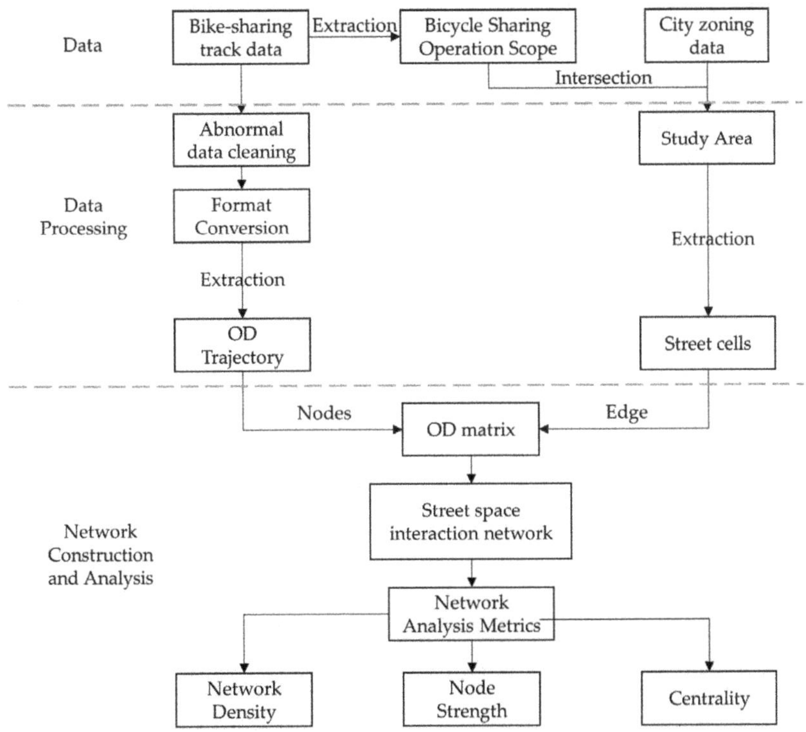

Figure 2. Technical approach.

3. Results

3.1. Spatial Distribution of Shared Bikes

We used kernel density estimation (KDE) to calculate and visualize the spatial distribution of shared bicycle data. Data from three working days in July, specifically July 6th, July 7th, and July 8th, were aggregated at four specific time points: 6:00 AM, 9:00 AM, 5:00 PM, and 8:00 PM. The calculation and visualization of kernel density values are shown in Figure 3. From the range of kernel density values, it is apparent that the KDE values at 9:00 AM and 5:00 PM are higher than those at 6:00 AM and 8:00 PM. Since kernel density reflects the density of data distribution in spatial dimensions, the analysis suggests that the spatial distribution of shared bikes is more concentrated at 9:00 AM and 5:00 PM. This indicates that in the morning, bicycle trip destinations are spatially concentrated, whereas in the evening, bicycle trip origins are more clustered. Considering that shared bicycles are commonly used as commuting tools during weekday peak hours, this observation has implications. On one hand, it suggests that the employment areas in the core of Beijing, as compared to residential areas, have a more concentrated geographical distribution. On the other hand, by comparing the kernel density value ranges between 6:00 AM and 8:00 PM and between 9:00 AM and 5:00 PM, we found that shared bicycle usage is higher and more concentrated in the morning than in the evening. This indicates that more people opt for shared bicycles as a mode of transportation in the morning and switch to other modes in the evening. This could be attributed to the morning trips being primarily for commuting, which are more concentrated. In the evening, trip purposes become more diverse, leading to a variety of transportation modes being used.

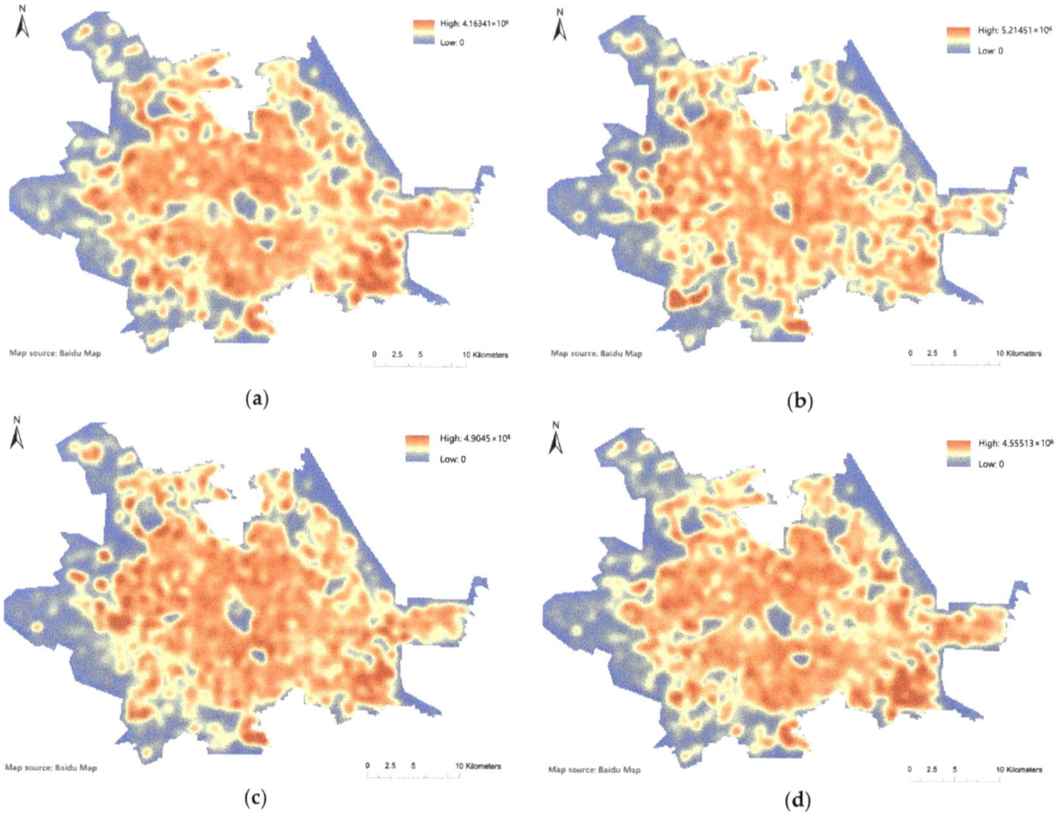

Figure 3. Shared bicycle space distribution: (**a**) 6:00; (**b**) 9:00; (**c**) 17:00; (**d**) 20:00.

3.2. Analysis of Spatial Interaction Network for Shared Bikes

3.2.1. Spatial Distribution of Shared Bikes

In this study, we adopted the principles of complex network theory to construct a spatial interaction network for shared bikes. The network was established by considering streets as nodes and representing the interactions between streets as directed edges. The weight of each edge was determined by the volume of shared bike usage, reflecting the intensity of interaction between streets. To build the network, we created a travel matrix that captures the travel relationships among streets within the urban central area. The network construction, analysis, and visualization were performed using the network library in Python, in conjunction with ArcGIS software. The resulting spatial interaction network is visualized in Figure 4, where the width of the lines corresponds to the magnitude of the flow between streets. In this case, the use of the early period means 6:00–9:00 and the use of the night period means 17:00–20:00. This visualization allows us to visually assess the patterns and dynamics of shared bicycle movements within the study area. Thicker lines indicate higher flow volumes, indicating streets with greater usage and stronger connections with other streets. To further analyze the characteristics of the spatial interaction network, we calculate the network density for each network. Network density quantifies the level of connectivity within a network by measuring the proportion of existing connections compared to the total number of possible connections.

Table 2 presents the network density values for the spatial interaction networks under investigation. Higher network density values indicate a denser network with stronger interconnectivity between streets, implying a higher potential for shared bicycle flow and interactions among streets.

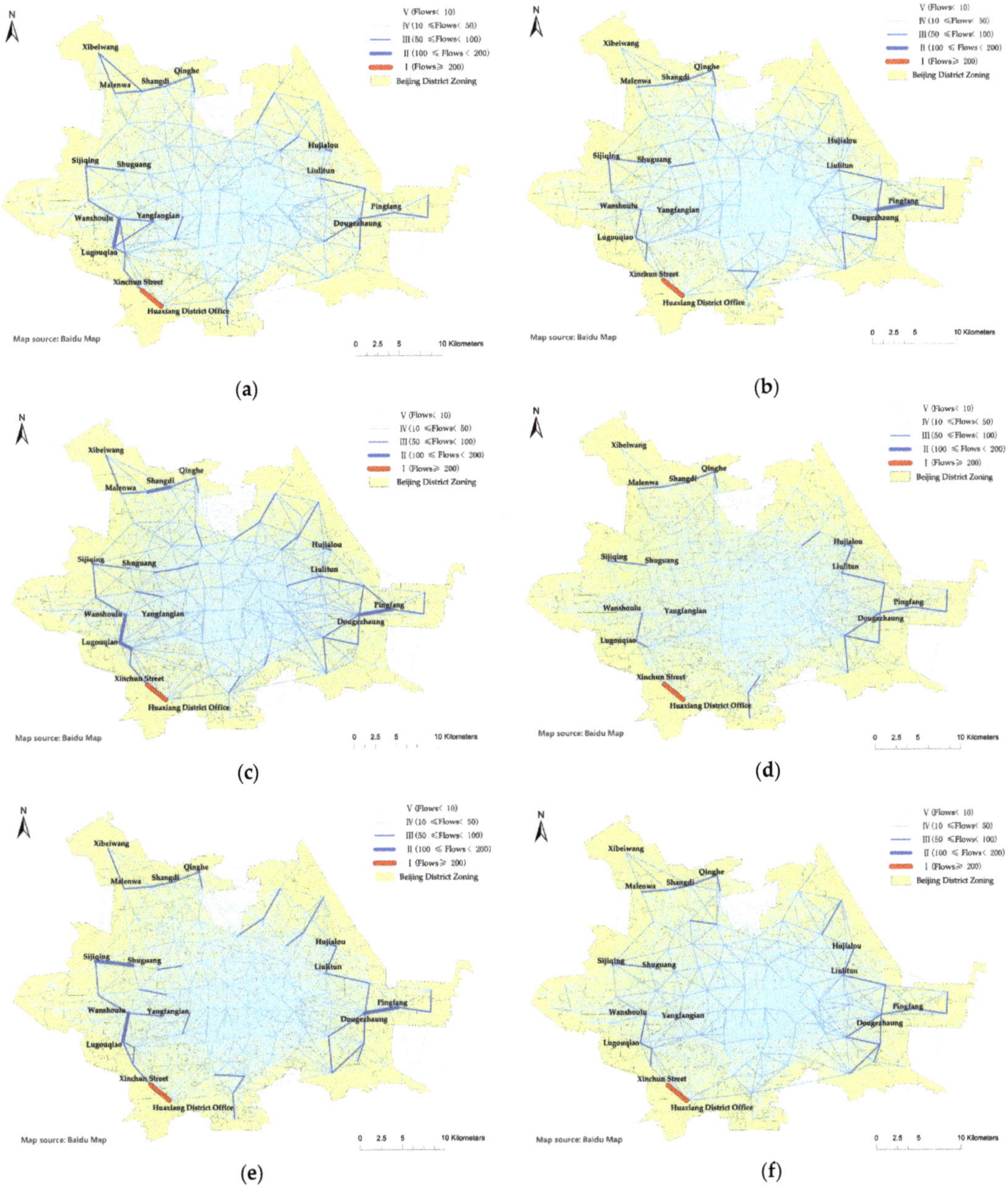

Figure 4. Visualization of spatial interaction networks: (**a**) July 6 early period network; (**b**) July 7 early period network; (**c**) July 8 early period network; (**d**) July 6 night period network; (**e**) July 7 night period network; (**f**) July 8 night period network.

Table 2. Network basic attribute statistics.

Time Period	Network Density		
	July 6	**July 7**	**July 8**
6:00–9:00	0.1661699	0.1715215	0.1764666
17:00–20:00	0.1793117	0.1650673	0.1791085

Upon examining the network density values presented in Table 2 and analyzing the spatial structure of the network depicted in Figure 4, it becomes evident that interactions between streets that are geographically distant from each other are relatively infrequent. Most of the interaction activities occur between adjacent streets, such as the substantial pedestrian flow observed daily between the Hua Xiang District Office and Xin Cun Street. This observation indicates that shared bicycles are primarily used for short-distance trips between neighboring streets, especially during weekdays. This finding is consistent with previous research results.

Furthermore, the visualization results also show that edges with higher flow volumes are primarily distributed on the periphery of the study area, while the central area of the study region exhibits fewer high-flow interactions. The reason for this travel pattern may be attributed to the distinct characteristics of the peripheral areas of Beijing's core region, where the boundaries between employment and residential zones are more pronounced. In contrast, the central area of Beijing's core region is characterized by a more complex mix of attributes, often resulting from historical or cultural factors. Employment units and residential communities are intermixed spatially in this central area. Such spatial attributes often lead to a certain degree of road congestion within the region.

3.2.2. Spatial Characteristics Analysis

Figure 5 illustrates the visualization and analysis of node strength within the spatial interaction network. In this analysis, dot size corresponds to the node strength, offering insights into the relative significance of various streets. Upon examining the figure, it becomes evident that streets with higher node strength can be categorized into two distinct groups. The first category encompasses streets with more substantial weighted edges. These streets, including Dougezhuang District Office, Pingfang District Office, Huaxiang District Office, Lugouqiao Street, Hujialou Street, and Xincun Street, boast a greater number of significant connections with other streets. Their node strength is primarily influenced by the presence of these strong connections, signifying their role in facilitating interactions within the network. The second category comprises streets with a higher degree of interaction with numerous other streets. Streets like Wanshou Road Street, Liulitun Street, and Yangfangdian Street fall into this category, characterized by their popularity and extensive connections to other streets within the network.

Based on the findings above, two recommendations can be made. Firstly, there is a need to enhance the infrastructure of important streets. Streets with higher node strength, such as Dougezhuang District Office, Pingfang District Office, Huaxiang District Office, Lugouqiao Street, Hujialou Street, and Xincun Street, play a crucial role in facilitating interactions within the network. Therefore, urban planners and policymakers should consider investing in bicycle infrastructure in these areas, including dedicated bike lanes and parking facilities. Ensuring an adequate supply of shared bikes during peak commuting hours is also vital. On the other hand, there is a need to expand the shared bike services. Streets with higher node strength, such as Wanshou Road Street, Liulitun Street, and Yangfangdian Street, stand out due to their popularity and extensive connections to other streets within the network. To promote equitable access to sustainable transportation modes, expanding shared bike services to neighboring areas could be considered, making these services more accessible to a broader range of people.

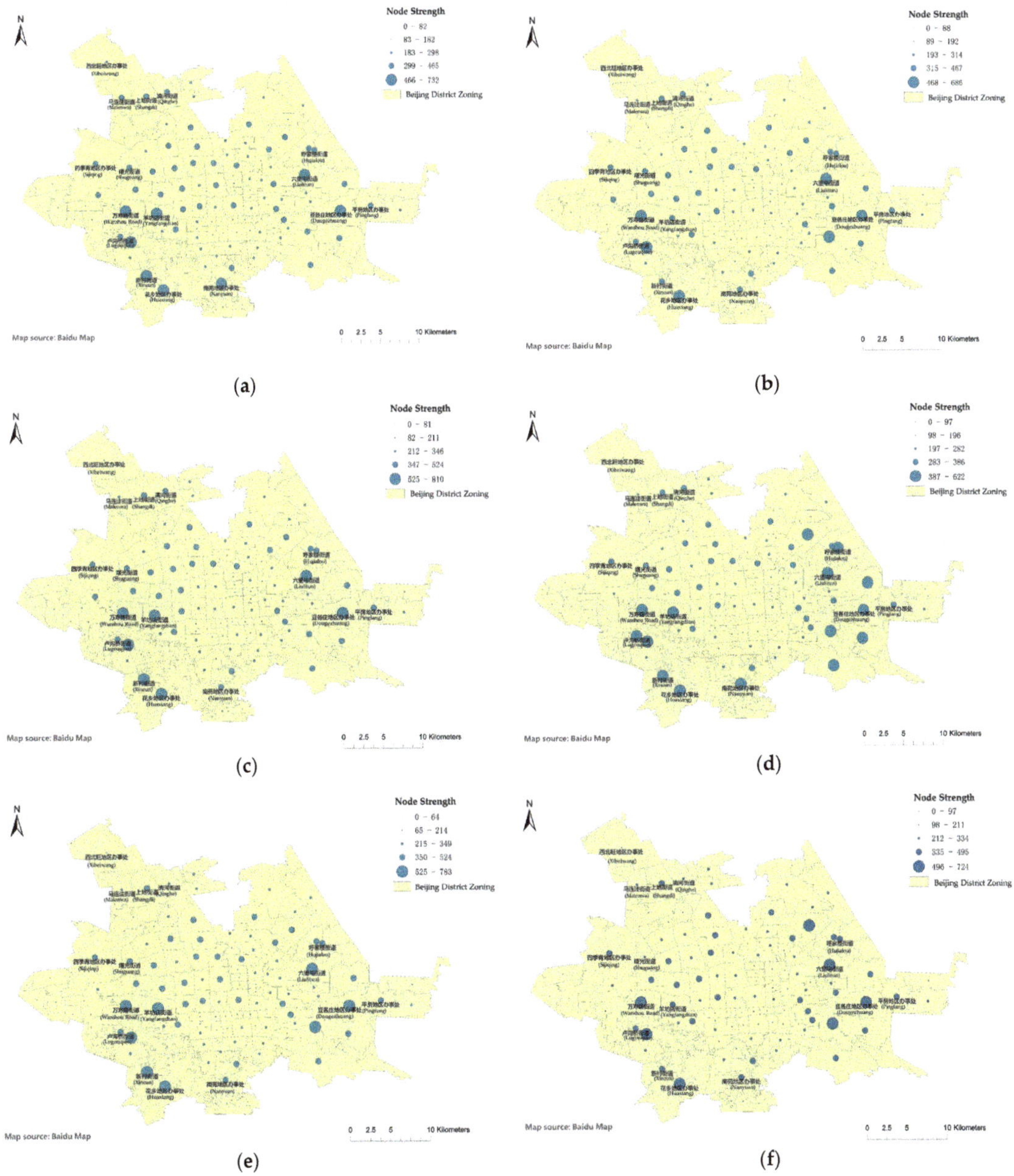

Figure 5. Node strength statistics: (**a**) July 6 early period; (**b**) July 7 early period; (**c**) July 8 early period; (**d**) July 6 night period; (**e**) July 7 night period; (**f**) July 8 night period.

The analysis of the spatial interaction network involves calculating the net flow ratio (NFR) based on the in-degrees and out-degrees of nodes. Figure 6 presents the visualization and analysis of the NFR, with dot size representing the magnitude of the NFR value. Positive NFR values are depicted in red, while negative values are depicted in green. Upon examining the figure, distinct patterns emerge during the early time period. Streets with NFR values greater than zero, indicating an increasing flow of shared bicycles, are primarily concentrated in the central area of the study region. Notable examples include Jinrong Street and Chongwenmenwai Street. These streets exhibit higher levels of activity and are characterized by a higher influx of shared bicycle usage during peak hours. Conversely, streets with NFR values less than zero, indicating a decreasing flow of shared bicycles, are primarily located near the periphery of the study region. This category includes streets such as Lugouqiao Street, Fangzhuang District Office, Huaxiang District Office, and Shibalidian District Office. The reduced NFR values in these areas suggest a lower demand for shared bicycles during the same time period.

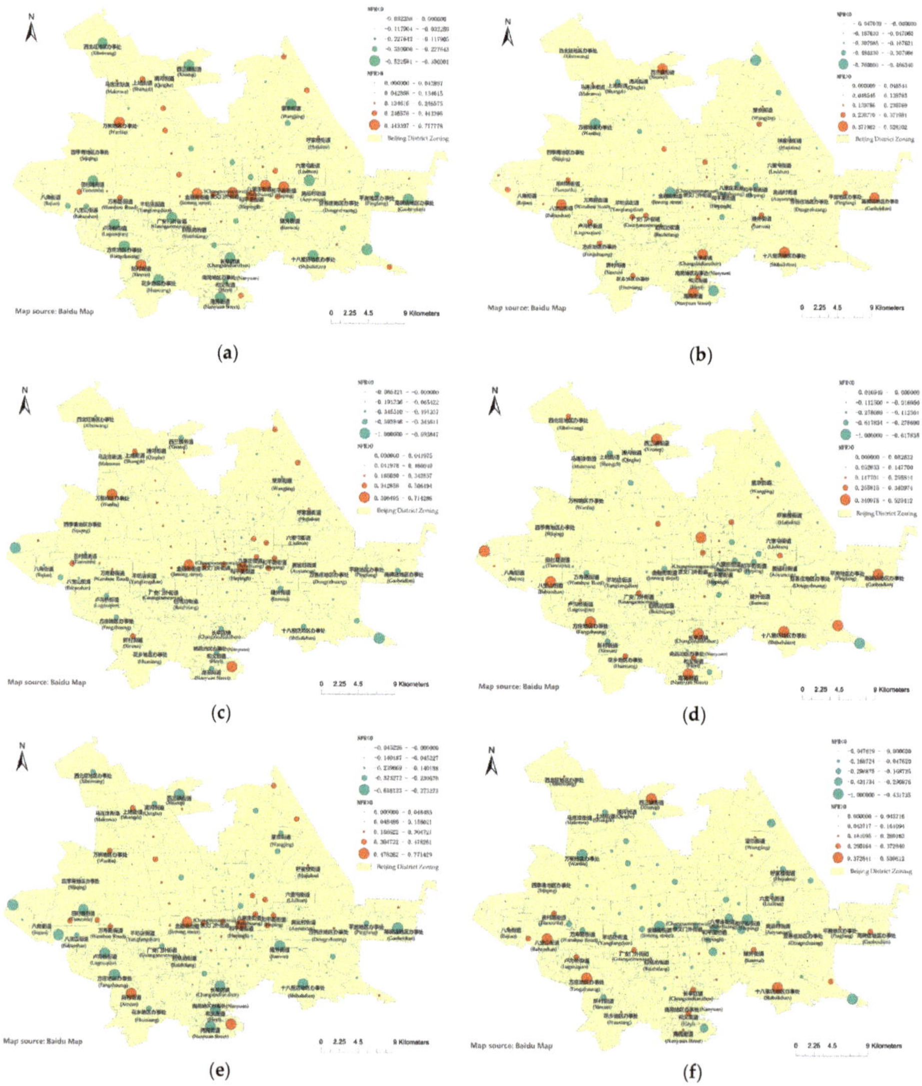

Figure 6. NFR analysis chart: (**a**) July 6 early period; (**b**) July 7 early period; (**c**) July 8 early period; (**d**) July 6 night period; (**e**) July 7 night period; (**f**) July 8 night period.

During the late time period, there is a reversal in the spatial distribution of activity within the study region. Central streets, which previously showed higher levels of activity, now exhibit decreased activity. Conversely, peripheral streets show increased activity, suggesting a higher demand for shared bicycles during this period. These observations are consistent with weekday commuting patterns when shared bicycles are frequently used for transportation to and from work. The central streets in the central district are more likely to be associated with employment areas, whereas the periphery mainly consists of residential areas.

The traffic flow on streets during the early time period corresponds to the flow in the opposite direction during the late time period. This phenomenon can be attributed to the temporal variations in travel patterns and the changing dynamics of shared bicycle usage. Moreover, the relationship between *Insurplus* and *Outsurplus*, as defined by Equations (9) and (10), indicates that they are opposite to each other. Leveraging this understanding, the present study focuses on comparing the *Insurplus* values during the early time period with the *Outsurplus* values during the late time period, as depicted in Figure 7. The findings consistently demonstrate that streets with positive *Insurplus* values exhibit a higher *Insurplus* during the early time period compared to the *Outsurplus* during the late time period. Conversely, for streets with negative *Insurplus* values, the opposite trend is observed. On one hand, they indicate that for the same street, the demand for shared bikes is generally higher during the early time period than during the late time period. This suggests a temporal preference for shared bicycle usage, potentially influenced by factors such as work commutes and daily activity patterns.

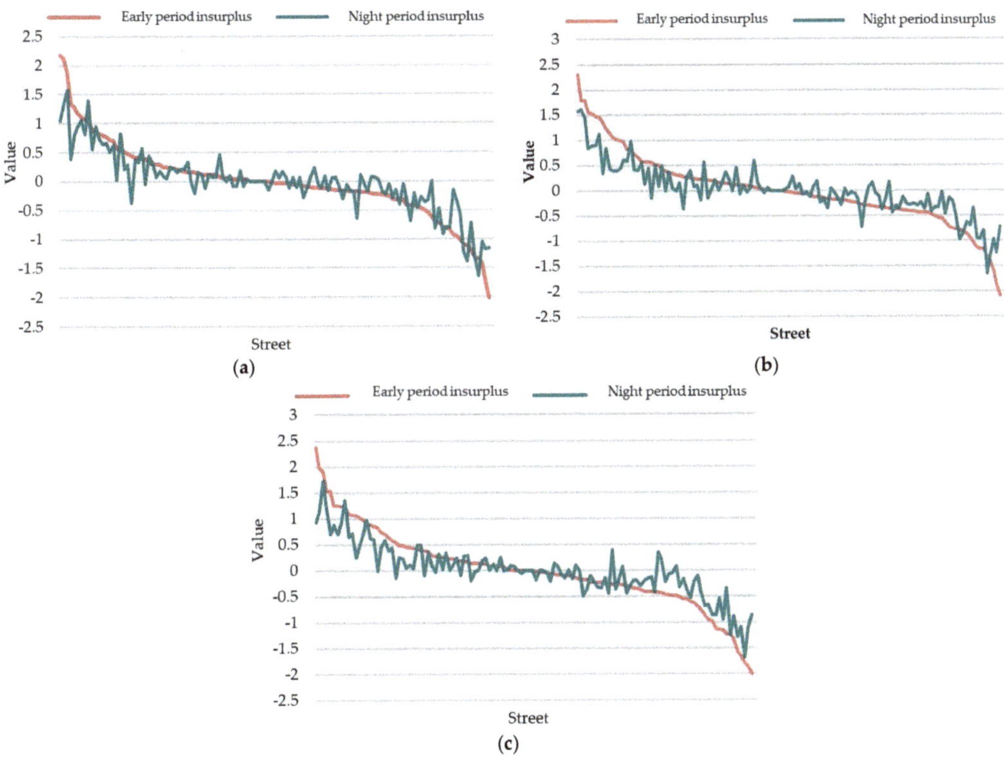

Figure 7. Degree centrality analysis plot: (**a**) July 6 degree centrality; (**b**) July 7 degree centrality; (**c**) July 8 degree centrality.

On the other hand, the results also indicate that streets with higher centrality in the network tend to experience a significant influx of shared bicycle usage during the early hours, followed by an efflux during the later part of the day. This observed pattern can be attributed to the distribution of employment centers and residential areas within the study region. Streets with greater centrality, often situated in the central area, typically function as primary employment hubs where commuters gather during peak hours, resulting in a heightened inflow of shared bicycle usage. Conversely, during the late hours, the outflow of shared bicycles from these central streets may be influenced by various factors, such as commuters returning home or engaging in recreational activities in peripheral areas.

4. Discussion and Conclusions

In this study, we applied complex network theory to construct and analyze a spatial interaction network of shared bicycles in the central urban area of Beijing. Our analysis focused on understanding the spatial interaction patterns of shared bicycle usage and their variations. We conducted an analysis of the spatial interaction characteristics of intra-regional traffic flow, providing valuable insights for urban planning and achievable transportation development. Several key findings have emerged from our investigation.

Regarding the spatial distribution of shared bikes, we employed kernel density estimation to visualize the concentration of bike usage during different time periods. Our results indicated that bike distribution was more concentrated during peak commuting hours, with destinations being more centralized in the early time period and origins being more centralized in the late time period. This suggests that shared bikes serve as a popular commuting option during weekdays, with employment areas exhibiting a higher concentration in the central region compared to residential areas.

Additionally, through the construction of a spatial interaction network, we conducted an analysis of the network structure and pinpointed key streets characterized by higher node strength and popularity. These streets can be categorized into two groups: those with substantial edge weights, signifying significant bicycle usage, and those engaged in interactions with a greater number of other streets, rendering them more favored within the network. In the network, streets with elevated node strength are typically situated in the central region, playing pivotal roles in fostering interactions within the network. These central streets generally bear a heavier load of shared bicycle usage during peak hours, which is associated with their positioning as employment hubs. In contrast, streets in peripheral areas undergo increased usage during the evening, reflecting distinct commuting patterns, with central areas being more employment-oriented and periphery areas primarily residential. Furthermore, the usage of shared bicycles exhibits pronounced variations across different time periods. There is a higher demand in the morning, likely influenced by work commutes and daily activity patterns. Conversely, during the evening, substantial outflows signify diverse travel needs, potentially influenced by factors such as commuters returning home or engaging in recreational activities in peripheral regions.

Based on the conclusions drawn above, given the concentration of shared bicycle usage during the morning peak hours, urban planners should consider enhancing bicycle infrastructure in areas with high employment and residential density. This may involve the addition of bike lanes, parking facilities, and ensuring an adequate supply of shared bicycles during peak commuting hours. Considering the complex mix of employment and residential areas within the central area of Beijing, urban planning policies should be implemented to encourage mixed-use zoning, thereby reducing traffic congestion. This can be achieved by promoting the development of mixed-use buildings, which can help reduce the need for long-distance commuting. Given the higher concentration of shared bicycles in peripheral areas, city authorities may contemplate expanding shared bicycle services to neighboring regions, promoting more equitable access to this sustainable mode of transportation. In conclusion, our study contributes to a deeper understanding of the spatiotemporal dynamics of shared bike usage in the urban central area. The findings emphasize the importance of considering both spatial and temporal factors in planning bike-sharing systems and urban transportation strategies. The insights gained from this research can inform policymakers and urban planners in optimizing the allocation of resources, improving transportation infrastructure, and promoting sustainable urban mobility.

There are certain limitations to this study. Firstly, the data used in this research were limited to just three days in the summer (6 July, 7 July, and 8 July 2021). Secondly, the variety of network indicators used in this study could be further expanded. Future research directions can build upon this study to further explore and address several aspects. Firstly, incorporating additional data sources, such as weather conditions and demographic information, could provide a more comprehensive understanding of the factors influencing shared bike usage patterns. Secondly, investigating the impact of bike-sharing systems on traffic congestion, air quality, and public health would contribute to a broader assessment of the benefits and challenges associated with promoting sustainable urban mobility. Additionally, exploring the spatiotemporal dynamics of shared bikes in different seasons and during special events or holidays could provide insights into the adaptability and resilience of bike-sharing systems. Lastly, integrating advanced analytical techniques, such as machine learning and predictive modeling, could enhance the accuracy of predicting bike usage demand and support more efficient resource allocation. By addressing these research directions, we can continue to advance our understanding of shared bike systems and contribute to the development of sustainable and efficient urban transportation networks.

Author Contributions: Conceptualization, D.T., Z.W., and Y.S.; data curation, D.T. and Z.W.; formal analysis, D.T. and Z.W.; funding acquisition, D.T.; investigation, Z.W.; project administration, D.T.; resources, D.T. and Y.S.; software, Z.W.; supervision, Y.S.; writing—original draft, D.T. All authors have read and agreed to the published version of the manuscript.

Funding: This work was supported by the Study on the Updating Assessment Technique of the Central City of Megacities under Multi-source Data–Beijing as an Example (2021-K-014), the Beijing Key Laboratory of Urban Spatial Information Engineering (20230110).

Institutional Review Board Statement: Not applicable.

Informed Consent Statement: Not applicable.

Data Availability Statement: Not applicable.

Acknowledgments: The authors would like to thank all of the reviewers for their valuable contributions to this work.

Conflicts of Interest: The authors declare no conflict of interest.

References

1. Fishman, E. Bikeshare: A Review of Recent Literature. *Transp. Rev.* **2016**, *36*, 92–113. [CrossRef]
2. Chen, D.; Hu, W.; Li, Y.; Zhang, C.; Lu, X.; Cheng, H. Exploring the temporary and spatial effects of city size on regional economic integration: Evidence from the Yangtze River Economic Belt in China. *Land Use Policy* **2023**, *132*, 106770. [CrossRef]
3. Wang, J.; Huang, J.; Dunford, M. Rethinking the Utility of Public Bicycles: The Development and Challenges of Station-Less Bike Sharing in China. *Sustainability* **2019**, *11*, 1539. [CrossRef]
4. Guo, K.; Cao, Y.; Wang, Z.; Li, Z. Urban and industrial environmental pollution control in China: An analysis of capital input, efficiency and influencing factors. *J. Environ. Manag.* **2022**, *316*, 115198. [CrossRef] [PubMed]
5. Shen, S.; Wei, Z.Q.; Sun, L.J.; Su, Y.Q.; Wang, R.C.; Jiang, H.M. The Shared Bicycle and Its Network—Internet of Shared Bicycle (IoSB): A Review and Survey. *Sensors* **2018**, *18*, 2581. [CrossRef] [PubMed]
6. Ling, S.; Ma, S.; Jia, N. Sustainable urban transportation development in China: A behavioral perspective Frontiers of Engineering. *Management* **2022**, *9*, 16–30.
7. Zhao, D.; Ong, G.P. Geo-Fenced Parking Spaces Identification for Free-Floating Bicycle Sharing System. *Transp. Res. Part Policy Pract.* **2021**, *148*, 49–63. [CrossRef]
8. Gao, K.; Yang, Y.; Li, A.; Li, J.; Yu, B. Quantifying Economic Benefits from Free-Floating Bike-Sharing Systems: A Trip-Level Inference Approach and City-Scale Analysis. *Transp. Res. Part Policy Pract.* **2021**, *144*, 89–103. [CrossRef]
9. Chen, Z.; van Lierop, D.; Ettema, D. Dockless Bike-Sharing Systems: What Are the Implications? *Transp. Rev.* **2020**, *40*, 333–353. [CrossRef]
10. Chen, X.; Ma, Z.; Ye, G.; Li, Z. Usage Behavior and Satisfaction Analysis of Free-Floating Bicycle Sharing System Service: Evidence from a Chinese University Campus. *Res. Transp. Bus. Manag.* **2022**, *43*, 100703. [CrossRef]
11. Hu, L.; Wen, Z.; Wang, J.; Hu, J. Spatial Interaction Analysis of Shared Bicycles Mobility Regularity and Determinants: A Case Study of Six Main Districts, Beijing. *ISPRS Int. J. Geo-Inf.* **2022**, *11*, 477. [CrossRef]

12. Li, Y.; Zhu, Z.; Guo, X. Operating Characteristics of Dockless Bike-Sharing Systems near Metro Stations: Case Study in Nanjing City, China. *Sustainability* **2019**, *11*, 2256. [CrossRef]
13. Garcia-Gutierrez, J.; Romero-Torres, J.; Gaytan-Iniestra, J. Dimensioning of a Bike Sharing System (BSS): A Study Case in Nezahualcoyotl, Mexico. *Procedia-Soc. Behav. Sci.* **2014**, *162*, 253–262. [CrossRef]
14. Torrisi, V.; Ignaccolo, M.; Inturri, G.; Tesoriere, G.; Campisi, T. Exploring the Factors Affecting Bike-Sharing Demand: Evidence from Student Perceptions, Usage Patterns and Adoption Barriers. *Transp. Res. Procedia* **2021**, *52*, 573–580. [CrossRef]
15. Parkes, S.D.; Marsden, G.; Shaheen, S.A.; Cohen, A.P. Understanding the Diffusion of Public Bikesharing Systems: Evidence from Europe and North America. *J. Transp. Geogr.* **2013**, *31*, 94–103. [CrossRef]
16. Wang, X.; Sun, H.; Zhang, S.; Lv, Y.; Li, T. Bike Sharing Rebalancing Problem with Variable Demand. *Phys. Stat. Mech. Its Appl.* **2022**, *591*, 126766. [CrossRef]
17. Dell'Amico, M.; Iori, M.; Novellani, S.; Subramanian, A. The Bike Sharing Rebalancing Problem with Stochastic Demands. *Transp. Res. Part B Methodol.* **2018**, *118*, 362–380. [CrossRef]
18. Chen, J.; Yang, Z.; Cheng, P.; Shu, Y. Rebalancing Bike-Sharing System with Deep Sequential Learning. *IEEE Intell. Transp. Syst. Mag.* **2020**, *13*, 92–98. [CrossRef]
19. Maas, S.; Nikolaou, P.; Attard, M.; Dimitriou, L. Spatial and Temporal Analysis of Shared Bicycle Use in Limassol, Cyprus. *J. Transp. Geogr.* **2021**, *93*, 103049. [CrossRef]
20. Faghih-Imani, A.; Eluru, N. Incorporating the Impact of Spatio-Temporal Interactions on Bicycle Sharing System Demand: A Case Study of New York CitiBike System. *J. Transp. Geogr.* **2016**, *54*, 218–227. [CrossRef]
21. Elhenawy, M.; Komol, M.R.; Masoud, M.; Liu, S.Q.; Ashqar, H.I.; Almannaa, M.H.; Rakha, H.A.; Rakotonirainy, A. A Novel Crowdsourcing Model for Micro-Mobility Ride-Sharing Systems. *Sensors* **2021**, *21*, 4636. [CrossRef] [PubMed]
22. Hu, J.-W.; Creutzig, F. A Systematic Review on Shared Mobility in China. *Int. J. Sustain. Transp.* **2022**, *16*, 374–389. [CrossRef]
23. Lin, P.; Weng, J.; Hu, S.; Alivanistos, D.; Li, X.; Yin, B. Revealing Spatio-Temporal Patterns and Influencing Factors of Dockless Bike Sharing Demand. *IEEE Access* **2020**, *8*, 66139–66149. [CrossRef]
24. Tao, F.; Wang, C. In Response to Green Travel, Full Coverage of Shared Bicycles in Beijing before the End of the Year. *Beijing Commercial Daily*, 23 September 2021.
25. Xing, Q.; Chen, Z.; Zhang, Z.; Huang, X.; Leng, Z.; Sun, K.; Chen, Y.; Wang, H. Charging Demand Forecasting Model for Electric Vehicles Based on Online Ride-Hailing Trip Data. *IEEE Access* **2019**, *7*, 137390–137409. [CrossRef]
26. Qiao, Z.; Xu, X.; Wu, F.; Luo, W.; Wang, F.; Liu, L.; Sun, Z. Urban Ventilation Network Model: A Case Study of the Core Zone of Capital Function in Beijing Metropolitan Area. *J. Clean. Prod.* **2017**, *168*, 526–535. [CrossRef]
27. Zacharias, J. Environmental Correlates of Dock-Less Shared Bicycle Trip Origins and Destinations. *J. Transp. Geogr.* **2021**, *92*, 103013. [CrossRef]
28. Harris, J.K. *Introduction to Exponential Random Graph Models*; Shanghai People's Publishing House and Gezhi Publishing House: Shanghai, China, 2016.
29. Li, X.; Li, W.; Zeng, M.; Zheng, R.; Li, M. Network-Based Methods for Predicting Essential Genes or Proteins: A Survey. *Brief. Bioinform.* **2020**, *21*, 566–583. [CrossRef]
30. Ryan, L.; D'Angelo, A. Changing Times: Migrants' Social Network Analysis and the Challenges of Longitudinal Research. *Soc. Netw.* **2018**, *53*, 148–158. [CrossRef]
31. Zhang, W.; Chong, Z.; Li, X.; Nie, G. Spatial Patterns and Determinant Factors of Population Flow Networks in China: Analysis on Tencent Location Big Data. *Cities* **2020**, *99*, 102640. [CrossRef]
32. Wang, X.; Ding, S.; Cao, W.; Fan, D.; Tang, B. Research on Network Patterns and Influencing Factors of Population Flow and Migration in the Yangtze River Delta Urban Agglomeration, China. *Sustainability* **2020**, *12*, 6803. [CrossRef]
33. Chong, Z.; Pan, S. Understanding the Structure and Determinants of City Network through Intra-Firm Service Relationships: The Case of Guangdong-Hong Kong-Macao Greater Bay Area. *Cities* **2020**, *103*, 102738. [CrossRef]
34. Zhou, Q. Hotspots Detection and Dynamic Analysis from Trajectory Data Based on Spatio-Temporal Data Field Clustering and Complex Network. Master's Thesis, Wuhan University, Wuhan, China, 2017.
35. Freeman, L.C. A Set of Measures of Centrality Based on Betweenness. *Sociometry* **1977**, *40*, 35–41. [CrossRef]
36. Gutiérrez-Moya, E.; Lozano, S.; Adenso-Díaz, B. Analysing the Structure of the Global Wheat Trade Network: An ERGM Approach. *Agronomy* **2020**, *10*, 1967. [CrossRef]
37. Borgnat, P.; Robardet, C.; Abry, P.; Flandrin, P.; Rouquier, J.-B.; Tremblay, N. A Dynamical Network View of Lyon's Vélo'v Shared Bicycle System. *Dyn. Complex Netw. Vol. 2 Appl. Time-Varying Dyn. Syst.* **2013**, *2*, 267–284. [CrossRef]
38. Chen, Q.; Wu, T.-T.; Fang, M. Detecting Local Community Structures in Complex Networks Based on Local Degree Central Nodes. *Phys. Stat. Mech. Its Appl.* **2013**, *392*, 529–537. [CrossRef]
39. Watts, D.J.; Strogatz, S.H. Collective Dynamics of 'Small-World' Networks. *Nature* **1998**, *393*, 440–442. [CrossRef] [PubMed]
40. Barabási, A.-L.; Albert, R. Emergence of Scaling in Random Networks. *Science* **1999**, *286*, 509–512. [CrossRef]

buildings

MDPI

Article

Entropy Change of Historical and Cultural Heritage in Traditional Tibetan Area of China Based on Spatial-Temporal Distribution Pattern

Xiwei Xu †, Junyu Zhang †, Shupeng Liu *,†, Jiaqi Liu, Zhen Zhang and Xiaoyuan Tian

Department of Urban Planning, School of Architecture, Tianjin University, Tianjin 300072, China; xuxiwei@tju.edu.cn (X.X.); zhang_jy@tju.edu.cn (J.Z.); liujiaqi0202@tju.edu.cn (J.L.); zhang_zhen@tju.edu.cn (Z.Z.); xiaoyuan.tian@tju.edu.cn (X.T.)

* Correspondence: shupeng_liu@tju.edu.cn; Tel.: +86-151-229-11068

† These authors contributed equally to this work.

Abstract: The traditional Tibetan area of China is an ethnically and culturally significant region with a historical geographical connection. This study investigates the spatial-temporal distribution patterns and entropy changes of historical and cultural heritage by examining the association between cultural heritage and socio-historical factors. It utilizes analytical methods such as information entropy and incorporates temporal, spatial, and typological information from the data obtained in the Third National Cultural Relics Census. The findings are as follows: (1) The three major regions in the Tibetan area of China alternately serve as development cores for the traditional Tibetan area, exhibiting a fluctuating "dispersion-aggregation" trend of historical and cultural heritage, which also displays notable regional variations. (2) The quantity and entropy change of historical and cultural heritage exhibit correlations between different periods, but there are also some intergenerational differences. (3) The spatial-temporal distribution pattern of historical and cultural heritage demonstrates an inter-era correlation, indicating that socio-historical development is a nonlinear process characterized by both "transition" and "accumulation". These findings are of significant importance for further understanding the social evolutionary process of human settlements in high-altitude areas and for the comprehensive protection of cultural heritage in ethnic regions.

Keywords: historical and cultural heritage; spatial distribution pattern; information entropy; traditional Tibetan area of China

1. Introduction

Since 2017, China has launched the Second Comprehensive Scientific Expedition to the Qinghai-Tibet Plateau (Qinghai-Tibet Scientific Expedition), which focuses on the human activity history and its impact on the Qinghai-Tibet Plateau. China has a civilization history of over two thousand years. Through long-term interaction and exchange among different ethnic groups, a diverse and unified Chinese civilization has been formed [1]. As an important ethnic region within the Chinese civilization framework, the traditional Tibetan area has produced a precious and unique historical and cultural heritage. Discussing the historical and cultural heritage of the traditional Tibetan area helps us better understand the movement and changes of human cultural activities in the plateau region throughout the history of Chinese civilization. It also provides valuable clues for the Qinghai-Tibet Scientific Expedition and cultural relics census.

In November 2023, the State Council of China issued a notice on conducting the Fourth National Cultural Relics Census, signaling the upcoming launch of a new round of historical and cultural heritage survey nationwide. According to incomplete statistics, China currently has 760,000 historical and cultural heritage sites, which not only include material remains such as ancient sites, buildings, and tombs, but also encompass natural resources with cultural significance, cultural landscapes involving human–nature interaction, and cultural exchange [2]. In 1984, the United States established its first national heritage area, considering interconnected historical and cultural heritage as special linear cultural landscape resources [3]. In recent years, cross-regional historical and cultural heritage resources such as canal waterways [4], railway roads [5], and commercial and cultural routes [6] have received widespread attention from academia.

Analyzing the spatial distribution patterns and formation processes of historical and cultural heritage has become an important scientific topic for reshaping regional civilization processes and strengthening local identity, as it allows for an exploration of the history of regional development and evolution. Historical and cultural heritage (HCH) is the explicit expression of national memory and cultural identity. It records the cumulated social history of spatial resource allocation and reflects social interactions and cultural transmission between nations or regions during certain historical periods [7]. The formation and evolution of historical and cultural heritage are influenced by various factors such as natural climate, geographical conditions, political orientation, and socio-economic factors. There is no consensus on how these factors in history influence historical and cultural heritage.

It can be inferred that social events and processes, cultural diffusion and dissemination, and other information are recorded to varying degrees in historical and cultural heritage. The political, economic, and cultural changes in different historical periods will have an impact on the distribution of heritage. Cultural heritage from different historical periods in the same region will overlap, merge, and transform, forming a new and organic landscape system and spatial order [8]. Although most cultural heritage in a region is expressed in point form, within the framework of territorial spatial planning, the spatial distribution pattern of historical and cultural heritage can be seen as a complex structural entity that involves social processes, cultural stratification, landscape systems, and human–land relationships [9].

Through literature review, we can find that there has been more attention given to the large-scale patterns of historical and cultural heritage [10,11] and cultural heritage routes [12], while there has been less discussion on the distribution of historical and cultural heritage in special regional areas such as ethnic regions, high-altitude areas, and cross-basin areas. Although the geographical characteristics of the integrity [13], systematicity [14], regionality [15], and scalability [16] of historical and cultural heritage have been widely recognized, there has been relatively limited research on the complexity [17], temporality [18], and multiplicativity of historical and cultural heritage. Although some studies have discussed the potential threats and interferences faced by current historical and cultural heritage [19], the impact mechanisms of past geographic factors on the current distribution patterns of historical and cultural heritage still need further investigation and verification [20,21]. Some studies have discussed the distribution patterns of cultural heritage of different periods and types, but little is known about the reasons for changes in the distribution patterns of cultural heritage [22].

Some scholars use time series models to analyze the advantages and changes in the declaration of world cultural heritage in different regions and periods, and assess the impact of heritage strategies on the number of heritage sites included in each region [23]. This interprets the spatial-temporal distribution patterns of historical and cultural heritage as guided by policies, excluding geographical factors. Some scholars have explained the social and historical role of cultural heritage and argue that cultural heritage is primarily a product of economic and social development [24], as well as the dissemination of religion [25], rather than a result of natural geography [23,26,27]. Therefore, the spatial-temporal distribution patterns of historical and cultural heritage should not be regarded as a simple system of geographical factors, but as a geographical phenomenon that can reflect the historical processes of regional societal development and human–environment relationships. It is necessary to consider cultural heritage as the richness of regional social culture [28] and judge the historical processes of regional society based on the complex distribution patterns of cultural heritage [29].

The study of China's historical and cultural heritage has shifted from a simple case to the study of the coupling between nature and humanity, the relationship between time-space and environment, and the social influence. Scholars have discussed the spatial distribution [30], temporal distribution [31,32], and cultural connotation [33] of historical and cultural heritage in different spatial scopes such as the world [34,35], regions [36,37], river basins [38], and cities [39], and have analyzed the influences of regional geography, national politics, ethnic differences, and other factors on the spatial-temporal distribution pattern of historical and cultural heritage [40,41].

Research on historical and cultural heritage in ethnic regions often concentrates on urban morphology patterns [42,43], ethnic culture [44], religious governance [45], and phenomenon analysis of cultural landscapes [46,47]. The research scope is mostly at the town level or within provinces in areas where ethnic minorities are concentrated, with less study on the spatial distribution and evolution of historical and cultural heritage across administrative boundaries in regions where ethnic minorities live [48]. Analyzing the spatiotemporal distribution characteristics and network structure relationships of historical and cultural heritage can promote human understanding of social changes, cultural succession, and the evolution of human–land relationships [49,50].

The majority of traditional Tibetan areas in China are located on the Qinghai-Tibet Plateau, which is the highest-elevated ethnic region in the world. Despite the analysis of the Prehistoric era in high-altitude areas through geological exploration [51], genetic testing [52,53], ecological footprint [54], and humanity relics [55], our understanding of human activities and social processes on the Qinghai-Tibet Plateau remains very limited [56–59]. Despite the evidence of human activities in the traditional Tibetan area dating back millions of years [60], including highly mobile, long-distance, and large-scale migration in the early human activities [61], our understanding of human activities and social processes in the traditional Tibetan area during the medieval period on the Qinghai-Tibet Plateau remains limited [62–64].

The continuity and development of Tibetan history and culture have been a topic of wide international interest, often criticized for marginalization or cultural extinction of Tibet due to various reasons [65,66]. Discussing the historical and cultural heritage of the traditional Tibetan area from a cultural perspective and incorporating it into the scope of anthropological research by utilizing temporal [67], spatial [68], and typological [69] information stored in the historical and cultural heritage can contribute to a better understanding of cultural exchange trends and spatial evolution mechanisms in high-altitude human settlements. It can also provide strong evidence that the traditional Tibetan area has been an integral part of China since ancient times and, more importantly, help better understand the close connection between the traditional Tibetan area and the central government in history, thereby reshaping the historical narrative discourse of the traditional Tibetan area [70].

In response to the issues of geographical determinism and causal inversion in the current research on the spatial and temporal distribution patterns of cultural heritage, the concept of the complexity of historical and cultural heritage is proposed [71,72]. It discusses the differences and connections between different types and periods of cultural heritage, with time as a potential driving force, forming a formative analytical framework to examine the diversity of historical and cultural heritage. By observing the changes in the diversity of historical and cultural heritage, the phenomenon of cultural exchange, dissemination, and integration that occurred in different historical periods within a region can be discussed [73,74]. Most historical and cultural heritage consists of inherited ancient buildings and structures. Discussing the diversity of historical and cultural heritage helps to bring cultural heritage research into the perspective of anthropology and regional history, which is of great significance for further understanding the unique historical and cultural processes of high-altitude ethnic regions [75].

Therefore, this study aims to analyze the spatiotemporal distribution patterns of 4367 historical and cultural heritages, revealing their distribution in different regions and time periods. This analysis not only helps us understand the background and evolution of these cultural heritages but also provides important reference for the study of regional history. Additionally, through studying their hierarchical patterns, we can uncover the interconnections and influences among different types of cultural heritages, contributing to a more comprehensive understanding of their value and significance. In this study, we will utilize tools such as information entropy and Pearson correlation for analysis. Information entropy assists in quantifying the complexity and diversity of the spatiotemporal distribution patterns of historical and cultural heritages, thus revealing their regularities and characteristics. Meanwhile, Pearson correlation helps explore the degree of correlation between historical and cultural heritages in different periods, further uncovering their mutual influence and mechanisms.

Through the implementation of this study, we hope to provide scientific basis and decision-making support for the protection and inheritance of historical and cultural heritages in traditional Tibetan areas of China. Additionally, we aim to promote the inheritance and promotion of traditional Chinese culture, enhance people's sense of identity and pride in historical and cultural heritages, and stimulate the diversified development and prosperity of culture [76].

2. Materials and Methods

2.1. Study Area

The research area of this study is the traditional Tibetan area (26°51′ N–39°12′ N, 78°23′ E–104°45′ E), with an area of approximately 2,613,819 km^2. The traditional Tibetan area in China consists of the regions of Weizang, Anduo, and Kangba, which cover the entire Qinghai and Xizang, as well as parts of Sichuan, Yunnan and Gansu Province. It is an extensive cultural region [77] inhabited mainly by the Tibetan ethnic group, with 19 prefecture-level cities and 161 districts and counties (Figure 1). The traditional Tibetan area is one of the highest-altitude regions in the world, with its steep terrain and complex natural geography nurturing a unique ethnic way of life and religious and social culture than other regions of China.

Figure 1. Study area. (Self-drawn. The Chinese map is produced based on the standard map with approval number GS (2020)4621 downloaded from the website of the Ministry of Natural Resources (http://www.mnr.gov.cn, accessed on 15 July 2023), and the boundaries of the base map have not been modified).

Since the Yuan Dynasty unified the Tibetan areas and established local governments, the three major Tibetan regions gradually formed. They include the Weizang region (religious domain) primarily in Tibet, the Anduo Tibetan region (pastoral domain) primarily in Qinghai, and the Kangba Tibetan region (human domain) primarily in western Sichuan and eastern Tibet. These three regions are not only dialectal regions in the Tibetan language but also different human and geographical regions. The Tibetan proverb says, "Weizang people are religious, Kangba people are warriors, and Anduo people are good at business". This reflects the basic understanding of the characteristics of the three major regions by Tibetans. The traditional Tibetan area has a large number and diverse types of historical and cultural heritage. For the purpose of convenience and understanding, this study uses the modern Chinese Pinyin system for place names. Terms such as Weizang (Ü-Tsang), Kangba (Kham), Anduo (Amdo), Rikaze (Shigatse), Changdu (Chamdo), and Lasa (Lhasa) are not written using the traditional Tibetan phonetic system.

2.2. Methods

2.2.1. Research Approach

The spatial-temporal distribution pattern of historical and cultural heritage refers to the structural distribution state of material remains with cultural functions within a certain time range and a certain geographical range, which is the result of long-term interaction between the natural environment and human activities. We know that the distribution pattern of historical and cultural heritage is the spatial expression of multiple interacting factors. However, the process of change for many of these factors is unclear. Therefore, by suspending other elements, we view the distribution pattern of historical and cultural heritage as a spatial system, hoping to find clues to the changes from the distribution pattern of historical and cultural heritage. By analyzing this spatial system, the complexity of traditional Tibetan historical and cultural heritage and the social-historical processes it reflects can be discussed (Figure 2).

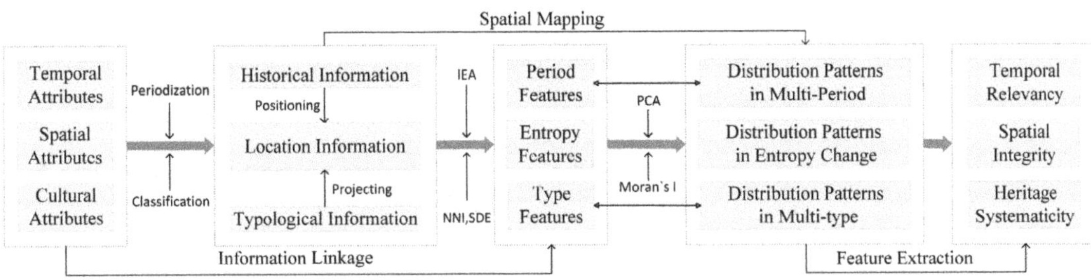

Figure 2. Technical route of this study (Self-drawn).

The distribution pattern of historical and cultural heritage is discussed using basic geographic analysis methods. The method of periodization is used to determine the temporal characteristics of cultural heritage, while the method of classification is used to determine the cultural characteristics of cultural heritage. The combination of temporal and typological features is used to discuss the entropy changes in historical and cultural heritage. By transforming the features of three dimensions into the distribution pattern of historical and cultural heritage, the temporal correlation, spatial integrity, and heritage system of traditional Tibetan historical and cultural heritage can be summarized. Analyzing cumulative temporal changes and entropy rules of historical and cultural heritage is of great significance for scientifically grasping regional development trends and optimizing the pattern of historical and cultural heritage.

2.2.2. Information Entropy Algorithm (IEA)

With technological advancements, new theories and technologies, including social spatial computing, cultural geospatial analysis, complex systems networks, and others, are gradually being introduced into the study of historical and cultural heritage. Methods such as exploratory spatial analysis have gradually been applied to the study of historical and cultural heritage [78–80], providing new possibilities for exploring the distribution patterns of historical and cultural heritage [81].

The spatial-temporal distribution patterns of historical and cultural heritage were gradually formed in ancient societal development and evolution. Although some studies have explored the conditions of the spatial-temporal distribution patterns of historical and cultural heritage using contemporary geographic data, such as construction [82], transportation [83], hydrology [30], and climate [84], and argue that these conditions have an impact on the distribution patterns of cultural heritage, the causal relationship between historical and cultural heritage and other geographic factors is still not clear [85,86], and often falls into the trap of geographical determinism.

The current mainstream methods of multi-factor exploration in academic research often result in logical problems of causal inference inversion, leading to temporal contradictions. Therefore, information entropy is used as the main technical method to describe the complexity changes in the distribution patterns of historical and cultural heritage. Within a certain period of time and spatial range, the more numerous and diverse the quantity and types of historical and cultural heritage, the more complex we consider them to be.

Information entropy is an index used to describe the degree of confusion of things. The distribution of historical and cultural heritage in the region presents local characteristics that change with time series, and this characteristic is composed of structure and randomness. In order to measure and predict historical and cultural heritage, Shannon's information entropy theory is used to perform information entropy calculation of historical and cultural space in different time phases.

$$H(x) \triangleq -P(x_i)\sum_{i=1}^{n} \log_2[P(x_i)] \ (P(x_i) = \frac{k_i}{N} \ i = 1, 2, \ldots, \mathrm{N}) \quad (1)$$

where $H(x)$ is the information entropy of historical and cultural heritage, $P(x_i)$ is the probability value of each type of cultural heritage, k_i is the frequency of a certain type of cultural heritage, and N is the total amount of cultural heritage. The calculation based solely on chronological and typological information may have theoretical shortcomings in terms of geographic factors and explanatory power of phenomena. However, it can still serve as a quantitative method to study the history of human societies within a specific region.

2.2.3. Basic Geographical Analysis

This paper analyzes the spatial information characteristics of historical and cultural heritage using the nearest neighbor index (NNI) and standard deviation ellipse (SDE) (Table 1). The diversity and complexity of the distribution of historical and cultural heritage are studied using the information entropy algorithm. Spatial Autocorrelation Analysis (Moran's I Index) and Pearson's Correlation Analysis (PCA) are applied to explore the interrelationships among historical and cultural heritage, providing a quantitative analysis of regional social-historical changes. This offers new insights into the complexity and diversity characteristics of historical and cultural spatial distribution.

Table 1. Basic geographic analysis technology of this study (Self-drawn).

Methods	Formula	Parameter	Role
Nearest Neighbor Index (NNI)	When the nearest neighbor index R = 1, it means that the point-like elements tend to be randomly distributed, which is random type; NNI < 1 tends to be agglomerative distribution; NNI > 1 tends to be uniform distribution. $NNI = \frac{\overline{r_1}}{\overline{r_E}}$ $\overline{r_E} = \frac{1}{2\sqrt{\frac{n}{A}}} = \frac{1}{2\sqrt{D}}$	r_1 is the average actual nearest neighbor distance value; r_E is the theoretical nearest neighbor distance; A is the area of the study area; N is the number of cultural heritage points; D is the density of cultural heritage points.	To judge the distribution pattern of point elements in space.
Standard Deviation Ellipse (SDE)	The larger the flatness, the more obvious the directionality of the data. Conversely, it indicates less obvious directionality; the shorter the short half axis, the more pronounced the centripetal force presented by the data. Conversely, the greater the degree of dispersion of the data. $C = \begin{bmatrix} var(x) cov(x,y) \\ cov(y,x) var(y) \end{bmatrix} = \frac{1}{n}\begin{pmatrix} \sum_{i=1}^{n} \overline{x_i}^2 \sum_{i=1}^{n} \overline{x_i}\overline{y_i} \\ \sum_{i=1}^{n} \overline{x_i}\overline{y_i} \sum_{i=1}^{n} \overline{y_i}^2 \end{pmatrix}$ $tan\theta = \frac{\left(\sum_{i=1}^{n} X_i^2 - \sum_{i=1}^{n} Y_i^2\right) + \sqrt{\left(\sum_{i=1}^{n} X_i^2 - \sum_{i=1}^{n} Y_i^2\right)^2 + 4\left(\sum_{i=1}^{n} X_i Y_i\right)^2}}{2\sum_{i=1}^{n} X_i Y_i}$ $\sigma_x = \sqrt{2}\sqrt{\frac{\sum_{i=1}^{n}(X_i cos\theta - Y_i sin\theta)^2}{n}}$ $\sigma_y = \sqrt{2}\sqrt{\frac{\sum_{i=1}^{n}(X_i sin\theta - Y_i cos\theta)^2}{n}}$	C is the coordinate of the standard deviation ellipse; $\overline{x}, \overline{y}$ represent the arithmetic mean center of the geographic feature; x_i, y_i are the spatial position coordinates of the geographic feature; θ represents the azimuth of the ellipse; X_i, $Y_i Y_i$ are the mean center deviations; σ_x, σ_y are the standard deviations of the x-axis and y-axis, respectively.	To describe the spatial distribution characteristics of geographical elements, including the distribution center, distribution range, density, development direction, etc.
Spatial Autocorrelation Analysis (Moran's I Index)	When I > 0 and Z > 0, it indicates a High-High cluster. When I < 0 and Z < 0, it indicates a Low-Low cluster. When I > 0 and Z < 0, it indicates a Low-High cluster. When I < 0 and Z > 0, it indicates a High-Low cluster. $I_i = \frac{n(x_i - \overline{x})\sum_{j=1}^{n} W_{ij}(x_j - \overline{x})}{\sum_{i=1}^{n}(x_i - \overline{x})^2}$ $(Z(I) = \frac{I - E(I)}{\sqrt{VAR(I)}})$	n represents the number of spatial objects; W_{ij} denotes the spatial weight matrix of unit i and j within the study area; *E (I)* represents the expected value of Moran's I statistic, *VAR (I)* represents the variance of Moran's I statistic.	Refers to the existence of relationships between variables within a certain spatial area and the same variables in surrounding areas.
Pearson's Correlation Analysis (PCA)	The correlation coefficient is not correlated or very weakly correlated between 0—0.2; weakly correlated between 0.2–0.4; moderately correlated between 0.4–0.6; strongly correlated between 0.6–0.8; and extremely strongly correlated between 0.8–1. $\rho_{x,y} = \frac{cov(x,y)}{\sigma_x \sigma_y} = \frac{\sum_{i=1}^{n}(x_i - \overline{x})(y_i - \overline{y})}{\sqrt{\sum_{i=1}^{n}(x_i - \overline{x})^2 \sum_{i=1}^{n}(y_i - \overline{y})^2}}$ $(\rho_{x,y} \in [-1,1])$	$\rho_{x,y}$ represents the correlation of two variables; x_i, y_i are the measurements of the random variables x and y, respectively; $\overline{x}$ and $\overline{y}$ represent the measurements of x and y, respectively; n is the number of sequences.	Pearson's correlation analysis is an indicator measure of the dependence of two variables.

2.3. Materials

This research is based on the cultural heritage resource list publicly released by national ministries and local governments, including the cultural heritage protection units (919 sites) and general cultural heritage sites (3448 sites) announced by cities, counties, and districts during the Third National Cultural Relic Census. The administrative division data are sourced from the Resource and Environmental Data Cloud Platform of the Chinese Academy of Sciences (http://www.resdc.cn/, accessed on 15 July 2023).

Based on the cultural heritage resources (Table 2), this study verifies the spatial location of cultural heritage by referencing local chronicles, cultural yearbooks, cultural relic bulletins, and government department files, and forms a cultural heritage database containing construction information, which is categorized by type and period. For cross-regional cultural heritage, multiple key areas are selected for placement. The cultural heritage data used in this study may have some duplication, such as some national key cultural heritage protection units including general heritage sites. The original lists are retained without deletion.

Table 2. Distribution statistics of the quantity statistics of cultural heritages (Self-drawn).

Regions	Cities	National Key Cultural Relics Protection Units	National Key Cultural Relics Protection Units	Municipal and County-Level Cultural Relics Protection Units	Ordinary Cultural Heritage	Total
WEIZANG	ALI	7	27	—	98	132
	HAIXI	5	14	6	57	82
	LASA	17	44	1	279	341
	LINZHI	2	10	8	109	129
	NAQU	2	13	24	128	167
	RIKAZE	17	24	16	366	423
	SHANNAN	17	33	9	246	305
KANGBA	ABA	33	18	88	599	738
	CHANGDU	11	15	33	148	207
	DIQING	6	5	19	9	39
	GANZI	17	22	35	91	165
	YUSHU	7	8	4	55	74
ANDUO	GANNAN	9	23	43	356	432
	GUOLUO	3	10	2	72	87
	HAIBEI	4	14	26	61	105
	HAIDONG	13	—	—	9	22
	HAINAN	9	60	43	527	639
	HUANGNAN	4	18	16	237	275
	XINING	5	—	—	1	6
Total		188	358	373	3448	4367

3. Results

3.1. The Distribution Pattern of Historical Cultural Heritage in Multiple Periods

3.1.1. Distribution Characteristics of Historical and Cultural Heritage in Multiple Periods

According to the principle of relative balance, the historical and cultural heritage of the traditional Tibetan region can be divided into eight periods (Table 3). The historical cultural heritage in the traditional Tibetan areas of China has to some extent preserved the process of civilization competition and cultural integration. The historical cultural heritage of the traditional Tibetan areas in China shows a significant trend of concentration in chronological order (Figure 2).

By calculating the nearest neighbor index (NNI) for cultural heritage from different periods, it can be observed that the traditional Tibetan cultural heritage in China exhibits significant clustering characteristics. Cultural heritage from the Stone Age shows the highest level of clustering, with an NNI of 0.265, while the cultural heritage from the

Qin-Han and Southern and Northern Dynasties is relatively dispersed, with an NNI of 0.451. Although the quantity of cultural heritage from the Ming Dynasty is similar to that of the Stone Age, the cultural heritage from the Ming Dynasty is more dispersed. Under the assumption of relatively stable natural conditions, it is evident that human activities during the Ming Dynasty were more extensive, indicating that the socio-economic aspects of the Ming Dynasty were more adaptable and resilient to the natural geographic environment. The number of cultural heritages from the Bronze Age is similar to the number of cultural heritages from the Qing Dynasty, while the distribution of cultural heritages from the Bronze Age is more concentrated, with an NNI of 0.272. The number of cultural heritages from the Qin-Han and Southern and Northern Dynasties is similar to the number of cultural heritages from the Liao, Song, and Yuan dynasties, but the cultural heritage sites from the Qin-Han and Southern and Northern Dynasties are more dispersed, with an NNI of 0.451.

Table 3. The NNI statistics of historical and cultural heritage in multiple periods (Self-drawn).

Periods	Era Range	Quantity	Average Actual Nearest Distance (m)	NNI	Z Value
The Stone Age (SA.)	Before BC 5000	593	8877.36	0.265	−34.26
The Bronze Age (BA.)	BC 5000–BC 221	722	22,513.85	0.272	−37.41
The Qin-Han and Southern and Northern Dynasties (QHSND.)	BC 221–AD 581	177	12,532.07	0.451	−13.96
The Sui-Tang, and Five Dynasties (STFD.)	AD 581–AD 960	820	8688.30	0.309	−37.85
The Liao, Song, and Yuan Dynasties (LSYD.)	AD 960–AD 1368	196	18,616.24	0.395	−16.22
The Ming Dynasty (MD.)	AD 1368–AD 1644	598	11,170.56	0.327	−31.47
The Qing Dynasty (QD.)	AD 1644–AD 1840	951	9386.84	0.346	−38.56
The Modern and Contemporary Period (MCP.)	AD 1840–AD 1949	310	15,909.95	0.358	−21.62

3.1.2. Distribution Pattern of Historical and Cultural Heritage in Multiple Periods

Overall, the distribution and aggregation characteristics of historical and cultural heritage in the traditional Tibetan areas have been continuously significant. With the passage of time, the core of heritage aggregation has shifted among the three major Tibetan areas and become stable (Figure 3). From the perspective of the distribution of cultural heritage, there is a trend of aggregation from west to east, which is particularly evident during the Bronze Age and the Qin-Han and Southern and Northern Dynasties.

The Stone Age historical and cultural heritage was concentrated in the eastern edge of the plateau and the surrounding area of the Yellow River basin, showing an overall uneven distribution pattern with denser distribution in the east and sporadic distribution in the west. This period was influenced by the altitude and natural environment, with the regions of the Gangdisi Mountains and the Hengduan Mountains being unsuitable for human activities, resulting in very few scattered cultural heritages. During the Bronze Age, historical and cultural heritage in the traditional Tibetan region gathered and appeared to have a trend of polarization in the Huangshui River basin and the foothills of the Laji Mountain due to the migration of Central Plains immigrants and the construction of military fortresses. During the Bronze Age to the Qin-Han and Southern and Northern Dynasties, the cultural heritage primarily gathered in the eastern part of the traditional Tibetan areas.

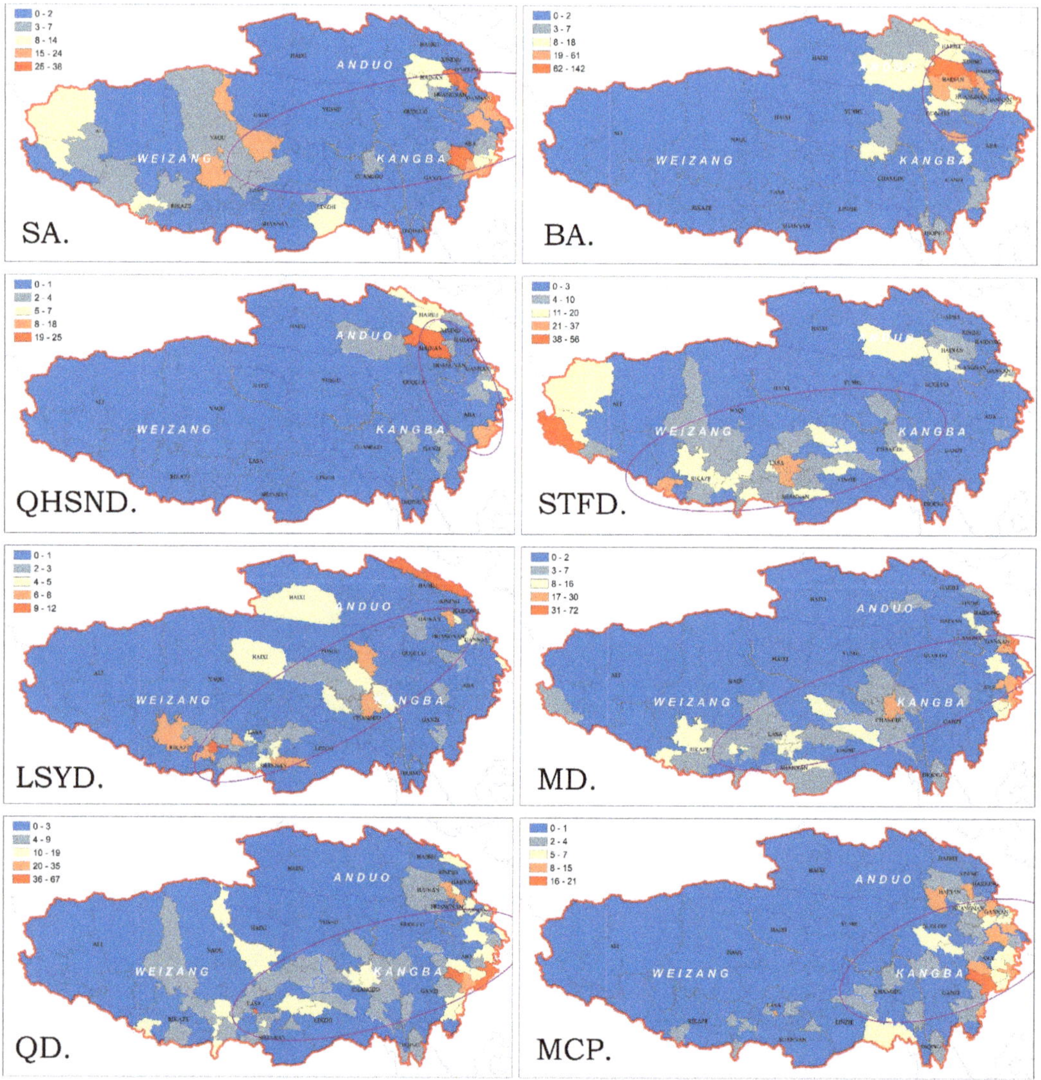

Figure 3. Overall distribution patterns of historical and cultural heritage in multiple periods. (Self-drawn. Some standard deviation ellipses exceed the drawing range, and the displayed standard deviation ellipses can reflect the overall distribution characteristics).

During the Qin, Han, and Southern and Northern Dynasties, as the central control area extended westward, the long-term confrontation between the Central Plains and the local areas led to a sharp decline in the population of ethnic minority groups such as the Wuhuan, Xiongnu, and Western Qiang. As a result, the cultural heritage also decreased. Government immigration and natural migration prompted population movement southward. At this time, the historical and cultural heritage in the traditional Tibetan areas showed a dual-core distribution pattern of Anduo and Kangba. Starting from the Sui and Tang Dynasties, cultural heritage gradually converged, showing a trend of moving closer to the east. The rise of the Tubo Dynasty prompted a spatial migration of the distribution center of historical and cultural heritage from east to west. The Anduo region experienced a peak of aggregation, while the Anduo and Kangba regions fell into a trough. At the same time, the economic and trade exchanges between Tubo and the central dynasties and the "Musk Route" also promoted the socio-cultural development of regions along the route such as Aba and Changdu, as well as the Ali region.

After the fall of the Tubo Dynasty (equivalent to STFD.), conflicts among various traditional tribes in the Tibetan region intensified, weakening the population and economic advantages in the Lhasa area. Branches of the Tubo tribe established their regimes in the Huanhe River basin in Qinghai Province, centered around Haidong (Ancient Miaochuan) and Xining (Ancient Qingtang), while Qiang tribes re-engaged in independent activities in the Yushu area after the fall of the Tubo Dynasty, leading to a significant increase in the quantity and density of cultural heritage in the Anduo and Kangba regions during the Liao, Song, and Yuan Dynasties. In the Ming Dynasty, inheriting the Tusi policy from the Yuan Dynasty, the distribution center shifted towards the eastern border of the Tibetan region, with no significant changes in the distribution range. Since the Qing Dynasty, with the continuous integration of the traditional Tibetan region and the mainland in politics, economy, and culture, the pattern of cultural heritage has shown a continuous movement of the center of gravity towards the Aba and Gannan regions, which are at the junction of the Han and Tibetan cultures, exhibiting characteristics of polarization.

Generally, the closer the heritage is to the present day, the higher the probability of its preservation. The longer the time span, the greater the number of preserved historical and cultural heritages. In terms of spatial distribution, the quantity and clustering degree of historical and cultural heritage in each period show fluctuating changes (Figure 4). Among them, four clusters of historical and cultural heritage have been formed from the Paleolithic Age to the Qin, Han, and Northern and Southern Dynasties periods, exhibiting the distribution characteristics of HH clusters against LL clusters. Among them, HH clusters and LH clusters are distributed alternately in the northeastern part of the traditional Tibetan region, while LL clusters are mainly distributed in the central and southern parts of the traditional Tibetan region, accompanied by a small number of LH clusters. The distribution ranges of each cluster expand with the passage of time. Since the Sui, Tang and Five Dynasties period, there have been obvious abrupt changes in the clustering distribution of the traditional Tibetan region.

HH clusters are distributed in the southern part of Changdu region and gradually shrink and disappear from the Sui, Tang to Qing Dynasties. LL clusters are mainly distributed in the central and northern parts of Kangba and the southern part of Anduo, with LH clusters showing a clear trend of migration along the Tanggula Mountains from southeast to northwest. In the Qing Dynasty and modern times, the clustering distribution characteristics are similar to those of the Paleolithic Age to the Qin, Han, and Northern and Southern Dynasties periods. The distribution areas of HH and LH clusters overlap more, and the clustering tendency continues to increase, reflecting the intermittent distribution pattern of LL clusters in a northeast-southwest direction in places like Rikaze, Naqu, and Haixi due to the accelerated process of modernization and Sinicization in this region.

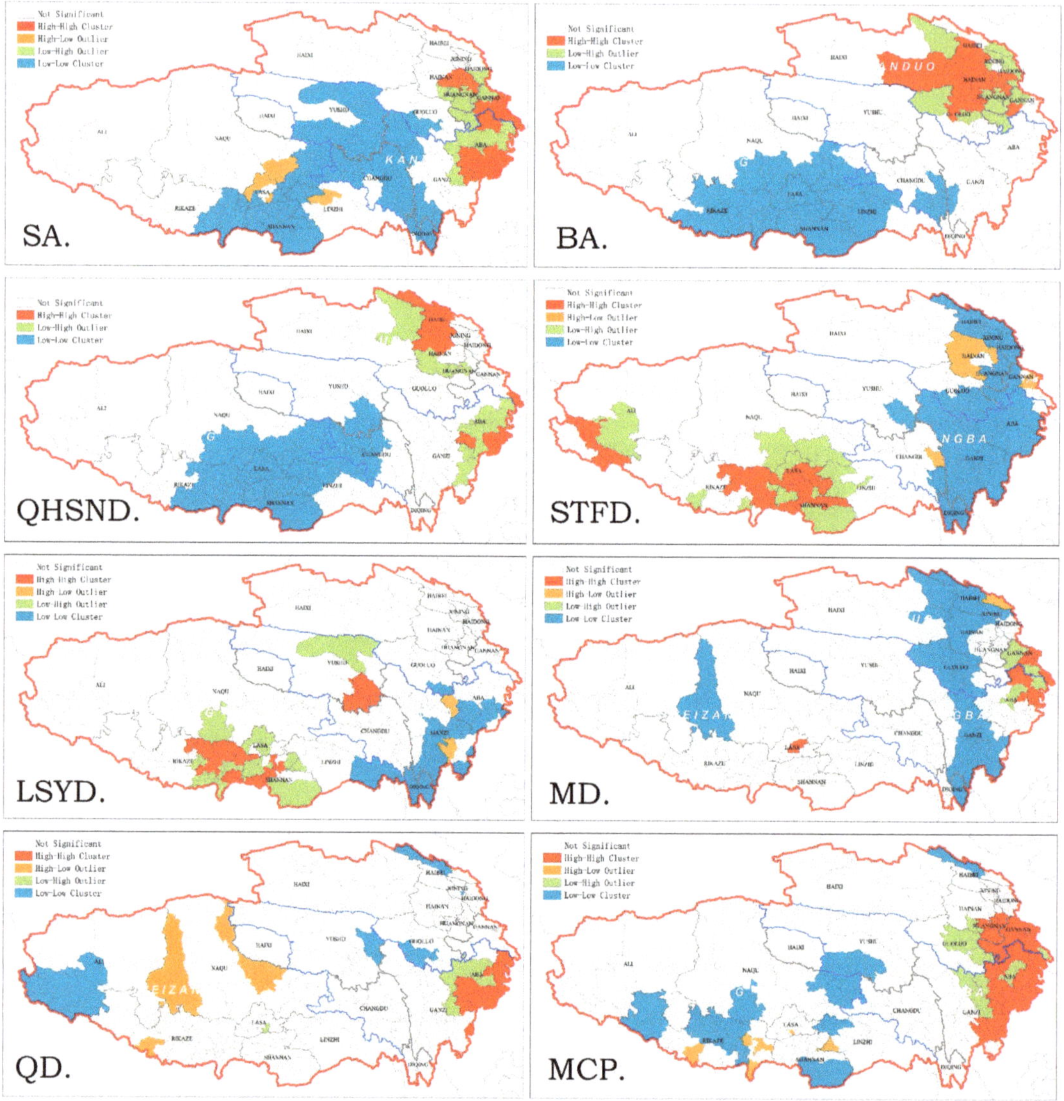

Figure 4. The autocorrelation distribution patterns of historical and cultural heritage in multiple periods (Self-drawn).

3.2. The Distribution Pattern of Multi-type Historical and Cultural Heritage

3.2.1. Distribution Characteristics of Multi-Type Historical and Cultural Heritage

The coverage of historical and cultural heritage in the traditional Tibetan areas is extensive, showing significant regional differences in types. According to the attributes of cultural heritage, it can be divided into six types, including the ancient ruins, the ancient tombs, temples and carvings, the ancient buildings, the Important Historical Sites from the Modern period, and other cultural relics (Table 4).

The quantity of ancient ruins is the highest, and their aggregation is also the most pronounced. The ancient ruins mainly include temple ruins, ancient city ruins, and military fortress ruins. These ancient ruins were abandoned in the course of history, and their high level of aggregation indicates the transition from more concentrated settlements to a broader geographic dispersion in ancient societies. The ancient tombs are generally found in burial clusters and have a strong spatial correlation with human settlements of different eras. Although the aggregation of ancient tombs is second only to ancient sites, their quantity is only one-third of that of ancient sites. The relatively high aggregation of ancient tombs provides a possibility for understanding human social behavior on the Qinghai-Tibet Plateau, suggesting a gradual transition from settlements in areas with better geographical and climatic conditions, such as valleys, to areas with poorer conditions, resulting in a shift from a concentrated pattern to a dispersed one. Temples and caves have slightly fewer numbers compared to ancient sites, and they are also more dispersed. Some temples after the Sui and Tang Dynasties were metaphorically referred to as "joints that suppress witches" in Tibetan legends. The dispersed pattern of temples and caves echoes this concept in the legends.

Table 4. The NNI statistics of multi-type historical and cultural heritage (Self-drawn).

Types	Features	Quantity	Average Actual Nearest Distance (m)	NNI	Z Value
The Ancient Ruins (AR.)	The ruins of ancient human activities, including villages, castles, and the like.	1876	5633.40	0.282	−59.52
The Ancient Tombs (AT.)	The above-ground and underground structures used for burying the deceased.	516	11,270.87	0.301	−30.36
Temples and Carvings (TC.)	Religious buildings, rock paintings, stone carvings.	1607	6839.41	0.314	−52.60
The Ancient Buildings (AB.)	Historically significant buildings before 1949.	196	19,518.05	0.365	−16.99
The Important Historical Sites from the Modern Period (IHS.)	Important historical relics related to modern China.	113	30,640.66	0.537	−9.41
The Other Cultural Relics (OCR.)	Unconventional heritage with historical, cultural, and artistic value.	59	43,374.68	0.565	−6.39
All Cultural Heritage Sites	——	4367	3766.29	0.283	−90.60

3.2.2. Distribution Pattern of Multi-Type Historical and Cultural Heritage

Overall, the historical and cultural heritage in the traditional Tibetan areas exhibits significant regional, temporal, and multi-polarization characteristics (Figure 5). High-density distribution areas are located at the intersection of the Anduo region and the eastern part of the Kangba region, while low-density distribution of historical and cultural heritage is found in the Changdu, Ganzi, and Yushu where the Hengduan Mountains are located. The distribution of ancient ruins, which mainly consist of early civilization settlements and town ruins, is influenced by the terrain factors of the interlacing between high mountains and valleys. They are densely distributed in the eastern Tibetan areas, such as the Hehuang Valley and Songpan Grassland, which have lower altitudes and are suitable for agriculture [87].

The ancient tombs are widely distributed in areas with dense population and cultural phenomena, and they highly overlap with the active range of the Tubo Dynasty. They form three core areas in Lasa of the Weizang region, Aba of the Kangba region, and Hainan, Huangnan of the Anduo region. The grotto temples and stone carvings are mainly distributed along the ancient Buddhist propagation route called "Tangfanni", forming a belt-shaped distribution. The main cores are in Lasa and Shannan, with multiple concentrations in places like Rikaze, Changdu, Aba, and Huangnan. With the rise of transportation methods such as Maritime Silk Road, there are rarely newly built grotto temples and stone carvings in the Tibetan areas.

Compared to other types of historical and cultural heritage, the number of ancient buildings is relatively small. They are concentrated in the relatively flat areas with a larger number of towns, such as the Aba region, and the secondary concentrated areas are in regions with relatively developed industry and commerce, such as Lasa, Xining, and Gannan. The Important Historical Sites from the Modern period mainly include battle memorial sites, administrative office sites, and industrial and commercial production sites. Therefore, they are concentrated in the eastern edge of the Tibetan areas where transportation conditions are better and resource development started earlier. Gannan, as an important node for the Red Army's Long March and the major defeat of the Nationalist Army, has a relatively large number of important modern and contemporary historical sites.

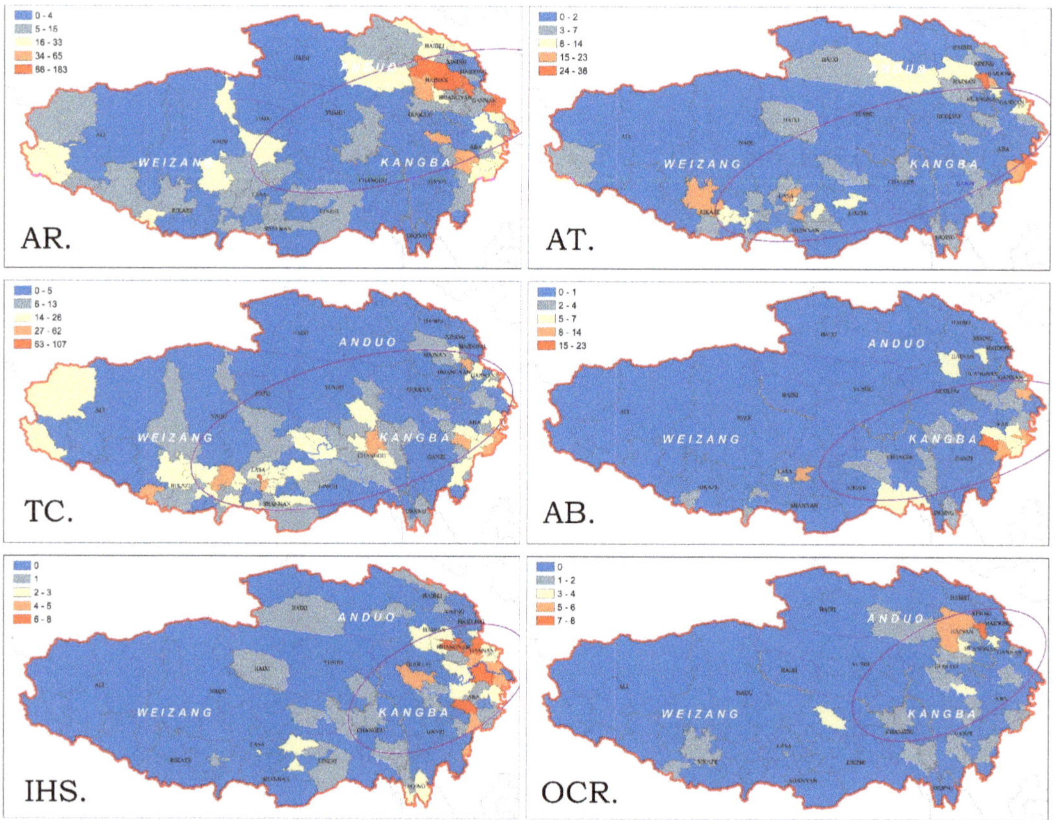

Figure 5. Overall distribution patterns of multi-type historical and cultural heritage. (Self-drawn. Some standard deviation ellipses exceed the drawing range, and the displayed standard deviation ellipses can reflect the overall distribution characteristics).

Different types of cultural activities and behaviors have shaped the diverse distribution patterns of historical and cultural heritage (Figure 6). The ancient ruins, the important Historical Sites from the Modern periods, and other cultural relics have similar distribution patterns. Natural conditions such as topography and terrain influence the clustering types of these types of historical and cultural heritage through their impact on human activities. HH clustering and LH clustering are widely distributed in the northeastern part of the traditional Tibetan area, indicating that the northeastern part is an important area where frequent production and living activities of traditional Tibetans occur. LL clustering is mainly distributed in the central and southern plateau regions, such as Rikaze and other areas with high mountains and deep valleys and higher utilization requirements, indicating that the production of historical and cultural heritage in the early stages of different development stages of agricultural activities and industrialization is limited by technological levels.

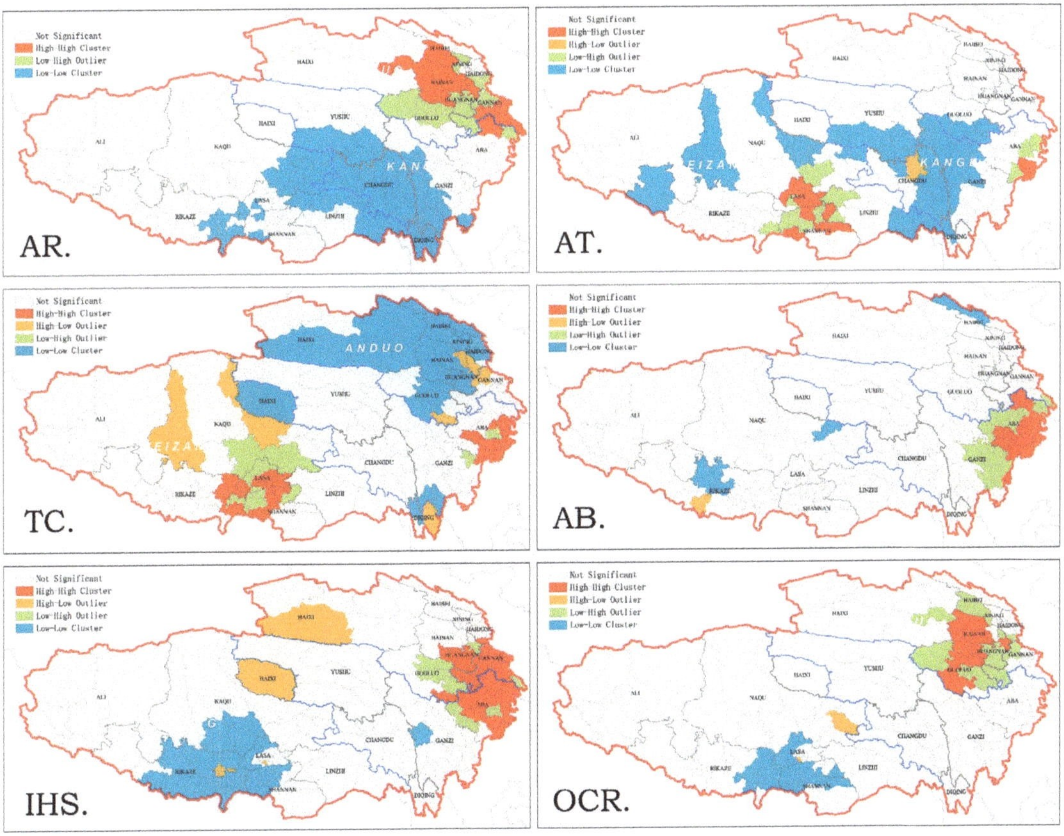

Figure 6. The autocorrelation distribution patterns of multi-type historical and cultural heritage (Self-drawn).

The clustering distribution of ancient buildings is more balanced. HH clustering is distributed in the eastern part of the traditional Tibetan area, which is also the region with the most favorable conditions in terms of topography, terrain, and transportation accessibility. The construction difficulty is relatively small, and the distribution of ancient buildings is more prominent in terms of quantity. The clustering of ancient tombs, grotto temples, and stone carvings is accompanied by HH clustering and LH clustering, mainly distributed in two major regions centered on Lasa and Aba, respectively. The clustering of LL is characterized by significant spatial complementarity.

The LL clustering of ancient tombs is mainly distributed in the Kangba region, while the LL clustering of grotto temples and stone carvings is mainly distributed in the Anduo region. This is closely related to the different functions developed over time in the three major regions of traditional Weizang, Anduo, and Kangba. The Anduo region, also known as the "Land of Horses", is an area where nomadic culture thrives. It is not primarily influenced by religion, hence there are relatively fewer temples in this region. The temples are densely distributed in areas where religious dissemination is widespread, while the ancient tombs are mainly found near settlements that developed in different periods. It can be inferred that the distribution patterns of different types of cultural heritage can reflect the cultural characteristics of a region.

3.3. The Distribution Pattern of Entropy Change in Historical and Cultural Heritage

3.3.1. Overall Distribution Pattern of Entropy Changes in Historical and Cultural Heritages

Entropy is a term used in physics to describe the degree of disorder or randomness in a system. In the study of cultural heritage, entropy is used to measure the diversity, complexity, and degree of change of traditional Tibetan cultural heritage in the Tibetan region. Entropy change analysis can provide a scientific basis and reference for the protection and inheritance of traditional Tibetan cultural heritage.

Entropy change analysis compares cultural heritage data from different time periods to calculate the entropy value of the heritage, thus understanding the dynamic process of heritage change. If a region has a greater variety of cultural heritage types and a more even distribution, its entropy value will be higher, indicating a higher level of cultural heritage diversity. Conversely, if the quantity of a certain heritage type decreases or disappears over different time periods, the entropy value will decrease, indicating that the heritage is facing the risk of loss or decline.

The clustering characteristics of entropy change in the traditional Tibetan areas of historical and cultural heritage in China are significant. The core of entropy increase has spread and migrated from the eastern part of Anduo region to the intersection of Weizang region and Kangba region, and then moved to the intersection of Anduo region and Kangba region (Figure 7). There has been no significant high-value aggregation of entropy change in the northern Tibetan Plateau. Prior to the Qin, Han, and Northern and Southern Dynasties, the high-value aggregation area of entropy change in the traditional Tibetan areas of China was mainly distributed in the surrounding areas of Gannan, Huangnan, and Haixi, and the range of high-value increase continuously expanded. The Weizang region centered on Lasa was a low-value aggregation area of entropy change, showing a significant spatial differentiation distribution pattern.

After the Tubo Dynasty (equivalent to STFD.) and the fusion of politics and ethnic groups, the unipolar entropy increase pattern centered around Hainan and Haibei gradually migrated southward, forming a distribution pattern centered around Gannan, Huangnan, and Aba regions. From the Sui, Tang, Five Dynasties to the Qing Dynasty, the areas where Changdu, Yushu, Linzhi, and Naqu are located maintained a long period of high-value aggregation of entropy increase. The high-value aggregation of entropy changes continuously moved northward and gradually showed a decreasing trend in aggregation degree. During this stage, there were no obvious low-value aggregation areas within the traditional Tibetan areas. In modern times, the high-value aggregation of entropy change in the historical and cultural heritage is concentrated in the intersection of Anduo region and Kangba region, where Aba, Gannan, and Guoluo are located, and there are a few low-value aggregation areas in the southeastern part of the Weizang area and the southern part of the Kangba region.

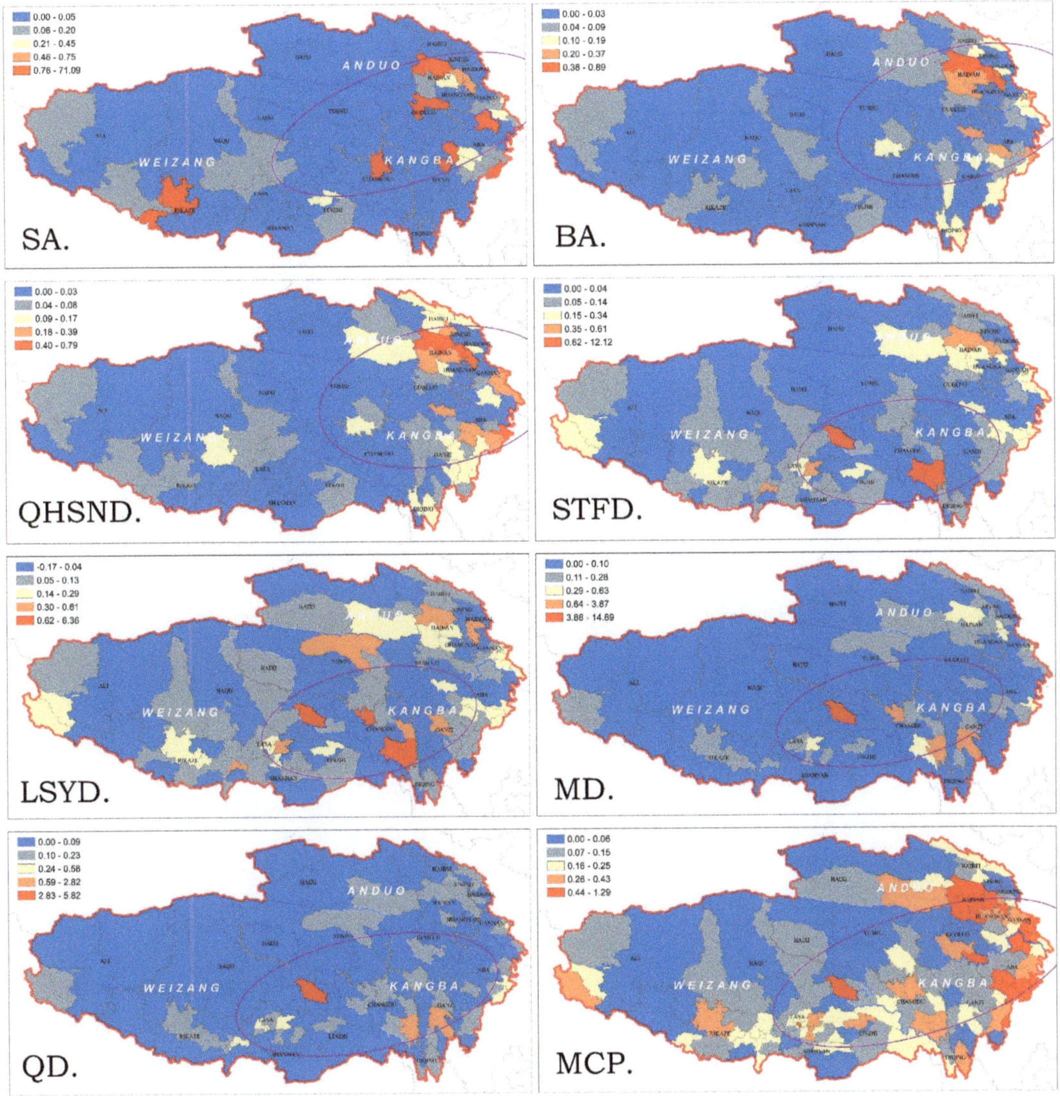

Figure 7. Overall distribution patterns of entropy change in historical and cultural heritage in multiple periods. (Self-drawn. Some standard deviation ellipses exceed the drawing range, and the displayed standard deviation ellipses can reflect the overall distribution characteristics).

3.3.2. The Autocorrelation Distribution of Entropy Changes in Historical and Cultural Heritage

The types of cultural heritage in the traditional Tibetan areas are gradually increasing and becoming more complex. They exhibit a trend of fluctuating growth and gradual expansion (Figure 8). During the Stone Age, the entropy increase of traditional Tibetan historical and cultural heritage in China was mainly characterized by low-value clustering, with a small amount of LH clustering distributed in the Qaidam Basin, where Haixi, Haibei, and Hainan are located. From the Bronze Age to the Qin, Han, and Northern and Southern Dynasties, the entropy changes of cultural heritage formed four distinct patterns with significant spatial differentiation. Among them, HH and LH clusters were distributed in the monsoon climate-dominated areas where Hainan, Gannan, and Aba are located, while LL clusters and a small amount of HL clusters were distributed in the plateau areas west of the Hengduan Mountains and north of the Himalayas.

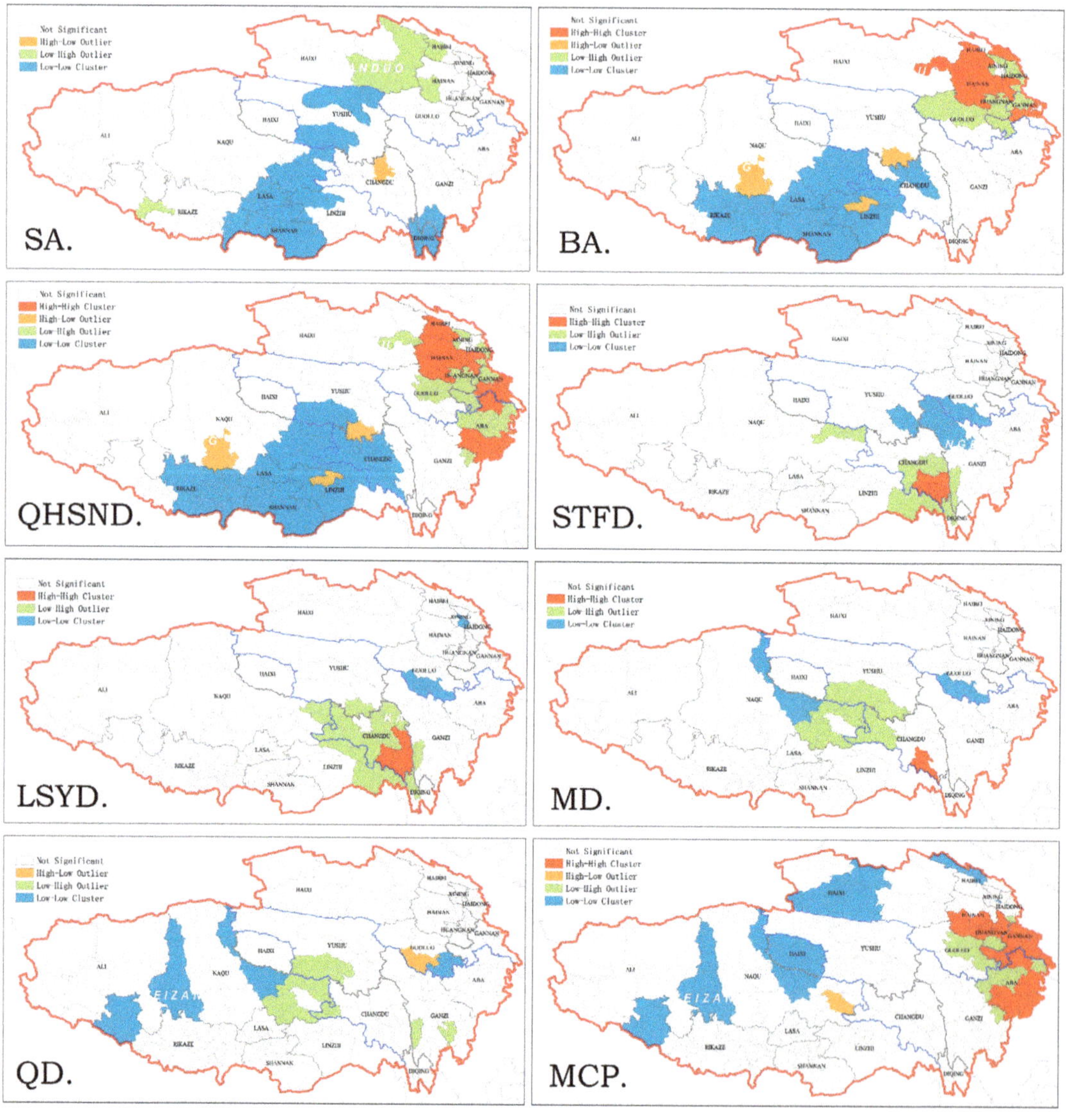

Figure 8. The autocorrelation distribution patterns of entropy change of historical and cultural heritage in multiple periods (Self-drawn).

The distribution of historical and cultural heritage from the Sui, Tang, Five Dynasties, Liao, Song, and Yuan Dynasties shows a significant leap, forming three clusters, with high-value and low-value clusters mainly distributed in the Kangba and Anduo region. The LH cluster is mainly located in the Changdu area, surrounded by the HH cluster, while a small number of LL clusters are distributed in the Ganzi and Guoluo. By the Ming and Qing Dynasties, the HH cluster disappeared, and the LH cluster moved northward to the border area of Naqu and Yushu, with a few scattered LL clusters appearing in the Weizang region. In modern times, the high-value cluster of historical and cultural heritage returned to the Aba, Gannan, Huangnan, and Hainan regions, while the LL cluster showed a southwest-to-northeast scattered distribution pattern in Rikaze, Naqu, Yushu, Haixi, and Haibei, indicating a shift in social, economic, and cultural activities towards towns intersecting with the mainland.

The increase in entropy of historical and cultural heritage is an indicator that measures the richness and disorder of cultural heritage in spatial distribution during different periods, and it explains the formation principles of the spatial distribution pattern of current historical and cultural heritage.

The distribution pattern of the increase in entropy of historical and cultural heritage, using the type information and age information of cultural heritage, depicts the formation process of social history in traditional Tibetan areas in China. In terms of the types and characteristics of historical and cultural heritage (Figure 9a), ancient sites are most closely associated with the overall distribution pattern of historical and cultural heritage, followed by ancient tombs, other types, temples, and caves. Although the number of other types of cultural heritage is the smallest, they are closely related to the overall distribution of historical and cultural heritage. Modern representative historical sites have the least influence on the overall distribution pattern of historical and cultural heritage. In addition, there is a high correlation between ancient sites and ancient tombs, temples and ancient buildings, and ancient tombs and temples, suggesting that the spatial distribution pattern of historical and cultural heritage is related to the cultural characteristics of human activities in different periods. Modern settlements have a significant spatial differentiation from ancient settlements, and there is also spatial differentiation between continuous ancient settlements and abandoned settlements. Temples play a crucial role in the development and evolution of settlements.

It is generally believed that "the closer the periods are, the stronger the correlation". Connected periods should have a strong correlation. According to the statistical analysis of the correlation between the distribution of the quantity of historical and cultural heritage in different periods and the distribution of entropy changes, it is found that the spatial distribution pattern of historical and cultural heritage has the dual characteristics of cumulation and transition. In terms of the quantity of historical and cultural heritage (Figure 9b), there is a more significant correlation between the Stone Age and the modern and contemporary, Qing Dynasty, and Ming Dynasty. This indicates that there is a discontinuous "pattern transition" in the long-term cumulation process of historical and cultural heritage. The Qin-Han and Southern and Northern Dynasties and earlier periods show strong correlation, while the Ming, Qing, and contemporary periods show strong correlation, and the values of the two stages of intergenerational superposition are closer.

The multigenerational relevance of the spatial distribution pattern of historical and cultural heritage reflects the social and historical development process of the regional space. In terms of the increase in entropy of historical and cultural heritage (Figure 9c), before the Sui and Tang Dynasties, the increase in entropy of historical and cultural heritage in the traditional Tibetan areas of China basically follows the cumulation characteristics. After the unification of the Tibetan area by the Tubo Dynasty, there is a sudden change in the increase in entropy of historical and cultural heritage, entering the second stage of linear superposition, and there is a significant negative correlation with the previous period in terms of entropy increase. The modern and contemporary period has a strong correlation with the entropy changes in the Sui and Tang Dynasties and earlier periods, reflecting a significant transition.

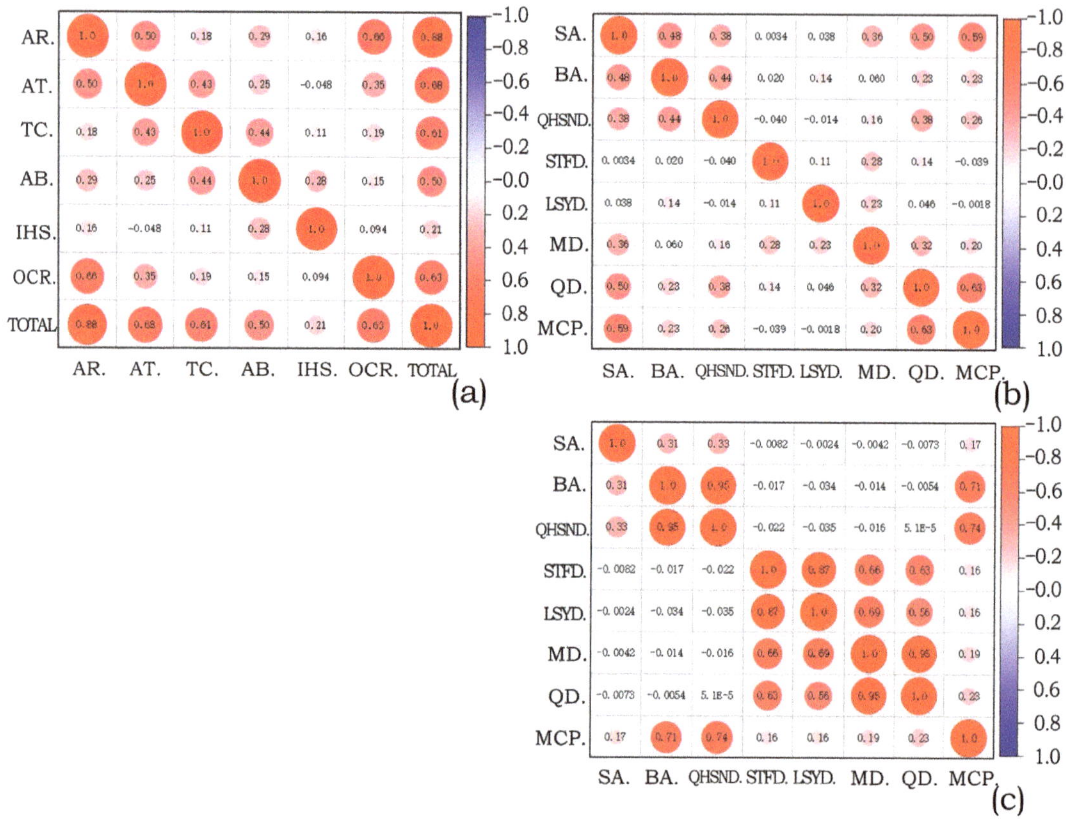

Figure 9. Correlation analysis of historical and cultural heritage: (**a**) Correlation of different types; (**b**) Correlation of different periods; (**c**) Correlation of entropy change in multiple periods (Self-drawn).

4. Discussion

The United Nations Educational, Scientific, and Cultural Organization (UNESCO) cultural heritage policies and Chinese state-directed tourism policies have played a significant role in the preservation of historical sites and the development of ethnic culture in contemporary Tibet [88]. This study of the temporal and spatial distribution patterns of historical and cultural heritage in traditional Tibetan can provide valuable support for the ongoing national cultural relics survey and comprehensive scientific expedition on the Qinghai-Tibet Plateau (Qinghai-Tibet Scientific Expedition). By analyzing the temporal and spatial distribution patterns and entropy trends of traditional Tibetan historical and cultural heritage, it is possible to discuss the process of social change in high-altitude regions from a regional perspective.

In recent years, social historical research at the regional level has gradually gained attention from scholars. In recent years, there has been increasing attention to cross-regional historical and cultural heritage resources, and researchers have been devoted to optimizing the allocation of regional cultural resources. The spatial distribution patterns of historical and cultural heritage reflect the structural distribution of cultural functions that have existed over a certain period of time and within a geographical range. In our previous research, we found that non-geographical factors such as politics, economy, and cultural changes are key driving factors affecting the distribution of heritage [89]. Therefore, we hope to verify the accumulative and transformative laws of regional social history by studying the distribution patterns of traditional Tibetan cultural heritage in China under the guidance of systems theory and entropy theory.

We will summarize the phenomena we discover into three aspects.

(1) The distribution patterns of traditional Tibetan historical and cultural heritage in China exhibit a "large dispersion-small aggregation" distribution characteristic and show fluctuating changes of "aggregation-dispersion-aggregation" during the process of socio-economic development. The historical and cultural heritage in traditional Tibetan areas of China is mainly distributed in high-altitude river valleys with suitable climate conditions [90] with the Weizang region, Anduo region, and Kangba region alternately becoming the center of the traditional Tibetan areas. The distribution patterns of historical and cultural heritage reflect the transfer of important cultural activities in the region, which is mainly influenced by geographical environment and socio-economic aspects.

According to investigations, the geographical environment of traditional Tibetan areas in China has remained relatively stable for thousands of years, with no drastic changes in geological landforms and other factors, while the distribution patterns of cultural heritage have undergone morphological evolution. It can be considered that although the distribution patterns of historical and cultural heritage are constrained by geographical conditions, they are mainly influenced by social changes and record the historical processes of cultural exchange and civilization confrontation. In the long run, the social and historical development of traditional Tibetan areas is influenced by political power and civilization, forming a series of unique political, military, economic, and social systems in the game between plateau civilization and valley civilization, Tibetan culture, Central Plains culture, and Buddhist culture [91]. This systemic nature can be confirmed through quantitative methods based on the distribution patterns of historical and cultural heritage.

(2) The distribution pattern of traditional Tibetan historical and cultural heritage in China exhibits a regional characteristic in typology. Over a long period of evolution, it has formed a pattern of diverse elements, adjacent distribution ranges, and deep cultural accumulation. The spatial distribution system is concentrated in the junction area of Xining, Hainan, Haidong, and Huangnan, supplemented by Lasa, Shannan, Gannan, and Aba as secondary cores, echoing the belt-like historical and cultural elements corridor formed by the southwest entry route to Tibet.

The three main regions of traditional Tibetan area exhibit significant differences in typological characteristics, reflecting the differences in cultural integration and dissemination processes in these regions [92]. The Weizang region has abundant ancient tombs, grotto temples, and stone carvings. Before the Qin-Han and Southern and Northern Dynasties, there were dense cultural heritage sites in this area, indicating that the Weizang region had a prosperous social civilization in history and wielded significant political and religious power. As a center for ancient sites and various cultural relics, the Anduo region exhibits a phenomenon of numerous archaeological sites but fewer tombs, indicating the presence of cultural discontinuities.

This may be attributed to ancient population migrations influenced by ethnic wars, natural disasters, and other factors. The Kamba region has a large number of ancient tombs, ancient buildings, and representative buildings from modern times, through which it can be inferred that although the Kamba area experienced decline in some historical periods, it has always been a core region for human activities in Tibet as an important communication passage between the Qinghai-Tibet Plateau and the Central Plains.

(3) The changing distribution pattern of historical and cultural heritage in traditional Tibetan areas of China reflects the cumulative and transition of social history, and even reflects the rise and fall of civilizations [93]. There is a significant correlation between the quantity of historical and cultural heritage from the Stone Age and the Modern, Ming Dynasty, and Qing Dynasty. From a spatial clustering perspective, the

clusters of historical and cultural heritage sites from the Stone Age, modern era, Qing Dynasty, and Ming Dynasty overlap, mainly in the eastern edge of the Tibetan region and the Tibetan-Han border area, while historical and cultural heritage in other areas is relatively scattered. Human and religious activities gradually shifted from simple ritual sites and caves to more complex buildings and settled tribal areas, and historical and cultural heritage gradually formed and accumulated during this process, thus the continuity of history is easily understood, manifested as linear accumulation [94] of history and cultural heritage.

However, the differences in the correlation of cross-era cultural heritage challenge the geographical continuity and the overall nature of history [95], mainly manifested in the significant correlation between the Stone Age and the Qing Dynasty, and the significant lack of correlation between the Sui and Tang Dynasties. Due to technological limitations, production and daily activities in the Stone Age were mainly concentrated in the low-altitude areas of the eastern edge of the Tibetan area [96]. Compared with the Stone Age, the Qing government established a large number of administrative institutions and military bases in the eastern edge of the Tibetan region in order to achieve practical control over the Tibetan region. Therefore, there is a significant correlation between the Stone Age and the Qing Dynasty.

Meanwhile, another phenomenon of transformative nature is the extremely weak correlation in the entropy changes of historical and cultural heritage before and after the Sui Dynasty. Before the Sui and Tang Dynasties, the complexity of historical and cultural heritage continued to increase, and spatial differentiation intensified, forming a high entropy value area centered on the Qinghai River Valley. After the Sui Dynasty, the spatial distribution pattern of historical and cultural heritage stabilized, and the high entropy growth area of historical and cultural heritage was mainly concentrated in the border towns between the Han and Tibetan regions.

The situation of local separatism in the Sui and Tang periods was broken, and with the unification of various tribes by the Tubo Dynasty, the central dynasty's discourse power and control significantly increased compared to the Han Dynasty. There was a great degree of exchange between the local and central dynasties in terms of economy, technology, and culture, and the social form of the traditional Tibetan area underwent a transformation, with a leapfrog change in production forms and cultural exchanges. The Tubo regime strengthened its control over religious beliefs, military, and politics, leading to a highly prosperous development of social and cultural aspects. Traditional Tibetan areas experienced a leap in social structure and technological advancement. Therefore, the entropy of historical and cultural heritage exhibits a significant weak correlation.

The above is research finding on the distribution pattern of historical and cultural heritage in traditional Tibetan areas. This is not just a case study. We hope to emphasize the importance of historical and cultural heritage in regional social and historical research in this study, and validate what we call the "historical leap" pattern. Although geographic factors can to some extent explain the changes in the distribution pattern of historical and cultural heritage, the diversity of cultural heritage exceeds the stability of geographic elements, and the social processes reflected by cultural heritage have not received sufficient attention. It is necessary to conduct comprehensive research on the differences and connections between different types and periods of cultural heritage, and explore the mutual influence between the diversity of historical and cultural heritage and cultural exchange and integration.

As seen in previous studies on cultural heritage, the combination of entropy weight and analytic hierarchy process can be used to evaluate heritage value and potential [97,98], and weigh the regional cultural stratification in urban structure using the entropy of spatial form [99]. It can discuss the inherent connections between place names, population, and ethnic groups [100], and evaluate the impact of endogenous and exogenous factors on culture using information matrices [101]. However, there is still limited quantification of the dispersed distribution of cultural heritage or ancient human activities. Dispersed distribution is an important concept in systems theory and can be measured by information entropy to assess the complexity of dispersed distribution explaining the cultural accumulation in the cycles of social development [102]. Compared with clustered distribution, dispersed distribution is more likely to have the resilience to survive in adverse conditions. In a vast region, different degrees of dispersed distribution reflect differences and connections in cultural diffusion patterns under the influence of settlement survival modes in different eras. The dispersed pattern of the Liao, Song, and Yuan Dynasties in the research on the historical and cultural heritage of the Beijing-Tianjin-Hebei region can also verify this viewpoint [89].

Some scholars have analyzed the cultural routes [103] formed by the traditional urban spatial composition in Tibet and summarized the basic types of traditional Tibetan spaces, such as single-core and multi-core. Although the idea of regional spatial integrity has been accepted in the theory and practice of cultural heritage protection, there is still a lack of discussion on the formation process of Tibetan cultural routes and traditional spaces in terms of chronology [104]. The continuity of time has not received sufficient attention in cultural heritage research [105–107]. Some studies have found that Tibetan cultural heritage exhibits high internal consistency and cultural identity [108].

In the historical process of China, the traditional Tibetan region is closely connected to the main ethnic regions of China. The traditional Tibetan areas of China experienced political struggles between the Tubo regime and the central regime during the Sui, Tang, and Five Dynasties period [108,109]. They underwent religious movements influenced by East Asian and Central Asian cultures during the Liao, Song, and Yuan Dynasties [110,111]. In the Ming Dynasty, they transitioned from a semi-independent mode managed by the Office of Pacification and Governance to a central governance mode managed by the Office of Military Affairs [112]. In the Qing Dynasty and modern times, they transitioned from a centralized system of combining politics and religion to a system of ethnic regional autonomy with the separation of politics and religion [113]. This has formed a distribution pattern of historical and cultural heritage characterized by multi-ethnic exchanges and religious integration.

The social history of a region is a process of fluctuation and dynamic development. Geographic and climate changes over thousands of years may have influenced the cultural forms of settlements, exhibiting similar characteristics in discontinuous periods. It should be emphasized that social change occurs faster than geographic change. The distribution pattern of historical and cultural heritage is closely related to the social historical process and requires exploration from the perspectives of social geography or anthropology. The distribution pattern of historical and cultural heritage is closely related to social-historical processes, requiring exploration from the perspectives of social geography or anthropology, rather than solely relying on geographical analysis of elements. The cultural heritage of each period has accumulated over time, forming the current distribution pattern of historical and cultural heritage.

Generally, the closer the cultural heritage is to the present, the higher the likelihood of its preservation; the longer the time span, the greater the quantity of preserved cultural heritage. Through the method of entropy, we found that although cultural heritage has grown in quantity and diversity in different regions over time, the fluctuating changes in the entropy of historical and cultural heritage are also influenced by the fluctuations of social history, and even exhibit cross-generational correlations. Further exploration of this cross-generational correlation will require the use of other technical methods. The types of historical and cultural heritage represent cultural characteristics, and different types of historical and cultural heritage are concentrated in different regions, implying different cultural features in these regions [114]. Overall, this study reveals the regional variations and temporal correlations within the interior of the Qinghai-Tibet Plateau.

5. Conclusions

Historical and cultural heritage refers to the material collection created by human activities, with functions of dissemination [115], cultural value [116], and historical significance [117]. This study focuses on the traditional Tibetan regions in China, and utilizes methods such as information entropy, correlation analysis, and historical periodization to explore the temporal complexity of the historical and cultural heritage patterns, and analyze the entropy change trends in these patterns. Through this research, we confirm that the spatial distribution pattern of historical and cultural heritage can effectively reflect the social and historical development process of the region [118,119]. Although the complexity changes in cultural heritage cannot fully restore the historical processes, it contributes to a scientific understanding of the local historical background and regional development trends.

The results show the following:

(1) The distribution pattern of historical and cultural heritage in traditional Tibetan areas of China exhibits fluctuating changes of "aggregation-dispersion-aggregation". This distribution characteristic is closely related to the process of social and economic development and is also influenced by geographical environment and socio-economic factors.

(2) The distribution pattern of historical and cultural heritage in traditional Tibetan areas of China has regional characteristics. Different regions have significant differences in the types of historical and cultural heritage, reflecting the differences in cultural integration and dissemination processes in each region. The Weizang region, Anduo region, and Kangba region each have distinct historical and cultural features and architectural heritage.

(3) The changes in the distribution pattern of historical and cultural heritage in traditional Tibetan areas in China reflect the cumulative and transformative nature of social history. The quantity and entropy change of historical and cultural heritage exhibit correlations between different periods, and there are also some intergenerational differences. The distribution pattern of historical and cultural heritage in contemporary times returns back to the characteristics of the Qin-Han and Southern and Northern Dynasties and the Northern and Southern Dynasties, forming the basic pattern of the current protection of historical and cultural heritage.

In previous studies, discussions on the spatial distribution pattern of cultural heritage mainly focused on the geographical and climatic causes [21,120–122], while the discussion on the temporal aspects of the distribution pattern of historical and cultural heritage was limited. The important perspective of this study is to combine the analysis of the types and time periods of cultural heritage to judge the complexity of cultural heritage in different historical periods, explaining the complexity of social culture and the systematic nature of historical landscapes. This can serve as effective evidence for the regional development process of social history.

This study on the spatiotemporal distribution pattern of historical and cultural heritage in the traditional Tibetan areas of China emphasizes the systematic relationship between historical cultural heritage and ancient societies. It further validates our theoretical understanding of regional social history, which is the combination of "transition" and "cumulation". In particular, the leapfrogging transition of history is of great value for a profound understanding of the laws of ancient social evolution.

In addition to these theoretical findings, our research also has practical value. From the perspective of a regional cultural heritage protection system, it is possible to develop integrated cultural heritage protection policies by utilizing large-scale cultural routes, which is a cross-administrative cultural strategy improving urban competitiveness [119]. The traditional Tibetan region in China can establish tourism service standards different from those in the general plain areas, creating a tourism system with bronze culture, Tubo culture, and Ming-Qing culture as the main features. The three major Tibetan regions can form cultural heritage exhibition systems with different themes. The Weizang region mainly showcases the splendid political and religious cultural history of Tubo, while the Amduo region and Kangba region mainly showcase the cultural exchange history between Tibet and the Central Plains. The vast expanse of the traditional Tibetan region is not conducive to the formation of a tourism network, but visitors can combine extensive sightseeing with in-depth experiences to better appreciate the unique culture of the traditional Tibetan region.

This study contributes to the preservation and inheritance of cultural heritage and the orderly utilization of historical heritage while also providing a reliable foundation for maximizing the protection of temporal and spatial information. This is of great significance for the rational optimization of the historical and cultural spatial pattern, the continuity of civilization, and the revitalization of culture. In addition, it is necessary to further explore the reasons for the complexity changes in cultural heritage and to investigate the causality across different eras using other technological methods. The classification of cultural heritage based on certain technical standards is insufficient to encompass the complete cultural characteristics and can also be categorized according to different social classes or professional attributes, which will be reflected in future research. The correlation between the spatial distribution of modern and early Tibetan historical and cultural heritages in terms of entropy values may be limited due to the limitations of archaeological work scope. Many early Tibetan activities still lack historical records, which may result in potential blind spots in this study. It is hoped that new evidence will be discovered in the upcoming Cultural Relics Census.

Author Contributions: Conceptualization, S.L., J.Z. and Z.Z.; methodology, S.L.; software, S.L.; validation, J.Z., Z.Z. and X.T.; formal analysis, J.Z. and S.L.; investigation, J.L.; resources, S.L.; data curation, J.L.; writing—original draft preparation, J.Z. and J.L.; writing—review and editing, S.L. and J.Z.; visualization, X.T. and Z.Z.; supervision, S.L.; project administration, X.X.; funding acquisition, X.X., S.L. and X.T. All authors have read and agreed to the published version of the manuscript.

Funding: This research was funded by "Ministry of Education Humanities and Social Science Fund Project", grant number 23YJA760101; "Tianjin Research Innovation Project for Postgraduate Students", grant number 2022BKY061; "Tianjin University Independent Innovation Fund Project", grant number 2020XSC-0034.

Data Availability Statement: All supporting data are cited within the Sencion 2 Materials and Methods. Restrictions apply to the availability of these data. The heritage data was obtained from http://www.ncha.gov.cn/ (accessed on 15 July 2023) and are available with the permission of the National Cultural Heritage Administration.

Conflicts of Interest: The authors declare no conflict of interest.

References

1. Quan, Z. The Concept of "National Unity" in Historical Records. *J. Weinan Teach. Coll. (Soc. Sci.)* **2003**, *3*, 18–19.
2. Yong, S.; Yubo, H. Planning and Governance of Regional Historical and Cultural Space Conservation and Development: A Case Study of the Loire Valley in France. *Urban Plan. Int.* **2022**, *37*, 111–121. [CrossRef]
3. Fei, L.; Tongqian, Z. National Culture Park: Logic, Origins and Implications. *Tour. Trib.* **2021**, *36*, 14–26. [CrossRef]
4. Gatta, G.; Bitelli, G. A HGIS for the study of waterways: The case of Bologna as ancient city of waters. *IOP Conf. Ser. Mater. Sci. Eng.* **2020**, *949*, 012051. [CrossRef]
5. Wenner, F.; Dang, K.A.; Hölzl, M.; Pedrazzoli, A.; Schmidkunz, M.; Wang, J.; Thierstein, A. Regional Urbanisation through Accessibility?—The "Zweite Stammstrecke" Express Rail Project in Munich. *Urban Sci.* **2020**, *4*, 2. [CrossRef]
6. Kostopoulou, S.; Sofianou, P.K.; Tsiokanos, K. Silk Road Heritage Branding and Polycentric Tourism Development. *Sustainability* **2021**, *13*, 1893. [CrossRef]
7. Liao, Y.; Cenci, J.; Zhang, J. Chinese Modern Architectural Heritage Resources: Perspectives of Spatial Distribution and Influencing Factors. *ISPRS Int. J. Geo-Inf.* **2023**, *12*, 358. [CrossRef]
8. Nicu, I.C. Cultural heritage assessment and vulnerability using Analytic Hierarchy Process and Geographic Information Systems (Valea Oii catchment, North-eastern Romania). An approach to historical maps. *Int. J. Disaster Risk Reduct.* **2016**, *20*, 103–111. [CrossRef]
9. Nicu, I.C. Tracking natural and anthropic risks from historical maps as a tool for cultural heritage assessment: A case study. *Environ. Earth Sci.* **2017**, *76*, 330. [CrossRef]

10. Schich, M.; Song, C.; Ahn, Y.-Y.; Mirsky, A.; Martino, M.; Barabási, A.-L.; Helbing, D. A network framework of cultural history. *Science* **2014**, *345*, 558–562. [CrossRef]
11. Jianguo, W. On professional problems of multi-scale protection of urban architectural heritage in China. *City Plan. Rev.* **2022**, *46*, 7–24.
12. Wu, Z.; Ma, J.; Zhang, H. Spatial Reconstruction and Cultural Practice of Linear Cultural Heritage: A Case Study of Meiguan Historical Trail, Guangdong, China. *Buildings* **2023**, *13*, 105. [CrossRef]
13. Feng, B.; Ma, Y. Network Construction for Overall Protection and Utilization of Cultural Heritage Space in Dunhuang City, China. *Sustainability* **2023**, *15*, 4579. [CrossRef]
14. Zhang, S.Y.; Liu, J.M.; Pei, T.; Chan, C.S.; Wang, M.D.; Meng, B. Tourism value assessment of linear cultural heritage: The case of the Beijing-Hangzhou Grand Canal in China. *Curr. Issues Tour.* **2023**, *26*, 47–69. [CrossRef]
15. Bertacchini, E.; Revelli, F. Kalos kai agathos? government quality and cultural heritage in the regions of Europe. *Ann. Reg. Sci.* **2021**, *67*, 513–539. [CrossRef]
16. Partarakis, N.; Zabulis, X.; Patsiouras, N.; Chatjiantoniou, A.; Zidianakis, E.; Mantinaki, E.; Kaplanidi, D.; Ringas, C.; Tasiopoulou, E.; Dubois, A.; et al. Multi-Scale Presentation of Spatial Context for Cultural Heritage Applications. *Electronics* **2022**, *11*, 195. [CrossRef]
17. Loren-Mendez, M.; Pinzon-Ayala, D.; Ruiz, R.; Alonso-Jimenez, R. Mapping Heritage: Geospatial Online Databases of Historic Roads. The Case of the N-340 Roadway Corridor on the Spanish Mediterranean. *ISPRS Int. Geo-Inf.* **2018**, *7*, 134. [CrossRef]
18. Spinney, L. Human cycles: History as science. *Nature* **2012**, *488*, 24–26. [CrossRef]
19. Ashrafi, B.; Kloos, M.; Neugebauer, C. Heritage Impact Assessment, beyond an Assessment Tool: A comparative analysis of urban development impact on visual integrity in four UNESCO World Heritage Properties. *J. Cult. Herit.* **2021**, *47*, 199–207. [CrossRef]
20. Qian, C.; Su-Pei, L. Geographical Distribution and Affecting Factors of the Intangible Cultural Heritage in China. *Sci. Geogr. Sin.* **2013**, *33*, 1166–1172.
21. Liu, C.; Kang, L. How Do Geographical Factors Affect the Distribution of Intangible Cultural Heritage: A Case Study of Xinjiang, China. *Sustainability* **2023**, *15*, 8943. [CrossRef]
22. Zhang, J. New functionalism: Rejuvenating historical and cultural heritage through urban revival. *Int. J. Anthropol. Ethnol.* **2019**, *3*, 8. [CrossRef]
23. Yongqi, L.; Ruixia, Y.; Pu, W.; Anlin, Y.; Guolong, C. A quantitative description of the spatial–temporal distribution and evolution pattern of world cultural heritage. *Herit. Sci.* **2021**, *9*, 80. [CrossRef] [PubMed]
24. Chen, J.; Zhou, Y.; Liu, D.; Zhu, A.; Liu, Q. Spatial distribution characteristics of religious architecture heritages in China and the influential factors. *J. Arid Land Resour. Environ.* **2018**, *32*, 84–90.
25. Miao, H.; Zhang, M. Spatial structure of intangible heritage in Northwest China based on GIS analysis. *J. Arid Land Resour. Environ.* **2014**, *28*, 179–186.
26. Agnoletti, M.; Tredici, M.; Santoro, A. Biocultural diversity and landscape patterns in three historical rural areas of Morocco, Cuba and Italy. *Biodivers. Conserv.* **2015**, *24*, 3387–3404. [CrossRef]
27. Li, M.; Wu, B.; Cai, L. Tourism development of World Heritage Sites in China: A geographic perspective. *Tour. Manag.* **2008**, *29*, 308–319. [CrossRef]
28. Galiakberova, A.A.; Mukhametshin, A.G.; Asratyan, N.M.; Kornilova, I.V.; Galiev, R.M. Categories and methods of civilizational approach in modern philosophical and historical discourse. *Rev. Amaz. Investig.* **2023**, *12*, 202–211. [CrossRef]
29. Hosner, D.; Wagner, M.; Tarasov, P.E.; Chen, X.; Leipe, C. Spatiotemporal distribution patterns of archaeological sites in China during the Neolithic and Bronze Age: An overview. *Holocene* **2016**, *26*, 1576–1593. [CrossRef]
30. Nie, X.; Xie, Y.; Xie, X.X.; Zheng, L.X. The characteristics and influencing factors of the spatial distribution of intangible cultural heritage in the Yellow River Basin of China. *Herit. Sci.* **2022**, *10*, 121. [CrossRef]
31. Sevilla, J.; Casanova-Salas, P.; Casas-Yrurzum, S.; Portales, C. Multi-Purpose Ontology-Based Visualization of Spatio-Temporal Data: A Case Study on Silk Heritage. *Appl. Sci.* **2021**, *11*, 1636. [CrossRef]
32. Kauppinen, T.; Mantegari, G.; Paakkarinen, P.; Kuittinen, H.; Hyvonen, E.; Bandini, S. Determining relevance of imprecise temporal intervals for cultural heritage information retrieval. *Int. J. Hum. Comput. Stud.* **2010**, *68*, 549–560. [CrossRef]
33. Yao, Y.; Wang, X.Y.; Lu, L.L.; Liu, C.S.; Wu, Q.; Ren, H.G.; Yang, S.; Sun, R.Q.; Luo, L.; Wu, K. Proportionated Distributions in Spatiotemporal Structure of the World Cultural Heritage Sites: Analysis and Countermeasures. *Sustainability* **2021**, *13*, 2148. [CrossRef]
34. Sesana, E.; Gagnon, A.S.; Ciantelli, C.; Cassar, J.; Hughes, J.J. Climate change impacts on cultural heritage: A literature review. *WIREs Rev. Clim. Chang.* **2021**, *12*, e710. [CrossRef]
35. Zheng, X.; Sun, W.; Huang, X.; Zang, X. Distribution of World Cultural Heritage and Human Urban Civilization. *Trop. Geogr.* **2018**, *38*, 424–431. [CrossRef]
36. Gao, H.; Wang, Y.; Zhang, H.; Huang, J.; Yue, X.; Chen, F. Spatial Distribution and Typological Classification of Heritage Buildings in Southern China. *Buildings* **2023**, *13*, 2025. [CrossRef]
37. Qiang, Z.; Wei, L. Heritage Area:A New Approach for the Preservation of Large-scale Cultural Landscape. *China Popul. Resour. Environ.* **2007**, *1*, 50–55.

38. Yuan, D.; Wu, R.; Li, D.; Zhu, L.; Pan, Y. Spatial Patterns Characteristics and Influencing Factors of Cultural Resources in the Yellow River National Cultural Park, China. *Sustainability* **2023**, *15*, 6563. [CrossRef]
39. Tong, M.; Li, B.; Li, Z. Research on the Spatial–Temporal Distribution and Morphological Characteristics of Ancient Settlements in the Luzhong Region of China. *Land* **2022**, *11*, 1579. [CrossRef]
40. Prieto, A.J.; Silva, A.; de Brito, J.; Macias-Bernal, J.M.; Alejandre, F.J. Multiple linear regression and fuzzy logic models applied to the functional service life prediction of cultural heritage. *J. Cult. Herit.* **2017**, *27*, 20–35. [CrossRef]
41. Pranskūnienė, R.; Zabulionienė, E. Towards Heritage Transformation Perspectives. *Sustainability* **2023**, *15*, 6135. [CrossRef]
42. Xie, Y.; Yang, R.; Liang, Y.; Li, W.; Chen, F. The Spatial Relationship and Evolution of World Cultural Heritage Sites and Neighbouring Towns. *Remote Sens.* **2022**, *14*, 4724. [CrossRef]
43. CAO Yongmao, L.H. Diachrony and Synchronism in the Protection of Historical Towns: Enlightenment and Thought of "Historic Urban Landscape". *Urban Dev. Stud.* **2019**, *26*, 13–20.
44. Deac, L.A.; Herman, G.V.; Gozner, M.; Bulz, G.C.; Boc, E. Relationship between Population and Ethno-Cultural Heritage—Case Study: Crișana, Romania. *Sustainability* **2023**, *15*, 9055. [CrossRef]
45. Chitty, G.; Smith, C. Principles into Policy: Assessing the Impact of Conservation Principles in Local Planning Policy. *Hist. Environ.-Policy Pract.* **2019**, *10*, 282–299. [CrossRef]
46. Liu, R.; Li, W.; Yin, X.; Zhang, W. Research on the Construction of Scenic Trails under the Background of Large Scale Homogeneous Landscape Patterns in Tibetan Regions: A Case Study of Gannan Tibetan Autonomous Prefecture. *Res. Dev.* **2012**, 109–112. [CrossRef]
47. Statuto, D.; Cillis, G.; Picuno, P. GIS-based Analysis of Temporal Evolution of Rural Landscape: A Case Study in Southern Italy. *Nat. Resour. Res.* **2019**, *28*, 61–75. [CrossRef]
48. Luan, F.; Xiong, H.; Wang, F.; Wang, Z. Spatial-temporal Changes and Man-land Relationship of Cultural Heritage Sites in Xinjiang. *Areal Res. Dev.* **2017**, *36*, 134–139.
49. Qing, W.; Gui, L.X.; Ming, Z. The Spatial Distribution and Relevant Factors of China's Different Types Intangible Cultural Heritages. *Econ. Geogr.* **2015**, *35*, 175–183. [CrossRef]
50. Giudici, E.; Mi Dahlgaard Park, S.; Melis, C.; Dessì, S.; Francine Pollnow Galvao Ramos, B. Is intangible cultural heritage able to promote sustainability in tourism? *Int. J. Qual. Serv. Sci.* **2013**, *5*, 101–114. [CrossRef]
51. Wang, Q.; Li, Z.; Yin, Y.; Yang, S.; Long, W.; Lai, F.; Hu, J. Distribution characteristics of typical geological relics in the Western Sichuan Plateau. *Open Geosci.* **2020**, *12*, 307–323. [CrossRef]
52. Chen, X.; Hou, G.; Jin, S.; Gao, J.; Duan, R. The Pollen Records of Human Activities in Qinghai-Tibet Plateau during the Middle and Late Holocene. *Earth Environ.* **2020**, *48*, 643–651.
53. Brantingham, P.J.; Rhode, D.; Madsen, D.B. Archaeology Augments Tibet's Genetic History. *Science* **2010**, *329*, 1467. [CrossRef] [PubMed]
54. Lu, H.; Chen, X.; Zhang, Z.; Tang, L.; Lemoine, X.; Wangdue, S.; Chen, Z.; Liu, X.; Frachetti, M.D. Early agropastoral settlement and cultural change in central Tibet in the first millennium BC: Excavations at Bangga. *Antiquity* **2021**, *95*, 955–972. [CrossRef]
55. Li, T.; Wu, Y.; Du, S.; Huang, W.; Hao, C. Age of Microlith human activities in Zhongba area, southwestern margin of Qinghai-Tibet Plateau. *J. Beijing Norm. Univ. Nat. Sci.* **2016**, *52*, 91–96.
56. Zhu, Y.; Hou, G.; Lan, C.; Gao, J.; Pang, L. GIS-based analysis of traffic routes and regional division of the Qinghai-Tibetan Plateau in prehistoric period. *Prog. Geogr.* **2018**, *37*, 438–449.
57. Zhang, D.D.; Li, S.-H. Comment on "Permanent human occupation of the central Tibetan Plateau in the early Holocene". *Science* **2017**, *357*, eaam9231. [CrossRef]
58. Hou, G.; Wei, H.; Chongyi, E.; Sun, Y. Human activities and environmental change in Holocene in the northeastern margin of Qinghai-Tibet Plateau: A case study of JXG2 relic site in Qinghai Lake. *Acta Geogr. Sin.* **2013**, *68*, 380–388.
59. Gao, X.; Zhou, Z.; Guan, Y. Human cultural remains and adaptation strategies in the Tibetan Plateau margin region in the late pleistocene. *Quat. Sci.* **2008**, *28*, 969–977.
60. Hou, G.; Cao, G.; Chongyi, E.; Ren, X.; Wuennemann, B.; Li, F. New evidence of human activities at an altitude of 4000 meters area of Qinghai-Tibet Plateau. *Acta Geogr. Sin.* **2016**, *71*, 1231–1240.
61. Hou, G.; Zhang, X.; Xiao, J.; Chongyi, E. Human Migration and Diffusion in Northeastern Margin of the Qinghai-Tibet Plateau since the Late Glacial Period. *Arid Zone Res.* **2013**, *30*, 149–155.
62. Song, J.; Gao, Y.; Tang, L.; Zhang, Z.; Tang, M.H.; Xu, H.; Wangyal, T.; Yuan, H.; Li, L.; Li, Y.; et al. Farming and multi-resource subsistence in the third and second millennium BC: Archaeobotanical evidence from Karuo. *Archaeol. Anthropol. Sci.* **2021**, *13*, 47. [CrossRef]
63. Madsen, D.B.; Ma, H.; Brantingham, P.J.; Xing, G.; Rhode, D.; Zhang, H.; Olsen, J.W. The Late Upper Paleolithic occupation of the northern Tibetan Plateau margin. *J. Archaeol. Sci.* **2006**, *33*, 1433–1444. [CrossRef]
64. Tang, L.; Lu, H.; Chen, X.; Xu, H.; Boivin, N.; Storozum, M.; Yang, F.; Li, S.; Liu, X.; Spengler, R.N., III. Prehistoric agricultural decision making in the western Himalayas: Ecological and social variables. *Antiquity* **2022**, *96*, 1214–1231. [CrossRef]
65. Hayk, P.; Melendez Carballido, R.; Alfaro Matos, M.; Carrion Leon, K.E. Cultural genocide and the question of Tibet. *Rev. Univ. Y Soc.* **2022**, *14*, 391–397.
66. Sinding-Larsen, A. Lhasa community, world heritage and human rights. *Int. J. Herit. Stud.* **2012**, *18*, 297–306. [CrossRef]

67. Blanco, A.; de Bustamante, I.; Pascual-Aguilar, J.A. Using old cartography for the inventory of a forgotten heritage: The hydraulic heritage of the Community of Madrid. *Sci. Total Environ.* **2019**, *665*, 314–328. [CrossRef]
68. Monna, F.; Rolland, T.; Denaire, A.; Navarro, N.; Granjon, L.; Barbe, R.; Chateau-Smith, C. Deep learning to detect built cultural heritage from satellite imagery.-Spatial distribution and size of vernacular houses in Sumba, Indonesia. *J. Cult. Herit.* **2021**, *52*, 171–183. [CrossRef]
69. Wagner, M.; Tarasov, P.; Hosner, D.; Fleck, A.; Ehrich, R.; Chen, X.; Leipe, C. Mapping of the spatial and temporal distribution of archaeological sites of northern China during the Neolithic and Bronze Age. *Quat. Int.* **2013**, *290*, 344–357. [CrossRef]
70. Park, M.; Ma, M. Sino-tibetian relationship in view of religion state and society: In a focus of national sovereignty disputes and correlative issues' historical meaning. *J. Mod. China Stud.* **2014**, *15*, 77–116. [CrossRef]
71. Strand, M.; Rivers, N.; Snow, B. The complexity of evaluating, categorising and quantifying marine cultural heritage. *Mar. Policy* **2023**, *148*, 105449. [CrossRef]
72. Gao, Q.; Jones, S. Authenticity and heritage conservation: Seeking common complexities beyond the 'eastern' and 'western' dichotomy. *Int. J. Herit. Stud.* **2021**, *27*, 90–106. [CrossRef]
73. Naguib, S.-A. Heritage in Movement: Rethinking Cultural Borrowings in the Mediterranean. *Int. J. Herit. Stud.* **2008**, *14*, 467–480. [CrossRef]
74. Yun, C.J. The Preservation and Utilization of Cultural Heritage in the Region of Haenama—Focused on the Ancient Cultural Heritage. *Baekje J.* **2018**, *26*, 103–127.
75. Lahdesmaki, T.; Ceginskas, V.L.A. Conceptualisation of heritage diplomacy in scholarship. *Int. J. Herit. Stud.* **2022**, *28*, 635–650. [CrossRef]
76. Cominelli, F.; Greffe, X. Intangible cultural heritage: Safeguarding for creativity. *City Cult. Soc.* **2012**, *3*, 245–250. [CrossRef]
77. Park, S. An Essay on the Tibetan's Intangible Cultural Heritage. *Asian Cult. Stud.* **2013**, *30*, 116–146.
78. Che, Y.J.; Cao, Y.; Wu, R.; Liu, J.; Chen, L.H.; Wu, J.K. Spatio-Temporal Pattern of World Heritage and Its Accessibility Assessment in China. *Sustainability* **2022**, *14*, 3033. [CrossRef]
79. Luo, W.; Hartmann, J.F.; Wang, F. Terrain characteristics and Tai toponyms: A GIS analysis of Muang, Chiang and Viang. *GeoJournal* **2009**, *75*, 93–104. [CrossRef]
80. Wright, W.C.C.; Eppink, F.V. Drivers of heritage value: A meta-analysis of monetary valuation studies of cultural heritage. *Ecol. Econ.* **2016**, *130*, 277–284. [CrossRef]
81. Marzeion, B.; Levermann, A. Loss of cultural world heritage and currently inhabited places to sea-level rise. *Environ. Res. Lett.* **2014**, *9*, 034001. [CrossRef]
82. Zhang, Y.; Zhang, H.; Sun, Z. Effects of Urban Growth on Architectural Heritage: The Case of Buddhist Monasteries in the Qinghai-Tibet Plateau. *Sustainability* **2018**, *10*, 1593. [CrossRef]
83. Yuewei, M.A.; Yongtao, Z.; Fubin, C.; Libo, L.A.N. Impact of Yaan-kangding highway on the world natural heritage of Sichuan giant panda sanctuaries. *Resour. Environ. Yangtze Basin* **2011**, *20*, 1017–1023.
84. Wu, P.-S.; Hsieh, C.-M.; Hsu, M.-F. Using heritage risk maps as an approach to estimating the threat to materials of traditional buildings in Tainan (Taiwan). *J. Cult. Herit.* **2014**, *15*, 441–447. [CrossRef]
85. Runze, Y. A study on the spatial distribution and historical evolution of grotto heritage: A case study of Gansu Province, China. *Herit. Sci.* **2023**, *11*, 165. [CrossRef]
86. Wang, X.; Zhang, J.; Cenci, J.; Becue, V. Spatial Distribution Characteristics and Influencing Factors of the World Architectural Heritage. *Heritage* **2021**, *4*, 2942–2959. [CrossRef]
87. Chen, F.H.; Dong, G.H.; Zhang, D.J.; Liu, X.Y.; Jia, X.; An, C.B.; Ma, M.M.; Xie, Y.W.; Barton, L.; Ren, X.Y.; et al. Agriculture facilitated permanent human occupation of the Tibetan Plateau after 3600 B.P. *Science* **2015**, *347*, 248–250. [CrossRef]
88. Shepherd, R. UNESCO and the politics of cultural heritage in Tibet. *J. Contemp. Asia* **2006**, *36*, 243–257. [CrossRef]
89. Liu, S.; Zhang, J.; Xu, X.; Dong, J.; Li, X. Distribution pattern of historical and cultural space and entropy variation in Beijing-Tianjin-Hebei region. *J. Arid. Land Resour. Environ.* **2023**, *37*, 84–93.
90. Rong, T.; Xu, S.; Lu, Y.; Tong, Y.; Yang, Z. Quantitative Assessment of Spatial Pattern of Geodiversity in the Tibetan Plateau. *Sustainability* **2023**, *15*, 299. [CrossRef]
91. Xia, H.; Zhang, D.; Wang, Q.; Wu, D.; Duan, Y.; Chen, F. A study of the construction times of the ancient cities in Ganjia Basin, Gansu Province, China. *J. Geogr. Sci.* **2020**, *30*, 1467–1480. [CrossRef]
92. Zhang, D.J.; Dong, G.H.; Wang, H.; Ren, X.Y.; Ha, P.P.; Qiang, M.R.; Chen, F.H. History and possible mechanisms of prehistoric human migration to the Tibetan Plateau. *Sci. China Earth Sci.* **2016**, *59*, 1765–1778. [CrossRef]
93. Dong, G.; Liu, H.; Yang, Y.; Yang, Y.; Zhou, A.; Wang, Z.; Ren, X.; Chen, F. Emergence of ancient cities in relation to geopolitical circumstances and climate change during late Holocene in northeastern Tibetan Plateau, China. *Front. Earth Sci.* **2016**, *10*, 669–682. [CrossRef]
94. Ding, M.; He, W.; Wang, T.; Shargan, W.; Zhang, M.; Cao, P.; Liu, F.; Dai, Q.; Fu, Q. A study of the mitochondrial genome of ancient inhabitants from the Latuotanggu cemetery, Tibet, China. *Acta Anthropol. Sin.* **2021**, *40*, 1–11.
95. Correia, F.; Merlo, G. Cross-temporal grounding. *Anal. Philos.* 2023; *early view*. [CrossRef]
96. Sun, L.; Fu, Y.; Yang, Y.; Wang, X.; Cui, W.; Li, D.; Yuan, X.; Zhang, Z.; Fu, Y.; Li, Y. Genomic Analyses Reveal Evidence of Independent Evolution, Demographic History, and Extreme Environment Adaptation of Tibetan Plateau Agaricus bisporus. *Front. Microbiol.* **2019**, *10*, 1786. [CrossRef]

97. Del, M.S.T.T.; Tabrizi, S.K. A methodological assessment of the importance of physical values in architectural conservation using Shannon entropy method. *J. Cult. Herit.* **2020**, *44*, 135–151. [CrossRef]
98. Meng, F.; Zhi, Y.; Pang, Y. Assessment of the Adaptive Reuse Potentiality of Industrial Heritage Based on Improved Entropy TOPSIS Method from the Perspective of Urban Regeneration. *Sustainability* **2023**, *15*, 7735. [CrossRef]
99. Brigatti, E.; Netto, V.M.; de Sousa Filho, F.N.M.; Cacholas, C. Entropy and hierarchical clustering: Characterizing the morphology of the urban fabric in different spatial cultures. *Chaos* **2021**, *31*, 113138. [CrossRef]
100. Zhao, F.; Fu, Y.; Luan, G.; Zhang, S.; Cai, J.; Ding, J.; Qian, J.; Xie, Z. Spatial-Temporal Characteristic Analysis of Ethnic Toponyms Based on Spatial Information Entropy at the Rural Level in Northeast China. *Entropy* **2020**, *22*, 393. [CrossRef] [PubMed]
101. Tabacchi, M.E.; Termini, S. Birkhoff's aesthetics, Arnheim's entropy. Some remarks on complexity and fuzzy entropy in arts. *Int. J. Comput. Intell. Syst.* **2015**, *8*, 1103–1115. [CrossRef]
102. Baciu, D.C. Cultural life: Theory and empirical testing. *Biosystems* **2020**, *197*, 104208. [CrossRef] [PubMed]
103. Wei, W.; Xia, J.; Wan, B. A Study on Morphological Characteristics and Types of the Traditional Urban Spaces in Tibet. *Urban Plan. Forum* **2016**, *4*, 102–111.
104. Wang, L. A Study on the Preservation of Linear Heritage Area of the Ancient Tea and Horse Road in Yunnan-Tibet. *Geogr. Geo-Inf. Sci.* **2012**, *28*, 101–105.
105. Ugwuanyi, J.K.; Itanyi, E.I.; Obieluem, U.H. Heritage Ontologies in Nigeria: An Analysis of How Heritage Connects and Disconnects People. *J. Afr. Archaeol.* **2021**, *19*, 131–145. [CrossRef]
106. Gultekin, N.T. Reading Administration Periods in Built Environment through the City of Lefke-Cyprus. *Iconarp Int. J. Archit. Plan.* **2020**, *8*, 461–479. [CrossRef]
107. Jiyoung, J. An argument about local cultural heritage and festival planning. *J. Soc. Korean Hist. Musicol.* **2020**, *65*, 525–542.
108. Luan, G.; Peng, Z.; Cai, J.; Fu, Y.; Song, L.; Shen, K.; Zhao, F. Spatio-Temporal Analysis on Evolution of Tibetan Human Activities in Qinghai from the Perspective of Geographic Names. *J. Geomat.* **2021**, *46*, 163–168.
109. Hu, Y. On the strategic position of Helong Area in Gyeonggidefense in the Tang Dynasty. *J. Northwest Univ. Nat. Sci. Ed.* **2017**, *47*, 769–775.
110. Chang, L.; Liu, R.; Jia, R. Homomorphism and isomorphism trait of Han and Tibet's culture about the peal-linked pattern. *J. Silk* **2019**, *56*, 72–78.
111. Lee, J. The Subsequent History of the Remaining Force of the Mongol Empire: Guan-xi Wei-suo of the Ming Dynasty. *Asia Rev.* **2022**, *12*, 375–428. [CrossRef]
112. Lau, H.-S. The Political Status of Tibet During the Ming Dynasty: An Analysis of Some Historical Evidence. *China Rep.* **2019**, *55*, 154–171. [CrossRef]
113. Chen, F. Transforming an Imperial Frontier: Japanese Knowledge and the Qing Empire's New Tibet Policy. *Asian Stud. Rev.* **2020**, *44*, 422–440. [CrossRef]
114. Zhu, H.; Liu, J.; Sang, Z.; Wei, W.; Wei, Z. Characteristic and contribution factors of different types of ethnic cultural resources: A case study of Gesar cultural and ecological protection experiment area. *Acta Geogr. Sin.* **2017**, *72*, 1118–1132.
115. Vodopianova, E. European Cultural Heritage as a Unity of the Traditional and the Innovative. *Contemp. Eur.* **2019**, *5*, 5–12. [CrossRef]
116. Lam-Knott, S. Contested Meanings of Urban Heritage in Hong Kong. *City Soc.* **2022**, *34*, 62–87. [CrossRef]
117. Guzel, O.G.; Kurnaz, H.A. Pomaks Protecting Their Cultural Heritage in Anatolia: A Research on Their Socio-Cultural Experiences. *Milli Folk.* **2023**, *18*, 199–213. [CrossRef]
118. Weinert, M.S. Reading world society phenomenologically: An illustration drawing upon the cultural heritage of humankind. *Int. Politics* **2017**, *55*, 26–40. [CrossRef]
119. Cerreta, M.; Daldanise, G.; Giovene di Girasole, E.; Torre, C.M. A Cultural Heritage Low Entropy Enhancement Approach: An Ex Post Evaluation of Creative Practices. *Sustainability* **2021**, *13*, 2765. [CrossRef]
120. Wang, F.; Mao, W.; Dong, Y.; Zhu, X. Implications for Cultural Landscape in a Chinese Context: Geo-analysis of Spatial Distribution of Historic Sites. *Chin. Geogr. Sci.* **2018**, *28*, 167–182. [CrossRef]
121. Wang, Q.; Bing, H.; Wang, S.; Xu, Q. Study on the Spatial Distribution Characteristics and Influencing Factors of Famous Historical and Cultural Towns or Villages in Hubei Province, China. *Sustainability* **2022**, *14*, 13735. [CrossRef]
122. Zhou, C.; Zhou, L.; Zhang, X. Spatial Differentiation and Driving Factors of Multi-category Intangible Cultural Heritages in China. *Geogr. Geo-Inf. Sci.* **2023**, *39*, 72–80.

buildings

MDPI

Article

An Experimental Study on Flexural-Shear Behavior of Composite Beams in Precast Frame Structures with Post-Cast Epoxy Resin Concrete

Peiqi Chen [1,2], Shuo Xu [1,2], Xiaojie Zhou [1,2,*] and Dezong Xu [1,2]

[1] Tianjin Key Laboratory of Civil Buildings Protection and Reinforcement, Tianjin 300384, China; cpq@tcu.edu.cn (P.C.); 17302290313@163.com (S.X.); m13525079386@163.com (D.X.)
[2] School of Civil Engineering, Tianjin Chengjian University, Tianjin 300384, China
* Correspondence: zhouxj@tcu.edu.cn

Abstract: Epoxy resin concrete has superior mechanical properties compared to ordinary concrete, and will play an increasingly important role in urban construction. In this paper, the application effect and prospect of epoxy resin concrete in precast composite frame structures are discussed. Taking the joint surface of the old and new concrete at the end of the composite beam as the research object, three specimens were devised and fabricated. Subsequently, a horizontal cyclic load test was conducted, and the seismic performance indices were analyzed. Multiple finite element models were established to assess the influence of precast concrete strength, the diameter of the longitudinal bar of the beam, the shear span ratio, and the epoxy resin concrete post-cast area, among other factors, on the seismic performance of the beam end. Four findings indicate the following: Firstly, epoxy resin concrete, characterized by its high performance attributes, can be used as a post-cast material in precast concrete structures. Secondly, when the strength of the post-cast epoxy concrete approximates or slightly exceeds that of the precast concrete, and the ratio of longitudinal reinforcement and shear span ratio are appropriately balanced, the operational performance of the composite beam frame structure is enhanced. In addition, when post-cast epoxy resin concrete is employed in the beam-column joint area, the mechanical performance of the composite beam end in the joint area matches or even surpasses that of the structure that was cast in situ. And subsequently, the expansion of post-cast area resulted in better mechanical performance. Finally, when the area of post-cast epoxy resin concrete is a non-node area, the mechanical properties of the composite beam end are worse than the former. However, the amount of epoxy resin concrete used will be greatly reduced, and as the precast node area expands, the bearing capacity of the beam end will increase and gradually approach the cast-in situ structure, indicating that this construction scheme also has advantages.

Keywords: monolithic precast composite frame structure; epoxy resin concrete; beam-column joints; seismic performance; simulation

1. Introduction

The advantages of precast structures comprise rapid production speed, consistent component quality, high construction efficiency, reduced labor resource consumption, and the promotion of sustainable development [1]. Therefore, they have progressively assumed a pivotal role in the burgeoning construction industry. Simultaneously, as individuals' demands for building functionalities continue to advance, the complexity and height of structures have escalated, thereby imposing more stringent requirements on the safety and reliability of building structures, including those of a precast nature.

Failures in prefabricated concrete structures frequently occur at the joints of each prefabricated component, thus rendering joint design a key point in structural design [2]. In recent years, numerous scholars have delved into the seismic performance of prefabricated concrete frame joints [3–6], as well as the seismic performance of bucket arch joints [7], yielding a wealth of findings. Refs. [8–11] mainly study the composite frame structure from the aspects of composite surface construction, shear resistance, and bending resistance of composite beams. However, so far, most of the studies on prefabricated frame structures focus on the post-cast area in beam-column joints. The non-node area is rarely involved as a post-cast area, and this is in fact also a commonly used construction scheme in engineering.

In addition, the global development of high-performance concrete [12] has prompted the gradual exploration of its application in precast structures. In 1993, Soubra et al. [13] conducted a study on the mechanical properties of post-cast steel fiber concrete precast beam-column members, revealing that the load-bearing capacity of steel fiber concrete-connected members surpassed that of ordinary concrete-connected members. In 2005, Gustavo et al. [14] employed high-performance fiber-reinforced cement composites (HPFRCC) in the post-cast joints of earthquake-resistant structures, resulting in significantly enhanced shear strength, ductility, and energy dissipation capacity when subjected to significant inelastic deformation. In 2009, Zheng et al. [15] proposed the utilization of ultra-high-performance concrete material (UHPC) for post-cast components of prefabricated members, demonstrating the favorable integrity performance of post-cast UHPC material in the prefabricated framework. In 2019, Deng et al. [16] substituted ordinary concrete with high ductility concrete (HDC) in nodal sections, thereby realizing the robust-nodal design principle and effectively enhancing the deformation and energy dissipation capacity of frame nodes. In terms of composite frame structures, in 2023, Wang et al. [17] proposed a new type of UHPC shear bond wet joint for connecting different prefabricated components of the precast composite beam, and evaluated the influence of different parameters of the UHPC shear bond on the shear performance of the wet joint through a series of launch tests. In 2016, Liang et al. [18] studied the important seismic characteristics of fiber-reinforced concrete (FRC) beam-column joints and RC beam-column joints under quasi-static reciprocating loads, and the results showed that the bearing capacity and deformation capacity of the beam-column joints were improved by post-cast FRC in the core area of the composite beams.

Epoxy resin concrete is a high-performance concrete—a combined material comprising epoxy resin, a curing agent, and a sand aggregate—that exhibits superior characteristics compared to conventional silicate concrete, including enhanced tensile strength, crack resistance, material uniformity, integrity, and freeze-thaw resistance [19]. Due to the illustrious interfacial bond strength between epoxy resin and concrete, as well as its commendable thermal stability, chemical resistance, and mechanical properties, it is commonly employed in reinforcement projects such as concrete crack repair [20–24]. Gil-Martin et al. [25] conducted a study comparing the mechanical properties of beam-column joints constructed with epoxy resin concrete and tire-doped rubber concrete. The findings revealed that the beam-column joints utilizing epoxy resin concrete exhibited satisfactory performance, whereas the tire-rubber doped concrete proved unsuitable for the core area of the joints. In addition, the mechanical properties of members of epoxy resin concrete were studied in Refs. [26–28]. This study provided rationale and direction for the application of epoxy concrete in building structures. However, there are very few studies on the application of epoxy resin concrete in building structures. At the same time, because the epoxy resin base has a good interfacial bond strength with concrete, if it is applied to the post-cast area of the monolithic precast concrete structure, it will greatly improve the bonding quality and shear strength of the new and old concrete bonding surface, and solve the key problems of prefabricated structures. However, there is almost no research in this area.

In light of the exceptional mechanical properties of epoxy resin concrete, this research aims to employ it as the post-cast material for a monolithic precast composite frame structure, takes the joint surface of new and old concrete at the end of the composite beam as the research object, and adopts the method of combining test and finite element analysis to study the influence of the type of post-cast concrete, the concrete strength, the ratio of longitudinal reinforcement of the beam, the shear-span ratio, and the scope and location of the post-cast area on the seismic performance of the composite beam end. Then, the bending and shear performance of the composite beam end with post-cast epoxy resin concrete was evaluated to provide data support for the popularization and application of epoxy resin concrete in precast building structures.

2. Test Profile

2.1. Test Piece Design

The research focused on the beam ends of the frame edge nodes, which served as the selected subjects for investigation. Three specimens were designed and produced, each utilizing different post-cast materials. Table 1 presents the fundamental particulars of these specimens. The dimensions of the (composite) beam section were $b \times h$ = 150 mm × 300 mm, with a post-cast layer thickness of 100 mm. The column section measured 300 mm × 300 mm. Both the beams and columns were reinforced with HRB400 longitudinal bars and stirrups. The joint interface between post-cast concrete and precast component of the composite beam was positioned 50 mm below the upper edge of the beam-column joint. The horizontal bonding surface was offered with a keyway, featuring a rough texture, while the vertical bonding surface also exhibited a rough texture. Figure 1 illustrates the geometry and reinforcement of the specimens.

Table 1. Basic information of specimens.

Specimen Number	Type of Post-Cast Concrete	Longitudinal Reinforcement	Column Stirrup	Beam Stirrup	Shear Span Ratio	Keyway Dimension/mm	Remark
WJ-RC	------	Φ16	Φ8@100	Φ8@80	1.5	---	cast-in situ specimen
WJ-CC	Ordinary concrete	Φ16	Φ8@100	Φ8@80	1.5	Width 90 mm, depth 30 mm	---
WJ-EC	Epoxy resin concrete	Φ16	Φ8@100	Φ8@80	1.5	Width 90 mm, depth 30 mm	---

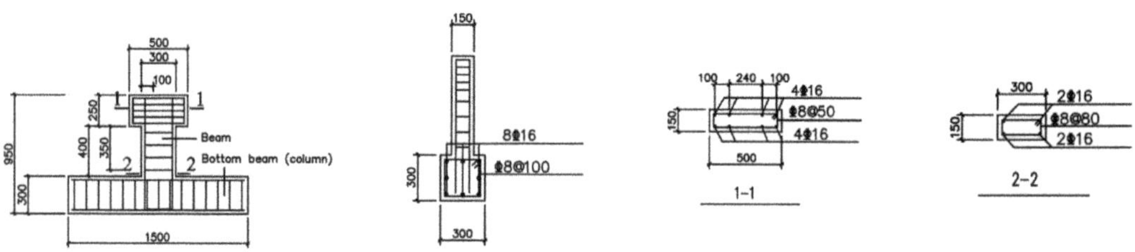

(a) Specimen WJ-RC

Figure 1. *Cont.*

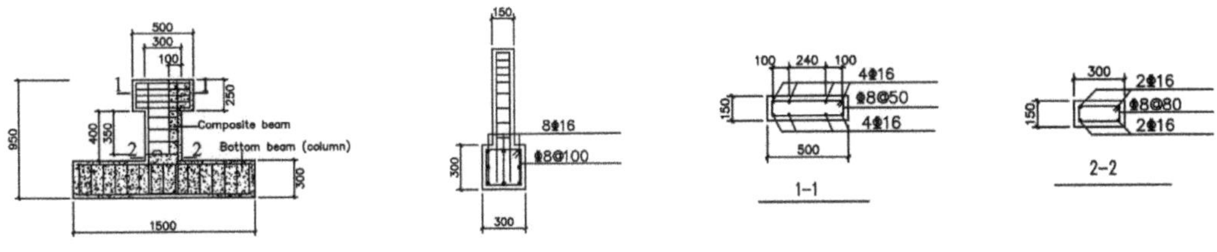

(**b**) Specimen WJ-CC

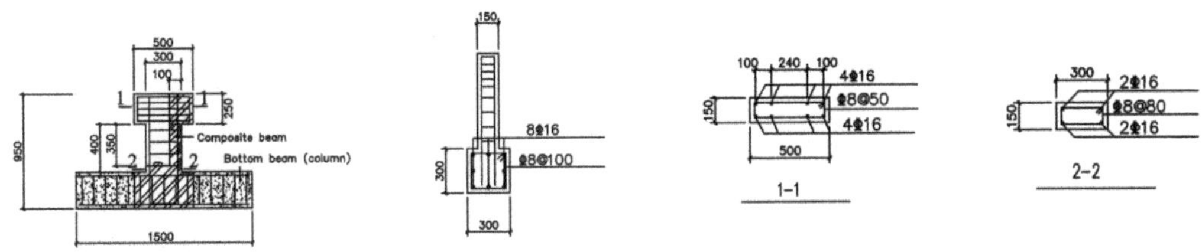

(**c**) Specimen WJ-EC

Figure 1. Specimens size and reinforcement. Note: Precast ordinary concrete; Cast- and Post-cast ordinary concrete; Post-cast epoxy resin concrete.

2.2. Basic Mechanical Properties of Materials

The average cubic compressive strength of the precast concrete was measured to be 30.9 MPa, whereas the average cubic compressive strength of the post-cast ordinary concrete amounted to 33.6 MPa. The mix ratio of the epoxy concrete is detailed in Table 2, with the average compressive strength of the measured cube reaching 55.4 MPa. The average yield strength of the 16 mm steel bar was determined to be 361.1 MPa, while the 8 mm steel bar exhibited an average yield strength of 369.5 MPa. In addition, the elastic modulus of the steel bar was measured to be 2.0×10^5 MPa.

Table 2. Mix ratio of epoxy resin concrete.

Epoxy Resin (kg/m^3)	Curing Agent (kg/m^3)	Cement (kg/m^3)	Cobble (kg/m^3)	Sand (kg/m^3)	Diluent (kg/m^3)
800	320	1200	2800	3200	80

2.3. The Production Process of the Specimens

The plywood was utilized as the formwork material. The cast-in situ specimen was constructed in one pouring process, while the other two specimens underwent two pouring processes. The key production process of specimens is depicted in Figure 2. Initially, the steel framework was assembled, and the formwork was produced. Subsequently, the mold was supported and the steel framework was positioned accordingly. At the joint of the new-old concrete, the polyethylene partition was temporarily supported and secured. Following this, the concrete (precast part) was poured for the first time. After the precast section had undergone curing and solidification, the temporarily fixed polyethylene partition was removed, and the formwork was cleansed and dried. Subsequently, a second concrete (post-cast part) pouring was conducted. Three days later, the mold was dismantled and stored in a cool environment for a duration of 28 days.

(a) Fixed partition and keyway die

(b) The first pouring

(c) The second pouring

Figure 2. Key production process of specimens.

2.4. Loading Device and Loading Scheme

The testing apparatus for loading is illustrated in Figure 3. The horizontal reciprocating load was applied at the midpoint of the terminal section of the loading beam. This process was divided into two loading stages: load control and displacement control. Prior to the specimen reaching its yield point, load control loading was implemented. Once the specimen had yielded, displacement control loading was employed. The load was regulated by a multiple of the yield displacement. The load cycle for each stage was repeated three times until the bearing capacity of the specimen declined to below 85% of the peak load, thereby concluding the test.

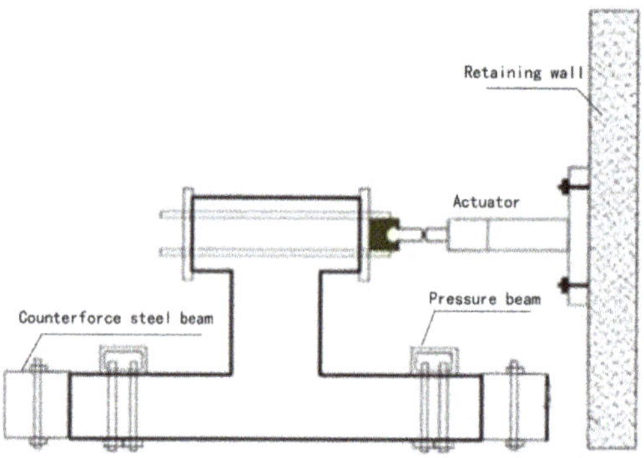

(a) Indication of loading device

(b) Loading device

Figure 3. Test loading device.

3. Test Process and Failure Characteristics of Specimen

(1) Specimen WJ-RC

Under 40 kN (push), the first horizontal crack emerges at the right side of the beam end, positioned 12 cm above the base. When the load is increased to 50 kN (push), the beam-column junction on the right side of the beam end experiences cracking. Upon applying a reverse load of 50 kN (pull), an oblique crack appears at the left side of the beam end, positioned 5 cm above the base. This crack extends diagonally towards the lower right, passing through the left root and causing the tensile steel bar to yield. The yield displacement, Δ_y = 1.21 mm. Subsequently, displacement control loading leads to the formation of oblique cracks at different heights along the beam end as the displacement increases. When the beam is loaded to $3\Delta_y$ (pull), the left oblique crack extends diagonally towards the lower right, intersecting with the right oblique crack to form a cross-shaped main crack. Additionally, a horizontal crack traverses the width of the beam section. Upon reaching $6\Delta_y$, the peak load of 130.8 kN is achieved. During the progressive reduction in bearing capacity, the shear oblique crack widens, causing slight compression of the concrete at the left root of the beam end. The right root of the beam end exhibits a slight tendency to lift, indicating its weakness. At a load of $9\Delta_y$, the load drops to 85% of the peak load, marking the end of the test. The failures of the specimen are depicted in Figure 4.

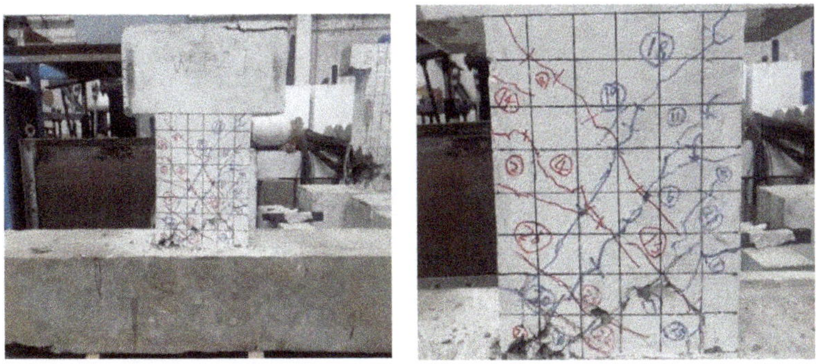

(**a**) Specimen WJ-RC (**b**) Damage pattern of beam end

Figure 4. Failure condition of specimen WJ-RC.

(2) Specimen WJ-CC

When subjected to a load of 30kN (pull), the first crack appears along the right side of the keyway on the horizontal bonding surface. Under a load of 40 kN (push), the right beam-column junction experiences cracking. Upon reaching a load of 41.3 kN (pull), the beam's longitudinal tendons yield, resulting in a yield displacement of Δ_y = 1.25 mm. Subsequent displacement-controlled loading leads to the emergence of horizontal cracks at various heights along the beam's end, with oblique cracks extending towards the keyway of the bonding surface as the displacement increases. When the beam is subjected to a load of $3\Delta_y$ (push), the horizontal cracks in the beam's width direction become interconnected. Conversely, when the beam is loaded to $3\Delta_y$ (pull), primary cross cracks form. The peak load is attained at $6\Delta_y$ (with an average peak load of 117.5 kN in both directions), beyond which the load reduces. Upon reaching a reverse load of $9\Delta_y$ (pull), slippage occurs at the interface between the new and old concrete, leading to crack development. Severe damage is observed on both sides of the keyway and the vertical joint, with the post-cast concrete on the right side of the beam experiencing near-shear failure. At a load of $11\Delta_y$, the load decreases to 85% of the peak load, marking the conclusion of the test. The failures of the specimen are depicted in Figure 5.

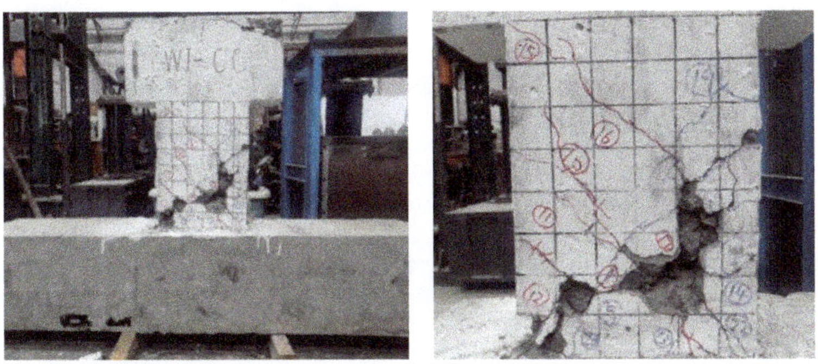

(**a**) Specimen WJ-CC (**b**) Damage pattern of beam end

Figure 5. Failure condition of specimen WJ-CC.

(3) Specimen WJ-EC

When subjected to a load of 40kN (push), the initial occurrence of a crack is observed on the left horizontal joint surface of the beam end, which subsequently propagates above the keyway. Upon further loading to 64.5kN (push), a portion of the tensile steel bar undergoes yielding, resulting in a yield displacement of Δ_y = 1.1 mm. Subsequently, displacement control loading is implemented, accompanied by the application of a reciprocating load. As a consequence, inclined cracks emerge in the prefabricated section, extending towards the keyway, while the post-cast epoxy resin concrete section remains devoid of evident cracks. Upon reaching a load of 6Δ_y, a series of subtle horizontal cracks occur in the tension area beneath the post-cast epoxy concrete on the right side of the beam end. Notably, these cracks do not exhibit significant propagation thereafter. Conversely, the cracks in the precast concrete, situated on the upper part of the keyway of the horizontal bonding surface, experience an increase in both magnitude and intersection. The peak load is attained at 10Δ_y (with an average value of 129.1 kN in both loading directions). Subsequently, the load gradually diminishes, resulting in severe concrete spalling above the keyway of the horizontal bonding surface. At a load of 13Δ_y, the load drops to 85% of the peak load, signifying the conclusion of the test. The failures of the specimen are depicted in Figure 6.

(**a**) Specimen WJ-EC (**b**) Damage pattern of beam end

Figure 6. Failure condition of specimen WJ-EC.

In summary, flexural-shear failure was observed in all three specimens. Among them, the integrity of cast-in situ specimens was found to be superior, exhibiting a greater number of inclined cracks and a more uniform distribution. The weakest section was identified as the beam end root section, which displayed characteristics of bending failure. The failure of the two composite beam specimens primarily occurred at the joint between the old and new concrete. In the case of specimen WJ-CC, the vertical joints and horizontal joints with keyway suffered severe damage, with the ordinary concrete pouring layer being nearly severed, resulting in poor integrity. Specimen WJ-EC, on the other hand, experienced concentrated failure on the horizontal bonding surface with keyway, while the vertical bonding plane remained largely intact. Notably, the precast part—i.e., the horizontal bonding plane—exhibited significant damage. In the post-cast epoxy concrete area, only a few small cracks with a uniform distribution were observed, indicating the participation and contribution of the post-cast epoxy resin concrete to improving the overall performance of the node area in the load-bearing process.

4. Experimental Results and Analysis

4.1. Hysteresis Curve

Figure 7 illustrates the *P*-*Δ* hysteresis curves of the three specimens. Due to a malfunction in the displacement meter, the hysteresis curve of WJ-RC is asymmetric, and then the negative hysteretic characteristics were observed and analyzed.

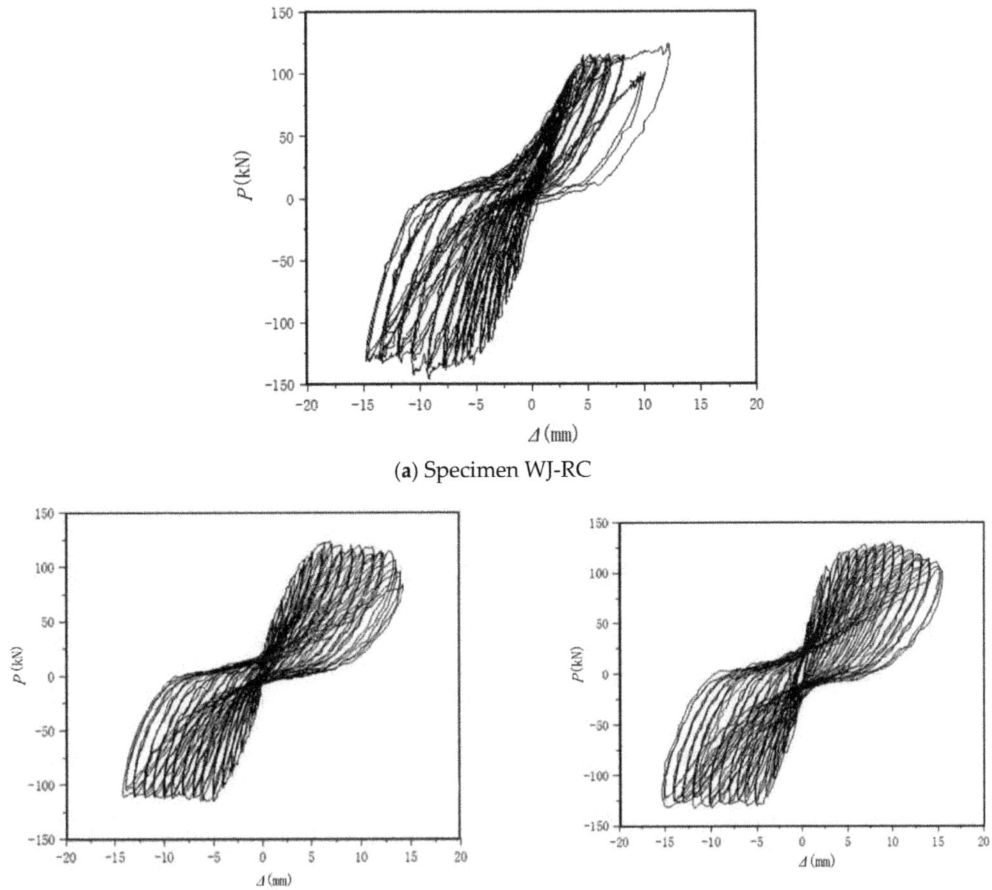

(**a**) Specimen WJ-RC

(**b**) Specimen WJ-CC

(**c**) Specimen WJ-EC

Figure 7. Hysteresis curve of specimens.

As depicted in Figure 7, the mechanical properties of the epoxy resin concrete render is superior to post-cast ordinary concrete specimen WJ-CC. The post-cast epoxy concrete specimen WJ-EC exhibits a decelerated strength degradation, a more complete hysteresis loop, and enhanced energy dissipation capacity. In addition, the hysteresis loop of WJ-EC surpasses that of WJ-RC, albeit marginally, thereby indicating that the hysteresis performance of composite beam frame joints in post-cast epoxy concrete is equivalent to, or even surpasses, that of fully-cast joints.

4.2. Skeleton Curve

Figure 8 illustrates the skeleton curves for each specimen, while Table 3 presents the measured data for each characteristic point on the skeleton curve. The following can be seen from Figure 8 and Table 3:

(1) During the early loading stage, the load value of the WJ-EC specimen exhibited the most rapid increase, indicating the highest stiffness. In comparison, the stiffness of the WJ-RC and WJ-CC specimens was slightly lower, with minimal difference between them. After reaching the ultimate load, the WJ-EC specimen demonstrated the slowest decrease in bearing capacity, followed by the WJ-RC specimen, while the bearing capacity of the WJ-CC specimen exhibited a more pronounced decline.
(2) The cracking load of the WJ-CC specimen was the lowest, whereas the cracking load of the WJ-EC specimen was marginally lower than that of the fully cast specimen, with no significant distinction between them. The initial crack in both the WJ-CC and WJ-EC specimens appeared on the horizontal joint surface between the old and new concrete. The cracking displacement of the two specimens was similar and smaller than that of the fully cast specimens.
(3) The overall trend of the skeleton curve for the WJ-EC specimen was higher, resulting in a 10% increase in bearing capacity compared to the WJ-CC specimen. The bearing capacity of the WJ-EC specimen was similar to that of the WJ-RC specimen, but the ductility of the WJ-EC specimen was 1.65 times greater than that of the WJ-RC specimen.

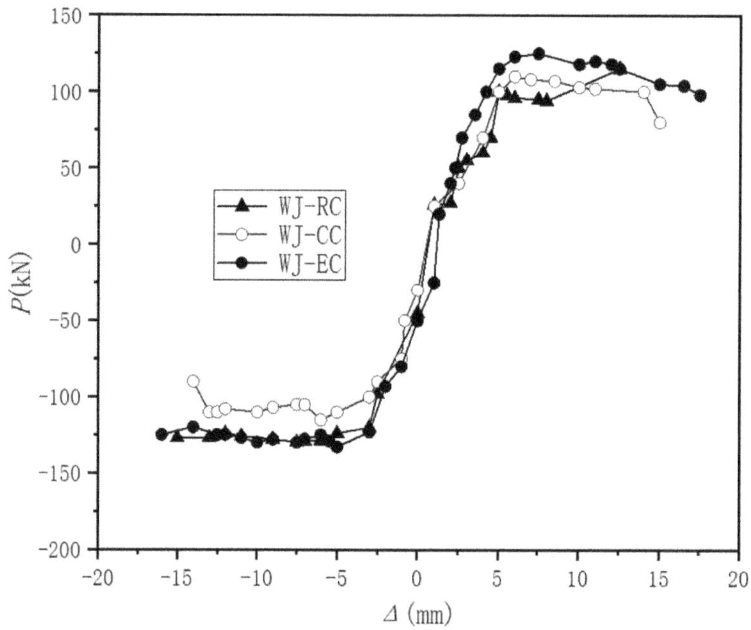

Figure 8. Skeleton curve.

Table 3. The measured data of skeleton curve feature points of each specimen.

Specimen Number	Cracking Load P_{cr}/kN	Cracking Displacement Δ_{cr}/mm	Yield Load P_y/kN	Yield Displacement Δ_y/mm	Peak Load P_{max}/kN	Peak Displacement Δ_{max}/mm	Failure Load P_u/kN	Failure Displacement Δ_u/mm	Ductility Coefficient $\mu = \Delta_u/\Delta_y$
WJ-RC	42.7	0.55	97.5	3.37	130.8	9.0	118	14.71	4.36
WJ-CC	28.1	0.43	80.9	2.76	117.5	6.0	92.8	14.08	5.07
WJ-EC	41.5	0.45	87.5	2.10	129.1	10.0	113.9	15.32	7.27

In summary, considering the cracking load, bearing capacity, stiffness, and ductility, the performance of post-cast ordinary concrete specimens is deemed the most inferior. Conversely, the performance of post-cast epoxy resin concrete specimens is either equivalent to or surpasses that of all the fuiiy-cast specimens.

5. Finite Element Analysis

Modeling and finite element analysis of concrete and epoxy resin concrete can be carried out from both micro and macro perspectives. Simulations at the micro scale are more accurate. For example, M.M. Shahzamanian [29] simulated the representative volume element (RVE) of a gelling system (Type I) containing fly ash (Class F) by using the voxel-based finite element analysis (FEA) method, and analyzed the homogenized macroscopic mechanical properties. Kamali-Bernard [30] proposed a three-dimensional multi-scale modeling method for the mechanical properties of cement-based materials—that is, to generate real 3D representative V-volume elements of cement-based materials at different scales, and then analyze and calculate their mechanical properties. The macroscopic perspective—that is, ignoring the microstructure of concrete and simulating its mechanical properties only from the macroscopic perspective—is more suitable for the analysis of concrete structures. In this paper, both concrete and epoxy resin concrete are simulated from a macroscopic perspective.

5.1. Constitutive Model of Materials

For the concrete, a plastic damage model is employed, while the constitutive relationship of ordinary concrete adheres to the uniaxial stress-strain curve recommended in the Code for Design of Concrete Structures [31]. As for epoxy resin concrete, its constitutive relation is established based on the comprehensive curve equation of compressive stress-strain, which has been fitted by our research group through tests (Equation (1)). The stress-strain curve of the rebar follows a double broken line model.

$$y = \begin{cases} ax + (4.9 - 4.23a)x^2 + (-4.67 + 6.67a)x^3 + (-0.27 - 4.74a)x^4 + (1.07 + 1.27a)x^5, & 0 \le x \le 1 \\ \frac{x}{b(x-1)^2 + x}, & x > 1 \end{cases} \quad (1)$$

a and b are undetermined parameters whose values range from:

$$\begin{cases} 0 < a < 1.0 \\ 1.0 < b < 10.0 \end{cases} \quad (2)$$

In this paper, a is 0.7 and b is 7.

5.2. Establishment of Model

The epoxy resin concrete is modeled using the C3D8R unit, while the rebar is modeled using the T3D2 unit. The rebar is fully embedded in the concrete. The interface between the new and old concrete is simulated using surface-to-surface contact, and the tangential action is based on the Coulomb friction model. According to Ref. [32], combined with EU Norms [33]: the friction coefficient μ is taken as 0.6 for new and old concrete bonding surfaces with untreated surfaces, μ is taken as 0.7 for bonding surfaces with a roughness not less than 3 mm and a width not less than 40 mm, and μ is taken as 0.9 for the cogging bonding surface. The friction coefficient μ of the interface between ordinary concrete in this paper is taken as 0.8, and after repeated calculations, the friction coefficient μ of the interface between epoxy resin concrete and ordinary concrete is taken as 0.95. Normal action is governed by "Hard contact", wherein the contact pressure is unrestricted when transmitted through the contact surface. If the contact pressure value becomes negative, the two contact surfaces separate, and the contact constraint on the node is released. The bottom beam is subjected to fixed constraints, and the loading mode is monotone loading.

5.3. Model Verification

The specimens WJ-RC, WJ-CC, and WJ-EC were subjected to finite element analysis individually, enabling an approximation of the damage conditions of the model through the distribution of the concrete compression damage factors, as depicted in Figure 9. It is evident from Figure 9a that the cast-in situ specimen WJ-RC exhibits superior integrity and a more uniform distribution of damage. Conversely, Figure 9b,c reveals noticeable differences in the damage conditions between the new and old concrete bonding surfaces. The vertical and horizontal joints of the WJ-CC specimens suffered severe damage, while the WJ-EC specimen primarily exhibited concentrated damage on the precast concrete of the upper side of the horizontal bonding surfaces, with the concrete on both sides of the vertical bonding surface remaining essentially intact. In summary, the numerical simulation results effectively approximate the actual failure state of the specimen.

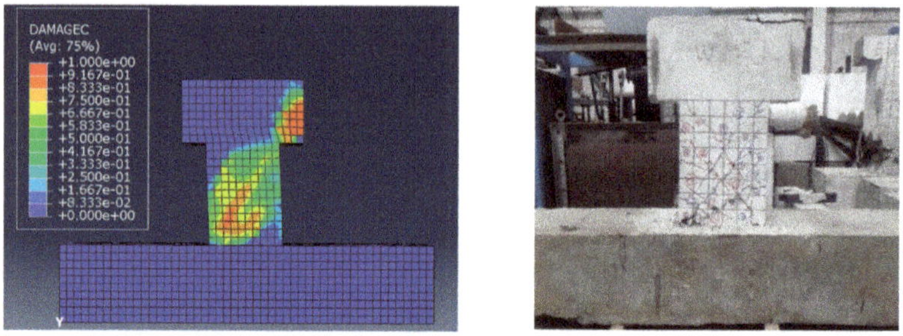

(**a**) WJ-RC

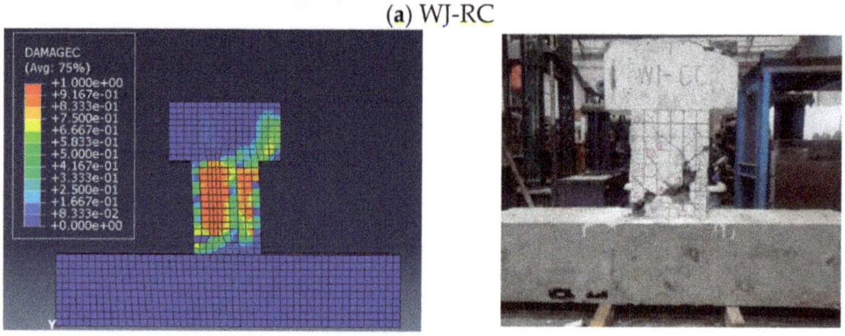

(**b**) WJ-CC

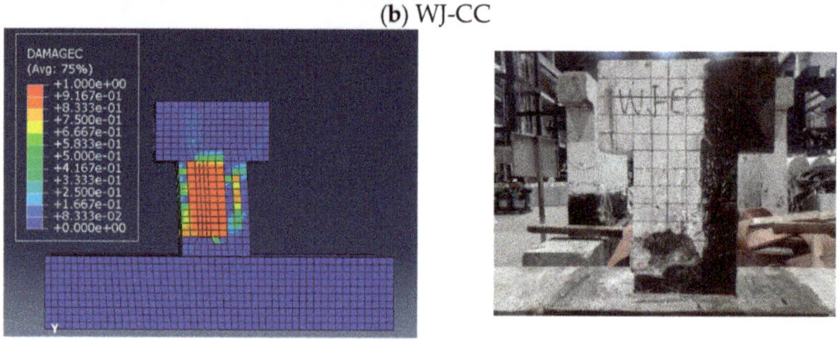

(**c**) WJ-EC

Figure 9. Comparison of failure forms of each specimen.

Figure 10 and Table 4 present a comparison between the test results and the simulation results of the skeleton curves for each specimen. The overall trend of the curves demonstrates a significant similarity between the two. The simulated value deviates from the test value of the peak load by approximately 10%, while the deviation between the simulated value and the test value of the peak displacement amounts to roughly 20%. The excessive initial slope of the simulated skeleton curve can be attributed to the finite element simulation's neglect of adverse factors such as residual stress and gap squeezing between the test loading devices. In general, the finite element model established in this study adequately approximates the mechanical properties of the beam ends in the joint region of the overlapping beams.

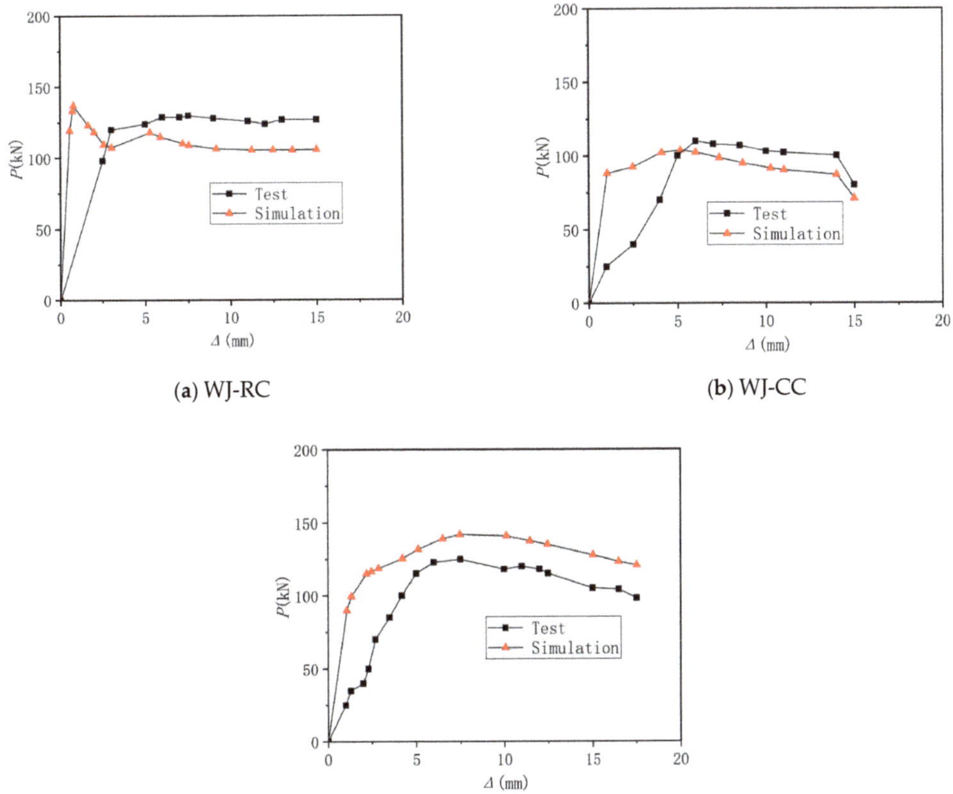

(**a**) WJ-RC (**b**) WJ-CC (**c**) WJ-EC

Figure 10. Comparison between simulated skeleton curve and test skeleton curve.

Table 4. Comparison of experimental and simulated shear strength.

Specimen Number	P_t/kN	Δ_t/mm	P_{nu}/kN	Δ_{nu}/mm	P_{nu}/P_t	Δ_{nu}/Δ_t
WJ-RC	130.8	9	139.0	1.2	0.90	0.67
WJ-CC	117.5	6	103.8	5.2	0.88	0.87
WJ-EC	129.1	10	139.4	8.8	1.07	0.88

Note: P_t and Δ_t are the peak load and peak displacement of the test, P_{nu} and Δ_{nu} are the peak load and peak displacement of numerical simulation.

5.4. Influence of Parameter Variation on Seismic Performance of the Composite Beam

Based on specimen WJ-EC, a finite element analysis was conducted, examining parameters such as the strength of precast concrete, the diameter of longitudinal reinforcement in beams and columns, the shear span ratio, and the post-cast area of epoxy resin concrete. Certain parameters remained unchanged, including the strength of post-cast epoxy resin concrete, the reinforcement ratio of stirrups, and the size of key slots. A total of 14 finite element models were established in 5 groups, as detailed in Tables 5 and 6. Among them, changing the post-cast area of epoxy resin concrete includes two cases. In the first case, the post-cast joint area with epoxy resin concrete was expanded by modifying the distance

y between the horizontal joint surface and the upper edge of the beam-column joint, resulting in an enlarged post-cast joint area (Figure 11a), as shown in specimens WJ-EC10 and WJ-EC11. The second case aimed to retain the integrity of the precast beam-column joints while changing the position of the precast joint area and horizontal joint surface in the composite beam (Figure 11b). Specimens WJ-EC12, WJ-EC13, and WJ-EC14 exemplify this scenario.

Table 5. Parameter setting of the model (The first case).

Specimen Number	Strength Grade of Precast Concrete	Beam and Column Longitudinal Reinforcement	Shear Span Ratio	y (mm)	Remark
WJ-EC	C30	Φ16	1.5	50	Compare specimens
WJ-EC1	C40	Φ16	1.5	50	---
WJ-EC2	C50	Φ16	1.5	50	---
WJ-EC3	C60	Φ16	1.5	50	---
WJ-EC4	C70	Φ16	1.5	50	---
WJ-EC5	C80	Φ16	1.5	50	---
WJ-EC6	C30	Φ18	1.5	50	---
WJ-EC7	C30	Φ20	1.5	50	---
WJ-EC8	C30	Φ16	1	50	---
WJ-EC9	C30	Φ16	2.5	50	---
WJ-EC10	C30	Φ16	1.5	200	---
WJ-EC11	C30	Φ16	1.5	300	---

Table 6. Parameter setting of the model (The second case).

Specimen Number	Strength Grade of Precast Concrete	Beam and Column Longitudinal Reinforcement	Shear Span Ratio	Post-Cast Materials	Horizontal Seam Thickness (mm)	y (mm)	Remark
WJ-RC	---	Φ16	1.5	---		---	Contrast specimen
WJ-CC	C30	Φ16	1.5	Ordinary concrete	---	---	Contrast specimen
WJ-EC12	C30	Φ16	1.5	Epoxy resin concrete	70	70	---
WJ-EC13	C30	Φ16	1.5	Epoxy resin concrete	70	165	---
WJ-EC14	C30	Φ16	1.5	Epoxy resin concrete	70	260	---

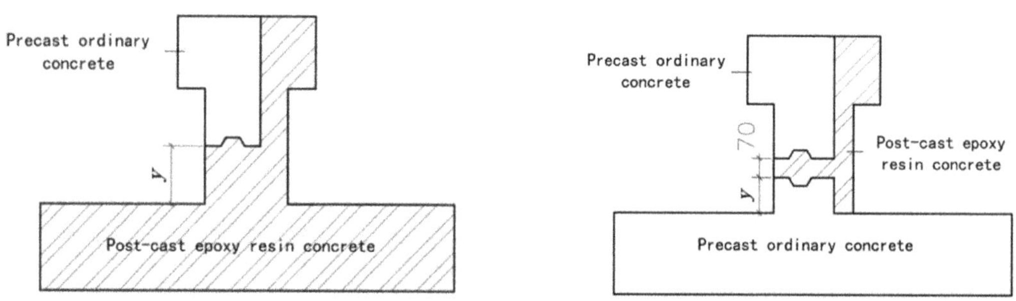

(**a**) Case 1: Expand the area of the post-cast node (**b**) Case 2: Retain the precast node area

Figure 11. The model diagram of change of epoxy concrete post-cast area.

5.4.1. Influence of Precast Concrete Strength

The stress nephogram and concrete compression damage factors distribution map of the typical model are presented in Figure 12. It is evident that the strength of the precast partial concrete exerts a minimal influence on the failure pattern of joints. In the limit state, the peak stress points of all specimens all appear near the joint surface of the precast section and the post-cast section, with the post-cast portion contributing significantly to the stress distribution. The P-Δ curve of each specimen is depicted in Figure 13. It is observed that the initial stiffness of the specimen remains largely unaffected by the concrete strength, whereas the bearing capacity exhibits a noticeable dependence on it. Specifically, when the concrete strength grade is elevated from C30 to C40, the bearing capacity of the specimen increases from 139.4 kN to 144.6 kN, corresponding to a growth rate of 3.7%. Subsequently, as the concrete strength grade is further increased from C40 to C50, the bearing capacity rises from 144.6 kN to 153.3 kN, exhibiting a growth rate of 6.0%. Similarly, when the concrete strength grade is raised from C50 to C60, the bearing capacity increases from 153.3 kN to 160.0 kN, demonstrating a growth rate of 4.0%. However, beyond this point, the rate of increase in bearing capacity reduces gradually with the enhancement of concrete strength. In summary, the optimal working performance of the specimen is achieved when the strength of the post-cast epoxy resin concrete is approximately equal to or slightly higher than that of the precast partial concrete. The main reason is that in this case, the interaction ability of the new and old concrete of the bonding surface including the shear bond is equivalent, and its overall working performance is the most fully exerted.

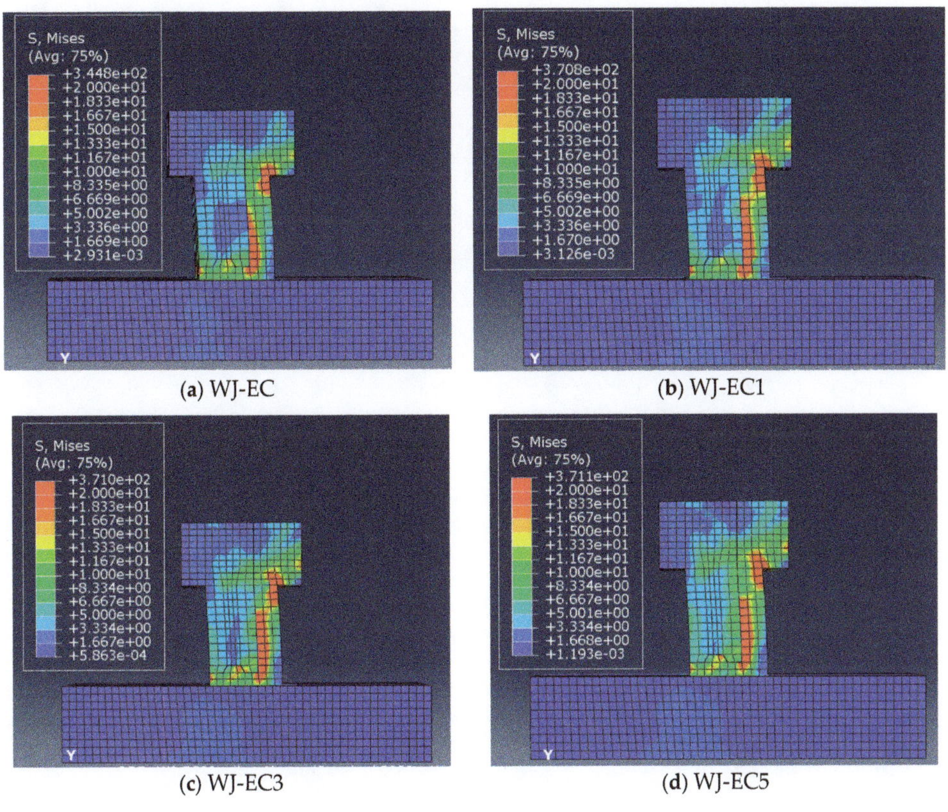

Figure 12. Stress nephogram and damage factors distribution map of specimens with different strength.

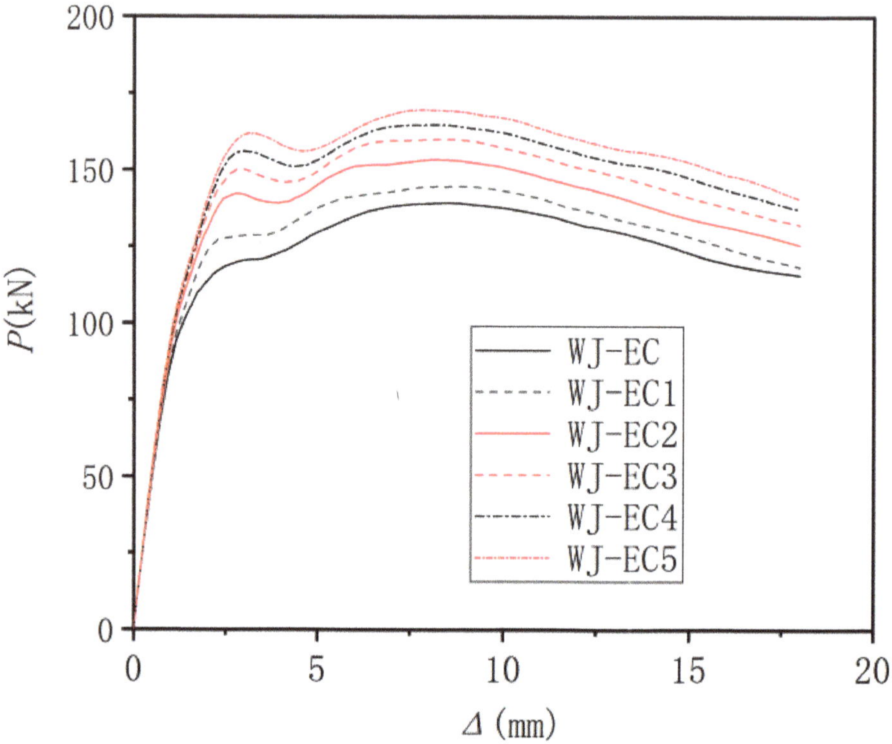

Figure 13. *P-Δ* curves of different concrete strength specimens.

5.4.2. Effect of the Longitudinal Reinforcement Ratio of the Composite Beam

Figure 14 presents the stress nephogram and concrete compression damage factors distribution map for each specimen of the composite beam, featuring varying longitudinal bar diameters. Figure 15 displays the *P-Δ* curves for each specimen. It is evident that the reinforcement ratio of the longitudinal reinforcement in the composite beam has a minimal influence on the failure pattern of the joint area, but significantly impacts the bearing capacity. During the initial loading stage, the *P-Δ* curves of all specimens exhibit significant overlap, indicating similar initial stiffness. As the plastic stage is reached, an increase in the longitudinal reinforcement ratio leads to an elevation in peak load, a reduction in peak displacement, and a steeper declining section of the curve, thereby accelerating the degradation of bearing capacity after reaching the peak. For instance, when the diameter of the longitudinal reinforcement increases from 16 mm to 18 mm, the bearing capacity rises from 139.4 kN to 153.2 kN, reflecting a growth rate of 9.8%. Similarly, when the diameter of the longitudinal reinforcement increases from 18 mm to 20 mm, the bearing capacity increases from 153.2 kN to 157.2 kN, with a growth rate of 2.6%. These findings demonstrate that at lower reinforcement ratios, the bearing capacity experiences a significant increase, with an increased reinforcement ratio. However, at higher reinforcement ratios, the bearing capacity growth slows down, and the ductility experiences a significant decrease. The main reason is that with the increase in the reinforcement ratio of longitudinal reinforcement, the shear bearing capacity of the joint surface gradually depends on the strength of the concrete, and the brittle failure property is also gradually obvious.

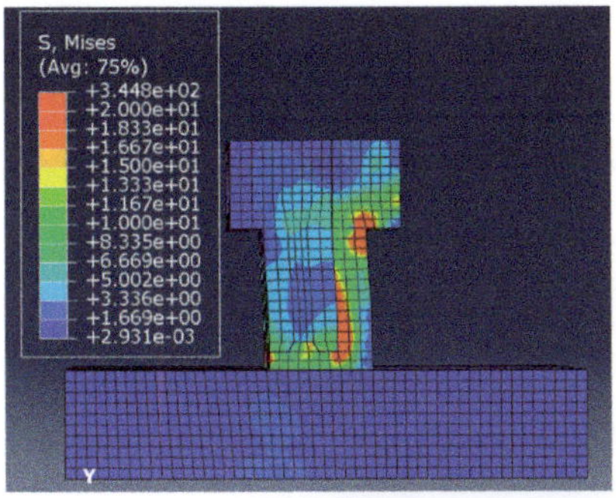

(**a**) WJ-EC

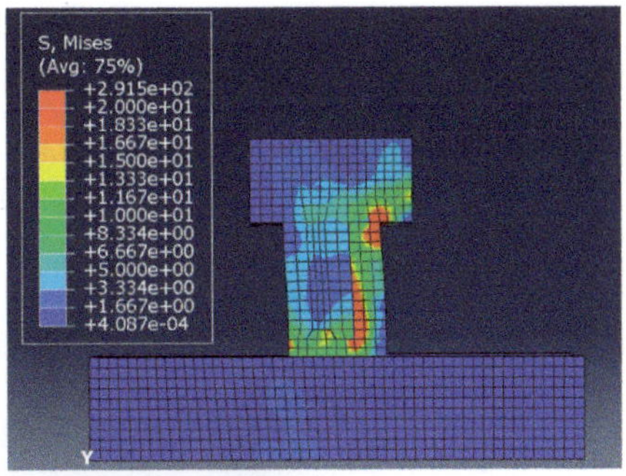

(**b**) WJ-EC6

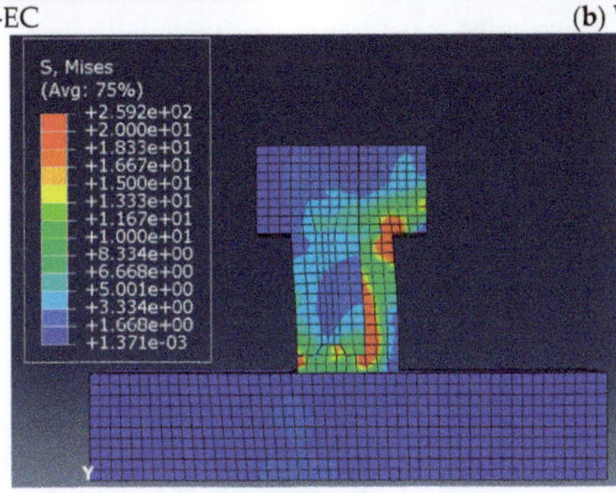

(**c**) WJ-EC7

Figure 14. Stress nephogram and damage factors distribution map of specimens with different longitudinal diameters.

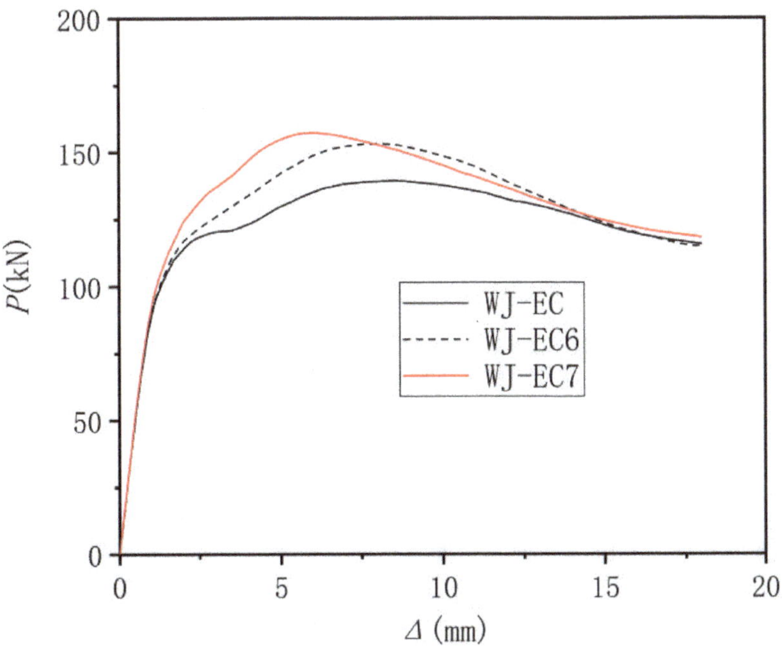

Figure 15. P-Δ curves of specimens with different longitudinal diameters.

5.4.3. The Influence of Shear–Span Ratio

Figure 16 presents the stress nephogram and concrete compression damage factors distribution map for each specimen, varying in shear–span ratio. Figure 17 displays the *P-Δ* curves for each specimen. It is evident that the damage observed on both sides of the joint surface in the three specimens is similar. Specifically, the prefabricated side exhibits more severe damage, while the after-cast epoxy concrete side experiences relatively lighter damage. As the shear–span ratio decreases, the shear failure characteristics of the specimen become more pronounced, and the right side of the vertical joint surface demonstrates a greater contribution to the shear force, thus enhancing the bearing capacity. Notably, specimens with a shear to span ratio of 2.5 exhibit a lower bearing capacity, with the right side of the post-cast section of the vertical joint plane exhibiting minimal force participation, while the left side of the prefabricated section displays evident bending failure characteristics. The bearing capacities of specimens with shear span ratios of 1.0, 1.5, and 2.5 are recorded as 195.7 kN, 139.4 kN, and 91.3 kN, respectively. Thus, an increase in the shear span ratio leads to a decrease in the specimen's bearing capacity, an increase in the peak displacement, a deceleration in the falling section, and an enhancement in deformation capacity. The main reason is that the shear-span ratio reflects the ratio of bending moment and shear force on the cross-section of the specimen, the shear force has an obvious effect when the shear-span ratio is small, and it has the characteristics of shear brittleness failure during failure. However, when the shear-span ratio is large, the bending moment has an obvious effect, and the ductile failure characteristics are obvious.

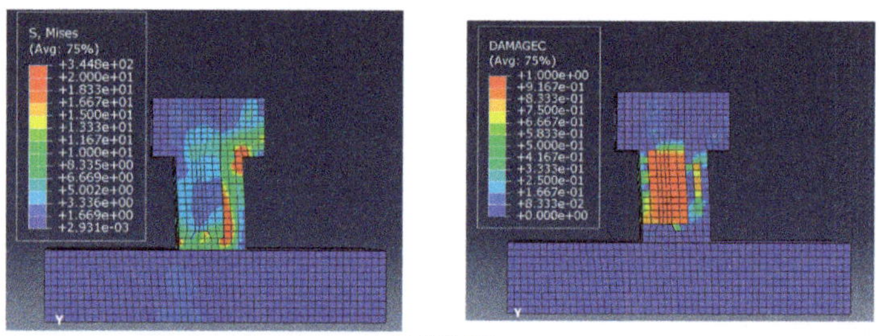

(**a**) WJ-EC

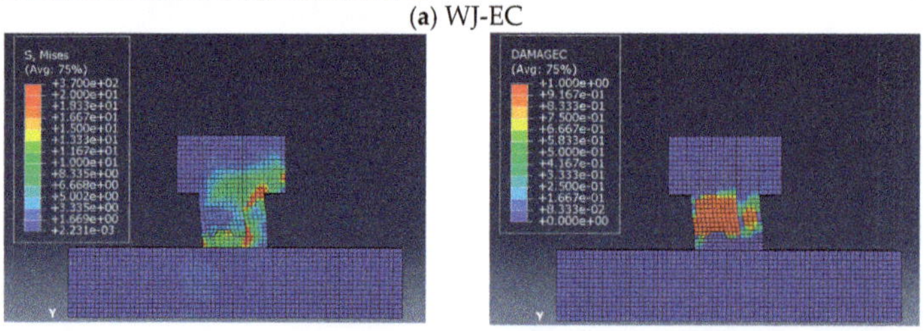

(**b**) WJ-EC8

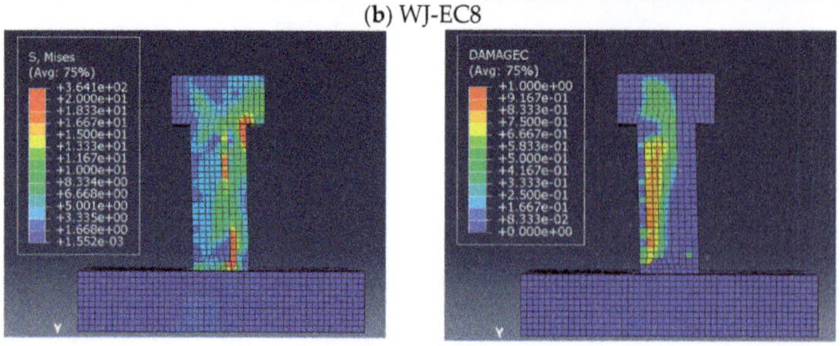

(**c**) WJ-EC9

Figure 16. Stress nephogram and damage factors distribution map of specimens with different shear–span ratios.

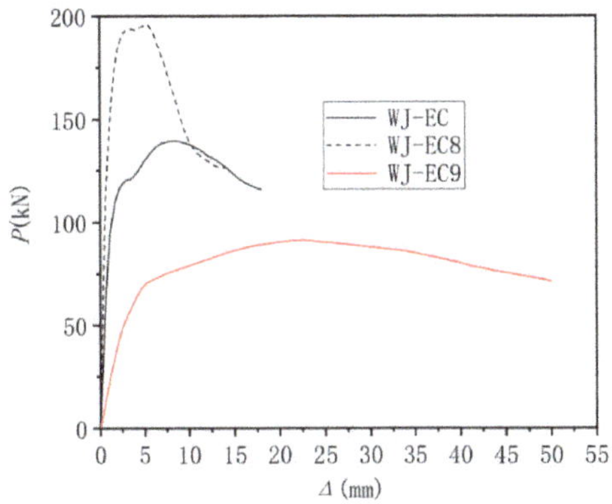

Figure 17. P-Δ curves of specimens with different shear–span ratios.

5.4.4. The Effect of the Epoxy Resin Concrete Post-Cast Area

(1) Case 1: Expand the area of the post-cast node

Premised on specimen WJ-EC, the post-cast area was expanded—specifically, the horizontal bonding surface with a keyway was positioned at distances of 200 mm and 300 mm above the beam-column joint edge, denoted as models WJ-EC10 and WJ-EC11, respectively (Figure 11a). The stress nephogram and concrete compression damage factors distribution map of each specimen are depicted in Figure 18. Additionally, Figure 19 illustrates the P-Δ curves for each specimen.

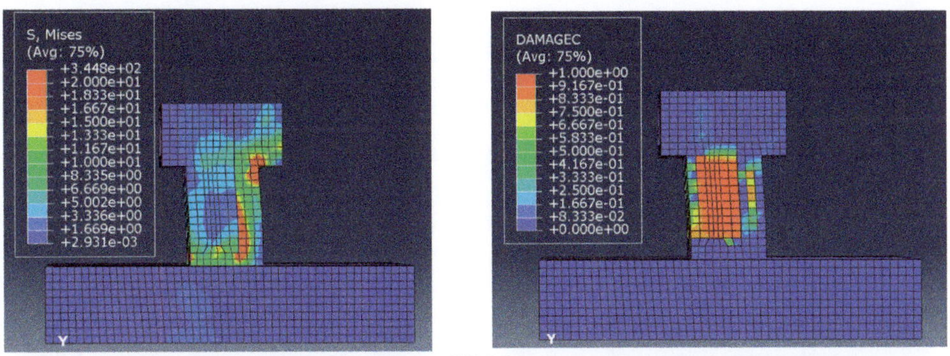

(**a**) WJ-EC

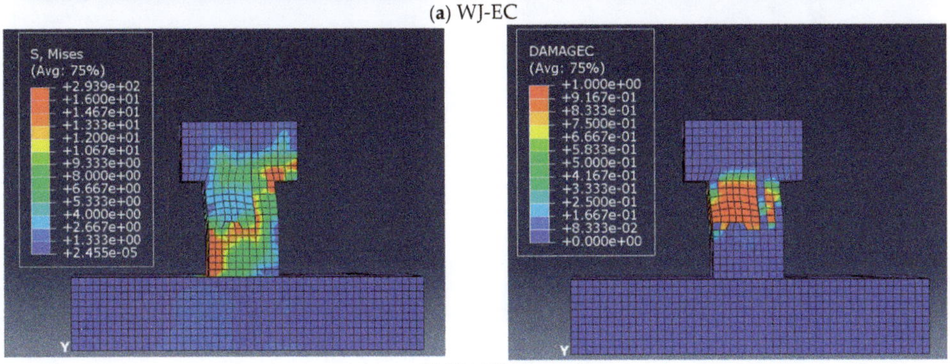

(**b**) WJ-EC10

Figure 18. *Cont.*

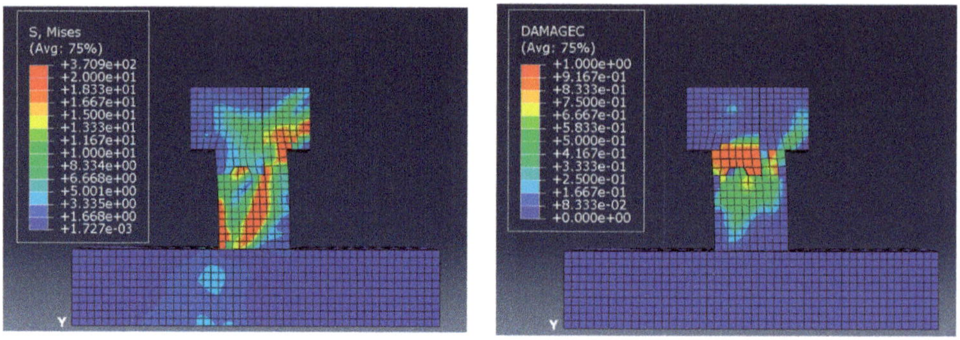

(c) WJ-EC11

Figure 18. Stress nephogram and damage factors distribution map of each specimen (Case 1).

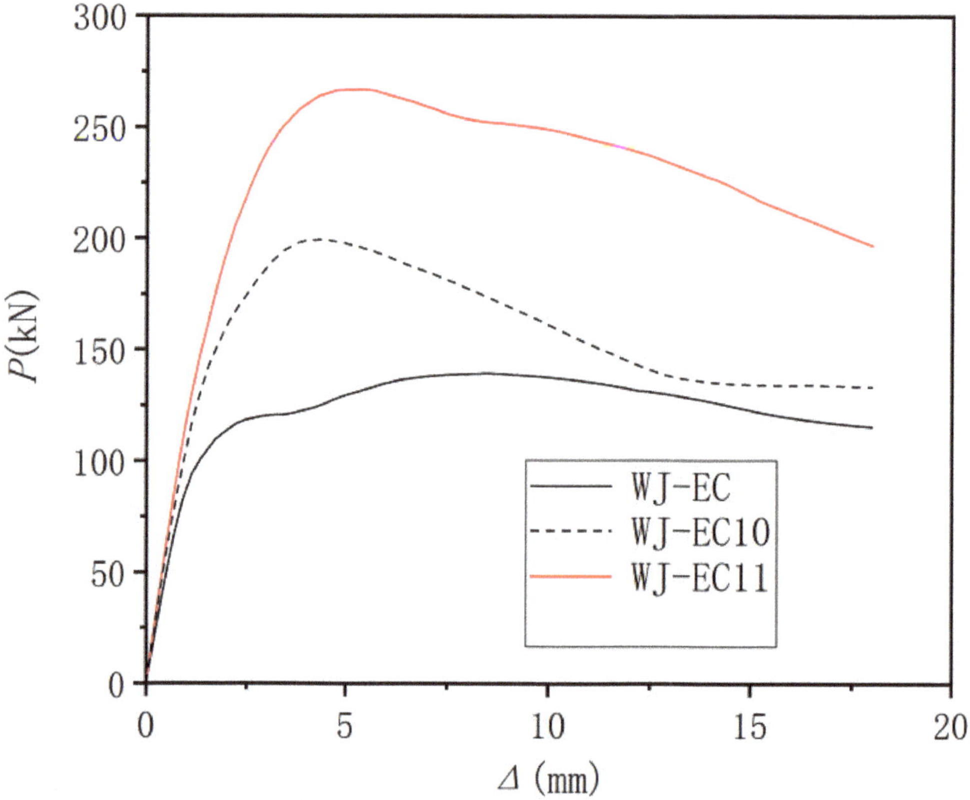

Figure 19. P-Δ curves of each specimen (Case 1).

As observed in Figure 18, the enlargement of the post-cast area resulted in an increased contribution of the post-poured epoxy concrete, with an excellent mechanical properties section for the specimen's stress, hence leading to a significant enhancement in its bearing capacity. The damage in all three specimens was predominantly concentrated in the precast section, characterized by lower material strength. Figure 19 reveals that the bearing capacity and initial stiffness of the specimen experienced a significant increase with the subsequent expansion of the post-cast area. Moreover, the declining section of the curve exhibited a slightly steeper slope, indicating a minor reduction in the specimen's deformation ability, albeit not significantly. The bearing capacities of specimens WJ-EC, WJ-EC10, and WJ-EC11 were determined to be 139.4 kN, 199.6 kN, and 267.0 kN, respectively. Therefore, compared to specimen WJ-EC, specimen WJ-EC10 exhibited a 43% increase in bearing capacity, while specimen WJ-EC11 demonstrated a remarkable 91% increase.

(2) Case 2: Reserve the precast node area

To ensure the integrity of the precast beam-column joints, the distance y from the joint surface of the precast node area and the composite beam to the edge of the node was set at 70 mm, 165 mm, and 330 mm, respectively. This resulted in an incremental increase in the precast joint area for specimens WJ-EC12, WJ-EC13, and WJ-EC14. The steel bar connection at the horizontal joint of each specimen was mechanically connected. Considering the sleeve length and construction requirements, the thickness of the horizontal joint was set at 70 mm (Figure 11b). The simulation results of these three specimens were compared with those of WJ-RC and WJ-CC. The stress nephogram and concrete compression damage factors distribution map of each specimen are shown in Figure 20, while the P-Δ curves of each specimen are presented in Figure 21.

As depicted in Figure 20, the stress distribution and damage distribution of the cast-in situ specimen WJ-RC were found to be the most uniform. The post-cast epoxy resin concrete played a significant role in bearing the stress of the specimen, with its contribution increasing as the y value increased. In contrast, the damage in the other specimens was concentrated in the precast ordinary concrete. With the expansion of the precast joint area, the participation degree and damage degree of the post-cast epoxy resin concrete increased, leading to an increase in the bearing capacity of the specimens. Figure 21 demonstrates that the bearing capacities of specimens WJ-RC, WJ-EC12, WJ-EC13, and WJ-EC14 were 139.0 kN, 117.0 kN, 125.0 kN, and 134.4 kN, respectively. It is worth noting that the bearing capacity of the specimens increased with the expansion of the precast joint area, but remained smaller than that of the cast-in situ specimens. This can be attributed to the fact that the post-cast epoxy resin concrete compromised the overall performance of the specimen, and its volume was relatively small, thereby underutilizing the superior mechanical properties of epoxy resin concrete. In addition, the bearing capacity of specimen WJ-CC was 103.8 kN, which was lower than that of the cast-in situ specimen and the post-cast specimen with epoxy resin concrete. This indicates that the prefabricated construction scheme, which preserves the integrity of the precast joint area, is advantageous.

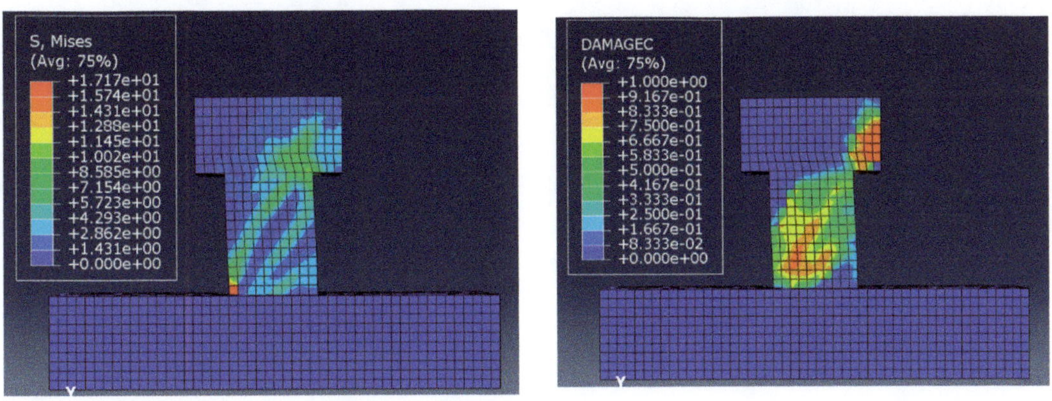

(a) WJ-RC

Figure 20. *Cont.*

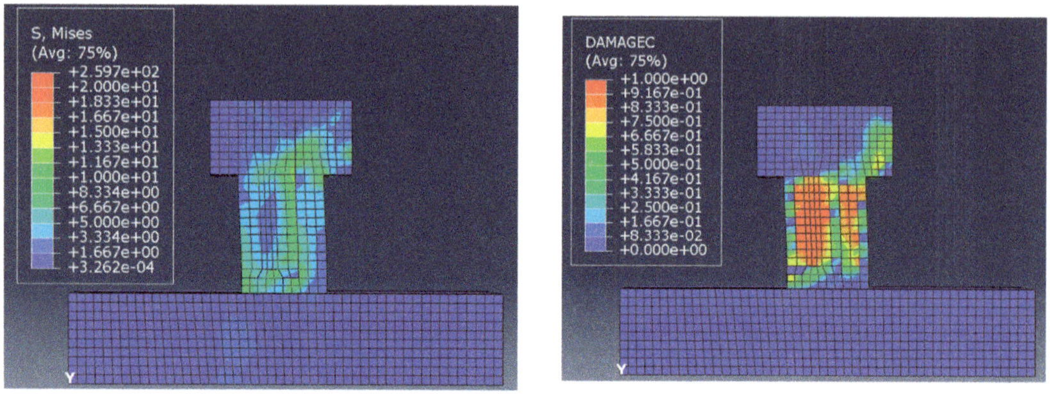

(**b**) WJ-CC

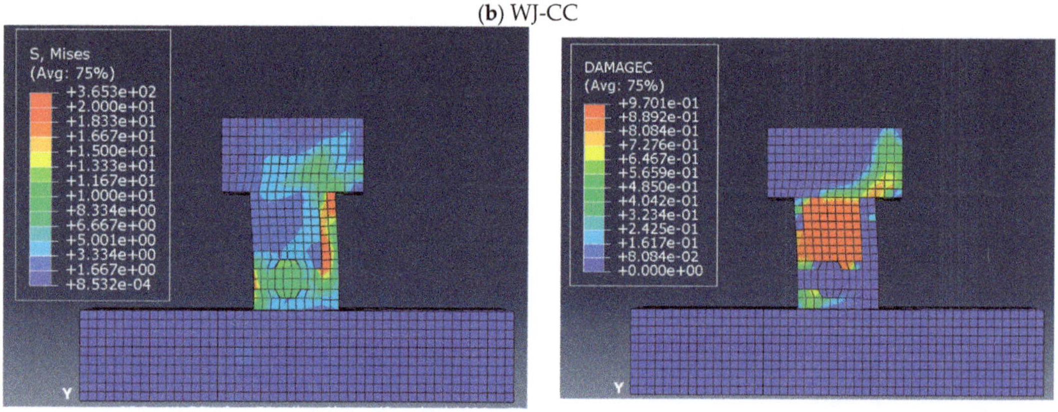

(**c**) WJ-EC12

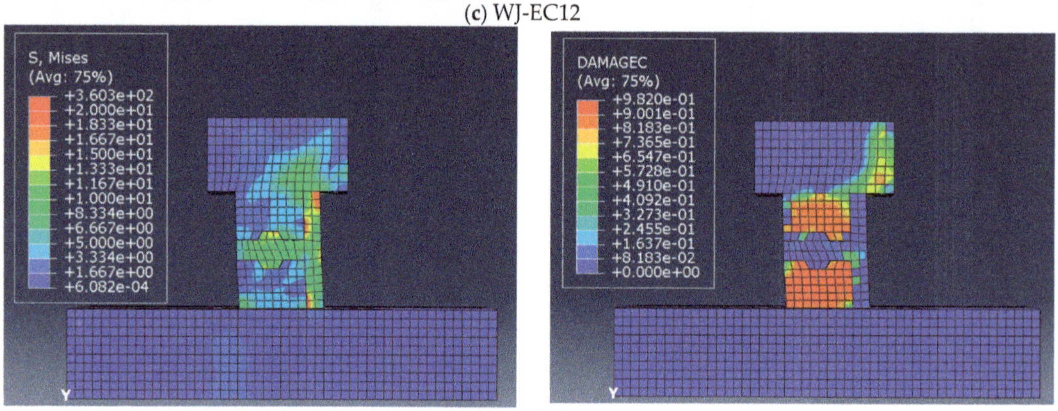

(**d**) WJ-EC13

Figure 20. *Cont.*

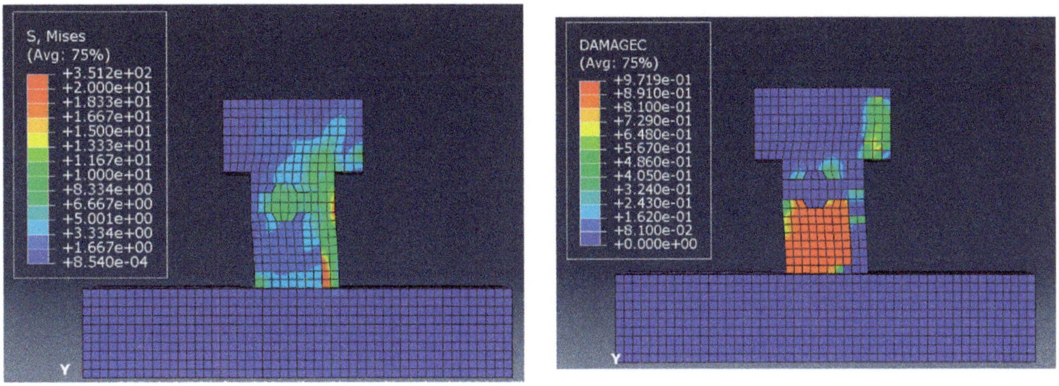

(**e**) WJ-EC14

Figure 20. Stress nephogram and damage factors distribution map of each specimen (Case 2).

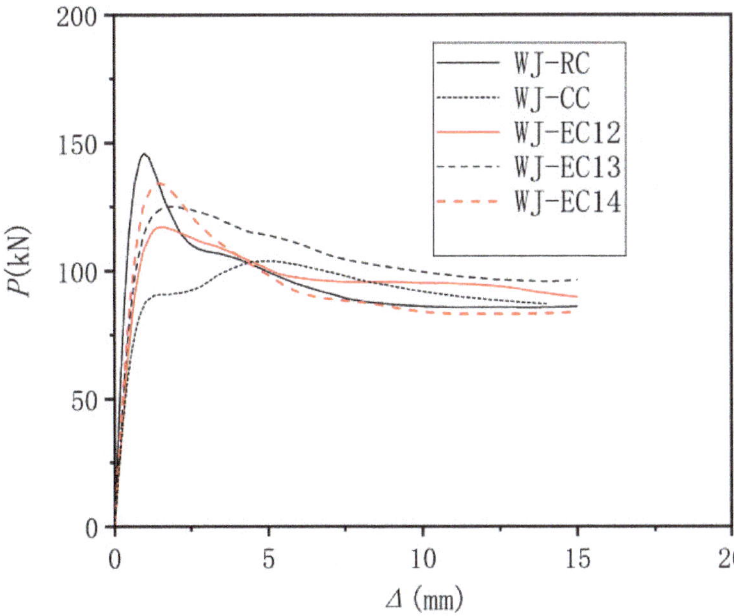

Figure 21. *P*-*Δ* curves of each specimen (Case 2).

6. Conclusions and Prospects

6.1. Conclusions

In this paper, ordinary concrete and epoxy resin concrete were utilized as post-cast materials for the purpose of conducting bending and shear performance tests, finite element simulation, and parameter analysis on the composite beam end of precast composite frame structures. The conclusions obtained can provide reference for the further application of epoxy resin concrete in urban construction.

(1) The composite beam frame structure, specifically in the beam-column joint area featuring post-cast epoxy resin concrete, exhibits a structural behavior that is essentially identical to that of the cast-in situ reinforced concrete frame structure. Moreover, it demonstrates superior load-bearing capacity and deformation performance.

Conversely, the composite beam frame structure with post-cast ordinary concrete in the joint area exhibits inadequate mechanical properties. Therefore, epoxy concrete, as a high-performance concrete, holds potential for applications in precast concrete structures.

(2) In the case of the composite beam frame structure with post-poured epoxy resin concrete in the joint area, an increase in the strength of precast concrete leads to an enhancement in the beam end's load-bearing capacity and ensures stable deformation performance. Notably, when the strength of post-cast epoxy concrete approaches or slightly surpasses that of precast concrete, the load-bearing capacity experiences the most significant improvement. In addition, an increase in the longitudinal reinforcement ratio results in an increase in the beam end's load-bearing capacity, albeit with a slight decrease in deformation performance. Conversely, as the shear span ratio decreases, the beam end's load-bearing capacity increases, but the deformation performance deteriorates significantly. Therefore, it is necessary to avoid excessively small shear–span ratios for composite beams, such as those lower than 1.5. Additionally, the expansion of the post-cast area allows for the full utilization of epoxy resin concrete's superior mechanical properties, leading to a significant increase in the beam end's load-bearing capacity and initial stiffness, albeit with a slight decrease in deformation capacity.

(3) The composite beam frame structure, which preserves the integrity of the precast joint area and is infused with epoxy resin concrete, exhibits an enhanced load-bearing capacity at the beam end due to the expansion of the prefabricated joint area. However, this capacity remains inferior to that of the cast-in situ reinforced concrete frame structure. Therefore, the aforementioned composite beam frame structure, as discussed in Conclusion (2), demonstrates superior mechanical properties. Nevertheless, it necessitates a substantial quantity of epoxy resin concrete, resulting in high costs.

(4) In contrast, the composite beam frame structure, which maintains the integrity of the precast joint area and is filled with ordinary concrete, exhibits enhanced load-bearing capacity and deformation performance at the beam end compared to the composite beam frame structure filled with ordinary concrete in the joint area. As the precast joint area expands, the load-bearing capacity at the beam end increases and gradually approaches that of the cast-in situ reinforced concrete frame structure. This observation highlights the advantages of the precast construction scheme that preserves the integrity of the precast joint area.

6.2. Prospects

From the perspective of mechanical properties, there is good bond strength between epoxy resin concrete and ordinary concrete, which is suitable for the connection joints of precast concrete structures. But other properties, such as durability, chemical stability, atmospheric stability, temperature effect, environmental protection, construction convenience, etc., need to be further studied, and the engineering applicability of epoxy resin concrete in construction engineering should be fully and comprehensively analyzed to promote its engineering application.

Author Contributions: Methodology, X.Z.; Software, S.X.; Investigation, P.C.; Data curation, D.X.; Writingoriginal draft, S.X.; Writing—review & editing, P.C.; Supervision, X.Z. All authors have read and agreed to the published version of the manuscript.

Funding: This research was funded by the National Natural Science Foundation of China, grant number 51678389.

Data Availability Statement: Data are contained within the article.

Conflicts of Interest: The authors declare no conflict of interest.

References

1. Sha, M.; Liu, P.; Gao, J.; Ma, H. Research status and application of prefabricated concrete frame joints. *E3S Web Conf.* **2021**, *260*, 03025. [CrossRef]
2. Lin, K.; Lu, X.; Li, Y.; Guan, H. Experimental study of a novel multi-hazard resistant prefabricated concrete frame structure. *Soil Dynam. Earthq. Eng.* **2019**, *119*, 390–470. [CrossRef]
3. Wang, C.; Liu, Y.; Zheng, X.; Wu, J. Experimental investigation of a precast concrete connection with all-steel bamboo-shaped energy dissipaters. *Eng. Struct.* **2019**, *178*, 298–308. [CrossRef]
4. Ghayeb, H.H.; Abdul Razak, H.; Ramli Sulong, N.H. Seismic performance of innovative hybrid precast reinforced concrete beam-to-column connections. *Eng. Struct.* **2020**, *202*, 109886. [CrossRef]
5. Ding, T.; Xiao, J.; Khan, A.-u.-R. Behavior of concrete beam-column frame joints with DfD connections: A simulation study with interface modelling. *Eng. Struct.* **2019**, *189*, 347–358. [CrossRef]
6. Gou, S.; Ding, R.; Fan, J.; Nie, X.; Zhang, J. Experimental study on seismic performance of precast LSECC/RC composite joints with U-shaped LSECC beam shells. *Eng. Struct.* **2019**, *189*, 618–634. [CrossRef]
7. Si, J.; Chen, J.; Li, F.; Wu, Z.; Zhou, M.; Ju, J. Experimental study on seismic performance of prefabricated concrete bucket arch joints. *Earthq. Eng. Eng. Dynam.* **2021**, *41*, 114–122. (In Chinese)
8. Hanson, N.W. Precast-prestressed concrete bridge horizontal shear connections. *J. PCA Res. Dev. Lab.* **1960**, *2*, 38–58.
9. Gohnert, M. Horizontal shear transfer across a roughened surface. *Cement Concrete Compos.* **2003**, *25*, 379–385. [CrossRef]
10. Lawrence, F.K.; Adam, S. Interface shear in high strength composite T-beams. *PCI J.* **2004**, *49*, 102–110.
11. Salih, M.N.; Tahir, M.M.; Mohammad, S.; Ahmad, Y.; Sulaiman, A.; Shek, P.N.; Abraham, A.; Firdaus, M.; Aminuddin, K.M. Experimental study on flexural behaviour of partially encased cold-formed steel composite beams using rebar as shear connector. *IOP Conf. Ser. Mater. Sci. Eng.* **2019**, *513*, 012038. [CrossRef]
12. Ozawa, K.; Maekawa, K.; Okamura, H. Development of high performance concrete. *J. Faculty Eng. Univ. Tokyo Ser. B (Jpn.)* **1992**, *41*, 381–439.
13. Soubra, K.S.; Wight, J.K.; Naaman, A.E. Cyclic response of fibrous cast-in-place connections in precast beam-column subassemblages. *Struct. J.* **1993**, *90*, 316–323.
14. Parra-Montesinos, G.J. High-Performance Fiber-Reinforced Cement Composites: An Alternative for Seismic Design of Structures. *Struct. J.* **2005**, *102*, 668–675.
15. Zheng, Q.; Liu, Y.; Long, L.; Chen, G.; Ma, Y. Experimental research on seismic behavior of precast concrete frame connected with UHPC. *Ind Constr.* **2019**, *49*, 85–91.
16. Deng, M.; Ma, F.; Ye, W.; Yin, P. Experimental study on aseismic behavior of locally used HDC assembled frame beam-column joints. *Eng. Mech.* **2019**, *36*, 68–78. (In Chinese)
17. Wang, H.; Wang, N.; Liu, X.; Yue, Q.; Yan, J.; Zhang, Y. Effects of UHPC shear key on strengthening shear performances of wet joint in prefabricated composite beams. *Eng. Struct.* **2024**, *299*, 117130. [CrossRef]
18. Liang, X.; Wang, Y.; Tao, Y.; Deng, M. Seismic Performance of Fiber-Reinforced Concrete Interior Beam-Column Joints. *Eng. Struct.* **2016**, *126*, 432–445. [CrossRef]
19. Qian, Y.; Li, Z.; Jin, Y.; Wang, R. Experimental Study on Axial Tension Members of a New Epoxy Resin Concrete. *Sci. Adv. Mater.* **2021**, *13*, 2005–2015. [CrossRef]
20. Xiang, Q.; Xiao, F. Applications of epoxy materials in pavement engineering. *Construct. Build. Mater.* **2020**, *235*, 117529. [CrossRef]
21. Rahman, M.M.; Akhtarul Islam, M. Application of epoxy resins in building materials: Progress and prospects. *Polym. Bull.* **2022**, *79*, 1949–1975. [CrossRef]
22. Zhang, W.; Guan, X.; Ren, J.; Gu, X. Experimental study on chloride permeability of concrete surface-treated with epoxy resin. *J. Build. Mater.* **2008**, *11*, 339–344.
23. El-Hawary, M.; Al-Khaiat, H.; Fereig, S. Fereing Effect of Sea Water on Epoxy-repaired Concrete. *Cement Concrete Compos.* **1998**, *20*, 41–52. [CrossRef]
24. Yu, T. The Experimental Study of Repairing Concrete Cracks with Epoxy Resin Grouting. Master's Thesis, Beijing University of Technology, Beijing, China, 2016. (In Chinese)
25. Gil-Martín, L.M.; Rodríguez-Suesca, A.E.; Fernández-Ruiz, M.A.; Hernández-Montes, E. Cyclic behavior of RC beam-column joints with epoxy resin and ground tire rubber as partial cement replacement. *Construct. Build. Mater.* **2019**, *211*, 659–674. [CrossRef]
26. Li, X. Study on the Behavior of the Axial Compression Members of the New Epoxy Resin Concrete. Master's Thesis, Jilin Jianzhu University, Jilin, China, 2019. (In Chinese)
27. Zhang, G. Study on the Mechanical Properties of New Epoxy Resin Concrete Axial Tensile Members. Master's Thesis, Jilin Jianzhu University, Jilin, China, 2019. (In Chinese)
28. Sun, Z. Study on Mechanical Behavior of New Epoxy Resin Concrete Beam. Master's Thesis, Jilin Jianzhu University, Jilin, China, 2019. (In Chinese)
29. Shahzamanian, M.M.; Basirun, W.J. Modeling of Cementitious Representative Volume Element with Additives. *J. Multiscale Modell.* **2017**, *29*, 1750003. [CrossRef]
30. Bernard, F.; Kamali-Bernard, S. Performance simulation and quantitative analysis of cement-based materials subjected to leaching. *Comput. Mater. Sci.* **2010**, *218*, 2262010. [CrossRef]

31. *GB 50010-2010*; Code for Design of Concrete Structures. China Architecture & Building Press: Beijing, China, 2015. (In Chinese)
32. Lv, X.; Yu, Z.; Shan, Z. Seismic behaviour of frame structures with assembly of prefabricated concrete beam. *J. Build. Eng.* **2021**, *40*, 102765. [CrossRef]
33. *EN1992-1-2*; European Standard. Euro-Code2: Design of Concrete Structures-Part1: General Rules and Rules for Buildings. European Committee for Standardization: Bruxelles, Belgium, 2021.

buildings

MDPI

Article

Environmental Sustainability Study of Urban Waterfront Landscapes Based on the LCA–Emergy–Carbon Footprint and Artificial Neural Network Method

Gang Jiang [1,*], Lanlan Zuo [2], Ashish T. Asutosh [3] and Junxue Zhang [4,*]

1 School of Architecture, Southeast University, Nanjing 210096, China
2 Hunan City University Design and Research Institute Co., Ltd., Changsha 401120, China; zll11072023@163.com
3 Construction Engineering & Management, Civil and Environmental Engineering, University of Delaware, Newark, DE 19716, USA; aasutosh@udel.edu
4 School of Civil Engineering and Architecture, Jiangsu University of Science and Technology, Zhenjiang 212100, China
* Correspondence: mj23h990112@163.com (G.J.); zjx2021@just.edu.cn (J.Z.)

Abstract: The ecological landscape design of urban rivers plays a crucial role in mitigating the urban heat island effect and preserving urban ecology. This study focuses on the construction process data of key landscape nodes along Nanjing's urban rivers. By employing a whole life cycle emergy approach and carbon emission method, the sustainable changes in the landscape system are quantitatively assessed. Furthermore, artificial neural networks have been used to conduct long-term sustainability analysis and predictions for the landscape system. The research findings reveal that over time, the maintenance investment in landscape projects gradually becomes dominant, increasing from 2% in the first year to approximately 75% after 30 years. This phenomenon signifies a decline in the efficiency of the landscape system. Sustaining the ecological balance of the landscape system necessitates continuous inputs of material flow, energy flow, and information flow. The major contributors to carbon emissions in the landscape engineering system are diesel fuel, cement, and steel. This highlights opportunities for sustainable improvement from a low-carbon perspective. To enhance the ecological sustainability of urban waterfront landscapes, three measures are proposed: sponge city construction concepts, coupled sewage treatment systems, and information flow monitoring systems. The effectiveness of these measures was preliminarily validated.

Keywords: sustainability; LCA–Emergy–Carbon footprint analysis; artificial neural network method; water landscape system

1. Introduction

In the face of the deteriorating global environment, an ecologically sustainable urban environment is crucial. The urban system involves numerous environmental impact factors, and rivers possess exceptional characteristics. Through waterfront landscape design, rivers can effectively mitigate the urban heat island effect and reduce carbon emissions in cities [1,2]. The purpose of this study is to provide a feasible approach for urban sustainable development through the ecological and low-carbon assessment and analysis of waterfront landscapes.

Urban waterfront landscapes can protect and restore local ecosystems. Wetlands and vegetation in these landscapes have the ability to filter water, absorb pollutants, and provide habitats for various flora and fauna. The vegetation in waterfront landscapes can absorb harmful gases from the air and release oxygen, thus improving the air quality of cities. Trees and plants also help reduce air temperature and mitigate the urban heat island effect. Urban waterfront landscapes can manage water resources through rainwater collection and treatment systems. These systems can collect and utilize rainwater, reducing the burden on city drainage systems and minimizing the demand for groundwater. Additionally, waterfront landscapes provide recreational activities and social spaces for residents. Contact with natural environments and engagement in outdoor activities are crucial for physical and mental well-being. Moreover, green waterfront landscapes alleviate psychological stress and enhance residents' quality of life. In conclusion, urban waterfront landscapes effectively enhance the ecological standards of cities through their roles in ecosystem protection, air quality improvement, water resource management, and the promotion of community health and well-being [3,4].

The ecological emergy theory is a promising method for conducting sustainable assessment research on urban waterfront landscapes. The theory of emergy, originally introduced in the field of ecology [5–7], has been extended to multiple domains including urban studies [8,9], industrial sectors [10,11], building systems [12,13], and economic fields [14]. Currently, many scholars have coupled the emergy theory with other methods, such as GIS approaches [15,16], carbon emission methods [17,18], ecological footprint methods [19–21], and life cycle assessment (LCA) evaluation [22].

Among them, the LCA–Emergy methodology creates a comprehensive approach that enables the calculation of emergy values throughout the entire lifecycle. This integration ensures the completeness of the assessment. It considers the entire lifecycle of a specific system, including stages such as resource acquisition, manufacturing, use, disposal, and waste management, when assessing its sustainability. By using emergy as a universal unit, this method calculates the emergy consumed and generated by the system at different stages. The key advantage of the LCA–Emergy method is its ability to quantify and comprehensively consider the energy consumption, material flows, and environmental impacts of the system, thereby providing more comprehensive and holistic assessment results. By integrating life cycle theory with the concept of emergy, it enables a better understanding and evaluation of the energy efficiency and environmental performance of the system at different stages, facilitating informed decision-making for sustainability [23–25]. The research on the integration of the LCA–Emergy method with the ecological sustainability of urban waterfront landscapes involves several aspects [26–28], such as: (1) Sustainability assessment: The LCA–Emergy method can serve as one of the tools to assess the ecological sustainability of urban waterfront landscapes. (2) Optimization of waterfront landscapes: By employing the LCA–Emergy method, the performance of different design options can be evaluated and compared in terms of energy, material, and ecological benefits, thereby providing decision support and design recommendations to achieve more sustainable waterfront landscapes. (3) Energy management: By analyzing the flow of emergy and energy conversion processes, critical areas of energy utilization can be identified and optimized to reduce energy waste and promote energy efficiency improvement in waterfront landscapes. (4) Evaluation of ecosystem services: The LCA–Emergy method can be used to assess the value of ecosystem services provided by urban waterfront landscapes. In summary, the LCA–Emergy method can support the design, assessment, and management of urban waterfront landscapes, facilitating the achievement of their ecological sustainability. At the same time, the ecological sustainability of waterfront landscapes can be quantified and analyzed through the application of the LCA–Emergy method, providing a scientific basis for decision-making.

In addition, the integration of LCA and carbon emissions is a recognized methodology for studying the sustainability of systems. The cross-methodology of LCA and carbon emissions is known as carbon footprint assessment, which combines LCA methods with carbon emission calculations to evaluate the carbon emissions generated by products, services, or systems throughout their entire life cycle. By combining LCA with carbon emission calculations, carbon footprint assessment comprehensively considers the carbon emissions associated with products, services, or systems, helping to identify and optimize high-carbon-emitting stages to achieve carbon reduction and sustainable development goals. It can not only be applied in corporate environmental management and product design, but also provides clear carbon information to consumers, promoting sustainable purchasing and consumption behavior [29–37].

Neural network models have potential applications in predicting the sustainable effects of waterfront landscape design to achieve intelligent waterfront landscape design [38–40]. The key steps include: (1) Data collection and feature selection: Collecting relevant data related to the sustainability of waterfront landscapes and performing feature selection for the prediction target. These data may include parameters of waterfront landscape design, environmental indicators, ecosystem service values, etc. (2) Model training and optimization: Utilizing the collected data to train and optimize the neural network model. This involves selecting an appropriate neural network structure, setting model hyperparameters, and training the model using the data to accurately predict the sustainability effects of waterfront landscape designs. (3) Prediction and evaluation: After model training, using the model to predict new waterfront landscape design scenarios. By inputting the corresponding design parameters and features, the model can output predictions such as ecosystem service values, energy utilization efficiency, etc. These predictions can help evaluate the sustainability effects of different design scenarios and support decision-making. (4) Model validation and improvement: To validate the accuracy and robustness of the model, independent test datasets can be used for model validation. If the model performs poorly, improvements can be made by adjusting the neural network structure or parameters, or introducing other enhancement methods.

To date, no scholars have explored the coupling of LCA–Emergy–Carbon with a neural network model to predict the ecological sustainability of waterfront landscapes. This study addresses this research gap and introduces an innovative approach by integrating the LCA, emergy analysis, carbon footprint method, and neural network model to comprehensively analyze the ecological sustainability of waterfront landscapes. This integrated analysis provides a more comprehensive and accurate assessment, revealing the long-term impacts of waterfront landscape design on ecosystems.

Traditionally, assessments of waterfront landscapes have primarily focused on the environmental benefits during the design phase, often overlooking the energy consumption and environmental impacts throughout the entire life cycle. By applying the LCA–Emergy–Carbon approach, this study considers the integrated effects of energy consumption and environmental loadings across various stages, including material acquisition, construction, operation, and decommissioning. The carbon footprint method serves as a measure of human activities' contribution to greenhouse gas emissions. By incorporating the carbon footprint analysis, this study quantitatively evaluates the carbon emissions associated with different waterfront landscape design scenarios, providing guidance for reducing carbon footprints. Additionally, the use of a neural network model enables predictions based on historical data and relevant features, allowing for inference on the impact of waterfront landscape design scenarios on ecological sustainability. This model can handle complex non-linear relationships and provide accurate prediction results.

By combining these three approaches, this study offers a more comprehensive, accurate, and innovative analysis in assessing the ecological sustainability of waterfront landscapes, thereby providing robust support for future waterfront landscape design and decision-making.

2. Materials and Methods

2.1. Research Framework

The study in this paper is based on the methodology of systems engineering to conduct a sustainable assessment of landscape systems. The application of systems engineering methodology and the study of ecological emergy and low-carbon cross-coupling in architecture can help categorize the various input factors of the building system, establish a modular framework, and define structural boundaries. This facilitates comprehensive evaluation and design for the entire building system. In order to establish a research model for the building system, the input types are divided into three categories: material flow, energy flow, and information flow. Figure 1 illustrates the research framework of the entire paper.

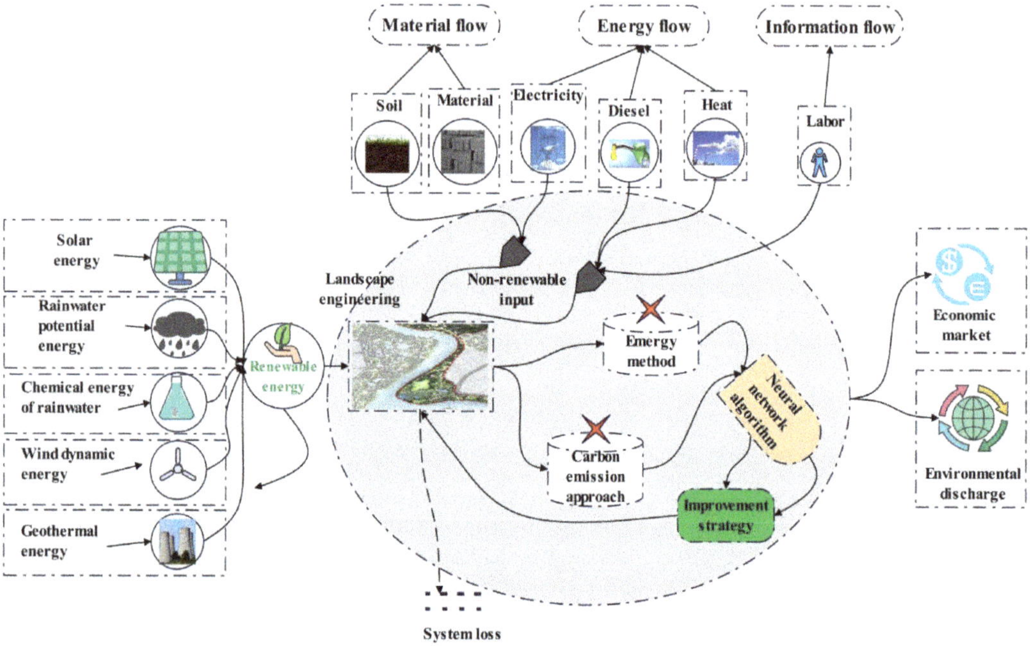

Figure 1. Research framework.

In addition, this paper takes a dynamic approach to the architectural system, involving not only static analysis but also the sustainable aspects and predictions of carbon emissions related to subsequent riverfront landscape projects. This contributes to the enhancement of efficiency for urban managers.

2.1.1. Hypothesis for Research

This article verifies the following three hypotheses through an ecological assessment of urban river landscapes:

(1) Ecological assessments of landscape transformation projects should consider the impact of information lag on sustainability outcomes. In general, information lag can have a negative effect on the overall sustainability performance of the building system, particularly concerning feedback from landscape designers.
(2) The influence of feedback structures should be evaluated to examine their impact on the entire building system. Feedback structures can be categorized into open-loop, closed-loop, and cross-feedback structures.
(3) The effect of water quality in rivers on the ecological sustainability of the entire river landscape, including emergy and carbon emission aspects.

2.1.2. Typical Feedback Structure

The feedback systems are evaluated based on two aspects: ecological energy value and carbon emissions. The three types of feedback systems are open-loop feedback system, closed-loop feedback system, and cross-feedback system. These systems are represented by measures such as energy value quantity, energy value indicators, and carbon emissions. A diagram of the feedback system pathways can be found in Figure 2 of this article.

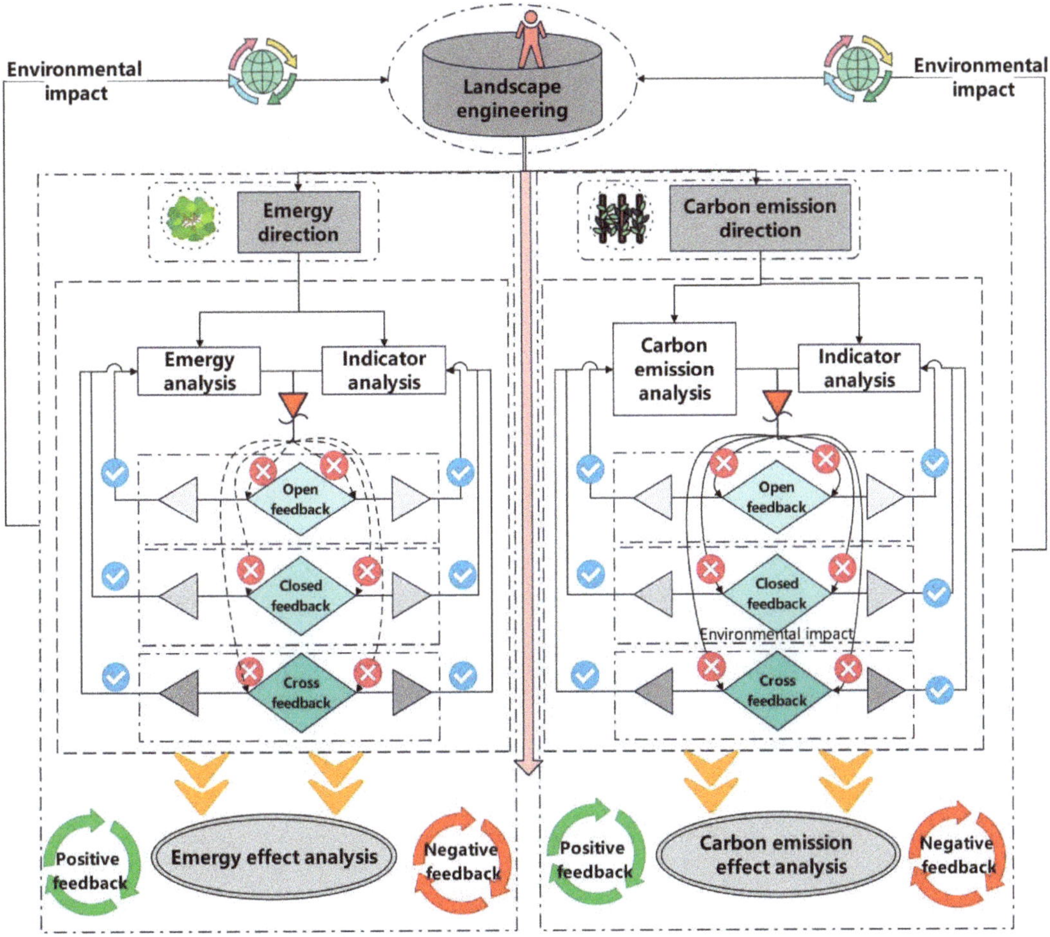

Figure 2. Typical feedback system path.

2.2. LCA–Emergy Model

2.2.1. Emergy Introduction

The ecological emergy method is an analytical approach used to assess and compare the impacts of different decision options on ecosystems. It is based on ecological principles and aims to quantify the value of ecosystems as energy values, helping decision-makers make sustainable choices between environmental conservation and economic development. The ecological emergy method takes into account the structure, functions, and services provided by ecosystems, as well as their contributions to human well-being. By converting these ecological values into emergy values, it becomes possible to compare the costs and benefits of different decision options for ecosystems and reveal their potential environmental benefits and risks. In the ecological emergy method, the first step involves assessing the ecosystem, including data collection and analysis of aspects such as species diversity, ecological processes, and natural resources. Then, by using appropriate indicators and models, the emergy of the ecosystem is quantified as emergy values and integrated into the decision analysis framework. One of the advantages of the ecological emergy method is its ability to quantify the value of ecosystems, allowing for comparisons with economic benefits. This helps raise awareness and appreciation for ecosystems, promoting decision-making that supports sustainable development. Additionally, the ecological emergy method can uncover the ecological risks and uncertainties associated with decision options, aiding decision-makers in better managing environmental issues. The UEVs consist of three types, which are sej/kg, sej/j, sej/USA$, respectively [41].

2.2.2. Emergy Indicators

The sustainability hierarchy can be realized based on a set of emergy indicators [42,43]. In this study, three are four critical indexes that have been displayed in Figure 3.

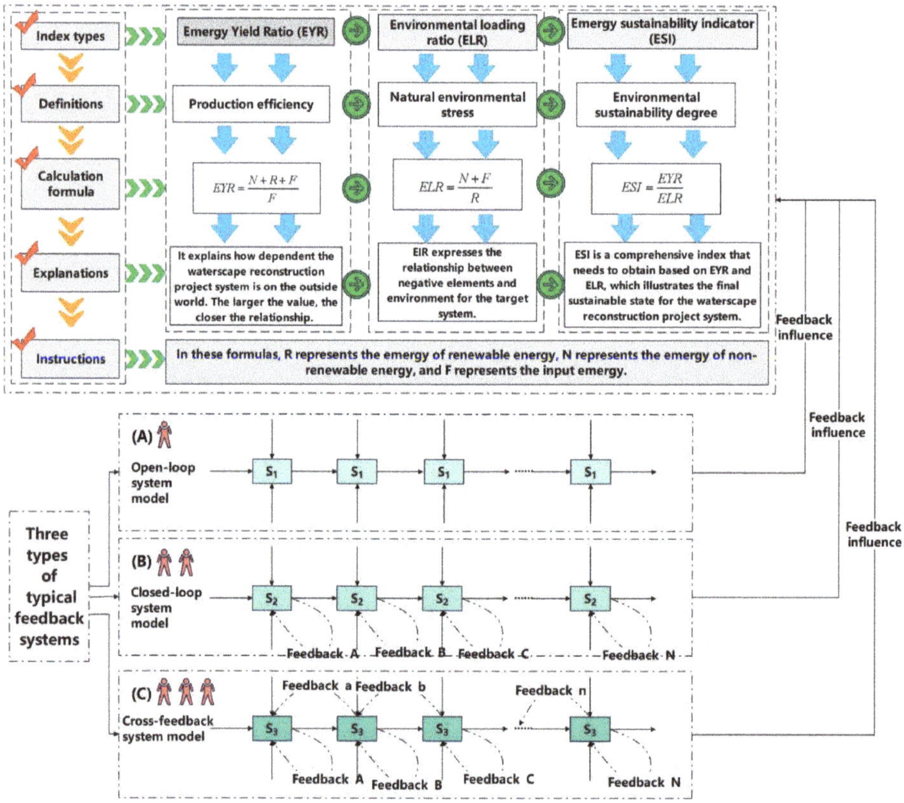

Figure 3. Sustainable index frame diagram of the coupled feedback system.

The environmental loading rate of a landscape system is an indicator that assesses the pressure exerted by the system on environmental resources in terms of resource consumption, energy utilization, waste generation, and pollution emissions. It helps in understanding the extent of the landscape system's impact on the environment.

The emergy yield rate measures the efficiency and output levels of a landscape system in providing ecological services, maintaining ecological functions, and supporting human well-being. It quantifies the system's contribution to energy flow, material cycling, and ecological processes, and evaluates the value it creates for the socio-economic system.

The environmental sustainability indicator comprises a set of indicators and factors used to assess the sustainability performance of a landscape system. These parameters consider factors such as resource use efficiency, ecological conservation capacity, carbon footprint, water footprint, and biodiversity protection, aiming to ensure the harmonious development of the landscape system with environmental protection and achieve long-term sustainable development goals.

These three categories of indicator collectively provide a comprehensive assessment of the landscape system. They consider its impact on the environment (ELR), evaluate its contribution to the socio-economic system (EYR), and also focus on its sustainability performance (ESI). This aids in guiding landscape planning and management, promoting a balance between ecological conservation and economic development.

2.3. LCA–Carbon Calculation Model

In accordance with the national carbon emission calculation standards [44], Figure 4 and Table 1 have been designed and presented. Figure 4 represents the carbon emission implementation pathway, while Table 1 depicts the carbon sink calculation pathway.

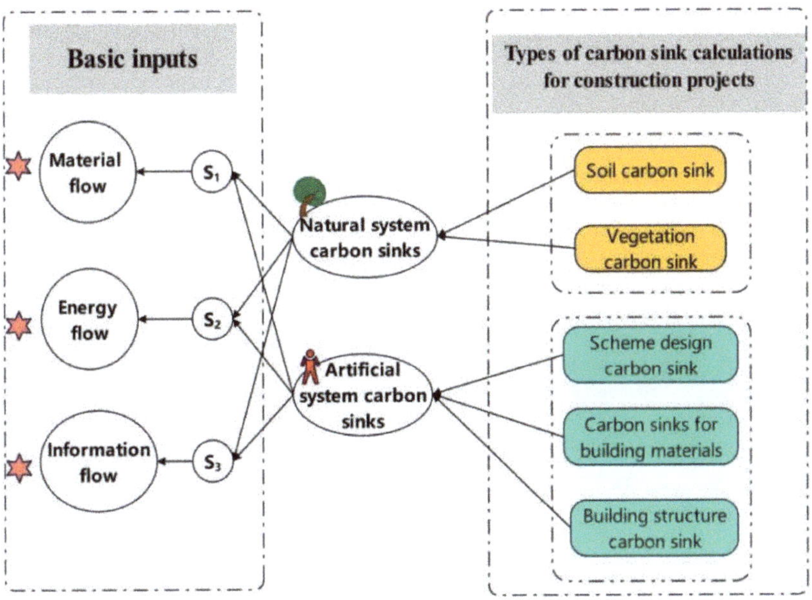

Figure 4. Landscape system carbon emission frame diagram.

Table 1. Carbon sink calculation list.

Method Types	Equations	Explanations
Soil type method	$T_{jd} = \sum_{i=1}^{k} \rho_i P_i D_i (1 - S_i)$ $M_d = \sum_{j=1}^{k} A_j T_{jd}$	ρ_i is the soil weight; P_i is the average organic carbon storage; D_i is soil thickness; S_i is the average gravel content. A_j is the area of a grid cell, T_{jd} is the unit mean organic carbon density; n is the total soil area grid units.
Life zone method	$B_D = b_0 + b_{1D} + b_2 \lg C_f$ $C = C_f + B_D(1 - \delta_2)V$	B_D is the soil weight; b_1, b_2, b_3 is the constant of soil weight and carbon density under different vegetation types; D is the depth from the surface to the center of the soil layer; C_f is the organic carbon mass fraction. δ_{2mm} is the gravel fraction; V is the soil layer volume.
Remote sensing technology method	$C_i = 0.58 S_i \sum (H_j Q_j W_j)$	i is the soil type; C_i is the soil organic carbon storage (t); 0.58 is the carbon storage conversion factor; S_i is the soil area; H_j is the mean soil thickness; Q_j is the average mass fraction of soil organic matter; W_j is the average soil weight.
Classical concrete carbonation theory	$d = \sqrt{\frac{2D_{CO_2}C_0}{m_0}} \cdot \sqrt{t}$	d is the concrete carbonation depth; D_{CO_2} is the effective diffusion coefficient of carbon dioxide in concrete; C_0 is the concentration of the concrete surface; m_0 is the carbon dioxide absorption per unit volume of concrete; t is the carbonization time.

2.4. Artificial Neural Network Method

Artificial neural network models can be used to predict and analyze the changing trends in ecological sustainability of landscape engineering. By training the model with historical data, it can learn patterns and trends related to the ecological sustainability of landscape engineering. Subsequently, the model can be utilized to forecast future trends and provide recommendations on how to improve or maintain ecological sustainability. It is important to note that the predictive capability of artificial neural network models is influenced by factors such as data quality, feature selection, and model parameters. Hence, during predictive analysis, ensuring the accuracy and representativeness of data and conducting appropriate tuning and validation of the model are crucial to enhance the reliability and accuracy of the prediction results.

Figure 5 illustrates the implementation steps and pathway of the neural network. Figure 6 represents the basic prediction model of the neural network.

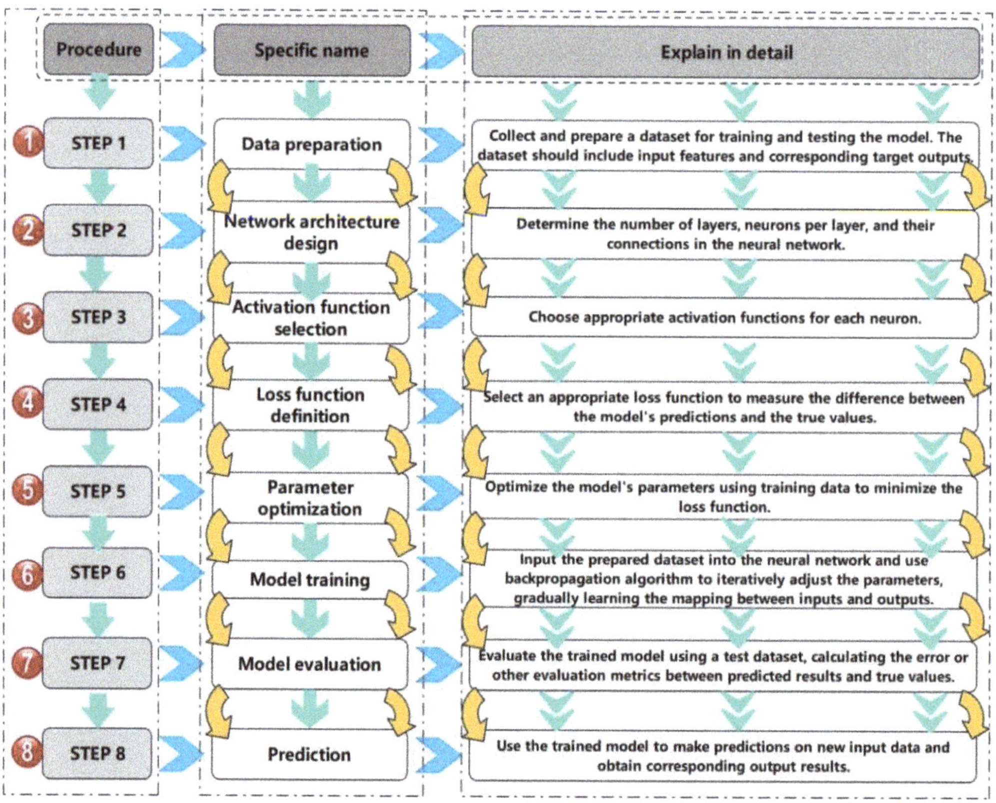

Figure 5. Artificial neural network implementation path diagram.

This study involves a significant amount of data statistics, which is crucial for the accuracy of the research methods. Analyzing from the perspectives of ecological emergy and carbon footprint methods, both require a comprehensive life cycle inventory of the research system, which poses some challenges for this study. Additionally, evaluating the accuracy of neural network prediction models is also a difficulty. As the number of hidden layers increases, the overall accuracy of the prediction model will vary, thus influencing the research results of this study.

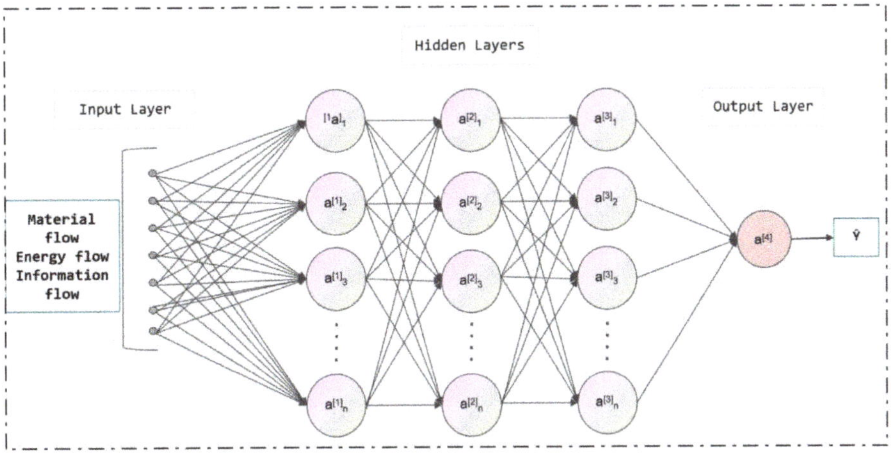

Figure 6. Neural network prediction model diagram.

2.5. Data Identification and Collection

The entire landscape renovation project involves multiple types of data, including material inventory data, energy inventory data, labor input data, etc. This article also involves information flow-related data, which has a crucial impact on analyzing the ecological transformation of the entire landscape. Figure 7 illustrates the process of data collection and selection.

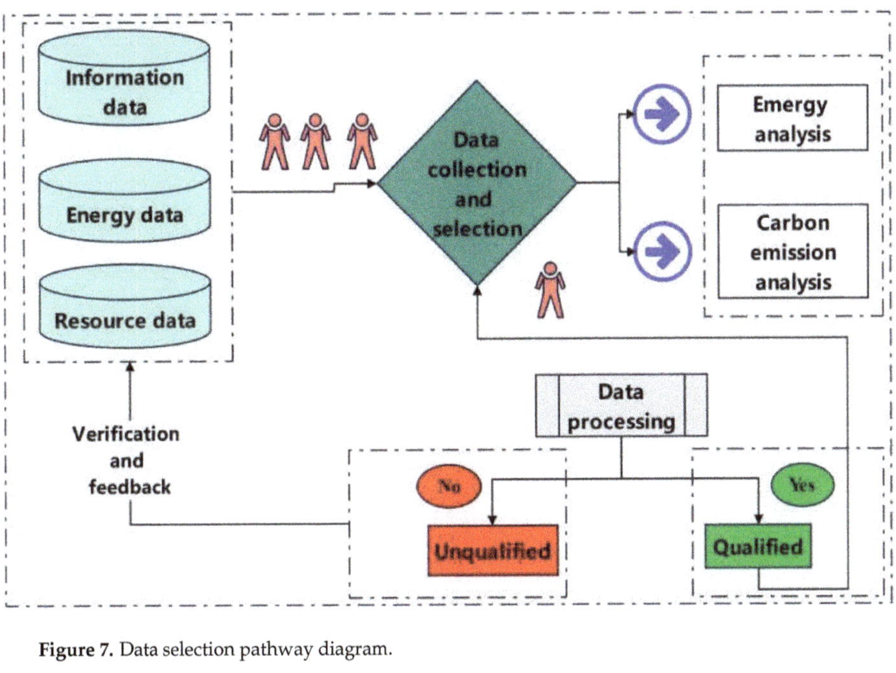

Figure 7. Data selection pathway diagram.

The importance of data filtering lies in its ability to help us extract useful information from a large volume of data, enabling decision-making and insights. For instance: improving data quality: data filtering eliminates invalid, erroneous, or duplicate data, ensuring the reliability and accuracy of the analyzed data set. Reducing noise and redundancy: by filtering data, it is possible to reduce the impact of noise and redundant information, leading to clearer and more credible analytical results. Saving time and resources: data filtering allows for the exclusion of irrelevant or unnecessary data, saving time and resources involved in processing and analyzing data. Discovering hidden patterns and trends: by filtering and focusing on specific subsets of data, it becomes easier to uncover hidden patterns, trends, and correlations within the data. Supporting decision-making: carefully filtered data can provide strong evidence for decision-making, reducing decision risks, and guiding organizational actions in market, operational, and strategic aspects. In conclusion, data filtering is a crucial step to ensure the effectiveness and accuracy of data analysis. It improves data quality, saves time and resources, and supports decision-making.

3. Case Study

Case Introduction

This case is located in Nanjing, Jiangsu Province, China. In order to enhance the urban living environment and promote ecological sustainability, a portion of the river landscapes has been redesigned. The selected rivers mainly focus on the main channel of the Jinchuan River, with a total length of approximately 50 km. The overall condition of the buffer zones on both sides of the river varies greatly, with limited space for landscape design and being mostly in a neglected state. From a landscape design perspective, the design of the riverbanks is monotonous, with a single type of planting, and lacks distinctive features. Table 2 and Figure 8 summarize the current status of waterfront greening in selected sections of the river (a total of nine river locations). Figure 9 shows a diagram of node changes in the river channel before and after landscape design.

Table 2. The current status of waterfront greening in selected sections of the river.

No.	Name	Length (km)	Waterfront Greening Status
1	Neijin River mainstream	2.88	Most of the waterfront areas have been restored as leisure green belts, with widths ranging from 2 to 10 m. The overall landscape greening quality is high, with lush vegetation along the riverbanks. Deciduous and shrub tree species are combined with riparian aquatic plants. Along the riverbanks, the use of stacked stones and faux wood pilings creates an environment for the growth of aquatic plants, achieving a unified combination of esthetics and ecology.
2	Neijin River east tributary	2.22	With the exception of small-scale riverside buildings in certain areas, most of the waterfront leisure green belts have been restored, with widths ranging from 2 to 5 m. The green belts along the riverbanks are generally connected, dominated by single rows of tall trees, with a linear and relatively simple hierarchy.
3	Neijin River central tributary	1.42	The green belts are discontinuous, with some sections lacking sufficient space along the river channels, resulting in the absence of waterfront green belts. Some sections of the rivers have abundant green spaces on both sides, with rich vegetation and pleasant walking experiences.
4	Neijin River central tributary	1.27	Apart from small-scale riverside buildings in specific locations, most of the waterfront leisure green belts have been restored. However, due to the dense surrounding residential areas and narrow landscape spaces along the river channels, it is not possible to form large-scale waterfront green belts. The width typically ranges from 2 to 5 m. Some sections of the rivers lack softscape due to their proximity to residential areas, parking lots, etc. The green belts along the mainstream of the Neijin River are intermittent, featuring linear and monotonous plant landscapes, and they lack connections with surrounding parks and green spaces.

Table 2. *Cont.*

No.	Name	Length (km)	Waterfront Greening Status
5	Northwest City Moat	1.16	The two sides of the waterfront leisure green belts are mostly connected, with widths ranging from 2 to 10 m. The combination of riverside landscapes and public green spaces creates an overall high-quality open space. While meeting the basic greening requirements along the entire line, attention is paid to the arrangement of tall trees, shrubs, and grass layers, creating a pleasant and high-quality walking and recreational experience.
6	North City Moat	1.13	The two sides of the waterfront leisure green belts are mostly connected, with widths ranging from 2 to 10 m. The overall quality of greening is high, with reasonable arrangements of tall trees, shrubs, flowers, and plants. The vegetation is well-distributed, creating a beautiful and pleasant riverside recreational space with rich layers, vivid colors, and distinct seasonal changes.
7	South Ten-Mile Long Ditch mainstream	0.93	Most of the waterfront leisure green belts have been restored, but there are significant differences in the current state of waterfront greening across different sections of the rivers. The river section in Yuanyuan Residential District has a purely hard landscape with a lack of softscape, while the river section near Hongshan Zoo Station has a good landscape environment with well-planned arrangements of tall trees, shrubs, flowers, and plants.
8	Zhang Wang Miao Ditch	0.86	There are significant differences in the current state of waterfront greening across different sections of the rivers. The section from Jianning Road to Mufu South Road is adjacent to a major road without any greenery. The section along Jinbi Road has a rich variety of green plants and shows healthy growth.
9	Waijin River	0.79	Most of the waterfront leisure green belts have been restored, with widths ranging from 2 to 10 m. The spaciousness of the waterfront leisure green belts allows for abundant vegetation with diverse colors that change noticeably throughout the seasons.

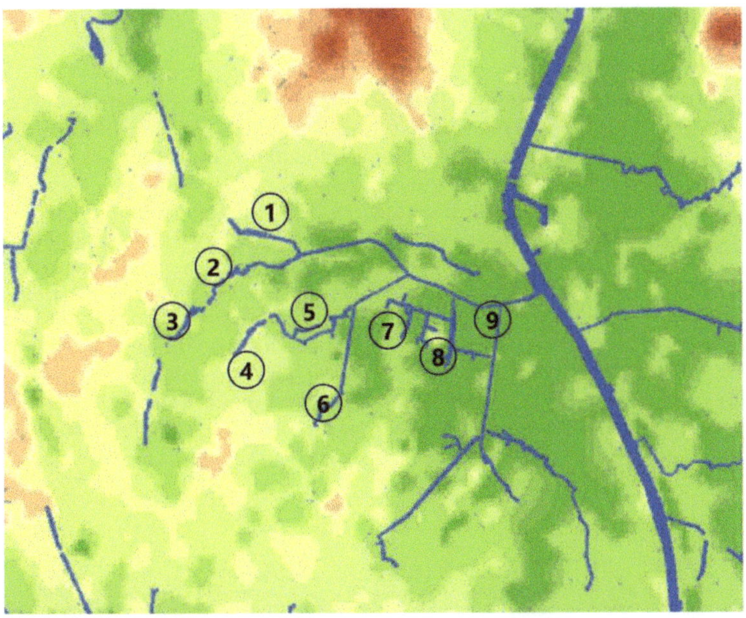

Figure 8. The research location of a river object.

Figure 9. Changes before and after landscape design.

4. Results and Discussion

The entire landscape engineering transformation process can be divided into five stages: the design stage, building material production stage, transportation stage, construction stage, and maintenance stage. By analyzing the emergy and carbon emissions throughout the five stages of the life cycle, the ecological sustainability of the entire landscape transformation system can be assessed. Additionally, by using a neural network prediction model, long-term ecological sustainability analysis of landscape engineering can be conducted.

Due to the large scale of the urban riverfront in the city, three specific locations have been selected to conduct a comprehensive life cycle analysis of emergy and carbon emissions for three small-scale landscape projects.

4.1. LCA–Emergy Analysis

4.1.1. Dominated Contributor

Taking the Neijin River mainstream landscape renovation project (Project No. 1) as an example, the emergy ratios across five stages are calculated and displayed for analysis. The time points chosen for analysis are the first year and the 20th year of operation. Figure 9 provides a clear view of the emergy changes in each stage.

From the perspective of the entire life cycle, Figure 10A depicts the emergy distribution of five stages of landscape engineering operations over one year. Material-related emergy has the highest contribution, accounting for approximately 49%. Construction comes next, representing 23% of the total. The other three stages have a smaller impact and are considered secondary. In Figure 10B, which simulates the operation of the landscape project over 20 years, it is evident that the emergy proportion during the entire operational phase reaches 75%, significantly higher than the other four stages. This indicates that the ecological sustainability of the entire landscape system requires continuous maintenance, including a constant input of material flow, energy flow, and information flow. Consequently, the proportion of emergy in the maintenance stage keeps increasing.

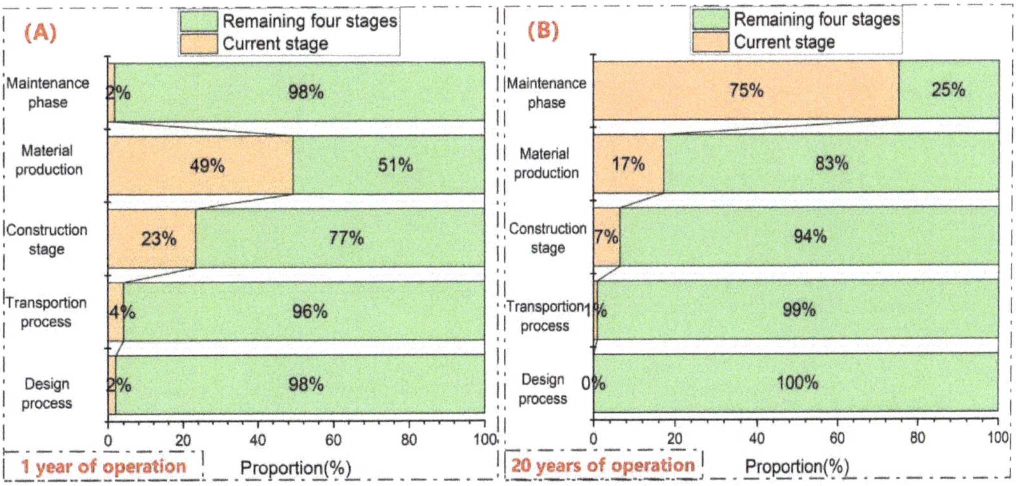

Figure 10. Contribution proportion of five stages. (**A**) Proportion of operation for one year. (**B**) Proportion of operation for twenty years.

4.1.2. Emergy Indicator Analysis

Using No. 1 project as an example (Table 2), the sustainability indicators were simulated for three periods: a 10-year cycle, a 20-year cycle, and a 30-year cycle. Figure 10 illustrates the changes in the indicators at these three time points. From Figure 11, it can be observed that the EYR (Energy Yield Ratio) decreases over time, indicating a decreasing emergy production efficiency of the entire landscape system and a tendency toward decline. Conversely, analyzing the Environmental Load Ratio, which increases continuously from 132.8 to 238.6, indicates a gradual increase in the environmental cost of operating the landscape system. Based on the assessment of the emergy production rate and environmental load ratio, the Energy Sustainability Index (ESI) is quantitatively calculated. The ESI for the landscape system operating for 10 years is 0.335, reducing to 0.212 after 20 years and further declining to 0.109 after 30 years. This implies that the landscape system requires reconstruction or major refurbishment.

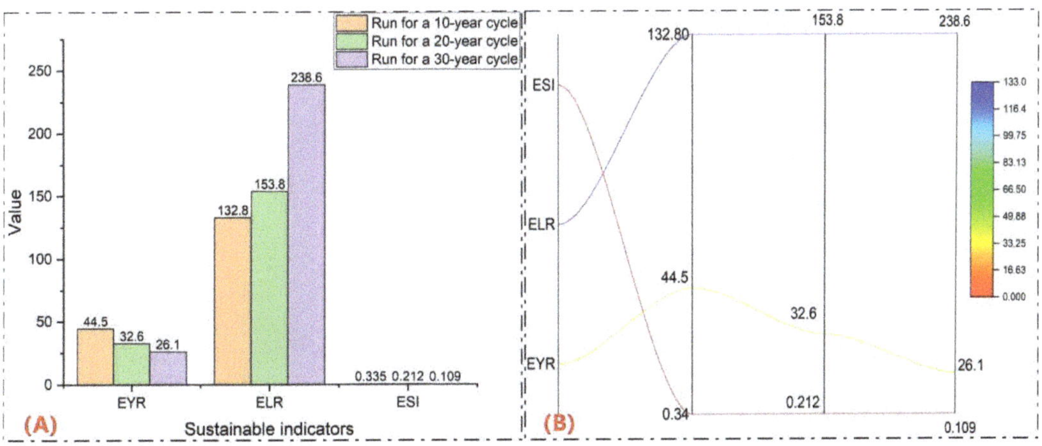

Figure 11. Sustainable indicator comparisons. (**A**) Sustainable indicators. (**B**) Comparative analysis.

4.1.3. Sensitivity Analysis

Data uncertainty analysis is a crucial step in understanding and assessing uncertainties within data. It aids decision-making, model evaluation, information communication, and risk management, thereby enhancing the accuracy and reliability of data analysis.

To quantitatively demonstrate the sensitivity of data using EYR (Economic Yield Ratio), ELR (Energy Loss Rate), and ESI (Emergy Sustainability Index) as core indicators, we can analyze the following hypotheses:

Hypothesis 1. *Basic data fluctuates by 10%, including material flow data, information flow data, energy flow data, etc.*

Hypothesis 2. *The energy conversion rate fluctuates by 5% to examine the sensitivity changes of the three indicators.*

Figure 11 presents the sensitivity trends of nine landscape engineering (Table 2) sustainability parameters.

Figure 12 presents the sensitivity results, showing that the first three types of landscape renovation project result in minor changes in sustainable parameters, as depicted in Figure 12 (No. 1/No. 2/No. 3). Taking the ESI parameter as an example, the fluctuation ranges are 8.47% and 4.31% for Scenario 1 and Scenario 2, respectively. However, the middle three items (No. 4/No. 5/No. 6) exhibit significantly increased fluctuations, with ESI changes of 11.6% and 7.29%, respectively. The last three landscape renovation projects (No. 7/No. 8/No. 9) have fluctuations ranging from 8% to 12%.

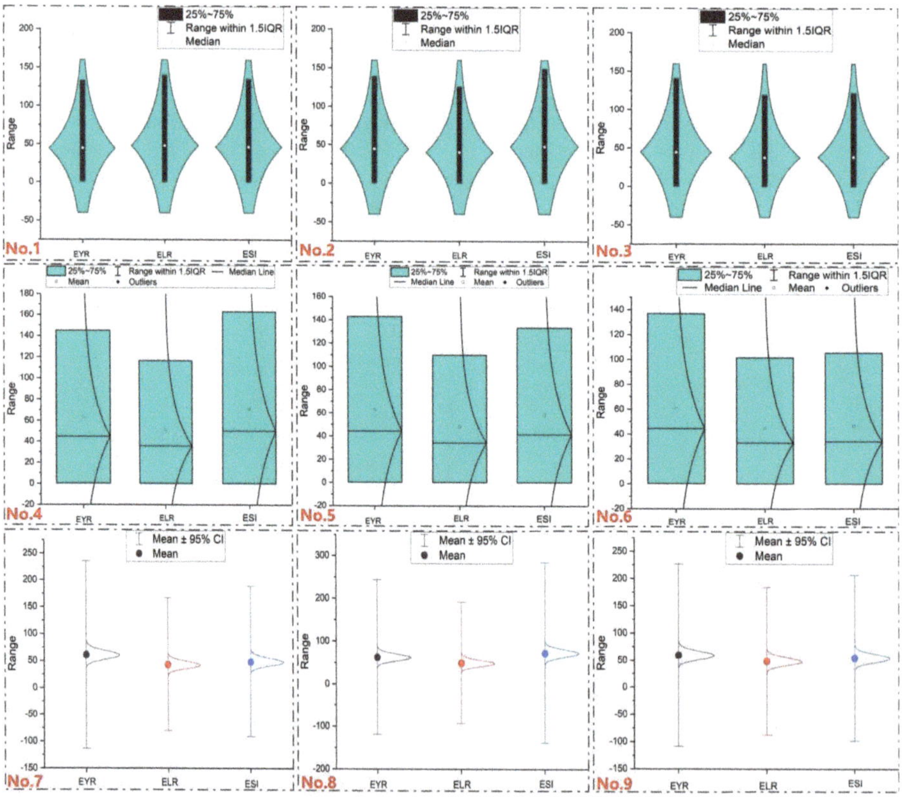

Figure 12. Sensitivity changes under Hypotheses 1 and 2.

As the proportion of assumed data fluctuations increases, the final variation in sustainable indicators also increases. Additionally, in situations with a smaller sample size, the sensitivity is more pronounced, and the impact received is greater.

4.2. LCA–Carbon Emission Analysis

This chapter consists of two parts: carbon emission calculations for nine landscape renovation nodes, and calculation and validation of carbon sequestration effects.

4.2.1. Carbon Emission Analysis

In this section, the energy consumption of transportation (diesel fuel) is accounted for in the materials, totaling eight items. Figure 13 illustrates the carbon emission calculation results for nine landscape renovation nodes. From Figure 13, it is evident that diesel fuel has the highest carbon emissions, followed by cement, steel, and gravel. The other four materials (brick, lime, sand, and wood) contribute less to carbon emissions. Analyzing the nine types of renovation project, Project 1 has the highest carbon emissions from diesel fuel at 33.8 tCO_2, followed by cement at 26.6 tCO_2 and steel at 18.7 tCO_2. Due to the largest length of landscape renovation in Project 1, it has the highest consumption of materials and energy.

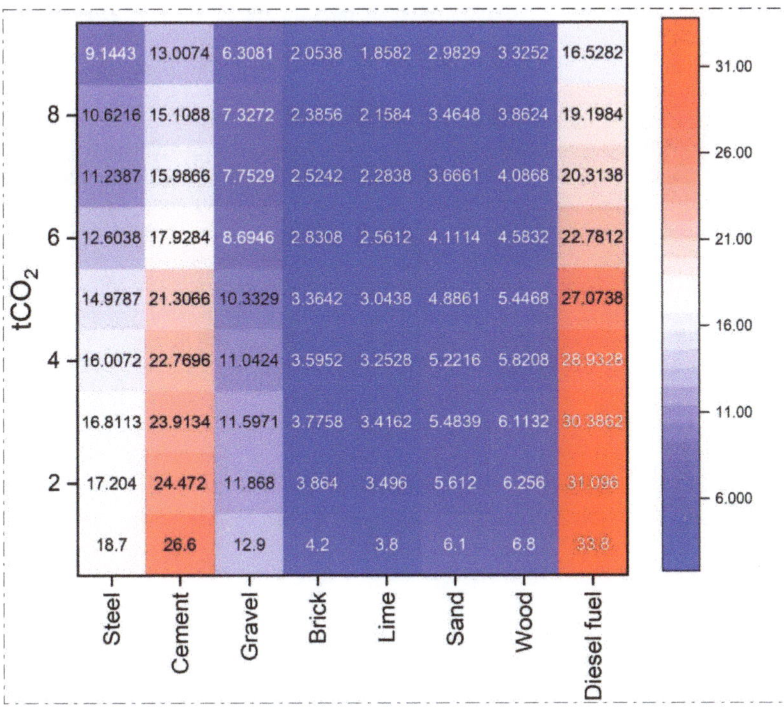

Figure 13. Presentation of carbon emissions from main materials and energy sources.

4.2.2. Carbon Sink Effect Analysis

Carbon sink calculations can help assess and determine the contribution of landscape renovation projects to carbon emission reduction targets. It quantifies the amount of carbon dioxide absorbed by projects through measures such as increasing vegetation coverage and improving soil quality, providing a basis for evaluating their environmental benefits. Carbon sink calculations can be used to evaluate the effectiveness of landscape renovation projects in adapting to climate change. By increasing vegetation and enhancing soil health, projects can enhance ecosystem resilience and reduce vulnerability to climate change.

Section 2.3 provides models for calculating four types of carbon sink, including soil carbon sinks and concrete material carbon sinks. Landscape systems are an important approach to mitigating climate change because they can absorb carbon dioxide through the elements planted within the landscape system, thereby reducing CO_2 levels in the environment.

Taking soil carbon sequestration as an example, the selected parameters are an average carbon concentration of 26.41 g/kg (1–20 cm) for riverbank soil and a carbon density of 3.98 kg/m^2. The regional variation coefficient is 0.48. Based on these parameters and the formulas from Section 2.3, the calculated annual carbon sequestration for soil is 2.46 tCO_2. Similarly, the carbon sink capacity of concrete (used for river embankments) can reach 5.63 tCO_2 per year. This demonstrates that the carbon sequestration potential of the entire landscape renovation project is significant and requires special attention and should not be ignored.

4.3. Neural Network Model Prediction Analysis

The neural network model can be used to predict and analyze the ecological sustainability of landscape engineering systems. In this section, taking the ESI index as an example, the predicted trends of the sustainability of the entire landscape engineering system over 10, 20, and 30 years are analyzed. Figure 14 illustrates the patterns of these changes.

Figure 14A shows the ESI indicator trends over a 10-year period, indicating an overall growth despite significant fluctuations in individual data points. This demonstrates that the landscape engineering system is in its early stage of development and is still in a sustainable state. Figure 14B presents the projected ESI trends over a 20-year period, showing a continued growth trend, although the rate of growth is diminishing. In Figure 14C, which represents the ESI indicator trends over 30 years, the landscape system is observed to be gradually declining. Although regular maintenance and small-scale renovations are implemented for the landscape system, the overall sustainability of the system decreases.

Neural network models can play a positive role in predicting the ecological emergy of building projects, similar to architectural engineering. While some scholars have conducted neural network evaluations for various aspects of building projects [43–46], research on the sustainable ecological energy value of landscape architecture is still in its early stages. This paper represents a forward-looking exploration of this aspect.

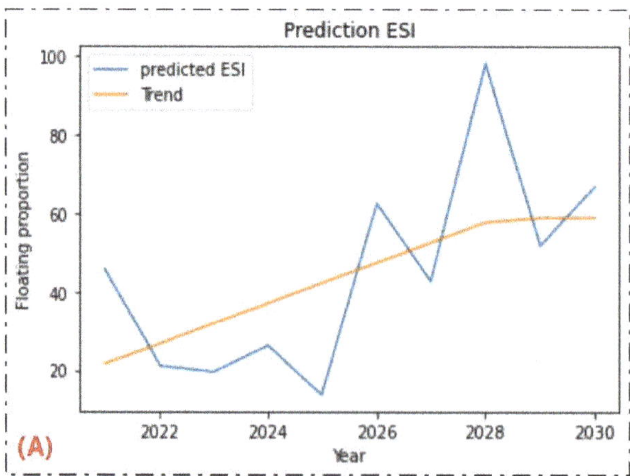

Figure 14. *Cont.*

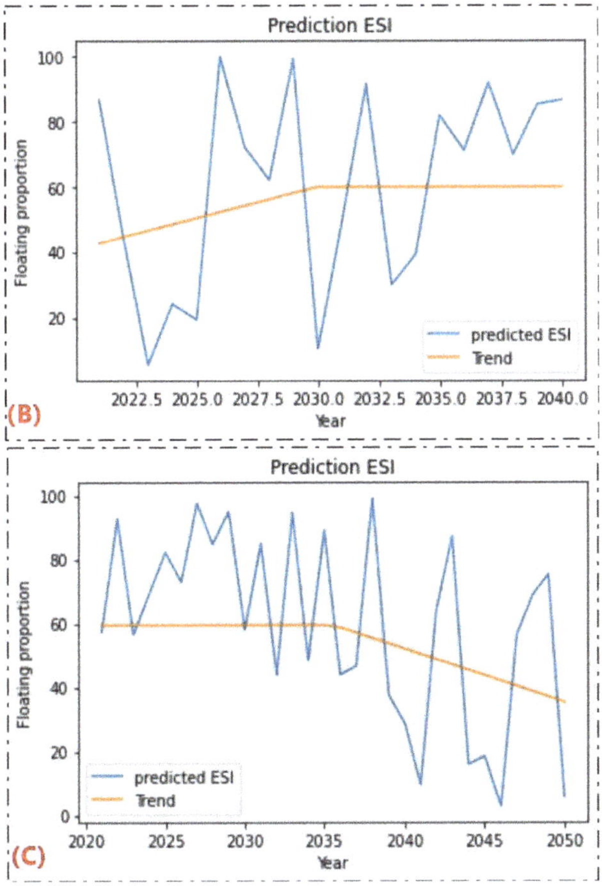

Figure 14. Predicted trends based on an ESI indicator. (**A**) Prediction of ESI parameters for 10 years. (**B**) Prediction of ESI parameters for 20 years. (**C**) Prediction of ESI parameters for 30 years.

Neural network systems can be used to predict the ecological emergy sustainability of landscape engineering. By collecting and analyzing relevant data, neural networks can learn and establish patterns to forecast the trends and sustainability of the ecological emergy in landscape engineering systems. These predictions can assist decision-makers in evaluating and optimizing the design, management, and maintenance of landscape engineering to achieve better ecological benefits and sustainable development. However, this requires appropriate training datasets, rational feature selection, and effective neural network architectures to ensure accurate prediction results and reliable decision support.

5. Landscape System Correction Strategy

According to the research in this article, the overall sustainability of the system needs to be improved from an ecological energy perspective. Furthermore, the efficiency of the entire system needs to be enhanced based on carbon footprint evaluation. This article attempts to improve the sustainability of the system from three aspects: sponge city system, coupling of sewage treatment subsystems, and optimization of landscape engineering system efficiency. By implementing these three enhancement measures, the sustainability of the waterfront landscape system can be improved to some extent, providing valuable reference for urban managers.

The concept of a sponge city Is a planning and design approach aimed at addressing water resource management and environmental issues in urban areas. This concept emphasizes transforming cities into sustainable entities that can absorb, store, and utilize rainwater by mimicking the functions of natural ecosystems. The advantages of a sponge city include reducing the risk of flooding, improving urban microclimates, enhancing water resource utilization efficiency, increasing urban green spaces and ecological functionality, and improving residents' quality of life. This concept has been widely applied in many cities and has become an important strategy for tackling climate change and promoting urban sustainability.

A wastewater treatment system is a facility and process used to treat wastewater generated from urban, industrial, and rural areas. Its purpose is to remove or reduce harmful substances in the wastewater to an acceptable level in order to protect the environment and public health. The implementation of wastewater treatment systems helps to reduce water resource pollution, improve environmental quality, and protect public health. This concept is widely applied in urban and rural areas to ensure sustainable water resource management and protection.

A landscape engineering information flow monitoring system is a facility and technology used to collect, analyze, and monitor information flows related to landscape engineering projects. The implementation of the landscape engineering information flow monitoring system helps landscape engineering teams and relevant stakeholders gain an in-depth understanding of public attitudes, feedback, and demands regarding projects for effective project management, public sentiment handling, and decision-making.

5.1. Coupled Design of the Sponge City Concept

The sponge city design approach enhances the sustainability of waterfront landscapes through several key elements. Firstly, it focuses on effective stormwater management by integrating green infrastructure such as permeable pavements, rain gardens, and bioretention areas to absorb and filter rainwater, reduce runoff, and improve water quality. Secondly, sponge city design promotes the protection and restoration of natural ecosystems in waterfront areas. This includes creating wetlands, riparian buffers, and green spaces that provide habitat for wildlife and serve as natural filtration systems for water bodies. Additionally, the design approach encourages the use of sustainable materials and technologies in construction and development. This includes adopting energy-efficient techniques, utilizing recycled materials, and implementing low-impact development practices to minimize environmental impact. By implementing these strategies, sponge city design enhances the resilience and long-term sustainability of waterfront landscapes, ensuring their ecological integrity and promoting a more sustainable and livable urban environment. Figure 15 illustrates the implementation pathway of the sponge city design approach.

The main improvement measures in this section involve selecting three types of sponge city practice. Firstly, the hierarchical landscape drainage system (Figure 15A–C) is implemented to intercept materials from reinforced cages filled with broken stones (naturally sourced) and set up water discharge gates with varying elevation differences using gravity. The reinforced cages can filter some large particle pollutants in wastewater, thereby improving water quality to a certain extent. At the same time, planting vegetation inside the reservoirs (Figure 15A,B,E,F) helps in automatically purifying the water and further enhancing the sustainability of the water system. Additionally, to filter the water entering the river from both sides, a cascading landscape design can be established along the riverbanks, serving the dual purpose of mitigating urban river overflow and purifying the water quality entering the river (Figure 15D).

To validate the specific effects, an emergy estimation was conducted for the landscape renovation project of Project 1 (Table 2). If the design approach of the sponge city is adopted, there will be a significant increase in the overall engineering quantity of the landscape system. Initially, it may require substantial investment. However, as the usage period increases, based on the ecological emergy index ESI calculation, the overall sustainability of the landscape project can be improved by approximately 15% (from ESI = 0.335 to 0.385). The efficiency of this aspect has been confirmed by relevant scholars. For specific details, please refer to references [45–47].

Figure 15. Exhibition of sponge city design elements. (**A**) Water purification system. (**B**) Single-layer water purification system. (**C**) Components of a natural water purification system. (**D**) Step-style purified water landscape. (**E**) Water purification collection pool. (**F**) Landscape pathway.

5.2. Sewage Purification System

Wastewater treatment systems can effectively purify the water of urban rivers. These systems employ a series of physical, chemical, and biological processes to remove suspended solids, organic substances, nutrients, and harmful substances from wastewater. Firstly, wastewater undergoes primary treatment where solid residues are removed and settled. Next, the wastewater enters secondary treatment, utilizing biological methods such as activated sludge or anaerobic digestion to eliminate organic matter and nutrients. During this process, bacteria and other microorganisms decompose organic matter, thereby reducing the pollutant content in the water. Finally, the wastewater undergoes advanced treatment to remove residual organic matter, nutrients, and microorganisms. Common methods for advanced treatment include biofiltration, biogas fermentation, and ozone oxidation. These processes further purify the water quality, ensuring compliance with environmental standards and sustainable development requirements. The water treated by the wastewater treatment system is then discharged into urban rivers, significantly improving their water quality. This helps protect aquatic ecosystems, provide clean water resources, and reduce adverse impacts on the environment and human health. However, continuous monitoring and effective maintenance are crucial to ensure the proper functioning and safety of wastewater treatment systems.

Figure 16 depicts the basic flow path and key nodes of a wastewater treatment system. Taking the ESI (Ecological Sensitivity Index) as an example, coupling wastewater treatment with landscape engineering systems significantly enhances ecological sustainability, increasing it from 0.335 to 2.41, representing an approximately 7.19-fold improvement. However, this calculation does not consider the various inputs required by the wastewater treatment system; it only focuses on the positive effects of integrating the wastewater treatment system into the landscape engineering.

Studies conducted by relevant scholars regarding the impact of wastewater treatment systems on urban river water quality have been validated and shown to have significant effectiveness in greatly enhancing the sustainability of urban rivers [48,49]. However, their shortcomings are also evident, namely the requirement for a substantial investment and the need for assessment as a subsystem of the entire urban system.

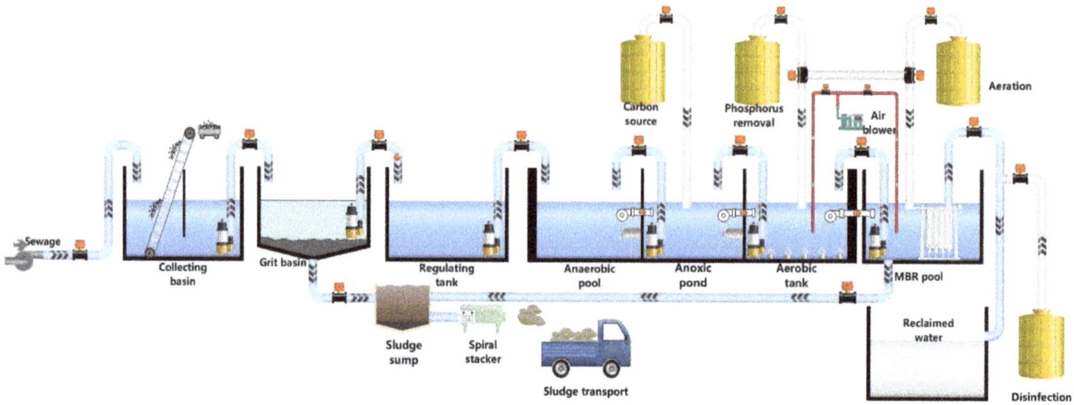

Figure 16. Sewage treatment system.

5.3. *Information Flow Optimization Control System*

In order to enhance the intelligent management level of landscape engineering, a complete monitoring, detection, and control topology diagram is designed and applied. Figure 16 represents the topology control system diagram. The entire process involves data collection from cameras, detectors, sensors, and probes, which is then collected and processed by regional servers. The data are uploaded to a central processing computer for data analysis, providing managers with appropriate recommendations.

The smart landscape topology diagram is a visual representation of the interconnected components and systems that make up a smart landscape. It illustrates the structure and relationships between various elements such as sensors, actuators, controllers, data networks, and management systems. At the center of the topology diagram is typically a central control hub or server that serves as the core of the smart landscape system. This hub receives data from different sensors distributed throughout the landscape, such as weather sensors, soil moisture sensors, light sensors, and motion sensors. The central control hub processes and analyzes the collected data in real time. Based on the analysis results, commands and instructions are sent to actuators and devices distributed across the landscape. These devices can include irrigation systems, lighting fixtures, water pumps, and other infrastructure components. The smart landscape topology diagram also depicts the communication infrastructure used for seamless data transmission and exchange between different components. This may involve wired or wireless networks, including local area networks (LANs), wide area networks (WANs), and Cloud-based platforms. Overall, the smart landscape topology diagram provides an overview of how different components within the system are connected and interact to create an intelligent and efficient management system for landscapes. It helps stakeholders understand the flow of information, control mechanisms, and the overall architecture of the smart landscape solution.

According to the design and application of the topology diagram in Figure 17, the information flow within the entire landscape system is more efficient, avoiding ecological landscape damage caused by information delays. Taking the ESI index as an example, the use of the complete monitoring system greatly increases the proportion of emergy in the information flow. In this study on landscape ecology in design engineering, ESI can improve efficiency by approximately 25% (from 0.335 to 0.268).

Similarly, there is a limiting negative factor. Due to the higher cost associated with the overall automated monitoring system, when converted into ecological costs, it has a significant negative impact in the early stages of landscape projects. The positive effects can only become evident as the lifecycle of the landscape project is extended. Relevant scholars have also studied the effectiveness of information flow topology [50,51].

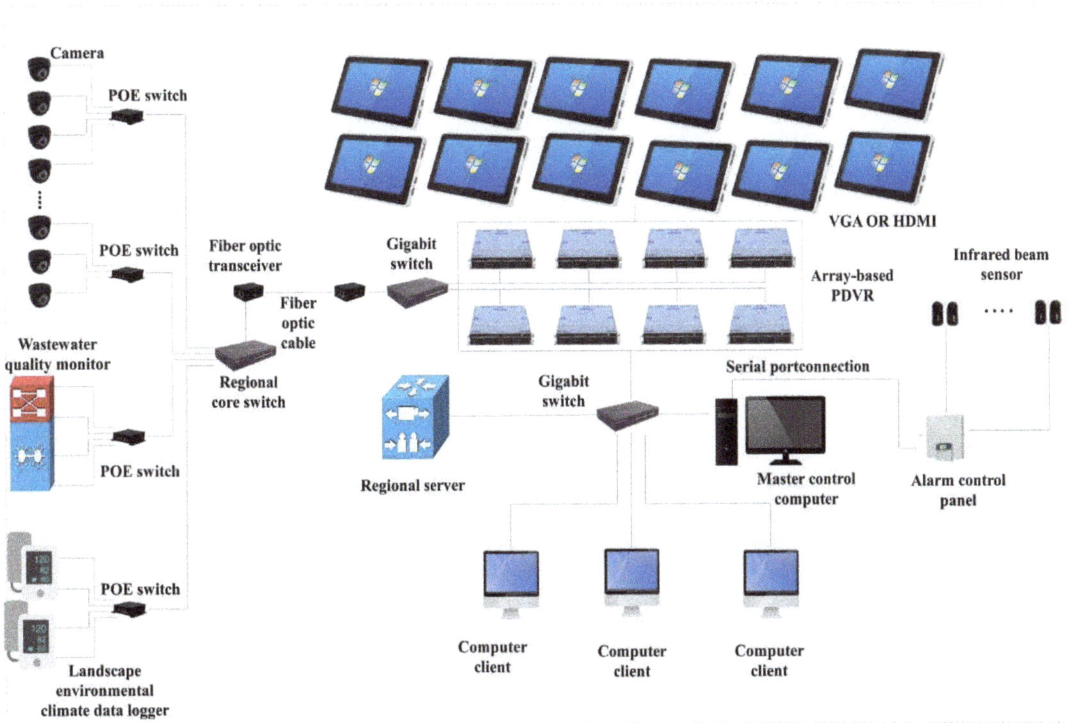

Figure 17. Coupling design of a landscape engineering monitoring system.

6. Discussion on the Comparison with Other Studies

Sustainability research on urban waterfront landscapes is a hot topic, especially regarding the influencing factors, including water bodies, landscapes, slopes, and management [52–56]. Specifically, different researchers have analyzed and discussed this type of research from various perspectives.

For example, some researchers investigate the energy efficiency of different design strategies for urban riverfront landscapes and propose sustainable approaches to reduce energy consumption [57]. Several authors evaluate the carbon emissions associated with urban riverfront development in two cities and identify potential mitigation measures to enhance sustainability [58]. By using life cycle assessment to analyze the environmental impacts of green infrastructure, interventions in urban riverfront redevelopment, focusing on energy use and carbon emissions, have been realized [59]. From the view of the importance of incorporating green spaces into urban planning for carbon reduction, the research [60] assesses the carbon sequestration capacity of riverfront parks in an urban context, emphasizing the importance of incorporating green spaces into urban planning for carbon reduction. A study examines energy-efficient design strategies applied to urban riverfront landscapes, highlighting successful practices and lessons learned from a case study in GHI city [61].

Compared to previous studies, this article conducts an assessment and research on the sustainability of waterfront landscapes from the perspective of the entire life cycle's emergy and carbon emissions. Specifically, focusing on urban river scenarios, an estimation is made of their ecological emergy and carbon emissions, providing new insights for urban ecological sustainability research.

7. Conclusions

To verify the ecological sustainability of urban river landscape systems, this study applies the whole life cycle emergy and carbon emission methods to landscape renovation projects. Additionally, through the prediction of artificial neural networks, a long-term assessment of the ecological sustainability of landscape projects is conducted.

From the perspective of ecological emergy, in the long term, landscape engineering's ecological maintenance emergy accounts for approximately 75% and is a major influencing factor. Over time, the three main indicators (EYR/ELR/ESI) show varying degrees of change, especially the emergy sustainability parameter ESI, which serves as a core reference indicator, exhibiting a decreasing trend. This indicates that as time goes on, the landscape system gradually becomes less efficient. Analyzing carbon emissions from nine types of landscape node, diesel fuel, cement, steel, and other factors were identified as the main contributors to carbon emissions. At the same time, a quantitative estimation of the landscape system's carbon sink was conducted, which helps to clearly identify and understand the carbon emissions of the entire landscape system. This is beneficial for urban managers in implementing low-carbon management.

To enhance the ecological sustainability of the entire landscape system, three types of improvement measure have been preliminarily designed and applied. These include sponge city design enhancement pathways, coupled wastewater treatment systems, and optimization of information flow. From the effectiveness perspective, these measures have a certain positive impact on the landscape's ecological sustainability as described in this article. However, all three improvement measures require relatively high investment costs, especially the coupling of wastewater treatment modes and information flow monitoring systems. The selection of these measures should be based on the type and pattern of different ecological landscapes.

In addition, this study needs to consider the continuity of waterfront landscapes, especially their ecological impact on the entire urban system. This is an area that requires in-depth research. Moreover, due to the significant carbon sequestration effects of water systems, precise calculations at the carbon sink level are necessary to enhance the overall accuracy of the study. Therefore, the next research step should focus on these two aspects.

Author Contributions: Conceptualization, G.J. and A.T.A.; investigation, L.Z.; formal analysis, G.J. and A.T.A.; methodology, G.J. and J.Z.; resources, G.J. and L.Z.; writing—review and editing, G.J. and A.T.A. All authors have read and agreed to the published version of the manuscript.

Funding: The work described in this paper was supported by the State Key Laboratory of Silicate Materials for Architectures (Wuhan University of Technology) (SYSJJ2022-16); the XJTLU Urban and Environmental Studies University Research Centre (UES) (UES-RSF-23030601). Major Projects of Philosophical and Social Science Research in Universities in Jiangsu Province (grant number 2023SJZD131).

Data Availability Statement: Data are contained within the article.

Conflicts of Interest: Author Lanlan Zuo was employed by the company Hunan City University Design and Research Institute Co., Ltd. The remaining authors declare that the research was conducted in the absence of any commercial or financial relationships that could be construed as a potential conflict of interest.

References

1. Beck, D.; Bridging, M.F. "Stakeholder Value Creation" and "Urban Sustainability": The need for better integrating the Environmental Dimension. *Sustain. Cities Soc.* **2023**, *89*, 104316. [CrossRef]
2. Hao, Z.; Wang, Y. Evaluation of socio-economic-ecological environmental benefits of urban renewal projects based on the coupling coordination degree. *Environ. Sci. Pollut. Res.* **2023**, *30*, 56946–56968. [CrossRef] [PubMed]

3. Li, D.; Yan, S.; Chen, G. Effects of Urban Redevelopment on Surface Urban Heat Island. *IEEE J. Sel. Top. Appl. Earth Obs. Remote Sens.* **2023**, *16*, 2366–2373. [CrossRef]
4. Ramsay, E.E.; Duffy, G.A.; Burge, K.; Taruc, R.R.; Fleming, G.M.; Faber, P.A.; Chown, S.L. Spatio-temporal development of the urban heat island in a socioeconomically diverse tropical city. *Environ. Pollut.* **2023**, *316*, 120443. [CrossRef]
5. Odum, H.T. *Environmental Accounting. Emergy and Environmental Decision Making*; John and Wiley and Sons: Hoboken, NJ, USA, 1996; pp. 32–34.
6. Mufan, Z.; Yong, G.; Hengyu, P.; Fei, W.; Dong, W. Ecological and socioeconomic impacts of payments for ecosystem services-A Chinese garlic farm case. *J. Clean. Prod.* **2021**, *285*, 124866. [CrossRef]
7. Asgharipour, M.R.; Amiria, Z.; Campbell, D.E. Evaluation of the sustainability of four greenhouse vegetable production ecosystems based on an analysis of emergy and social characteristics. *Ecol. Model.* **2020**, *424*, 109021. [CrossRef]
8. Bergquist, D.; Garcia-Caro, D.; Joosse, S.; Granvik, M.; Peniche, F. The Sustainability of Living in a Green Urban District: An Emergy Perspective. *Sustainability* **2020**, *12*, 5661. [CrossRef]
9. Zhang, J.; Ma, L. Urban ecological security dynamic analysis based on an innovative emergy ecological footprint method. *Environ. Dev. Sustain.* **2021**, *23*, 16163–16191. [CrossRef]
10. Santagata, R.; Zucaro, A.; Fiorentino, G.; Lucagnano, E.; Ulgiati, S. Developing a procedure for the integration of Life Cycle Assessment and Emergy Accounting approaches, The Amalfi paper case study. *Ecol. Indic.* **2020**, *117*, 106676. [CrossRef]
11. Zhang, J.; Zhang, H.; Asutosh, A.T.; Sun, N.; Fu, X.; Wang, H.; Li, X. Ecological sustainability assessment of building glass industry in China based on the point of view of raw material emergy and chemical composition. *Environ. Sci. Pollut. Res.* **2023**, *30*, 40670–40697. [CrossRef]
12. Cristiano, S.; Gonella, F. To build or not to build? Megaprojects, resources, and environment: An emergy synthesis for a systemic evaluation of a major highway expansion. *J. Clean. Prod.* **2019**, *223*, 772–789. [CrossRef]
13. Yadegaridehkordi, E.; Hourmand, M.; Nilashi, M.; Alsolami, E.; Samad, S.; Mahmoud, M.; Alarood, A.A.; Zainol, A.; Majeed, H.D.; Shuib, L. Assessment of sustainability indicators for green building manufacturing using fuzzy multi-criteria decision making approach. *J. Clean. Prod.* **2020**, *277*, 122905. [CrossRef]
14. Yang, L.; Wang, C.; Yu, H.; Yang, M.; Wang, S.; Chiu, A.S.F.; Wang, Y. Can an island economy be more sustainable? A comparative study of Indonesia, Malaysia, and the Philippines. *J. Clean. Prod.* **2020**, *242*, 118572. [CrossRef]
15. Wanga, X.; Liu, G.; Cosciemec, L.; Giannettiad, B.F.; Haoab, Y.; Zhang, Y.; Brown, M.T. Study on the emergy-based thermodynamic geography of the Jing-Jin-Ji region: Combined multivariate statistical data with DMSP-OLS nighttime lights data. *Ecol. Model.* **2019**, *397*, 1–15. [CrossRef]
16. Leea, D.J.; Brown, M.T. Estimating the Value of Global Ecosystem Structure and Productivity: A Geographic Information System and Emergy Based Approach. *Ecol. Model.* **2021**, *439*, 109307. [CrossRef]
17. Fan, Y.; Meng, J.; Ye, H.; Wang, P.; Wang, Y.; Wang, Y. Sustainability and ecological efficiency of low-carbon power system: A concentrating solar power plant in China. *J. Environ. Manag.* **2021**, *290*, 112659. [CrossRef] [PubMed]
18. Hou, M.; Braham, W.W. Land development in an urban megalopolis: A holistic environmental assessment of housing on Chongming Eco-island, China. *J. Clean. Prod.* **2021**, *312*, 127617. [CrossRef]
19. Xie, B.; Zhang, X.; Lu, J.; Liu, F.; Fan, Y. Research on ecological evaluation of Shanghai port logistics based on emergy ecological footprint models. *Ecol. Indic.* **2022**, *139*, 108916. [CrossRef]
20. Du, Y.-W.; Wang, Y.-C.; Li, W.-S. Emergy ecological footprint method considering uncertainty and its application in evaluating marine ranching resources and environmental carrying capacity. *J. Clean. Prod.* **2022**, *336*, 130363. [CrossRef]
21. Zadehdabagh, N.; Monavari, S.M.; Kargari, N.; Taghavi, L.; Pirasteh, S. Sustainability of agroecosystems by indices: A comparative study between indicators of ecological footprint sustainability and emergy analysis; a case study in Dez catchment, Iran. *Ecol. Model.* **2022**, *474*, 110165. [CrossRef]
22. Xie, Y.; Li, B.; Geng, Y. Life cycle assessment and emergy synthesis for evaluating the environmental performance of residential buildings in China. *J. Clean. Prod.* **2015**, *109*, 228–237.
23. Wang, Y.; Liu, G.; Cai, Y.; Giannetti, B.F.; Agostinho, F.; Almeida, C.M.V.B.; Casazza, M. The Ecological Value of Typical Agricultural Products: An Emergy-Based Life-Cycle Assessment Framework. *Front. Environ. Sci.* **2022**, *10*, 824275. [CrossRef]
24. Tam, V.W.; Zhou, Y.; Illankoon, C.; Le, K.N. A critical review on BIM and LCA integration using the ISO 14040 framework. *Build. Environ.* **2022**, *213*, 108865. [CrossRef]
25. Xu, J.; Teng, Y.; Pan, W.; Zhang, Y. BIM-integrated LCA to automate embodied carbon assessment of prefabricated buildings. *J. Clean. Prod.* **2022**, *374*, 133894. [CrossRef]
26. Wang, Y.; Qi, T.; Su, M. Sustainable Design Strategies for Waterfront Landscapes: A Review. *Sustainability* **2023**, *15*, 496.
27. Chen, J.; Liu, X.; Zhang, Y.; Li, J. Assessing the Ecological Sustainability of Urban Waterfront Development: A Case Study in Shanghai, China. *Environ. Manag.* **2022**, *69*, 166–180.
28. Liao, K.; Wang, C.; Yang, Y.; Ma, Y. Evaluating the Ecological Sustainability of Waterfront Landscape Development: A Comprehensive Assessment Framework. *J. Clean. Prod.* **2023**, *309*, 167127.
29. Huo, T.; Xu, L.; Liu, B.; Cai, W.; Feng, W. China's commercial building carbon emissions toward 2060: An integrated dynamic emission assessment model. *Appl. Energy* **2022**, *325*, 119828. [CrossRef]

30. Zhang, N.; Luo, Z.; Liu, Y.; Feng, W.; Zhou, N.; Yang, L. Towards low-carbon cities through building-stock-level carbon emission analysis: A calculating and mapping method. *Sustain. Cities Soc.* **2022**, *78*, 103633. [CrossRef]
31. Gan, L.; Liu, Y.; Shi, Q.; Cai, W.; Ren, H. Regional inequality in the carbon emission intensity of public buildings in China. *Build. Environ.* **2022**, *225*, 109657. [CrossRef]
32. Wang, X.; Du, Q.; Lu, C.; Li, J. Exploration in carbon emission reduction effect of low-carbon practices in prefabricated building supply chain. *J. Clean. Prod.* **2022**, *368*, 133153. [CrossRef]
33. Liu, W.; Zuo, B.; Qu, C.; Ge, L.; Shen, Q. A reasonable distribution of natural landscape: Utilizing green space and water bodies to reduce residential building carbon emissions. *Energy Build.* **2022**, *267*, 112150. [CrossRef]
34. Fabrizio; Amoruso, M.; Sonn, M.-H.; Schuetze, T. Carbon-neutral building renovation potential with passive house-certified components: Applications for an exemplary apartment building in the Republic of Korea. *Build. Environ.* **2022**, *215*, 108986. [CrossRef]
35. Chen, C.; Bi, L. Study on spatio-temporal changes and driving factors of carbon emissions at the building operation stage-A case study of China. *Build. Environ.* **2022**, *219*, 109147. [CrossRef]
36. Gan, L.; Ren, H.; Cai, W.; Wu, K.; Liu, Y.; Liu, Y. Allocation of carbon emission quotas for China's provincial public buildings based on principles of equity and efficiency. *Build. Environ.* **2022**, *216*, 108994. [CrossRef]
37. Grinham, J.; Fjeldheim, H.; Yan, B.; Helge, T.D.; Edwards, K.; Hegli, T.; Malkawi, A. Zero-carbon balance: The case of HouseZero. *Build. Environ.* **2022**, *207*, 108511. [CrossRef]
38. Zhang, Y.; Li, X.; Zhang, J.; Chen, W. Sustainable Design of Waterfront Landscapes Using Neural Network Models: A Case Study in Guangzhou, China. *Environ. Impact Assess. Rev.* **2023**, 105335.
39. Wu, H.; Deng, Y.; Wu, X.; Li, Q.; Yang, C. Forecasting the Environmental Performance of Waterfront Landscape Designs using Neural Networks. *J. Environ. Manag.* **2022**, *299*, 113512. [CrossRef]
40. Liu, S.; Zeng, C.; Liu, L.; Gou, Z. Predicting the Sustainability Effects of Waterfront Landscape Design using a Neural Network Model and Life Cycle Assessment. *J. Clean. Prod.* **2021**, *304*, 127030. [CrossRef]
41. Zhang, J.; Ravi; Srinivasan, S.; Peng, C. A Systematic Approach to Calculate Unit Emergy Values of Cement Manufacturing in China Using Consumption Quota of Dry and Wet Raw Materials. *Buildings* **2020**, *10*, 128. [CrossRef]
42. Zhang, X.; Chen, Y.; Yang, Q. A comparative analysis of life cycle energy, carbon emissions, and emergy evaluation of green roofs in China. *Energy Build.* **2019**, *185*, 25–35.
43. Brown, M.T.; Campbell, D.E.; De Vilbiss, C.; Ulgiati, S. The geobiosphere emergy baseline: A synthesis. *Ecol. Model.* **2016**, *339*, 92–95. [CrossRef]
44. Chu, L.K. The role of energy security and economic complexity in renewable energy development: Evidence from G7 countries. *Environ. Sci. Pollut. Res.* **2023**, *30*, 56073–56093. [CrossRef] [PubMed]
45. Wan, S.; Xu, L.; Qi, Q.; Yang, H.; Zhou, Y. Building a multi-objective optimization model for Sponge City projects. *Urban Clim.* **2022**, *43*, 101171. [CrossRef]
46. Sun, Y.; Jiang, F. Garden Landscape Design of Sponge City Residential Area Based on Digital Technology. *Mob. Inf. Syst.* **2022**, *2022*, 3684422. [CrossRef]
47. Yang, J.; Zheng, B. Spatial structure planning and optimization strategy of sponge city in coastal area. *J. Coast. Res.* **2020**, *103*, 561–565. [CrossRef]
48. Zhou, K. Study on application of fuzzy control model in phosphate removal system of sewage water. *Appl. Water Sci.* **2022**, *12*, 34. [CrossRef]
49. Zhang, J.; Ma, L. Environmental Sustainability Assessment of a New Sewage Treatment Plant in China Based on Infrastructure Construction and Operation Phases Emergy Analysis. *Water* **2020**, *12*, 484. [CrossRef]
50. Harush, U.; Barzel, B. Dynamic patterns of information flow in complex networks. *Nat. Commun.* **2017**, *8*, 2181. [CrossRef]
51. Zheng, Y.; Li, S.E.; Wang, J.; Cao, D.; Li, K. Stability and Scalability of Homogeneous Vehicular Platoon: Study on the Influence of Information Flow Topologies. *IEEE Trans. Intell. Transp. Syst.* **2015**, *17*, 14–26. [CrossRef]
52. Smith, J. Sustainable urban waterfront development: A comparative study. *J. Urban Plan. Dev.* **2019**, *145*, 123–135.
53. Jones, A.; Brown, K. The role of green infrastructure in sustainable waterfront revitalization. *Landsc. Urban Plan.* **2018**, *176*, 45–56.
54. Chen, L. Sustainable design strategies for urban waterfront regeneration: A case study of XYZ city. *Sustain. Cities Soc.* **2017**, *35*, 123–134.
55. Wang, Q. Integrating social, economic, and ecological aspects for sustainable waterfront planning and design. *J. Clean. Prod.* **2016**, *112*, 4328–4339.
56. Turner, M. Towards sustainable waterfront developments: An integrated approach. *Sustain. Dev.* **2015**, *23*, 88–98.
57. Li, X. Assessing the energy performance of urban riverfront landscape design: A case study of XYZ city. *Sustain. Cities Soc.* **2020**, *55*, 102043.
58. Zhang, Y. Carbon footprint analysis of urban riverfront development: A comparative study of ABC and XYZ cities. *J. Environ. Manag.* **2019**, *245*, 234–243.
59. Wang, H. Life cycle assessment of green infrastructure in urban riverfront redevelopment projects. *J. Clean. Prod.* **2018**, *183*, 394–404.

60. Liu, J. Exploring the carbon sequestration potential of urban riverfront parks: A case study of DEF city. *Landsc. Urban Plan.* **2017**, *160*, 12–21.
61. Chen, S. Energy-efficient design strategies for urban riverfront landscapes: Lessons from GHI city. *Sustain. Dev.* **2016**, *24*, 345–356.

buildings

MDPI

Review

Greenhouse Gas Emissions in Railways: Systematic Review of Research Progress

Diogo da Fonseca-Soares [1,2,3,*], Sayonara Andrade Eliziário [4], Josicleda Domiciano Galvincio [2] and Angel Fermin Ramos-Ridao [3]

1 Brazilian Company of Urban Trains, Square Napoleao Laureano, 01 Varadouro, João Pessoa 58010-150, Brazil
2 Department of Geographical Sciences, Federal University of Pernambuco, Avenue Professor Moraes Rego 1235, Cidade Universitária, Recife 50670-901, Brazil; josicleda.galvincio@ufpe.br
3 Civil Engineering Department, University of Granada, Campus de Fuentenueva, s/n. E.T.S. Ingenieros de Caminos, Canales y Puertos, 18071 Granada, Spain; ramosr@ugr.es
4 Department of Renewable Energy Engineering, Federal University of Paraíba, Campus I, Castelo Branco III, João Pessoa 58051-900, Brazil; sayonara@cear.ufpb.br
* Correspondence: diogosoares@cbtu.gov.br; Tel.: +55-83-98701-6734

Abstract: Rail transportation plays a crucial role in reducing carbon emissions from the transportation system, making a significant contribution to environmental impact mitigation due to the efficiency of passenger and freight rail transportation. Accurate assessment of carbon emissions resulting from rail transit is essential to quantify the positive impact of this mode of transportation on overall urban transport emission reduction. Given that measuring carbon emissions throughout the lifecycle of rail transportation involves a wide array of factors, adopting a systematic framework for analyzing these aspects is crucial. This study conducts a comprehensive review of existing research related to carbon emissions in rail transportation and its mitigation. Initially, the distinct characteristics of carbon emissions associated with rail transportation are identified, along with the complexity involved in accurately measuring these emissions. Subsequently, a comparison and analysis are conducted regarding various models for measuring carbon emissions in rail transportation. Finally, the study examines some greenhouse gas emission measurement research within the railway system. Redirecting research efforts toward measuring carbon emissions in the rail transportation system is essential to help the development of robust and effective public policies. This measure will play a crucial role in emission reduction, climate change mitigation, and the promotion of more sustainable transportation. Furthermore, the identified results propose which LCA methodology offers a valuable framework improving the quality of railway transportation emissions for future generations.

Keywords: railway; sustainability; GHG emission; LCA; climate change

1. Introduction

The increase in concentrations of greenhouse gas (GHG) emissions in the atmosphere has led to an amplified greenhouse effect, resulting in higher average temperatures worldwide. This global warming causes a range of adverse impacts on various aspects of the biosphere, from natural ecosystems to human activities [1].

Climate change represents one of the most pressing and complex challenges that humanity faces in the 21st century. It reflects significant changes in global climate patterns over time, primarily driven by human activity and its emissions of greenhouse gases such as carbon dioxide (CO_2), methane (CH_4), and nitrous oxide (N_2O) [2]. International cooperation plays a crucial role in addressing climate change, as evidenced by the Paris Agreement, a global milestone aimed at limiting global warming well below 2 °C compared to pre-industrial levels. Individual efforts, governmental actions, and corporate commitments are also essential to tackle this challenge [3].

Europe, for instance, is significantly impacted by climate change, presenting a series of environmental, social, and economic challenges for the region. These impacts are observed in various areas and have profound implications for European countries and the continent as a whole [4,5]. To address these challenges, the European Union and its member states are implementing policies and actions to mitigate and achieve climate neutral by 2050, through green technology, sustainable industry and clean transport. Investments in sustainable technologies, promotion of energy efficiency, protection of natural ecosystems, and public awareness are key strategies to reduce greenhouse gas emissions [6,7]. This involves decarbonization of industrial activities by transitioning to renewable energy sources, the electrification of vehicle (including trains, buses, planes, boats and cars) and investing in fleets powered by other alternative energy sources, such as hydrogen, efuels or biofuels. Furthermore, adaptation is equally crucial, involving the implementation of strategies to address ongoing impacts and the development of resilient infrastructures [8].

Europe plays an active role on the international stage, working together with other countries to build a more sustainable and resilient future in the face of climate change [9,10]. However, it's important to emphasize that climate change knows no borders, and international cooperation is essential to address this global challenge. In light of this panorama, climate change mitigation has become a global priority.

Furthermore, climate change disproportionately affects vulnerable populations, such as low-income communities, indigenous peoples, and densely populated coastal regions. Lack of access to resources, adequate infrastructure, and early warning systems increases their susceptibility to extreme weather events, exacerbating existing inequalities and creating new socio-economic challenges. This reality underscores the urgency of addressing climate change not only as an environmental issue but also as a matter of social justice and human rights [11,12]. Awareness, education, and cooperation are essential to confront this threat and ensure a healthy and habitable planet for present and future generations [12–14].

Therefore, the study of greenhouse gas (GHG) emissions and their life cycle is a crucial research area for understanding the environmental impact of human activities and developing strategies for climate change mitigation [13]. The life cycle approach is a valuable tool for analyzing environmental burdens and impacts of GHG emissions across all stages of a process, product, or system, from raw material extraction to final disposal. This includes production, transportation, use, and disposal of goods and services. Life cycle assessment (LCA) involves collecting detailed data on material flows, energy, and emissions at each stage, enabling a comprehensive analysis of environmental impacts [15,16].

Estimates of GHG emissions impacts into the atmosphere are intrinsically linked to the modernization of transportation system, due this sector still heavily relies on fossil fuels, such as gasoline and diesel, currently accounting for over 20% of global CO_2 emissions. The impact of GHG from the transportation system is complex and encompasses a variety of transportation modes and related practices [17,18]. Studies show that factors such as population growth and concentration, economic growth, technological development, and motorization rates increase energy intensity and CO_2 emissions. According to Tang and Jiang, the effects of energy intensity, population urbanization, industry scale, and energy structure collectively exert negative driving influences because propelling the augmentation of carbon emissions [19].

In the context of studying GHG emissions and their life cycle, it is common to assess different alternatives to identify the most sustainable options. For example, when comparing different transportation systems, one can analyze not only the direct emissions from vehicles but also emissions associated with the manufacturing, maintenance, and operation of infrastructure such as roads, rails, or airports [19]. Furthermore, life cycle analysis can highlight opportunities to reduce emissions at different stages of the cycle. This might involve improving energy efficiency in vehicle production, using more sustainable materials, promoting public transportation, expanding bike lane networks, and implementing policies that encourage the adoption of the mode/service with the lowest carbon footprint [20].

In this context, railways transportation has the potential to assume a central role in long-term transport sector decarbonization and provide huge opportunities with the modernizing of infrastructure, contributing to the reduction of GHG emissions and the development of more efficient and environmentally responsible transportation systems [19]. The life cycle analysis in the railway context is an approach that enables a complete understanding of the common environmental impact hotspots of railway operations, from the design and construction of infrastructures to the operation and maintenance of railway systems. This comprehensive methodology is essential for accurately and holistically assessing emissions associated with rail transportation and identifying mitigation opportunities [21,22].

Greenhouse gas emissions from the transportation system, including railways, pose one of the major challenges in addressing climate change. Furthermore, this approach allows for identifying critical emission points throughout the life cycle where interventions can be made to formulating policies for low-carbon development and to reduce overall emissions. This might involve adopting cleaner technologies, improving energy efficiency, changing consumption patterns, or implementing more sustainable practices [23,24]. The adoption of cleaner technologies, promotion of sustainable transportation alternatives, and smart urban planning are essential to mitigate the environmental impacts of the transportation sector and contribute to a more sustainable future [25].

In this context, this research provides a comprehensive approach to the study of investigation focusing on the analysis and measurement of greenhouse gas emissions in rail transportation. By doing so, it assists researchers in gaining a thorough understanding of the current state of research in this area and the persistent issues. Furthermore, it seeks to provide guidelines and recommendations for further in-depth investigation in this field in the future. Additionally, this research aims to address the following questions regarding studies of greenhouse gas emissions in the global railway system:

Q1. What are the carbon emission characteristics of railways?

Q2. What bibliographic trends exist in relation to studies of greenhouse gas emissions in railway systems?

Q3. What calculation models are used in research on carbon emissions from rail transportation?

Q4. What are the complexities and limitations of measuring carbon emissions from railway system?

Q5. What research trends exist regarding the mitigation of greenhouse gas emissions in the railway system?

2. Materials and Methods

This article proposes a systematic literature review focused on the analysis and measurement of greenhouse gas emissions in the context of railway transportation. The scope of this review primarily encompasses life cycle assessment (LCA) methods published between the years 2010 and 2023. A systematic review is defined as a critical synthesis of research evidence aimed at identifying, selecting, and synthesizing all published research on a specific topic or issue [11,17,26–29].The execution of this review follows a formal and rigorous methodological approach to achieve its objectives.

The systematic literature review process comprises three phases: Planning, Implementation, and Reporting. At the outset of this work, research questions were carefully formulated. Subsequently, a comprehensive research protocol was developed, outlining specific resources such as selected databases for the search, identification of key publications to be considered, and definition of a set of keywords to be used in the search process. The aim of this protocol was to ensure a systematic and exhaustive retrieval of relevant literature for our study.

The article search for the systematic literature review took place in September 2023 and was divided into four stages, as illustrated in Figure 1.

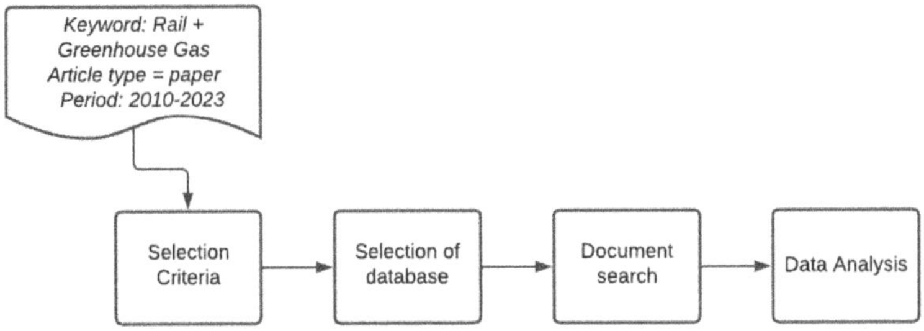

Figure 1. Stages of the articles search for the systematic review.

Selection Criteria: In this stage, the initial criteria for the search were defined, focusing on articles in the research area "Rail and Greenhouse Gas" from 2010 to 2023.

Choice of Database: While there are numerous types of databases for academic documents, Scopus was chosen for this study due to being a scientific database containing a large number of publications, authors, and journals that meet peer-reviewed scientific quality standards [26,29–34]

Document Search: The first search using the terms "Rail + Greenhouse Gas" resulted in 665 documents for the period 2010 to 2023. Subsequently, additional filters were applied, resulting in 327 documents that met the established criteria, as presented in Table 1.

Table 1. Summary of searches performed on the Scopus database.

Searches	Keyword	Filter	Result	Comment
1st	Rail + Greenhouse Gas		665	Not very discriminatory
2nd	Rail + Greenhouse Gas	Year: Between 2010 and 2023	561	A 13-years dataframe was chosen as indicated by most authors for analysis in this field of science
3rd	Rail + Greenhouse Gas	Document type: Articles	351	Articles passes through peer-review, which indicates higher quality
4th	Rail + Greenhouse Gas	Publication stage: Final	327	Documents under publishing processes were not included.

During the implementation phase, suitable research works were identified from the recruited resources based on relevant implementation criteria. Resources that met the implementation criteria were collected, and relevant information pertaining to the research questions was extracted. In the reporting stage, a systematic review article was prepared after investigating the research questions using the collected information.

3. Carbon Emission Characteristics of Railways

As carbon emission characteristics in the railway context are of great importance when considering the environmental footprint of this mode of transportation. Carbon emissions, primarily in the form of GHG, play a significant role in global climate change. Railways, as a mode of transportation, possess distinct characteristics that influence their carbon emissions and contribution to global warming [35].

One of the distinctive features of railways is their energy efficiency compared to other modes of transportation, such as roads and airplanes. Generally, rail transportation is considered one of the most efficient and environmentally friendly modes in terms of carbon emissions. Trains have a high capacity for carrying freight and passengers, resulting in relatively higher energy efficiency per unit of cargo or passenger transported [11,36]. Furthermore, electrification of railways, when electricity is sourced from renewable sources, can further reduce local carbon emissions. This leads to railway transportation emitting less compared to other modes [9].

Road transportation, which includes cars, trucks, and buses, generally has a higher carbon footprint compared to the railway system. This is partly due to the lower energy efficiency of these vehicles, especially when carrying smaller loads or fewer passengers [37]. Internal combustion engines commonly used in road vehicles are less efficient in terms of energy conversion while in motion, resulting in higher fuel consumption and consequently higher GHG emissions [38]. Moreover, a study on the impact of cargo modal transfer policy on carbon emissions in China concluded that road transportation is the most polluting mode among various transportation modes, while rail transportation has lower carbon emissions [39].

However, buses don't always emit more GHGs than trains. The case of the Sheppard subway line in Toronto, Canada, highlights the complexity of analyzing greenhouse gas emissions in transportation systems and how results can vary over time. In this specific example, the subway line initially produced more GHGs per passenger-kilometer than the buses it replaced during the first six years of operation. However, after this period, the carbon reduction benefits of the subway line began to become more evident [40].

This case underscores the importance of considering the evolution over time when assessing the environmental benefits of transportation projects. While the initial impact may not be immediately positive, the potential for significant reductions in GHG emissions can materialize as the system matures and becomes more efficient. This also highlights the need for a comprehensive and long-term approach when planning and evaluating transportation projects with the goal of mitigating climate change and reducing carbon emissions.

If transportation, on the other hand, is known to have a significantly higher carbon footprint compared to rail and road systems. Airplanes consume large amounts of fuel during flight, resulting in substantial GHG emissions per passenger or cargo transported. Additionally, GHG emissions from aviation occur at higher altitudes, which can have an amplified impact on the climate. Although there have been advances in improving the efficiency of jet engines and the adoption of biofuels, air transport is still considered one of the most polluting modes in terms of carbon emissions [25,41–43].

Comparing carbon emissions between rail, road, and air systems underscores the importance of energy efficiency and the type of fuel used. The railway system stands out for being more efficient and often cleaner due to electrification. The road system generally occupies an intermediate position in terms of emissions, while the air system is recognized for its relatively high emissions. This comparison emphasizes the need to consider environmental sustainability when choosing a mode of transportation and highlights the importance of policies and innovations aimed at reducing GHG emissions across all modes of transportation [9]. Although railway systems are generally considered a more sustainable option in terms of carbon emissions compared to other modes of transportation, it is crucial to acknowledge that emissions may still occur, regardless of the type of energy used.

The electrification of railways in many regions contributes to carbon emission reduction. Electric trains directly eliminate emissions from the combustion of fossil fuels, making them a cleaner option in terms of local emissions. However, the carbon footprint of electrification depends on the energy mix used to generate electricity. If electricity comes from renewable sources, carbon emissions associated with railway operation can be further reduced [44,45].

The study analyzing the four typical urban railway lines (lines 6, 9, 10, and 15 of the subway) in Beijing in 2014 yielded important results regarding carbon emission reduction and the influence of various factors in this context. The results demonstrated the complex interconnection between urban rail transit, carbon emissions, and specific factors affecting these emissions [46].

The study showed that urban railway lines had a positive impact on carbon emission reduction in Beijing. This suggests that the implementation of these lines contributed to a more sustainable environment and the mitigation of greenhouse gas emissions. Carbon emissions from urban rail transit were strongly correlated with the carbon emission factor of the electricity used to power the system. This underscores the importance of the energy source behind rail transportation, with electrification and the use of clean energy sources being crucial to maximize environmental benefits [46].

The study also identified a correlation between carbon emissions and the proportion of passenger trips using modes of transportation prior to the opening of urban railway lines. This suggests that passenger adoption and switching to rail transportation had a direct impact on emissions reduction. The sensitivity analysis conducted in the study helped identify which factors significantly influenced carbon emissions. This is crucial to guide future planning decisions and public policies, aiming to optimize the environmental efficiency of urban railway lines [46].

Collectively, these findings highlight the importance of urban railway transportation as a more sustainable alternative to previous modes of transportation, contributing to carbon emission reduction in urban areas. Furthermore, they underscore the need for integrated approaches that consider not only the railway system itself but also influencing factors such as electricity sources and passenger behavior. This can inform transportation and energy policies aimed at continuous reduction of greenhouse gas emissions in urban areas.

An investigation in China identified railway electrification as a key means to reduce carbon emissions and optimize the transportation energy structure in the country. The results indicated that railway electrification using the current energy generation mix could reduce carbon emissions by 8.9%. However, utilizing a generation mix similar to that of the United Kingdom could help achieve a maximum reduction in carbon emissions of 65.4% [47]

A fundamental aspect highlighted by the research was the role of the energy generation mix in determining the effectiveness of railway electrification in reducing carbon emissions. The composition of energy sources used for electricity generation plays a crucial role, as cleaner sources such as renewables result in lower emissions associated with the operation of electric trains. The study demonstrated that adopting an energy generation mix similar to that of the United Kingdom, which has a substantial proportion of renewable energy in its energy mix, could lead to a much more significant reduction in carbon emissions in the Chinese railway system [47].

This research underscores the importance of considering not only the electrification technology itself but also the origin of the electricity used to power railway systems. It also demonstrates the potential for international cooperation, as experiences and best practices from other countries can be leveraged to maximize the benefits of railway electrification in terms of carbon emission reduction. In summary, the research highlights railway electrification as a promising strategy to achieve emission reduction goals and optimize the energy structure in transportation systems, with positive implications for the environment and the global climate [47].

Railway electrification is recognized as a crucial strategy for making transportation systems more sustainable and with a lower carbon footprint. By replacing diesel locomotives with electric trains powered by renewable or cleaner energy sources, it's possible to significantly reduce greenhouse gas emissions. This shift not only contributes to climate change mitigation but also improves air quality and reduces noise pollution in urban areas near railway lines [48].

It's worth noting that carbon emissions are also related to operational practices and proper maintenance of tracks and trains [49–51]. Improvements in fleet management, regular maintenance, and route optimization can result in enhanced efficiency, which in turn can reduce carbon emissions. Furthermore, challenges related to infrastructure, network modernization, and expansion can also influence carbon emissions in different railway systems.

Carbon emissions in the railway sector are influenced not only by the energy sources used to power the trains but are also intrinsically linked to operational practices and proper maintenance of railway infrastructure and trains. The efficiency and sustainability of the railway system depend on a combination of factors, ranging from daily operations management to preventive maintenance strategies [35,52].

Operational practices play a pivotal role in determining carbon emissions. For instance, optimizing train scheduling and operations can result in lower energy consumption and, therefore, reduced emissions. The adoption of regenerative braking strategies, which capture and reuse part of the kinetic energy during train deceleration, can contribute to energy efficiency and emission reduction [53].

Moreover, proper maintenance of railway tracks and trains is crucial to minimize energy losses and ensure the efficient operation of the system. Well-maintained tracks reduce friction and allow smoother operation, which in turn reduces energy consumption. Similarly, regular maintenance of trains, including component cleaning and lubrication, ensures they operate efficiently, avoiding energy waste and unnecessary emissions [54–57].

Furthermore, innovative technologies such as more efficient propulsion systems and lightweight materials also play a role in reducing carbon emissions in the railway sector. Continuous research and development in these areas can contribute to improving energy efficiency and reducing emissions associated with railway operation [58].

The carbon emission characteristics of railways are shaped by their energy efficiency, electrification, and operational practices. While it offers advantages in terms of carbon emission reduction compared to other modes of transportation, it's essential to consider the energy source and implement sustainable practices to maximize its environmental benefits. Understanding these characteristics is crucial for guiding efforts in carbon emission mitigation in the railway sector and contributing to the sustainability of the transportation system as a whole [10].

In another recent study, an estimation of carbon emissions generated by the daily operation of the metro system was conducted. The results obtained from this analysis allowed for the calculation of carbon emissions per kilometer traveled or per passenger trip. This approach provided a solid theoretical basis for the government to consider implementing policies such as carbon taxes for citizens and the creation of a carbon offset mechanism [59].

By determining the carbon emissions associated with the day-to-day operation of the metro system, this study contributed to understanding the specific contributions of this mode of transportation to greenhouse gas emissions. Quantifying emissions per kilometer or per passenger trip offered valuable information to underpin environmental and economic policy formulation [59].

Particularly, the study results served as a foundation for considering measures such as carbon taxes. This strategy aims to incentivize carbon emission reduction by imposing a tax on greenhouse gas emissions. By establishing such a tax, the government seeks not only to promote environmental awareness but also to create a financial incentive for adopting low-emission alternatives, such as the use of public transportation such as the metro, over more polluting modes [59].

Furthermore, the study paved the way for the implementation of a carbon offset mechanism. This refers to an approach in which individuals or organizations can invest in carbon mitigation projects, such as reforestation or clean technologies, to offset their own carbon emissions. This approach plays a significant role in combating climate change by encouraging actions that result in the neutralization or net reduction of greenhouse gas emissions [59].

The study conducted by Yu [59] played a pivotal role in providing a solid theoretical foundation for policy formulation such as carbon taxes and the implementation of a carbon offset mechanism. These strategies are crucial for promoting environmental awareness, encouraging the adoption of sustainable modes of transportation, and contributing to the mitigation of greenhouse gas emissions.

In summary, railways stand out as a mode of transportation with favorable carbon emission characteristics due to their energy efficiency, electrification options, and potential for ongoing emissions reduction through sustainable operational practices. These aspects make railways an attractive alternative for mitigating GHG emissions in the transportation sector and contributing to the pursuit of more sustainable solutions in addressing climate change.

4. Trends in Greenhouse Gas Emission Studies in Railway Systems

The trends in studies concerning greenhouse gas emissions (GHG) in railway systems align with the global need to address climate change. The pursuit of more sustainable and efficient solutions in the railway transport sector continues to evolve, reflecting the growing importance of mitigating emissions and promoting greener and more responsible transportation. Studies are increasingly focused on understanding and reducing GHG emissions in railway systems as part of global efforts to combat climate change. There is a growing awareness of the importance of cleaner and more efficient transportation systems.

There is a rising interest in comparing GHG emissions from railway systems with other modes of transportation, such as highways and aviation. This helps identify the advantages and disadvantages of each mode and informs sustainable transportation policies [9,39]. The electrification of railways, along with the use of renewable energy sources, is gaining prominence. Studies are exploring how electrification can reduce GHG emissions and how the incorporation of clean energy can further enhance environmental efficiency [45]. Research is investigating innovative technologies, such as energy regeneration systems, aerodynamic improvements, and operational optimization, that can contribute to the reduction of GHG emissions in railway systems [55].

Public policies related to climate change and sustainability have influenced studies on GHG emissions in railway systems. Researchers are assessing how policies incentivizing electrification, infrastructure investment, and regulations can impact emissions [11]. Studies are using modeling to project future GHG emissions in railway systems based on different context of growth, electrification, and technology adoption. This helps identify emissions reduction pathways. The analysis of GHG emissions in railway systems is increasingly considering regional and global specificities, including differences in energy matrices, population density, travel patterns, and government policies [60].

5. Calculation Model Used in the Research on Carbon Emissions in Railway Transportation

There are two main approaches to calculate carbon emissions in urban railway transportation. The first is based on energy consumption at the end use and associated carbon emission factors. The second involves direct measurement of CO_2 emission factors, considering travel distance and the mode of transportation used.

In the first method, carbon emissions are calculated by considering the amount of energy consumed by the railway system during its operation. This includes electricity consumption for moving trains, station lighting, air conditioning systems, among others. The consumed energy is then related to carbon emission factors, which indicate the amount of carbon dioxide released into the atmosphere per unit of energy consumed. This approach considers the carbon footprint of the entire system and is often used in life cycle analysis studies [19,61].

In the second method, direct measurement of CO_2 emission factors is carried out based on travel distance and the mode of transportation. This involves collecting data about the trips taken, including the distance traveled and the type of transportation used, such as subway, train, or tram. From this data, CO_2 emissions associated with each mode of transportation are calculated, taking into account the specific operational characteristics of each. This approach is more direct and can provide immediate insights into carbon emissions in terms of distance traveled [62].

Both approaches are valuable for assessing carbon emissions in urban railway transit and are often used in comparative studies between different modes of transportation and policy analyses for emission reduction. The choice between these methods depends on data availability, study objectives, and the desired accuracy in estimating carbon emissions.

Table 2 summarizes and analyzes some of the research work on rail GHG emissions analysis and measurement in the literature.

Table 2. GHG emissions measurement model of rail in literature.

Country	City	Research Boundary	Summary	Reference
United States	Texas	LCA	Study of greenhouse gas emissions from metro construction, operation, and maintenance	[18]
Brazil	Pernambuco	Operation and maintenance	Study of greenhouse gas emissions from metro operation and maintenance	[61]
Korea	Gwangju	Construction and building	Study of greenhouse gas emissions from railway construction	[63]
China	shanghai	LCA	Study of greenhouse gas emissions from metro construction, operation, and maintenance	[64]
Austria	Vienna	LCA	Ignoring carbon emission from maintenance, dismantling, and recycling phases	[65]
United States	Los Angeles	LCA	Two different LCA frameworks are used	[66]
United States	Phoenix	LCA	Developed an integrated transportation and land use LCA framework	[67]
China	Beijing	Construction and building	A quota-based GHG emissions quantification model for metro station construction is proposed	[68]
Canada	Toronto	Operation and maintenance	Study of GHG emissions from the construction and reconstruction of the Spadina streetcar route	[38]
United Kingdom	London	Operation and maintenance	Analyzed the impact of changes in passenger travel behavior on GHG emissions from metro	[69]
United States	New Jersey	Construction and building	Different material inputs were evaluated during the construction of the railroad project	[70]
Italy	Rome	LCA	The use of data sourced from metro operators reduces the uncertainty of the results	[71]
Turkey	Kayseri	LCA	Integrating environmental, economic, and social factors with the LCA approach	[13]
China	Beijing	Operation and maintenance	No carbon emission by default for power generation methods other than thermal power	[24]
Índia	Delhi	Operation and maintenance	The impact of transport mode shift due to the introduction of the metro on carbon emission is considered	[72]
China	Baoji	Operation and maintenance	Developed a passenger demand-based carbon emission model	[73]

5.1. Lifecycle-Based Greenhouse Gas Emission Calculation

Most studies that utilize the approach based on final energy consumption and carbon emission factors focus on measuring carbon emissions during the operational phase of urban railway systems. In this method, the calculation of greenhouse gas (GHG) emissions is conducted by multiplying the electric energy consumption of urban railway systems by the associated carbon emission factor for electricity. This approach takes into account the total electric energy consumption necessary to operate the railway system, including train movement, lighting systems, ventilation, and other components. The energy consumption is then multiplied by the carbon emission factor of electricity, which represents the amount of CO_2 released into the atmosphere per unit of consumed electric energy. This multiplication provides an estimate of the carbon emissions associated with the operation of the railway system.

Life Cycle Assessment (LCA) in the railway system is a comprehensive approach for evaluating the emission of GHGs throughout all phases of the life cycle of a railway system, from raw material extraction to final disposal. LCA considers the environmental aspects related to material production, construction, operation, maintenance, and eventual disposal of a railway system [19]. LCA is a valuable tool for quantifying and evaluating GHG emissions and other environmental impacts associated with the railway system. It enables an objective comparison among different design, operation, and maintenance alternatives, identifying key areas where improvements can be made to reduce environmental impact [20].

The study conducted on the Pernambuco metro aimed primarily to analyze greenhouse gas (GHG) emissions associated with the operation and maintenance of this transportation system. The study found that the total GHG emissions over the life cycle of the Pernambuco railway system, excluding the construction phase, amounted to 6170.54 metric tons of carbon dioxide equivalent (tCO2e). This value represents the overall impact of GHG emissions from the operation and maintenance of the system, as the study did not consider the construction phase due to a lack of data [61].

Based on an estimated system lifespan of 50 years, the study also projected future emissions, estimated to be equivalent to 308,550 tCO2e. This underscores the importance of considering the complete life cycle of a railway system when assessing its environmental impact, as emissions are not limited solely to the operational phase but also include the construction phase and potential maintenance over the years [61].

The railway system in Pernambuco, Brazil, consists of a section of trains powered by diesel traction and another section powered by electricity. As a result, emissions from electricity account for 61% of the total emissions, while emissions from diesel traction represent about 23%. This differs from most railway systems around the world, which are predominantly powered by electricity, Figure 2 [61].

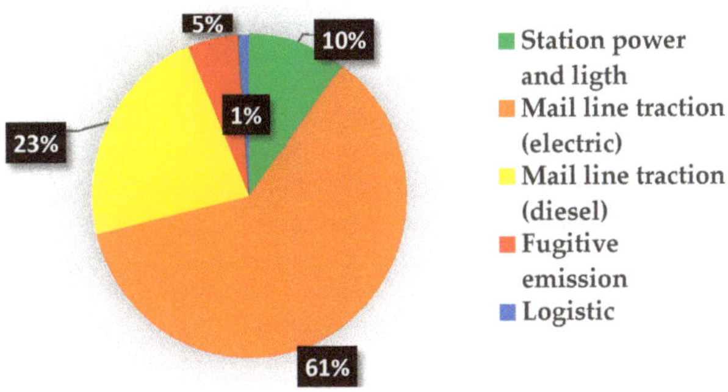

Figure 2. Composition of consumption and emissions of Pernambuco metro; reproduced from [61].

These results are crucial for understanding the role of GHG emissions in the sustainability of the railway system in Pernambuco and for guiding potential mitigation strategies. The study provides valuable information for making informed decisions to reduce emissions over time, whether through improvements in energy efficiency, adoption of cleaner energy sources, or implementation of more sustainable maintenance practices. This research contributes to raising awareness about the environmental impact of railway transportation and provides a starting point for the implementation of measures aimed at making the system more environmentally responsible and contributing to climate change reduction.

A study conducted in China employed the life cycle assessment (LCA) methodology to define the life cycle system of the Shanghai Metro, along with compiling observed real-world data on resource inputs and generated emissions. The study also conducted a comparative analysis of greenhouse gas (GHG) emissions from various urban railway systems worldwide [64].

The results indicate that the total GHG emissions over the life cycle, considering the complete construction extension of the Shanghai Metro with a 50-year lifespan, amount to 109,642.81 metric tons of CO_2 equivalent. Material production, material transportation, on-site construction, operation, and maintenance were responsible for approximately 4.1%, less than 0.1%, 0.4%, 92.1%, and 3.4% of the total emissions, respectively. While GHG emissions per passenger-kilometer traveled of the Shanghai Metro present global competitiveness, the study points out that there is significant energy-saving potential during the operational phase, especially in railway stations with more sustainable designs [64].

The preliminary conclusions of this study provide an understanding of the potential emissions reduction in urban railway transit systems, contributing to emissions reduction goals in China. Furthermore, the results can serve as a valuable source of information and data for future life cycle assessments (LCAs), aiding in the formulation of emissions mitigation strategies in similar transportation systems.

A study in Korea employed LCA to quantify GHG emissions originating from construction modules encompassing various stages such as earthwork, civil engineering activities (including tunnel, viaduct, and bridge construction), railways, passenger stations, and power transmission and telecommunication systems for all HSR lines. Additionally, the study identified the primary activity within each module. The study also analyzed the interdependence between GHG emissions and civil engineering works. This study provided an estimate of GHG emissions during the construction phase of a high-speed railway (HSR) infrastructure connecting Osong to Gwangju, South Korea, inaugurated in 2015. Total GHG emissions reached approximately 3.7 million metric tons of carbon dioxide equivalent (MtCO2eq.), with about 92% of the life cycle GHG emissions was due to vehicle operation and maintenance [63].

The study conducted in Belgium offered a comprehensive assessment of the environmental impact of freight railway transportation in the country, using the life cycle assessment (LCA) methodology. This study analyzed various aspects related to railway transportation, including different types of trains and energy sources used for traction, in order to understand the environmental impact associated with each of these options. The LCA adopted in this study considered not only direct processes related to transportation activities, such as energy consumption and direct emissions, but also other factors involved in the railway system, such as energy production, manufacturing of railway equipment, and the necessary infrastructure for operation. The study divided the freight railway transportation system into three distinct subsystems: operation, railway equipment, and infrastructure [22].

The results indicated that electric trains exhibit superior environmental performance compared to diesel trains in Belgium. The use of electric trains, driven by the Belgian electricity supply mix of 2012, led to a significant reduction in climate change-related environmental impact, reaching a reduction of 26%. Furthermore, the study suggested that increasing the use of electric trains in the future will have a significant impact on the environmental and energy aspects of freight railway transportation. The growing adoption of sustainable electric energy for the operation of electric trains can lead to a more efficient and environmentally friendly freight railway system [22].

The study conducted in the Houston-Dallas corridor (I-45) employed a life cycle assessment (LCA) approach to predict and evaluate net changes in environmental impact associated with the potential development of a high-speed railway (HSR) system along this corridor. The primary objective was to estimate the environmental impact, particularly in terms of carbon dioxide (CO_2) emissions and GHG emissions per vehicle/passenger-kilometer traveled (V/PKT) over the entire life cycle of the system. The study adopted a comprehensive approach, considering various phases of the railway system's life cycle, including material extraction and processing, infrastructure construction, vehicle manufacturing, system operation, and end-of-life. The Ecoinvent 3.4 inventory database was used as a data source for the analysis [18].

The study's results highlighted some important conclusions. Vehicle operation and maintenance were identified as the primary contributor to the total potential global warming, accounting for about 93% of the life cycle. Regarding the infrastructure component, the phase of material extraction and processing was identified as the major contributor, representing approximately 56.76% of GHG emissions, with around 23.75 kg CO_2 eq./VKT. In comparison, the study revealed that various emissions associated with the high-speed railway system throughout its life cycle, except for particulate matter (PM), are significantly lower than those emitted by passenger cars [18]

Another important study conducted a comprehensive assessment of the life cycle sustainability of a light rail transportation system in Kayseri, Turkey. The assessment incorporated environmental, economic, and social aspects to provide a holistic understanding of the system's sustainable performance. For the environmental assessment, a Life Cycle Assessment (LCA) approach was employed using SimaPro 8.4.1 software, in accordance with ISO 14040 and 14044 standards. Nine categories of environmental impact were considered to assess the environmental performance of the light rail transportation system, with a functional unit of 1 passenger-kilometer. The results indicated that the global warming potential and abiotic depletion potential of the light rail transportation system per passenger-kilometer were 2.4×10^{-2} kg CO_2 eq. and 2.7×10^{-1} MJ, respectively, considering a 50-year lifespan [13]

Economic assessment was conducted through life cycle costing, using the functional unit of 1 US dollar per 1 passenger-kilometer. The total life cycle cost of the light rail transportation system was calculated as 0.046 USD per 1 passenger-kilometer. The results highlighted that energy cost was the primary contributor to the total life cycle cost, accounting for about 92% of the total cost. Regarding the social performance assessment, the study identified that the light rail transportation industry demonstrates good performance in terms of benefits for society, the local community, and workers. However, social performance in relation to consumers was evaluated as weaker due to a less effective feedback mechanism [13]

In summary, life cycle assessment (LCA) studies of greenhouse gas emissions in railway systems are essential for promoting the transition to more sustainable transportation systems, enabling the adoption of informed strategies to reduce environmental impacts, optimize efficiency, and promote environmentally friendly practices.

5.2. Calculation of Carbon Emission Factors Based on Distance Traveled and Mode of Transportation

The calculation of carbon emission factors based on distance traveled and mode of transportation is a fundamental approach to estimating GHG emissions associated with transportation activities. This methodology involves determining the amount of CO_2 equivalent emitted per unit of distance traveled in different modes of transportation, enabling a direct assessment of the environmental impact of these activities [74].

First and foremost, the distance traveled is a key parameter in this calculation, representing the length of the journey undertaken. This distance can vary from short urban trips to longer routes between cities or regions. Next, the mode of transportation used is also a crucial factor, as different transportation modes have different energy efficiencies and consequently different GHG emissions. Modes such as gasoline, diesel, electric vehicles, or rail and road public transport have different emission pattern [63].

Carbon emission factors are expressed in terms of the mass of CO_2 equivalent emitted per unit of distance, such as kilometers traveled. These factors can be obtained through measurement data, fuel consumption analyses, direct emissions, or life cycle modeling, depending on the desired accuracy [75].

The combination of these elements—distance traveled and mode of transportation—results in a quantitative estimate of GHG emissions related to transportation activities. These calculations are vital for understanding the environmental impact of different mobility options, aiding in the formulation of sustainable policies, smart urban planning, and individual transportation choices that contribute to carbon emissions reduction and climate change mitigation [9,76].

In a study in Baoji, the assessment of the environmental impact of urban rail transportation was addressed, focusing on carbon emissions reduction. The study uses the prediction of passenger demand in the city of Baoji in China as a basis to analyze the environmental benefits of the rail transportation system. Scenarios of travel are outlined considering the use of rail transportation and its absence, taking into account the extent of demand generated by different modes of transportation. Carbon emissions from trips are calculated by multiplying the per capita emission of each mode of transportation by the total number of passengers on the target routes. The study highlights the importance of low-carbon urban rail transportation as a sustainable alternative to reduce carbon emissions related to urban transport [73].

A study on low-carbon in Beijing discussed the creation and use of software designed to calculate energy consumption and greenhouse gas emissions in the context of transportation. This software is developed based on research and principles of low-carbon transportation in Beijing, with the main objective of providing a comprehensive tool to assess the environmental impact of various modes of transportation [77].

The process of software development is outlined, including the incorporation of data related to vehicle fleets, energy sources, distances traveled, and other relevant factors. The authors emphasize the flexibility of the software, allowing customization based on specific urban contexts, transportation systems, and policy objectives. It also considers real-time data, increasing its applicability to assess dynamic transportation scenarios. The application of the software is demonstrated through case studies and simulations, focusing on various modes of transportation in the metropolitan area of Beijing. These studies illustrate how the tool can assess the impact of different transportation policies and interventions, providing insights into emission reduction and energy efficiency improvements. This can guide urban planners, policymakers, and researchers in identifying strategies to promote sustainable and low-carbon transportation systems [77].

Furthermore, the article discusses the potential of the developed software to contribute to broader urban sustainability goals, such as reducing traffic congestion, improving air quality, and enhancing overall quality of life. It underscores the importance of integrating energy consumption and emissions analysis into urban planning and transportation management [77].

These studies are important for understanding the environmental impact of the operational phase of urban rail systems and for assessing GHG emissions over time. However, it is crucial to consider that this approach may not capture all emissions associated with the complete life cycle of the system, such as the manufacturing, construction, and maintenance of rail infrastructures. Therefore, a comprehensive analysis of carbon emissions may require the combination of different approaches and the consideration of all stages of the life cycle.

6. The Measurement of Carbon Emissions in Transportation Faces Complexities and Limitations

The life cycle assessment (LCA) is an essential approach to understanding the environmental impact of systems and products throughout all stages of their life cycle, from raw material extraction to disposal. However, the full application of LCA to the railway system may face significant challenges due to the lack of complete and detailed data, especially in the construction stage of railways [78].

The lack of comprehensive data on the construction of older railways can create a gap in the life cycle analysis, particularly when it comes to calculating greenhouse gas emissions (GHG) associated with the construction phase. Many studies that aim to assess the environmental impact of railways might focus on operational and maintenance phases where data is more accessible. The absence of detailed construction data can hinder the accurate estimation of GHG emissions in the construction stage, which, in turn, can lead to an incomplete understanding of the overall impact of railways on climate change. This can be an obstacle to obtaining a holistic view of the life cycle and conducting accurate comparisons between different modes of transportation or sustainability interventions [79].

To address this limitation, researchers can adopt strategies such as using indirect data, modeling, and simulation to estimate GHG emissions in the construction phase. Furthermore, it's important to acknowledge that LCA is not limited solely to the construction phase; operational and maintenance phases also play a crucial role in evaluating the environmental impact of railway systems. Although the lack of comprehensive construction data may present a challenge, it's essential to consider that LCA can still provide valuable insights into the sustainability of railways throughout their lifespan. Focusing on stages where data is available, along with creative approaches to fill information gaps, can enable researchers to gain a more complete understanding of the environmental impact of railways, even when construction data is limited [22].

In the research conducted on the railway system in Pernambuco, the assessment of greenhouse gas emissions (GHG) exclusively focused on the maintenance and operational phases. This is due to the fact that, due to the age of the railway, construction-related data was no longer available. As a result, the life cycle of GHG emissions ends up being incomplete in this aspect [61]. Previous studies, such as those addressing gas emissions in the Shanghai metro, indicate that the portion of GHG emissions associated with the construction phase is not as significant compared to operational and maintenance phases. In many cases, this portion accounts for only about 5% of the entire GHG emissions cycle of the system. Therefore, while the life cycle assessment (LCA) approach may not be fully comprehensive due to the lack of construction data, this limitation does not seem to have a substantial impact on the final results [64].

This suggests that while the lack of construction data may create a limitation in the comprehensive life cycle analysis of GHG emissions, the focus on operational and maintenance phases, which generally have a larger contribution to the total emissions, still offers valuable insights into the sustainability of the railway system. Therefore, even though the lack of comprehensive construction data may influence the LCA approach, its impact on the final results is relatively limited, especially when considering the proportion of emissions from different life cycle phases.

The quality and comprehensiveness of data used in calculating greenhouse gas emissions (GHG) in the railway system play a fundamental role in the reliability of results obtained by assessment models. The accuracy of estimates depends directly on the availability of reliable and detailed data sources related to the railway transport sector (Manzo et al., 2018). Collecting reliable data is essential to ensure that the calculation model accurately reflects the real conditions of the railway system. Detailed, up-to-date, and comprehensive statistics are necessary to provide a complete view of railway operations, including factors such as distance traveled, type of locomotive used, energy consumption, and operational characteristics. Otherwise, estimates can be inaccurate and either underestimate or overestimate actual emissions [22,78,80].

Furthermore, it is crucial for data sources to be transparent and verifiable. The lack of transparency or unavailability of data used in the calculation can raise questions about the credibility of the results obtained. Obtaining data directly from responsible railway authorities or trustworthy research institutions can enhance the reliability of estimates. Another challenge is ensuring the consistency and updating of data over time. The railway transport sector is subject to changes, such as the introduction of more efficient technologies or the expansion of networks. Therefore, calculation models should be periodically reviewed and adjusted based on updated data to maintain the accuracy of estimates [63].

In a study conducted in China, which presents a comprehensive analysis of the transportation system and evaluates the effectiveness of current strategies aimed at sustainable transportation development, the authors use modeling techniques to assess various transportation scenarios. These scenarios consider factors such as modal choice, infrastructure development, and policy interventions. By simulating different strategies, they analyze their potential impacts on energy consumption, emission reduction, and overall transportation efficiency.

In the study conducted in China, significant findings related to oil consumption in the Chinese national transportation system were identified. They observed that oil consumption in the national transportation system accounted for 57% of the country's total oil consumption, a much higher proportion compared to the official statistics of 38%. This phenomenon was mainly attributed to the discrepancy between the Chinese energy statistics system and international standards. A relevant explanation for this discrepancy is related to the differences in data collection and calculation methods adopted by the Chinese energy statistics system compared to international standards. The variation in energy accounting systems can result in substantial differences in reported numbers of oil consumption in the transportation sector, which, in turn, affects the accurate understanding of this sector's contribution to total energy consumption and greenhouse gas emissions [81].

In the computational models used to estimate carbon emissions, many of these computational models have limitations in terms of dynamism and adaptability, as many components related to carbon emissions are defined statically. This characteristic can result in inaccuracies in emission calculations over time and in different contexts. Computational models used to estimate emissions should be developed with more dynamism and flexibility, allowing fluctuations in the transportation system over time to be captured more accurately in order to better reflect the operational reality of the transportation system [82].

7. Trends in Research for Greenhouse Gas Emissions Mitigation in the Railway System

As trends in greenhouse gas (GHG) emissions mitigation research within the railway system are evolving towards increasingly comprehensive and innovative approaches as awareness of climate change and the importance of reducing GHG emissions grows, researchers are exploring a variety of areas and strategies to enhance the sustainability of the railway system. This movement reflects the growing need to find innovative and comprehensive solutions to address environmental challenges, underscoring a commitment to building a more environmentally responsible railway system.

A significant trend is the transition to electrified railway systems or other systems with more sustainable fuels, reducing dependence on fossil fuels. This involves adopting renewable energy sources, such as electricity from wind, solar, hydro, or fuels such as hydrogen and biodiesel to power trains. Shifting to a more sustainable fuel helps reduce GHG emissions by eliminating direct emissions from trains and optimizing energy efficiency [83].

A study in the railway system of northeastern Brazil addressed the application of solar photovoltaic systems on light rail vehicles (LRVs). The study explored the feasibility and benefits of incorporating solar technology to assist in the electric power supply of LRVs. The research considers the scenario of sustainable public transportation and the potential advantages of solar energy generation to reduce dependence on conventional sources and mitigate GHG emissions [84].

Another energy efficiency study presented a strategy to optimize water and energy use in public buildings, using a Brazilian railway company as a case study. The study addresses the importance of resource efficiency in public facilities, considering both cost reduction and environmental benefits. The research details the approach taken, including the implementation of measures such as rainwater harvesting, the installation of water-saving devices, and the use of renewable energy systems. The specific case study in the Brazilian railway company demonstrates the results obtained from the implementation of these measures, both in terms of financial savings and the reduction of natural resource consumption and GHG emissions [85].

Studies on the use of hydrogen in trains have gained prominence due to the search for cleaner and more sustainable transportation solutions. Hydrogen is considered a promising energy source for trains, as its combustion or conversion into electricity through fuel cells produces only water as a byproduct, eliminating pollutant emissions and GHGs. Overall, studies on hydrogen use in trains aim to provide comprehensive information on the technical, environmental, and economic feasibility of this energy alternative, contributing to informed decision-making by transport authorities and promoting the transition to more sustainable railway systems [83].

A study on urban light hydrogen-powered rails explores the modeling, analysis, and practical implementation of an urban light rail powered by hydrogen fuel cells. The study focuses on exploring the feasibility and potential benefits of using hydrogen as an alternative energy source for urban rail transport. The study discusses various scenarios, considering different methods and hydrogen production technologies. This includes assessing emissions throughout the lifecycle and energy consumption of the entire system, from hydrogen production to train operation. The authors also consider the economic viability of implementing such a system, taking into account factors such as initial investment costs, operational expenses, and potential benefits in terms of reduced emissions and energy efficiency [86].

In conclusion, this article presents a comprehensive analysis of implementing a light urban hydrogen-powered rail system. By addressing technical, economic, and environmental aspects, it provides valuable insights into the feasibility and potential benefits of adopting this system. The study's modeling, analysis, and practical considerations contribute to advancing sustainable transportation solutions and encourage further research and development in the field of hydrogen fuel cell technology for urban rail systems [86].

Practical implementation is a crucial aspect as it aims to bridge the gap between theoretical analysis and real-world application. Exploring the technical challenges associated with integrating hydrogen fuel cells into existing rail systems, ensuring safety, reliability, and efficiency. Discussing potential barriers and regulatory considerations that need to be addressed for successful implementation is important for the evolution of sustainable railway transportation. The potential benefits of hydrogen rail systems, including reduced greenhouse gas emissions, improved air quality, and reduced dependence on fossil fuels, are also discussed [86,87].

Another significant research involving hydrogen in the railway system addresses the use of dual-fuel diesel engines operating with diesel injection and the introduction of fumigants such as hydrogen, gasoline, and ethanol. The main objective is to investigate how different diesel injection timing affects efficiency and emissions during multi-fuel combustion. The results show how varying the timing of diesel injection influences combustion efficiency, heat release rate, and emissions. These findings can offer valuable insights for optimizing the operation of dual-fuel engines, contributing to the reduction of pollutant emissions and the search for cleaner and more efficient alternatives in the transportation sector [88].

The use of biodiesel in the railway system is an approach aimed at partially or fully replacing conventional petroleum-derived diesel with a renewable and more sustainable fuel. Biodiesel is a biofuel produced from renewable sources such as vegetable oils, animal fats, or food waste through transesterification processes. Adopting biodiesel in the railway system aims to reduce pollutant emissions and contribute to climate change mitigation, as biodiesel tends to generate fewer carbon dioxide (CO_2) emissions and other atmospheric pollutants compared to fossil diesel. Additionally, biodiesel is biodegradable and produces less particulate matter, contributing to improving air quality in urban areas [20,89,90].

However, implementing biodiesel in the railway system also faces technical and logistical challenges. Ensuring that train engines are compatible with biodiesel blends is necessary, as biodiesel has different combustion properties compared to conventional diesel. Additionally, the availability and quality of biodiesel can vary, which can affect the efficiency of the railway system [91].

Ongoing studies and research are assessing the impacts of biodiesel use in the railway system, such as evaluating nitrogen oxide (NOx) and nitrogen dioxide (NO_2) emissions during the cold start of a diesel engine when fueled with different proportions of diesel and biodiesel blends. The main aim of the study is to investigate how biodiesel blends affect NO_2 and NO_x emissions during the cold start phase of a diesel engine. Cold start is a critical period in terms of emissions, as engine temperatures are low and combustion is not ideal, which can lead to increased pollutant emissions. The results indicate how different biodiesel blend ratios affect NO_2 and NO_x emissions, highlighting impacts during cold start. This can provide valuable insights for the automotive industry and environmental policy formulation, aiming to reduce pollutant emissions and develop cleaner and more efficient fuels [92].

Another biodiesel research investigates the spray behavior of a common rail diesel injector using a blend of biodiesel with 5% gasoline. The analysis provides information about the spray pattern and how different operational factors impact this process, contributing to a comprehensive understanding of the use of biofuel blends in diesel engines. The results offer insights into the angle, shape, and dispersion of spray droplets, considering different operating conditions such as injection pressure and flow velocity [93].

Researchers are also focusing on improving the energy efficiency of railway systems, such as designing lighter and more aerodynamic trains, energy regeneration systems during braking, and route optimization to reduce energy consumption [30,94,95].

A study addresses computational simulation of the contact response between wheels and tracks in a straight rail scenario. The research focuses on analyzing the behavior of wheel-track contact, taking into consideration different variables such as material characteristics, component geometry, and applied forces. The study utilizes dynamic simulations to understand how forces, deformations, and pressure distribution occur during the contact between wheels and tracks. The simulation results provide valuable information about the distribution of forces and stresses at the contact point, allowing for a deeper understanding of the system's performance and dynamic behavior. This has significant implications for the design and maintenance of railway systems, contributing to operational optimization and the reduction of greenhouse gas emissions [96].

In railway systems, real-time traffic management allows for monitoring and controlling train movement in real-time, identifying potential congestion, delays, or operational issues. This immediate responsiveness enables corrective measures to be adopted, such as adjusting train speed, optimizing stop schedules, and coordinating traffic flow, thus minimizing delays and improving service regularity. The integration of intelligent operation systems into railway systems offers a range of notable advantages. These systems can optimize energy efficiency by determining the best way to use available energy and intelligently managing power distribution. This is particularly valuable in electrified railway systems, where electrification is common, contributing to reduced energy consumption and pollutant emissions [30,97].

Effective integration among different modes of transportation (rail, road, cycling, etc.) allows for a more holistic approach to transportation system planning, reducing overall emissions by promoting more sustainable options.

A study involving an inference model for passenger itineraries in congested urban rail networks employs data from ticketing and train schedule information to estimate the routes and itineraries followed by passengers in congested urban rail transportation systems. Through the analysis of ticketing data and train schedules, the proposed model allows for estimating the most frequently used passenger routes, identifying major traffic flows and congestion points. This information can be used to develop congestion mitigation strategies, such as train redistribution, schedule adjustments, or the implementation of measures to avoid bottlenecks at key stations. This contributes to improving passenger experience and optimizing operations during high-demand situations, aiding in greenhouse gas emissions reduction [60].

In summary, research trends in greenhouse gas emissions mitigation within the railway system reflect a comprehensive approach involving electrification, energy efficiency, multimodal integration, sustainable technologies, and environmentally friendly policies. These trends aim to make railway transportation a more sustainable option and contribute to reducing greenhouse gas emissions in the transportation sector as a whole.

8. Conclusions

The approach of this research focuses on advancements in studies related to carbon emissions measurement in the railway transportation sector, providing a comprehensive analysis of existing research. Currently, there is maturity in analyses that seek to assess the potential reduction of carbon emissions in railway transportation, as well as in defining research scopes, utilizing measurement methods, and developing models. However, this maturity coexists with challenges posed by the inherent complexity of the railway environment, including electricity generation, urban rail expansion, system longevity, and interactions with other transportation modes in a multimodal context.

Within this complex scenario, ensuring the accuracy of carbon emissions measurements emerges as a critical aspect. Appropriate determination of carbon emission factors, judicious selection of measurement steps, and verification of the accuracy and validity of employed data are crucial aspects that need to be considered in future research. The application of life cycle assessment (LCA) methodology in the context of railway emissions is explored, and several articles are used as examples to illustrate this process. Despite the progress made, unresolved issues still remain. For instance, even though the use of LCA has become more systematic and refined, many studies fail to encompass the entire railway life cycle, especially in the initial construction phase.

Measuring carbon emissions in urban railway transportation systems presents multiple challenges and considerations. The precision and comprehensiveness of the required data for these measurements often encounter obstacles, and the very definition of carbon emissions can be ambiguous, complicating the task of accurately and simultaneously measuring emissions from this urban transport.

In summary, despite notable advancements, this research underscores that the intrinsic complexity of the railway transport context and the need for accurate measurements remain significant challenges in carbon emissions investigations within this sector. The application of LCA methodology offers a valuable framework, but the absence of data in certain life cycle phases and the ambiguity in emission definitions represent challenges to overcome in achieving a thorough and accurate understanding of the environmental impact of railway transportation.

To guide future research, it is crucial to direct efforts towards several key areas to enhance the understanding and approach to carbon emissions within the urban railway transportation system:

1. Data Collection and Quality Enhancement: An essential focus should be to ensure the accuracy and comprehensiveness of data detection and collection necessary for analyzing carbon emissions in urban railway transportation systems. This requires improved data acquisition methods and a more rigorous approach to data verification and validation, to enhance the overall quality of estimations.
2. Implementation of Monitoring Systems: Establishing a digital system that tracks greenhouse gas (GHG) emissions throughout the entire life cycle of the railway transportation system is fundamental. This digital twin system can integrate data from various life cycle phases, offering a holistic model and facilitating visualization and monitoring of railway transportation GHG emission measurements.
3. Integration of Multimodal Transport Modes: Most current studies focus on isolated analysis of urban railway traffic, often separating it from multimodal transportation systems. Thus, it is crucial for research to advance towards a more comprehensive analysis, considering the integration of railway transportation with other transportation modes in integrated urban systems, which can more realistically reflect the dynamics of carbon emissions.

Directing research efforts according to the outlined points will lead to significant advancements in the understanding and measurement of carbon emissions in the urban railway transportation system. This will provide essential information for well-founded and effective public policy formulation, contributing to emission reduction, climate change mitigation, and the promotion of more sustainable transportation. This approach will also have a significant impact on improving the quality of life for future generations.

Author Contributions: Conceptualization, D.d.F.-S. and A.F.R.-R.; methodology, D.d.F.-S.; software, D.d.F.-S.; validation, D.d.F.-S. and A.F.R.-R.; formal analysis, D.d.F.-S.; investigation, D.d.F.-S.; resources, D.d.F.-S.; data curation, D.d.F.-S.; writing—original draft preparation, D.d.F.-S.; writing—review and editing, A.F.R.-R., J.D.G. and S.A.E.; visualization, A.F.R.-R., J.D.G. and S.A.E.; supervision, A.F.R.-R., J.D.G. and S.A.E.; project administration, A.F.R.-R., J.D.G. and S.A.E. All authors have read and agreed to the published version of the manuscript.

Funding: This research received no external funding.

Data Availability Statement: The data presented in this study are available in this paper.

Conflicts of Interest: Author Diogo Da Fonseca-Soares was employed by the company Brazilian Company of Urban Trains. The remaining authors declare that the research was conducted in the absence of any commercial or financial relationships that could be construed as a potential conflict of interest.

References

1. Tian, Y.; Fleitmann, D.; Zhang, Q.; Sha, L.; Wassenburg, J.A.; Axelsson, J.; Zhang, H.; Li, X.; Hu, J.; Li, H.; et al. Holocene climate change in southern Oman deciphered by speleothem records and climate model simulations. *Nat. Commun.* **2023**, *14*, 4718. [CrossRef]
2. Kwakwa, P.A.; Adjei-Mantey, K.; Adusah-Poku, F. The effect of transport services and ICTs on carbon dioxide emissions in South Africa. *Environ. Sci. Pollut. Res.* **2023**, *30*, 10457–10468. [CrossRef]
3. Ogunkunbi, G.A.; Meszaros, F. Preferences for policy measures to regulate urban vehicle access for climate change mitigation. *Environ. Sci. Eur.* **2023**, *35*, 42. [CrossRef]
4. Christian, J.I.; Martin, E.R.; Basara, J.B.; Furtado, J.C.; Otkin, J.A.; Lowman, L.E.L.; Hunt, E.D.; Mishra, V.; Xiao, X. Global projections of flash drought show increased risk in a warming climate. *Commun. Earth Environ.* **2023**, *4*, 165. [CrossRef]
5. Sasse, J.P.; Trutnevyte, E. A low-carbon electricity sector in Europe risks sustaining regional inequalities in benefits and vulnerabilities. *Nat. Commun.* **2023**, *14*, 2205. [CrossRef]
6. Poppe Terán, C.; Naz, B.S.; Graf, A.; Qu, Y.; Hendricks Franssen, H.J.; Baatz, R.; Ciais, P.; Vereecken, H. Rising water-use efficiency in European grasslands is driven by increased primary production. *Commun. Earth Environ.* **2023**, *4*, 95. [CrossRef]
7. Nainggolan, D.; Abay, A.T.; Christensen, J.H.; Termansen, M. The impact of climate change on crop mix shift in the Nordic region. *Sci. Rep.* **2023**, *13*, 2962. [CrossRef] [PubMed]
8. Praticò, F.G.; Fedele, R. Economic Sustainability of High–Speed and High–Capacity Railways. *Sustainability* **2023**, *15*, 725. [CrossRef]
9. Aminzadegan, S.; Shahriari, M.; Mehranfar, F.; Abramović, B. Factors affecting the emission of pollutants in different types of transportation: A literature review. *Energy Rep.* **2022**, *8*, 2508–2529. [CrossRef]
10. Fageda, X. Do light rail systems reduce traffic externalities? Empirical evidence from mid-size european cities. *Transp. Res. Part D Transp. Environ.* **2021**, *92*, 102731. [CrossRef]
11. Chen, Z. Socioeconomic Impacts of high-speed rail: A bibliometric analysis. *Socio-Econ. Plan. Sci.* **2022**, *85*, 101265. [CrossRef]
12. Ren, X.; Chen, Z.; Wang, F.; Dan, T.; Wang, W.; Guo, X.; Liu, C. Impact of high-speed rail on social equity in China: Evidence from a mode choice survey. *Transp. Res. Part A Policy Pract.* **2020**, *138*, 422–441. [CrossRef]
13. Gulcimen, S.; Aydogan, E.K.; Uzal, N. Life cycle sustainability assessment of a light rail transit system: Integration of environmental, economic, and social impacts. *Integr. Environ. Assess. Manag.* **2021**, *17*, 1070–1082. [CrossRef]
14. Dolinayova, A.; Kanis, J.; Loch, M. Social and Economic Efficiency of Operation Dependent and Independent Traction in Rail Freight. *Procedia Eng.* **2016**, *134*, 187–195. [CrossRef]
15. Hausberger, L.; Cordes, T.; Gschösser, F. Life Cycle Assessment of High-Performance Railway Infrastructure, Analysis of Superstructures in Tunnels and on Open Tracks. *Sustainability* **2023**, *15*, 7064. [CrossRef]
16. Keiser, D.; Schnoor, L.H.; Pupkes, B.; Freitag, M. Life cycle assessment in aviation: A systematic literature review of applications, methodological approaches and challenges. *J. Air Transp. Manag.* **2023**, *110*, 102418. [CrossRef]
17. Da Fonseca-Soares, D.; Galvinicio, J.D.; Eliziário, S.A.; Ramos-Ridao, A.F. A Bibliometric Analysis of the Trends and Characteristics of Railway Research. *Sustainability* **2022**, *14*, 13956. [CrossRef]
18. Chipindula, J.; Du, H.B.; Botlaguduru, V.S.V.; Choe, D.; Kommalapati, R.R. Life cycle environmental impact of a high-speed rail system in the Houston-Dallas I-45 corridor. *Public Transp.* **2021**, *14*, 481–501. [CrossRef]
19. Ramos da Silva, T.; Moura, B.; Monteiro, H. Life Cycle Assessment of Current Portuguese Railway and Future Decarbonization Scenarios. *Sustainability* **2023**, *15*, 11355. [CrossRef]
20. Leichter, M.; Hackenhaar, I.; Passuello, A. Public Bus Transportation System Environmental Impact Projections Regarding Different Policy Scenarios—A LCA Study. *Infrastructures* **2021**, *6*, 169. [CrossRef]

21. De Bortoli, A.; Bouhaya, L.; Feraille, A. A life cycle model for high-speed rail infrastructure: Environmental inventories and assessment of the Tours-Bordeaux railway in France. *Int. J. Life Cycle Assess.* **2020**, *25*, 814–830. [CrossRef]
22. Merchan, A.L.; Belboom, S.; Léonard, A. Life cycle assessment of rail freight transport in Belgium. *Clean Technol. Environ. Policy* **2020**, *22*, 1109–1131. [CrossRef]
23. Gebler, M.; Cerdas, J.F.; Thiede, S.; Herrmann, C. Life cycle assessment of an automotive factory: Identifying challenges for the decarbonization of automotive production—A case study. *J. Clean. Prod.* **2020**, *270*, 122330. [CrossRef]
24. Kaewunruen, S.; Sresakoolchai, J.; Peng, J. Life cycle cost, energy and carbon assessments of Beijing-Shanghai high-speed railway. *Sustainability* **2020**, *12*, 206. [CrossRef]
25. Trevisan, L.; Bordignon, M. Screening Life Cycle Assessment to compare CO2 and Greenhouse Gases emissions of air, road, and rail transport: An exploratory study. *Procedia CIRP* **2020**, *90*, 303–309. [CrossRef]
26. Kumar, S.; Marrone, M.; Liu, Q.; Pandey, N. Twenty years of the International Journal of Accounting Information Systems: A bibliometric analysis. *Int. J. Account. Inf. Syst.* **2020**, *39*, 100488. [CrossRef]
27. Torraco, R.J. Writing Integrative Literature Reviews: Guidelines and Examples. *Hum. Resour. Dev. Rev.* **2005**, *4*, 356–367. [CrossRef]
28. Ciani, L.; Guidi, G.; Patrizi, G. Human reliability in railway engineering: Literature review and bibliometric analysis of the last two decades. *Saf. Sci.* **2022**, *151*, 105755. [CrossRef]
29. Radu, V.; Radu, F.; Tabirca, A.I.; Saplacan, S.I.; Lile, R. Bibliometric Analysis of Fuzzy Logic Research in International Scientific Databases. *Int. J. Comput. Commun. Control.* **2021**, *16*. [CrossRef]
30. Kostrzewski, M.; Melnik, R. Condition monitoring of rail transport systems: A bibliometric performance analysis and systematic literature review. *Sensors* **2021**, *21*, 4710. [CrossRef] [PubMed]
31. Wang, X.; Xu, Z.; Škare, M.A. bibliometric analysis of Economic Research-Ekonomska Istraživanja (2007–2019). *Econ. Res. Ekon. Istraz.* **2020**, *33*, 865–886. [CrossRef]
32. Merigó, J.M.; Pedrycz, W.; Weber, R.; de la Sotta, C. Fifty years of Information Sciences: A bibliometric overview. *Inf. Sci.* **2018**, *432*, 245–268. [CrossRef]
33. Munik, J.; Moura, L.F.; Treinta, F.T.; Cestari, J.M.A.P.; Deschamps, F.; Lima, E.P.D.; Costa, S.E.G. Bibliometric analysis of performance measurement systems in Non-Profit organizations: Cocitation analysis. In Proceedings of the 2017 International Annual Conference of the American Society for Engineering Management, APSEM 2017, Huntsville, AL, USA, 18–21 October 2017.
34. Qaiser, F.H.; Ahmed, K.; Sykora, M.; Choudhary, A.; Simpson, M. Decision support systems for sustainable logistics: A review & bibliometric analysis. *Ind. Manag. Data Syst.* **2017**, *117*, 1376–1388.
35. Jasti, P.C.; Vinayaka Ram, V. Estimation of CO_2 emission savings from a metro rail system using different methodologies: A case study of Mumbai, India. *Eur. Transp. Trasp. Eur.* **2021**, *81*.
36. Di Gangi, M.; Russo, F. Design of Hybrid Rail Services on Conventional and High-Speed Lines. *Int. J. Transp. Dev. Integr.* **2023**, *7*, 113–121. [CrossRef]
37. Liljenstrom, C.; Toller, S.; Akerman, J.; Bjorklund, A. Annual climate impact and primary energy use of Swedish transport infrastructure. *Eur. J. Transp. Infrastruct. Res.* **2019**, *19*, 77. [CrossRef]
38. Makarchuk, B.; Saxe, S. Temporal Assessment of the Embodied Greenhouse Gas Emissions of a Toronto Streetcar Line. *J. Infrastruct. Syst.* **2019**, *25*, 11. [CrossRef]
39. Chen, S.; Wu, J.; Zong, Y. The Impact of the Freight Transport Modal Shift Policy on China's Carbon Emissions Reduction. *Sustainability* **2020**, *12*, 583. [CrossRef]
40. Saxe, S.; Miller, E.; Guthrie, P. The net greenhouse gas impact of the Sheppard Subway Line. *Transp. Res. Part D Transp. Environ.* **2017**, *51*, 261–275. [CrossRef]
41. Kim, S.H.; Kim, J.H.; Chun, H.Y.; Sharman, R.D. Global response of upper-level aviation turbulence from various sources to climate change. *Npj Clim. Atmos. Sci.* **2023**, *6*, 92. [CrossRef]
42. Russo, F.; Musolino, G. (Eds.) Methodologies for Sustainable Development of TEN-T/RFC Corridors and Core Ports: Economic Impacts Generated in Port-Related Areas. In *Computational Science and Its Applications—ICCSA 2023 Workshops*; Springer Nature: Cham, Switzerland, 2023.
43. Russo, F.; Sgro, D.; Musolino, G. (Eds.) Sustainable Development of Railway Corridors: Methods and Models for High Speed Rail (HSR) Demand Analysis. In *Computational Science and Its Applications—ICCSA 2023 Workshops*; Springer Nature: Cham, Switzerland, 2023.
44. Han, S.; Wang, X.; Yang, C.; Xie, G.; Qiu, Z.; Wang, C. Optimization and Performance Analysis of Rail-Train Coupling System with Inerters. *Complexity* **2021**, *2021*, 9974803. [CrossRef]
45. Sekasi, J.; Martens, M.L. Assessing the contributions of urban light rail transit to the sustainable development of addis ababa. *Sustainability* **2021**, *13*, 5667. [CrossRef]
46. Chen, F.; Shen, X.; Wang, Z.; Yang, Y. An Evaluation of the Low-Carbon Effects of Urban Rail Based on Mode Shifts. *Sustainability* **2017**, *9*, 401. [CrossRef]
47. Xu, X.; Kent, S.; Schmid, F. Carbon-reduction potential of electrification on China's railway transport: An analysis of three possible future scenarios. *Proc. Inst. Mech. Eng. Part F J. Rail Rapid Transit* **2020**, *235*, 226–235. [CrossRef]

48. Lu, H.; Liu, Y.; Ali, A.; Tian, R.; Chen, Y.; Luo, Y. Empirical analysis of the impact of China–Japan–South Korea transportation infrastructure investment on environmental degradation and the validity of the environmental Kuznets curve hypothesis. *Front. Psychol.* **2022**, *13*, 977466. [CrossRef]
49. Hu, Y.; Yang, J.; Hu, N. Experimental study and optimization in the layouts and the structure of the high-pressure common-rail fuel injection system for a marine diesel engine. *Int. J. Engine Res.* **2021**, *22*, 1850–1871. [CrossRef]
50. Duraisamy, K.; Ismailgani, R.; Paramasivam, S.A.; Kaliyaperumal, G.; Dillikannan, D. Emission profiling of a common rail direct injection diesel engine fueled with hydrocarbon fuel extracted from waste high density polyethylene as a partial replacement for diesel with some modifications. *Energy Environ.* **2021**, *32*, 481–505. [CrossRef]
51. Zhu, Q.; Chen, G.X.; Kang, X.; Dong, B.J. Study on the Effect of the Modeling Method of Railway Fastener on Rail Corrugation Prediction Model. *Tribol. Trans.* **2021**, *65*, 180–191. [CrossRef]
52. Sinha, M. Harnessing land value capture: Perspectives from India's urban rail corridors. *Land USE Policy* **2021**, *108*, 105526. [CrossRef]
53. Rudin-Brown, C.M.; Harris, S.; Rosberg, A. How shift scheduling practices contribute to fatigue amongst freight rail operating employees: Findings from Canadian accident investigation. *Accid. Anal. Prev.* **2019**, *126*, 64–69. [CrossRef]
54. Wu, D.; Ding, W.; Guo, F.; Ma, L. Effects of harmonic wear of wheels on creep characteristics of a wheel-rail system. *Zhendong Yu Chongji/J. Vib. Shock* **2021**, *40*, 1–9.
55. Yang, Y.; Liu, Z.; Gao, X.; Ling, L.; Wang, K.; Zhai, W. Analysis on Essential Characteristics of the Polygonal Wear of Locomotive Wheels and Its Effect on Wheel/Rail Dynamic Impact. *Jixie Gongcheng Xuebao/J. Mech. Eng.* **2021**, *57*, 130–139.
56. Bai, Y.; Lan, Q.; Fan, L.; Ma, X.; Liu, H. Investigation on the fuel injection stability of high pressure common rail system for diesel engines. *Int. J. Engine Res.* **2021**, *22*, 616–631. [CrossRef]
57. Wang, Z.; Lei, Z. Analysis of influence factors of rail corrugation in small radius curve track. *Mech. Sci.* **2021**, *12*, 31–40. [CrossRef]
58. Guo, F.; Tang, Y.; Ren, L.; Li, J. A novel permanent magnetic rail for HTS levitation propulsion system. *Phys. C Supercond. Its Appl.* **2009**, *469*, 1825–1828. [CrossRef]
59. Yu, W.; Wang, T.; Xiao, Y.; Chen, J.; Yan, X.A. Carbon Emission Measurement Method for Individual Travel Based on Transportation Big Data: The Case of Nanjing Metro. *Int. J. Environ. Res. Public Health* **2020**, *17*, 5957. [CrossRef]
60. Zhu, Y.; Koutsopoulos, H.N.; Wilson, N.H.M. Passenger itinerary inference model for congested urban rail networks. *Transp. Res. Part C Emerg. Technol.* **2021**, *123*, 102896. [CrossRef]
61. Da Fonseca-Soares, D.; Eliziário, S.A.; Galvinicio, J.D.; Ramos-Ridao, A.F. Life-Cycle Greenhouse Gas (GHG) Emissions Calculation for Urban Rail Transit Systems: The Case of Pernambuco Metro. *Appl. Sci.* **2023**, *13*, 8965. [CrossRef]
62. Wei, P.; Pan, H. Research on individual carbon dioxide emissions of commuting in peri-urban area of metropolitan cities—An empirical study in Shanghai. *Transp. Res. Procedia* **2017**, *25*, 3459–3478. [CrossRef]
63. Lee, J.Y.; Lee, C.K.; Chun, Y.Y. Greenhouse gas emissions from high-speed rail infrastructure construction in Korea. *Transp. Res. Part D-Transp. Environ.* **2020**, *87*, 12. [CrossRef]
64. Li, Y.; He, Q.; Luo, X.; Zhang, Y.R.; Dong, L. Calculation of life-cycle greenhouse gas emissions of urban rail transit systems: A case study of Shanghai Metro. *Resour. Conserv. Recycl.* **2018**, *128*, 451–457. [CrossRef]
65. Gassner, A.; Lederer, J.; Kanitschar, G.; Ossberger, M.; Fellner, J. Extended ecological footprint for different modes of urban public transport: The case of Vienna, Austria. *Land USE Policy* **2018**, *72*, 85–99. [CrossRef]
66. Houston, D.; Dang, A.; Wu, J.; Chowdhury, Z.; Edwards, R. The cost of convenience; Air pollution and noise on freeway and arterial light rail station platforms in Los Angeles. *Transp. Res. Part D Transp. Environ.* **2016**, *49*, 127–137. [CrossRef]
67. Kimball, M.; Chester, M.; Gino, C.; Reyna, J. Assessing the Potential for Reducing Life-Cycle Environmental Impacts through Transit-Oriented Development Infill along Existing Light Rail in Phoenix. *J. Plan. Educ. Res.* **2013**, *33*, 395–410. [CrossRef]
68. Liu, K.; Wang, M.; Cao, Y.; Zhu, W.; Wu, J.; Yan, X. A Comprehensive Risk Analysis of Transportation Networks Affected by Rainfall-Induced Multihazards. *Risk Anal.* **2018**, *38*, 1618–1633. [CrossRef] [PubMed]
69. Ball, R.J.; Langdon, N.J.; Giles, D.P. Impact of the Calder's Wharf development on Docklands Light Railway tunnels, London, UK. *Proc. Inst. Civ. Eng. Geotech. Eng.* **2021**, *174*, 419–429. [CrossRef]
70. Hanson, C.S.; Noland, R.B.; Porter, C.D. Greenhouse gas emissions associated with materials used in commuter rail lines. *Int. J. Sustain. Transp.* **2016**, *10*, 475–484. [CrossRef]
71. Del Pero, F.; Delogu, M.; Pierini, M.; Bonaffini, D. Life Cycle Assessment of a heavy metro train. *J. Clean. Prod.* **2015**, *87*, 787–799. [CrossRef]
72. Gopal, K.; Shin, E.J. The impacts of rail transit on the lives and travel experiences of women in the developing world: Evidence from the Delhi Metro. *Cities* **2019**, *88*, 66–75. [CrossRef]
73. Zhang, N.; Wang, Z.; Chen, F.; Song, J.; Wang, J.; Li, Y. Low-Carbon Impact of Urban Rail Transit Based on Passenger Demand Forecast in Baoji. *Energies* **2020**, *13*, 782. [CrossRef]
74. Soni, A.R.; Chandel, M.K. Assessment of emission reduction potential of Mumbai metro rail. *J. Clean. Prod.* **2018**, *197*, 1579–1586. [CrossRef]
75. Shao, L.M.; An, S.J.; Chang, H.B. Calculation and analysis of combustion and emission of a diesel engine with high pressure common rail system. *Neiranji Gongcheng/Chin. Intern. Combust. Engine Eng.* **2005**, *26*, 1–4.
76. Steffen, W.; Percival, T.; Flannery, D. Greenhouse gas emission reductions from Canberra's light rail project. *Aust. Plan.* **2015**, *52*, 337–345. [CrossRef]

77. Shen, W.; Chen, F.; Wang, Z. (Eds.) *The Development and Application of Transport Energy Consumption and Greenhouse Gas Emission Calculation Software Based on the Beijing Low-Carbon Transport Research*; LTLGB 2012, 2013 2013//; Springer: Berlin/Heidelberg, Germany, 2013.
78. Olugbenga, O.; Kalyviotis, N.; Saxe, S. Embodied emissions in rail infrastructure: A critical literature review. *Environ. Res. Lett.* **2019**, *14*, 21. [CrossRef]
79. Ingrao, C.; Scrucca, F.; Matarazzo, A.; Arcidiacono, C.; Zabaniotou, A. Freight transport in the context of industrial ecology and sustainability: Evaluation of uni- and multi-modality scenarios via life cycle assessment. *Int. J. Life Cycle Assess.* **2021**, *26*, 127–142. [CrossRef]
80. Boschiero, M.; Zanotelli, D.; Ciarapica, F.E.; Fadanelli, L.; Tagliavini, M. Greenhouse gas emissions and energy consumption during the post-harvest life of apples as affected by storage type, packaging and transport. *J. Clean. Prod.* **2019**, *220*, 45–56. [CrossRef]
81. Liu, W.; Lund, H.; Mathiesen, B.V. Modelling the transport system in China and evaluating the current strategies towards the sustainable transport development. *Energy Policy* **2013**, *58*, 347–357. [CrossRef]
82. De Bortoli, A.; Christoforou, Z. Consequential LCA for territorial and multimodal transportation policies: Method and application to the free-floating e-scooter disruption in Paris. *J. Clean. Prod.* **2020**, *273*, 122898. [CrossRef]
83. Xu, Z.; Zhao, N.; Hillmansen, S.; Roberts, C.; Yan, Y. Techno-Economic Analysis of Hydrogen Storage Technologies for Railway Engineering: A Review. *Energies* **2022**, *15*, 6467. [CrossRef]
84. Carneiro, M.A.; Soares, D.D.F. Solar photovoltaic assistance system study for a Brazilian light rail vehicle. *U Porto J. Eng.* **2020**, *6*, 35–45. [CrossRef]
85. Carneiro, M.A.; Da Fonseca-Soares, D.; Pereira, L.H.M.; Ramos-Ridao, A.F. An Approach for Water and Energy Savings in Public Buildings: A Case Study of Brazilian Rail Company. *Sustainability* **2022**, *14*, 15921. [CrossRef]
86. Ciancetta, F.; Ometto, A.; D'ovidio, G.; Masciovecchio, C. Modeling, Analysis and Implementation of an Urban Electric Light-Rail Train Hydrogen Powered. *Int. Rev. Electr. Eng.* **2019**, *14*, 237–245. [CrossRef]
87. Ibrahim, M.M.; Ramesh, A. Experimental investigations on a hydrogen diesel homogeneous charge compression ignition engine with exhaust gas recirculation. *Int. J. Hydrogen Energy* **2013**, *38*, 10116–10125. [CrossRef]
88. Fang, W.; Huang, B.; Kittelson, D.B.; Northrop, W.F. Dual-*Fuel* diesel engine combustion with hydrogen, gasoline, and ethanol as fumigants: Effect of diesel injection timing. *J. Eng. Gas Turbines Power* **2014**, *136*, 81502. [CrossRef]
89. Kurczyński, D. Effect of the rme biodiesel on the diesel engine fuel consumption and emission. *Commun. Sci. Lett. Univ. Zilina* **2021**, *23*, B308–B316. [CrossRef]
90. Rajeswari, S.; Saravanan, P.; Kumaraguru, K.; Jaya, N.; Rajeshkannan, R.; Rajasimman, M. The Scientometric Evaluation on the Research of Biodiesel Based on HistCite and VOSviewer (1993–2019). *Biomass Convers. Biorefinery* **2021**, *13*, 4093–4103. [CrossRef]
91. Thongchai, S.; Lim, O. Investigation of the combustion characteristics of gasoline compression ignition engine fueled with gasoline-biodiesel blends. *J. Mech. Sci. Technol.* **2018**, *32*, 959–967. [CrossRef]
92. Zare, A.; Stevanovic, S.; Jafari, M.; Verma, P.; Babaie, M.; Yang, L.; Rahman, M.M.; Ristovski, Z.D.; Brown, R.J.; Bodisco, T.A. Analysis of cold-start NO_2 and NOx emissions, and the NO_2/NOx ratio in a diesel engine powered with different diesel-biodiesel blends. *Environ. Pollut.* **2021**, *290*, 118052. [CrossRef] [PubMed]
93. Thongchai, S.; Lim, O. Macroscopic spray behavior of a single-hole common rail diesel injector using gasoline-blended 5% biodiesel. *Energies* **2020**, *13*, 2276. [CrossRef]
94. Ma, C.; Gao, L.; Zeng, Q.; Cui, R. Study on Modal Characteristics and Influencing Factors of Wheel-rail Coupling Vibration of High-speed Railway. *Tiedao Xuebao/J. China Railw. Soc.* **2021**, *43*, 85–93.
95. Rong, K.J.; Xiao, Y.L.; Shen, M.X.; Zhao, H.P.; Wang, W.J.; Xiong, G.Y. Influence of ambient humidity on the adhesion and damage behavior of wheel–rail interface under hot weather condition. *Wear* **2021**, *486–487*, 204091. [CrossRef]
96. Zhou, X.; Jing, L.; Ma, X. Dynamic finite element simulation of wheel–rail contact response for the straight track case. *Adv. Struct. Eng.* **2021**, *24*, 856–869. [CrossRef]
97. Aydin, Ö.L.; Kuşkapan, E.; Çodur, M.Y. Investigation of adequacy of light rail systems route in erzurum province in terms of ground safety. *Jeol. Muhendisligi Derg.* **2021**, *45*, 41–66. [CrossRef]

buildings

MDPI

Review

Strategies for the Preservation of Historic Areas within Existing Middle Eastern Cities: The Case of Historic Jeddah

Mohammed Aloshan [1,*], Islam Elghonaimy [2], Essam Mesbah [3], Moustafa Gharieb [4], Khaled Mahmoud Heba [5] and Mohammed Humaid Alhumaid [1]

1 Department of Architectural Engineering, Imam Mohammad Ibn Saud Islamic University (IMSIU), Riyadh 11432, Saudi Arabia; mhmalhumaid@imamu.edu.sa
2 Department of Architecture and Interior Design, The College of Engineering, University of Bahrain, Sakhir 32038, Bahrain; eelghonaimy@uob.edu.bh
3 College of Engineering, Department of Architectural Engineering, University of Jeddah, Jeddah 21959, Saudi Arabia; emosbah@uj.edu.sa
4 Department of Architecture, College of Architecture and Planning, Qassim University, Buraydah 52571, Saudi Arabia; mg.abdou@qu.edu.sa
5 Department of Architecture, Faculty of Engineering, Al-Azhar University, Cairo 11371, Egypt; khaled_heba@azhar.edu.eg
* Correspondence: mamoshan@imamu.edu.sa; Tel.: +966-54-249-9915

Abstract: During the recent founding of Saudi Arabia in 1932, there was no interest in the urban and architectural heritage that Jeddah possesses. As a result, historic Jeddah was exposed to many problems similar to many historical Middle Eastern cities. For example, the historic city wall and many heritage buildings were demolished. With the economic recovery during the 1970s, the original dwellers abandoned the city altogether. They were soon replaced by a class of ex-patriate workers who would inhabit the area, and the city's distinct heritage fell into neglect. At the beginning of the 1980s, attempts to preserve that area began with the expert Robert Matthew, who studied the remaining historical buildings and proposed strategies for preservation that were based on three main directions: restoration, replacement, or new construction. The issuance of "Saudi's 2030 vision" included programs to develop Saudi cities, including Jeddah. This program includes the removal of unplanned and slum areas around historic Jeddah without developing a clear master plan for what will replace them. In addition, it includes the complete expropriation of historic Jeddah, without announcing a specific plan for what the area will be used for. Therefore, this study seeks to review current preservation strategies, with the aim of reaching results related to the challenges facing the area; demonstrating the effectiveness of the preservation strategies used; and deducing what could contribute in the future to the development, preservation, and optimal exploitation of the area, without losing its value as a result of the modernization processes currently taking place.

Keywords: the world heritage list; historic Jeddah; strategies; preservation; historical areas

1. Introduction

Historic Jeddah has a distinctive architecture as a result of the many changes in its architecture it underwent throughout various historical eras.

It was influenced by Mamluk architecture and then the Ottoman architecture prevalent in most of the historical cities of the Middle East, whether in Egypt, Iraq, the Levant, or elsewhere [1].

As a result of several factors, the historical area of Jeddah witnessed urban deterioration, especially with the rapid urban expansion beginning after the founding of Saudi Arabia in 1932 with the lack of awareness of the importance of such heritage at that time. The demolition of its historical wall in 1947 [2] in particular led to an acceleration in its urban deterioration as its historical urban fabric began to be mixed with the urban fabric of the new areas. This led to a loss of its historical privacy, in addition to the loss of many heritage buildings, whether through neglect or to claims of modernization and development, without a vision or strategy to preserve the area or an awareness of its historical importance, until the end of the 1970s [3].

In addition to the above, changes in its demographic composition with the emigration of its original dwellers to more modern urban areas and their replacement by a class of foreign expatriate workers, with their lack of interest in maintenance operations for its buildings, led to further deterioration. This is the case with many historical areas in existing cities, especially in Saudi Arabia and the Arab Gulf countries [4].

With the renewed interest in this historical area, its urban fabric, and its heritage buildings, in addition to its inclusion on the World Heritage List in 2014 [5], it is necessary to review and analyze the conservation strategies already drawn up. The problems that historic Jeddah currently faces or may face in the future should be studied with the aim of achieving sustainable urban preservation, especially in the context of the comprehensive urban development projects taking place in the city within the framework of "Saudi's 2030 vision" [6] since 2021. Such developments include the removal of the unplanned districts adjacent to historic Jeddah and their conversion to other uses without the announcement of a clear master plan and without presenting such a plan to public opinion for discussion, especially since this demolition and removal of these slums and unplanned districts around the historical area aims, as was announced, to support the historical area. Therefore, this study aims to draw attention to the selection of appropriate cultural and tourism activities to replace that removed area in order to support and preserve the historical area.

1.1. Study Purpose

This study aims to review and analyze the urban preservation strategies currently implemented in historic Jeddah and review whether or not they are sufficient. It also ad-dresses the most important challenges facing the area and the impacts related to these strategies. This comes as a result of a large project being implemented under the pretext of developing Jeddah, which includes the demolition of the unplanned districts surrounding the historic area and the expropriation of the historic area without a clear plan in place for what the area will become.

Therefore, the study examines development in the area and then proposes a set of recommendations that can be directed to those in charge of those projects and everyone who bears responsibility for preserving that area, with the aim of ensuring that any preservation is primarily in the economic, social, and urban interest of the historic area.

In addition to drawing attention to the necessity of ensuring that the development project for the historical area and its surroundings is carried out in a way that contributes to preserving and supporting the historic area by choosing appropriate activities, whether cultural, touristic, or other, it draws attention to the necessity to replace the removed slums around the historical area.

1.2. Study Significance

The importance of this study lies in the need to ascertain the effectiveness of the strategies drawn up for the urban preservation of historic Jeddah so far in addition to addressing the most important problems and challenges facing the historic area as a world heritage site.

It also highlights the necessity of defining international standards that are approved and applied globally in urban preservation operations, especially in the context of the large project aiming to develop the historic area and its surroundings, which began in 2021. The details of this project have not yet been announced, other than the removal of some of the surrounding districts and the expropriation of the historical area.

A specific and clear plan for what the region will become next was not announced, in addition to no plan being presented to the public and specialists in all fields for discussion. Therefore, it is important for this study to put forward some recommendations that may benefit decision makers in preserving the historical area and converting the surrounding area to uses that are compatible with it as an additional buffer zone. All of this is to be achieved within the framework of developing the historical region without the region as a whole losing its historical identity.

2. Materials and Methods

The study follows a theoretical and descriptive method and deals with the concepts of cultural heritage, including the characteristics and patterns of archaeological and historical areas and their distinctive character according to "the world heritage convention" from 1972 and others [7].

It also reviews how to deal with historical areas and buildings and the urban problems facing them by presenting the most important approved strategies and naming and explaining the most important international charters and institutes that sponsor them.

Then, the study moves on to present and review urban preservation plans and strategies for the historic Jeddah area as a case study of historical areas within existing Middle Eastern cities, for which the first detailed master plan to classify their heritage buildings and historical urban fabrics was prepared by appointing a specialized international consulting body between 1978 and 1979 [8,9].

After that, the study addresses the establishment of the department for the protection of historic Jeddah in 1993 and the launch of Prince Majid bin Abdelaziz's project to take care of historic Jeddah [9]. Then, the efforts of "the Saudi commission for tourism and national heritage" to develop historic Jeddah since its establishment in 2000 are discussed, thanks to which a special file was prepared to register the area on the UNESCO World Heritage List in 2011. This effort was initially met with rejection by the international organization due to several considerations related to the current state of the area at that time, which forced the Saudi government to withdraw the nomination [10].

However, this rejection motivated the Saudi Commission to develop many urgent and future plans to implement specific projects to care for, maintain, and preserve historic Jeddah's urban heritage. These efforts succeeded in achieving Jeddah's inclusion on the World Heritage List in 2014 [8,9].

This brings us to the current stage, in which the city of Jeddah is undergoing, through "Saudi's 2030 vision", several projects to develop its various districts. This involves the implementation of some modern development projects in Jeddah starting in 2021, which include the expropriation of historic Jeddah and the removal of its neighboring districts as a buffering zone.

Therefore, this study seeks to draw the attention of decision makers by making recommendations on the importance of choosing strategies that support the historical area when developing the general plan for the surrounding areas currently being dealt with, in addition to the historical area itself after its expropriation.

2.1. Definition of the Cultural Heritage

"The world heritage convention" 1972 defines cultural heritage in "Article 1" as including the following [7]:

- Monuments: architectural works, works of monumental sculpture and painting, elements or structures of an archaeological nature, inscriptions, cave dwellings, and combinations of features, which are of outstanding universal value from the point of view of history, art or science;
- Groups of buildings: groups of separate or connected buildings which, because of their architecture, their homogeneity, or their place in the landscape, are of outstanding universal value from the point of view of history, art or science;
- Sites: works of man or the combined works of nature and of man, and areas including archaeological sites which are of outstanding universal value from historical, aesthetic, ethnological, or anthropological points of view.

2.2. *Characteristics and Patterns of Archaeological and Historical Areas*

The identity of archaeological and historical areas is formed by many components. Its buildings possess a quantitative and qualitative richness of architectural and historical styles, combining tangible features such as the urban fabric, buildings, and monuments and intangible features such as social traditions, popular culture, commercial activities, etc. The combination of all these factors becomes a multi-dimensional civilizational system. These areas also gain moral value due to the events they witnessed and the meanings associated with them. Therefore, reviving the meanings and events associated with these areas aims to reaffirm their importance in the conscience of its people and visitors [11].

Archaeological and historical areas consist of a group of urban fabrics and foci that record specific periods of the city's stages. These foci may be in the form of markets, distinctive gates, or archaeological or historical buildings with a distinctive architectural character. They also gain their value or lack thereof based on the state of the urban fabric. Historical and archaeological areas have some other features that distinguish them, including the distinctive and authentic urban environment and the architectural features of buildings with a unique character that have the ability to carry and influence culture. In addition, distinguished urban fabrics contain a large concentration of archaeological and historical buildings characterized by a common architectural vocabulary that confirms the character of these areas [11,12].

2.3. *Dealing with Archaeological and Historical Areas and Buildings in Historic Cities*

2.3.1. The Urban Problems That Face Historical Areas and Buildings

The historic centers of cities are exposed to many dangers as a result of a number of factors, including increasing rates of population growth, internal and external migration to historic areas, and the increase in the number and size of modern means of transportation and their penetration into areas that were not historically designed for them. This leads to the emergence of other problems related to expanding old existing roads by demolishing historical buildings, as well as providing places to wait for these modern means of transportation [13].

In addition, the high profitability to be gained from constructing modern buildings in historical areas has led to an increase in the size and number of buildings, which exacerbates the problem of transportation and movement, such as the need to build special parking places for vehicles such as cars. In turn, increased space for transportation affects the climates and environments of these historical centers and leads to an increase in environmental pollution. These factors have led to the degradation and destruction of historical areas of high heritage value [13].

The state of imminent danger threatening these historical buildings has also been exacerbated by the neglect of maintenance and care and the community's lack of awareness of their historical, heritage, and artistic value, and even their economic value [13,14].

Therefore, realistic scientific strategies and plans must be developed to protect and preserve buildings of historical and artistic value. This process is known as an integrated and comprehensive maintenance plan, which is followed by a repair and rehabilitation plan for buildings of historical and heritage value [14,15].

2.3.2. Rehabilitation of Historical Areas and Buildings

The rehabilitation and maintenance process essentially stems from the data on the environment in which the historical areas and buildings to be rehabilitated and maintained are located, taking into account the changes that result from the interactions of economic and social activities and the impact of these changes on the environment and buildings. Accordingly, any practical plan must give great attention to these buildings. Therefore, the process of maintenance and rehabilitation must be given preference and precedence over all other options [16], including the development option itself in its simple, abstract concept, as this process of preservation is considered in itself a development process [17]. A lack of attention to the process of regulating uses at the level of the urban environment or deficiencies in maintenance procedures lead to the destruction of valuable buildings and thus the destruction of historical areas [14].

2.3.3. Considerations for the Rehabilitation and Maintenance of Historic Areas and Buildings

It is important to understand that the maintenance process of historical or heritage buildings does not limit or reduce the standard of living of their residents. There must also be some kind of preparation to meet the requirements of the modern lifestyle. For example, if it is necessary for residents of historical areas to have modern vehicles, then providing parking places for these vehicles within an acceptable distance must be considered, in addition to providing shaded pedestrian paths or using internal means of transportation that do not pollute the surrounding environment, such as bicycles, electric cars, etc. [16].

Infrastructure providing high-quality and appropriate technical methods should be provided in a way that does not conflict with the fact that it is located in an environment that represents a distinctive urban character. The historical area must also be supplied with all utilities and services that have become necessities of modern life. It is also necessary to supervise some necessary public activities and facilities that may be a source of disturbance to the privacy and distinction of such historical areas, such as public restaurants and cafes, which must be placed under surveillance and supervision by the relevant authorities [18].

Perhaps one of the most fundamental problems facing the rehabilitation process is how to create new functions and roles for historical buildings. The development planner must be aware of the extent of the problems and shortcomings that the building or historical center suffers from, and the architect must also know the areas within the scope of the historic area in which historical buildings can be repurposed. Through the integration of these efforts between architects, planners, administrators, and other members of the team, these historical centers can be saved from a state of neglect and decay and transformed into a state in which they can effectively serve their historical and cultural role as pieces of invaluable heritage and cultural wealth [16].

2.4. Strategies and Policies for Preserving Historic Areas and Buildings

The process of preserving the urban and architectural heritage of valuable historical, archaeological, and heritage buildings and areas is carried out by following many strategies and policies. The most important of these methods are listed below [19,20]:

- Restoration.
- Protection.
- Preservation.
- Conservation.
- Adaptation–Reuse.
- Clearance.
- Replacement.
- Renewal.
- Developing.
- Upgrading.

The method used depends on the individual circumstances and conditions of each area or building [15,19].

2.5. International Conventions Related to the Preservation of Historical Areas and Buildings

Many agreements have been sponsored by UNESCO for the maintenance and preservation of historical areas and buildings following the adoption of the Venice Charter [21] in 1964, which was issued in thirteen documents, the most important and famous of which was the first document, known as "International charter for the conservation and restoration of monuments and sites" [21]. This document is considered the first international charter upon which most international conventions were based on, and it is considered one of the most important documents related to the preservation of archaeological and historical areas and buildings, as it has become the main reference for restoration and maintenance operations.

This charter established important definitions used in this field [21]. The charter led to the issuance of the "Convention concerning the protection of the world cultural and natural heritage" in 1972, which approved a program aimed at classifying, naming, and preserving sites of special importance to mankind, namely the "World Heritage List" [22], as well as the "Nairobi Recommendations/Recommendation concerning the safeguarding and contemporary role of historic areas" in 1976 [23], which defined protection measures of historical areas into the categories of renewal, prevention, restoration, maintenance, and revitalization of historical or traditional areas and their environments; thus, protection includes all possible methods of intervention [23].

In addition, the 1979 "Burra Charter" was adopted for places of cultural importance in 2013 to be a guide for best practices in preserving and managing cultural heritage, as it emphasizes that preservation processes are required in sites of important cultural value and that, during the preservation process, caution must be taken to minimize change in archaeological buildings [24].

The "Washington Charter" was published for the preservation of historic cities and urban areas in 1987 [25] and indicated that new functions and activities in a historical area should be compatible with their historical nature [25]. There are many other relevant frameworks, charters, agreements, and recommendations related to the processes of preserving the urban and cultural heritage and protecting historical areas and other cultural properties of humankind [8].

3. Historic Jeddah (the Study Case)

3.1. Location and Historical Importance of Jeddah

Jeddah is considered one of the governorates of the "Mecca Region" and is located in the west of Saudi Arabia on the coast of the Red Sea. Its population was about 4,697,000 in 2021 [26]. Jeddah is considered the economic capital and the first tourist destination in Saudi Arabia and is the second largest city after the country's capital. It is also considered a gateway to "Mecca", which has unique value to all Muslims. It has the largest seaport on the Red Sea and is thus considered a financial and business center in Saudi Arabia and is a major port for exporting non-oil goods and importing local needs [26].

Jeddah has a long history [27] entwined with the emergence and spread of Islam. The historical transformation of Jeddah took place during the reign of Caliph Othman bin Affan in 647 when he that ordered the city be transformed into a port for sea pilgrims heading to perform the Hajj in Mecca [1]. To this day, Jeddah remains the main crossing for pilgrims and remains under the influence of successive Islamic disputes, starting with the Umayyads and including the Abbasids, Ayyubids, and then the Mamluks [1,27].

3.2. Historical Urbanization in Jeddah (Its Development and the Problems It Faces)

Jeddah has a distinctive historical architecture that differs in quality from the rest of the urban and architectural product present in various regions of Saudi Arabia. This is a result of the diversity of changes in its construction that the city witnessed during various eras and times. It was greatly influenced by Mamluk architecture and then the Ottoman architecture that prevailed in the Levant and Egypt at that time [1], and we can clearly distinguish the features of this distinct urbanism in the remaining historical part of the city, where the traces and evidence of urban products such as buildings, mosques, markets, and others are still witness to a different culture, distinguished by their use of local building materials that were commonly used at that time in building houses [8,26]; it is this cultural evidence which gave the historical part of Jeddah a distinction that earned it inclusion on the UNESCO World Heritage List [5].

For a long time, this distinctive historical heritage was neglected, especially with the modernization and development processes that the city has witnessed since the founding of the "Third Saudi State" in 1902 by King Abdulaziz Al Saud, who subjugated Jeddah to his kingdom in 1926; the Kingdom of Saudi Arabia was later established in 1932 [28]. As a result of this modernization, the historical parts of the city were neglected, and the historic city wall and many of its gates were demolished in 1947 with the aim of expanding and modernizing the city without taking into account the distinctive historical heritage that the city possesses. The last quarter of the twentieth century, however, witnessed a rise in interest in the historical part of the city, and attempts were made to develop it in order to preserve it from neglect and extinction [8].

3.2.1. The Historic City Wall of Jeddah and Its Old Gates

During the Mamluk era (1250–1517), Sultan Al-Ashraf Qansuh al-Ghuri, the last Mamluk sultan in Egypt, built the Jeddah Wall in 1509 to protect it from Portuguese raids [1,2]. The wall includes seven gates: Bab Mecca, Bab Al-Madina, Bab Sharif, Bab Al-Nafi'a, Bab Al-Mughrabi, Bab Al-Sabbah, and Bab Sarif. These gates were built in stages according to necessity. Because of the narrowness of the city gate, an eighth double gate with sufficient width to allow cars to pass through it easily was added on the northern side of the wall at the beginning of the Saudi era, known as Bab Jadid. The entire wall was demolished in 1947 [2] due to the increasing urban expansion of the city [29] (Figures 1 and 2).

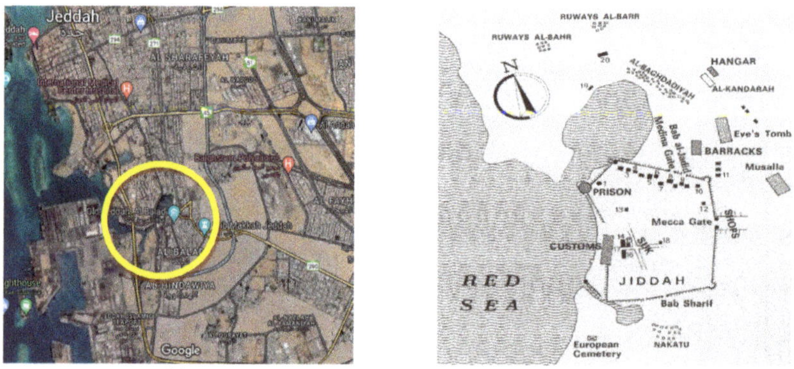

(a) (b)

Figure 1. (**a**) A current map of Jeddah after the huge expansions it witnessed during last eighty years, showing the size of the historical part of the city, which was located within its old wall before its demolition in 1947 [30]. (**b**) Map of Jeddah as drawn by the Italian orientalist Carlo Alfonso Nallino in 1938. It shows the historic city wall [31].

(a) (b)

Figure 2. (**a**) Jeddah as shown in an aerial photograph taken in 1938, where its urban mass appears inside the historic city wall, which represents the historical area of the current city [32]. (**b**) "Bab Jadid"—New Gate—which was built at the end of the 1930s. It is considered the last gate built on the wall before its demolition and was a double gate that could accommodate cars [33].

3.2.2. Jeddah's Historic Districts (Harrat) and Buildings

Historic Jeddah is divided into four main districts; these districts are called "Harah". In the north is "Haret Al-Sham" and in the south is "Haret Al-Yemen", "Haret Al-Bahar", and "Haret Al- Al-Mazloum". These districts, called "Harrat", contain many distinctive historical houses and buildings that are considered valuable pieces of human heritage, where construction methods and building materials were used depending on the materials available in the local environment [29,34].

The most famous and oldest buildings existing to date are the houses of "Al-Baeshen", "Al Sheikh", and "Al-Shafi'i Mosque", which are located in "Haret Al-Mazloum", as well as the houses of "Al-Banaja", "Al-Zahid", and "Al-Pasha Mosque" in "Haret Al-Sham"; the houses of "Al-Nassif" and "Nour Wali" in "Haret Al-Yaman"; and the houses of "Al-Sharbatli" and "Al-Nimr" in "Haret Al-Bahr", among others [34] (Figure 3).

(a)

(b)

Figure 3. (**a**) House of "Al-Baeshen". It was built in 1857 and is considered an example of a historical house whose owners are still keen on maintaining and caring for it [35]. (**b**) House of "Al-Nassif", which was built in 1872 and is considered one of the area's most important heritage landmarks, gaining historical importance since King Abdulaziz resided there in 1925; it has now been transformed into a museum for Jeddah [36].

3.3. Problems of Urban Deterioration in Historic Jeddah

With the rapid development that Jeddah witnessed at the beginning of the second half of the twentieth century, there was rapid growth in the urban population, which is considered one of the most important causes of urban deterioration in the area. As the social composition of historic Jeddah changed, its people moved to live in new, more modern districts, renting the buildings they owned in the historical area to expatriate workers of different nationalities, which led to the deterioration of the area due to the absence of the necessary maintenance and severe neglect from the migrant workers who occupied these historic buildings. Such neglect exposed the historic area more than once to the danger of fire as a result of high electrical loads, lack of safety measures, etc. [4] (Figure 4a).

In addition, natural factors such as a high level of humidity also contributed to urban deterioration. This led to the erosion of wood and damage to the facades, as well as a rise in the groundwater level, which damaged the foundations of the buildings and led to corrosion and cracking and the walls being saturated with moisture. Incorrect restoration operations that did not respect the character of the buildings also led to changing the heritage features of many of these buildings [37,38] (Figure 4b).

(a)

(b)

Figure 4. (**a**) Historic Jeddah suffered from fires for long periods as a result of the neglect of the area and the poor condition of its infrastructure [39]. (**b**) An example of the incorrect restoration of the historic Al-Shafi'i Mosque in 2015, which did not respect the original building materials and turned it into what resembles a white plastic block from the outside, with glass doors used on the inside on the courtyard due to air-conditioning and other additions not related to the original building being added to the mosque [40].

4. Strategies and Policies for Urban Preservation of Historic Jeddah

4.1. The Beginnings of Interest in the Urban and Architectural Heritage of Historic Jeddah

Interest in the historic Jeddah area began with the establishment of "the old Jeddah municipality" in 1977. Many steps have been taken that aim to preserve the urban and architectural heritage of the old city of Jeddah, including the following [9]:

- Preparing the first detailed master plan to preserve and develop historic Jeddah and classify its heritage buildings and its historical urban fabric by the British consultant Robert Matthew during the period from 1978 to 1983 [37].
- Establishing "the historic Jeddah protection department" in 1993 with the aim of protecting heritage buildings from demolition [9].
- Establishing the "Prince Majid bin Abdelaziz's project to take care of historic Jeddah", which included the restoration of "Nassif house" and turning it into a museum, revealing parts of old city walls, as well as partial restoration of 12 heritage buildings [9,38].
- Forming a "Higher committee for fire extinguishing project", where the first phase of the fire extinguishing network was implemented in 2011 and which is now completed [8].
- Beginning the implementation of the "King Abdulaziz project for the preservation of historic Jeddah", starting in 2005, adopting policies for dealing with heritage buildings and establishing "the urban development department in historic Jeddah" [8,9].

4.2. Efforts of the Saudi Commission for Tourism and National Heritage to Develop Historic Jeddah

The Saudi commission for tourism and national heritage" was established in 2000 [8] and was transformed into the "Ministry of tourism" in 2020. The commission developed the "historic Jeddah development project", through which it worked in cooperation with the Jeddah municipality to preserve, rehabilitate, and develop its historical sites architecturally, culturally, and economically in a way that highlights its distinctive features and urban heritage. In addition, the commission encouraged the owners of that urban heritage to protect, maintain, and repurpose their properties in a manner commensurate with their historical and cultural value [38].

The authority has also played a role in the restoration of many heritage buildings and others and, in cooperation with the Jeddah municipality, has worked to set a specific timetable for the completion of many projects in historic Jeddah. These efforts have the aim of highlighting the area's historical and heritage importance and value. Among these efforts, the authority has implemented many tasks, including the following [9]:

- Contracting international consultants to prepare a protection and management plan for the area, within the framework of completing its file for registration on the World Heritage List, which was achieved in 2014.
- Collaborating with the Jeddah municipality in determining a priority plan for the projects required to develop the area, and coordinating on the status of the area and the studies needing to be implemented. Among the most important of these projects is the project to revive tourist axes and paths in historic Jeddah [41]. The most important of these historical and tourist paths and axes are the "Al-Nada axis", "Alawi axis", and "Abu Anaba Axis", on which pedestrians can cross the area in a walking time of fifteen minutes. This last axis is considered a distinguished tourist route; in addition to the distinguished historical buildings it contains, it is flanked on both sides by many outlets displaying the crafts and handicrafts of productive families and many stores that serve popular foods [29,41] (Figure 5).

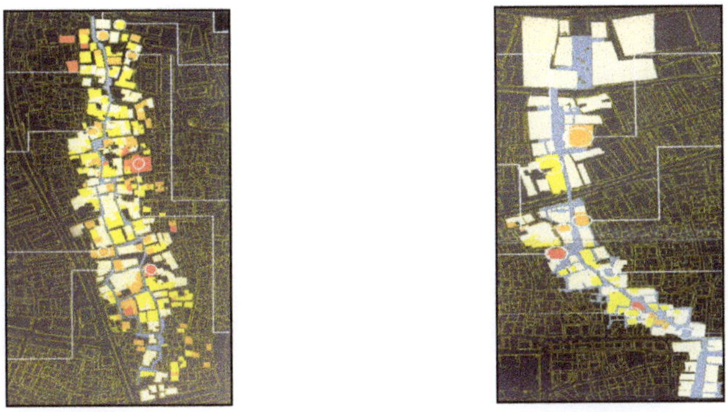

(a) (b)

Figure 5. The project to revive tourist axes and paths in historic Jeddah. (**a**) Path: "Abu Anaba axis," on which many historical buildings are located. It has important historical and heritage houses, and pedestrians can cover it in a walking time of fifteen minutes [41]. (**b**) Path: "Al-Nada axis," which represents the market in the old city of Jeddah [41].

4.3. *Strategies for Preserving the Historic Jeddah*

The British consultant Robert Matthew is considered the first to develop a masterplan to preserve the historical area from 1978 to 1983. In his plan, he identified more than a thousand valuable historical buildings, which were classified into three levels, historical, artistic, and spatial, based on several criteria related to the uniqueness of the building's architectural style and its value, in accordance with adopted standards in the UNESCO classification, as described below [37,42] and illustrated in Figure 6:

- First Class: buildings of national importance;
- Second Class: buildings of regional importance;
- Third Class: buildings of local importance.

In addition, Matthew's classification also includes modern high-rise buildings that were constructed in recent periods, as well as vacant lands, paths, corridors, etc.

Matthew's plan also proposed adopting a "sustainable preservation" strategy [2] by re-using the historical buildings according to their degree of classification, as elaborated below:

- First-level buildings: buildings being used institutionally, whether as governmental, administrative, cultural, or educational buildings [42];
- Second- and third-level buildings: buildings being used for hospitality, residential, office, or commercial activities [42] within a specific zone of the historic area called the "Nominated Property" [42,43] (Table 1).

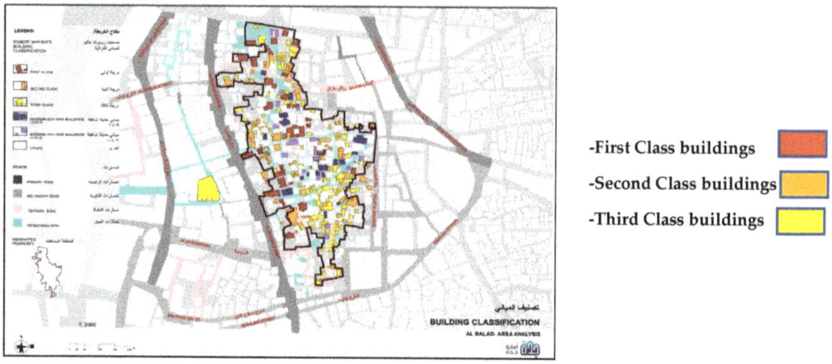

Figure 6. Matthew's classification for the historical buildings in historic Jeddah according to the uniqueness of the building's architectural style and its value. The classes are historical, artistic, and spatial, in accordance with adopted standards in the UNESCO classification [42].

Table 1. Classification of heritage buildings in historic Jeddah and proposals for their reuse *.

Classification	Reuse Proposal	Architectural Processing and External Facades Processing	Architectural Processing and Interior Design Processing
First Class	- The optimal reuse is chosen according to the master plan, provided that it does not conflict with the nature and privacy of the building. The alternatives for proposed uses are as follows: governmental, administrative, cultural, and educational (Provided that the optimal proposal for reuse is subject to the approval of the relevant authorities).	- The presence of building materials and systems used in constructing and finishing the facades of heritage buildings, which must be carried out in the traditional manner and in accordance with the original drawings of the building.	- The distribution and design of the building's interior spaces, as well as all interior design elements, must adhere to the building's original drawings. - Perfect match between the proportions of the interior spaces and the original design.
Second and Third Class	- The optimal reuse is chosen according to the master plan, provided that it does not conflict with the nature and privacy of the building. The alternatives for proposed uses are as follows: governmental, administrative, cultural, and educational (provided that the optimal proposal for reuse is subject to the approval of the relevant authorities).	- The presence of building materials and systems used in constructing and finishing the facades of heritage buildings, which must be carried out in the traditional manner and in accordance with the original drawings of the building. - Preserving the original proportions of the facade design of the heritage building.	- Alterations should range from 30% to 50% of the total elements and components of the original interior design.

* (Jeddah Municipality, 2014. adapted).

The "Nominated Property" is the protection area that was registered in 2014 on the UNESCO World Heritage List [5], and it includes more than about 250 classified historical and heritage buildings. It also covers an area of 18 hectares. It is located within another circle called the "Buffer Zones", which are a group of urban areas that surround the "Nominated Property" on all sides and cover an area of 113.53 hectares over the remaining parts of the historic area and other surrounding areas [43].

The Buffer Zones are divided into four conservation areas with the aim of playing an important role in preserving the "Nominated Property" heritage conservation area and enabling it to continue to flourish in light of the changes and developments taking place in the city as a whole. The "Historic Jeddah building requirements system" is followed in this area, which comprises building regulations established by the Jeddah municipality [29,42] (Figure 7).

With time and the acceleration of the modernization movement, in addition to the various factors leading to deterioration in the historical area, many historical and heritage buildings have been lost. There are now only about 600 buildings remaining out of the approximately 1000 buildings inventoried by Robert Matthew in the 1980s distributed between the protection and conservation areas [4].

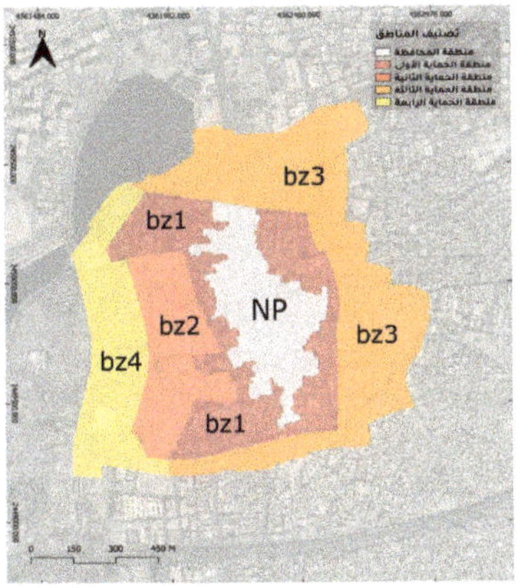

Figure 7. The historic Jeddah protection area "Nominated Property" (NP)—shaded in white—which is the area that was registered in 2014 on the UNESCO World Heritage List, followed by the four "Buffer Zones" (bz1–bz4), which are a group of conservation areas surrounding the historic Jeddah protection area from all sides [43].

This means that the area has lost about 40% of its buildings in less than 40 years. Therefore, the "Jeddah municipality" has set a system of building regulations—the "Historic Jeddah building requirements system"—to demonstrate strategies for urban preservation in the historic area through three main paths, as follows [4]:

- Restoration of historical and heritage buildings;
- Replacement (by rebuilding) of heritage buildings that already exist but cannot be restored due to their poor structural and urban conditions;
- Constructing new buildings on vacant lands or replacing modern buildings that have no value and distort the urban and visual landscape of the area.

However, it is absolutely forbidden to demolish any historical building, as all buildings located within the "Nominated Property" area are subject to a protection policy that prevents removal, change, modification, addition, or other work [42,43] (Figure 8).

With the continuing state of urban deterioration in the historic area and the increase in the number of buildings on the verge of collapse, there was a need to adopt a number of clear and specific strategies to deal with historical and heritage buildings depending on the condition of each building individually. The Jeddah municipality classified historical and heritage buildings based on their structural and urban conditions, based on what was stated in Robert Matthew's classification of heritage buildings [43].

Buildings were thus classified into categories A, B, and C, in which the buildings are classified as having either minor damage (A), stable damage (B), or unstable damage(C).

Historical buildings represent 82% of the total building stock in the historical area, 9% of which are category (A) buildings, 54% are category (B) buildings, and 19% are category (C) buildings. These types of buildings require the use of a "Restoration" strategy, in order to prepare them for another strategy, such as the "Adaptation/Reuse" strategy, according to the relevant proposal in Robert Matthew's proposal of reusing heritage buildings, as reviewed and approved by the Jeddah municipality [4,43].

The rest of the buildings in the area are classified into categories D or E due to their poor structural and urban conditions, classified as having major damage in which there is no point in restoration, including cases of partial or complete collapse. These buildings represent 14% of the total building stock in the historical area. These types of buildings require the use of a "Replacement" strategy and rebuilding, after which they can be used to meet many purposes compatible with the nature of the historical area [4,43].

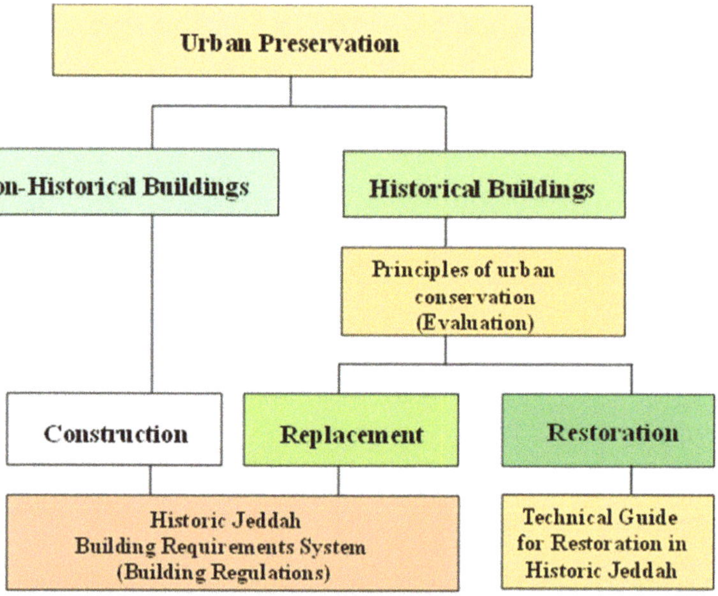

Figure 8. The strategies for urban preservation in historic Jeddah [43].

As for vacant lands, they are classified as category (F), and their percentage does not exceed 4% of the area's size. These lands are used for the purposes of new construction that is compatible with the historical area and supports it in accordance with the established architectural rules and urban requirements (Figure 9).

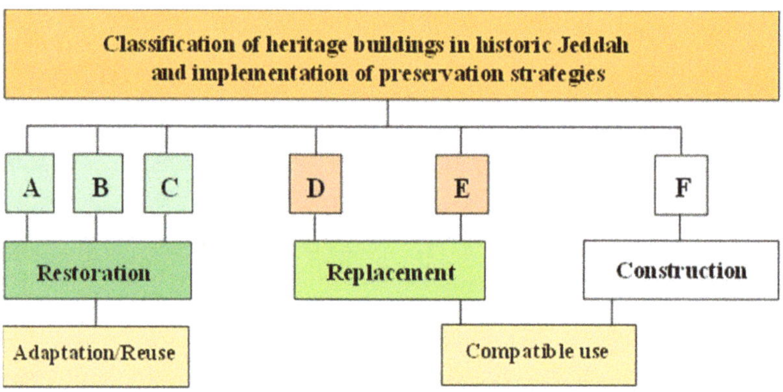

Figure 9. Classification of heritage buildings in historic Jeddah according to their structural and urban conditions and the preservation strategies followed [The Researchers].

In general, historical and heritage buildings are subject to an evaluation process carried out by a specialized committee of experts in the fields of architectural and structural preservation before proceeding to determining the most appropriate strategy and then setting the policy for preservation. The evaluation process includes several criteria considering the urban and structural conditions of the building, in addition to an assessment of the surrounding buildings and a study of the soil, foundations, walls, etc., bearing the building's weight [37,42,43].

From our analysis of the strategies and policies adopted for the urban preservation of the historic Jeddah area, we find that they were represented by several main strategies, as described in Table 2.

Table 2. Strategies that were adopted for the urban preservation of the historic Jeddah *.

Strategy	Scope of Use
Restoration	This is for historical buildings whose structural and urban conditions are classified within classifications A, B, and C. These restoration operations are subject to the requirements and standards of UNESCO for the preservation of historical and heritage buildings.
Replacement	This is for historical buildings whose structural conditions are classified within classifications D and E, including collapsed historical and heritage buildings. Under this strategy, buildings are rebuilt by returning to the original internal and external documentation of the building until it once again conforms to its former architectural character [43]
New Construction by following "Clearance & Renewal" strategies	This is for vacant lands located in the area, which are developed in accordance with the established architectural rules and urban requirements, provided that they are used to serve and develop the area. In addition, the area also contains a number of relatively modern structural buildings that were constructed after the demolition of the city wall in 1947 and during the following periods, especially in the 1960s and 1970s. These were mostly buildings that did not adhere to the height limits set by Robert Matthew in his masterplan, which led to urban and visual distortion. Therefore, the new construction policy is used to provide clearance for their renewal with buildings appropriate within the development scope of the historical area [43].
Adaptation/Reuse	This is for historical and heritage buildings that are being restored and in accordance with Robert Matthew's proposal and classification in this regard, as reviewed and approved by the Jeddah municipality [42].
Developing and Upgrading	This is enacted through some of the procedures followed in the area, such as preventing the entry of vehicles except for public service vehicles such as fire engines and ambulances, restricting the area to pedestrians only, providing parking areas for cars in the surrounding areas, providing the area with various necessary facilities and services, and adopting specific systems and procedures for cleaning and maintenance operations, in addition to adopting fire-fighting and extinguishing systems and other development and upgrading works.

* (The Researchers).

5. The Challenges Facing the Process of Preserving Historic Jeddah while Implementing Modern Development Decisions for the City (2021 until Now)

With the change in political leadership in Saudi Arabia in 2015, a new king of Saudi Arabia assumed power. The King chose, for the first time in the history of Saudi Arabia, a young crown prince in his thirties. The new crown prince was assigned to head the Council of Ministers, as Saudi Arabia was witnessing a radical change represented by openness to the contemporary world. For the first time, several decisions were made that may seem simple to most people in the world, but for Saudi Arabia, they amounted to lifting prohibitions, such as allowing women to drive cars, establishing cinemas, organizing film and theater festivals, and organizing concerts, attracting famous artists to visit Saudi Arabia and present their art there. In addition, sports of all kinds for both genders were promoted, and regional and international sports competitions were organized. These changes encouraged tourism, expanded archaeological excavations, established tourist resorts, etc. Mixing between men and women in restaurants and cafes, which was previously prohibited, was also allowed, among many other changes, in an attempt to change the face of Saudi society, which was accomplished in just a few years.

5.1. Saudi's 2030 Vision

In the same context, a massive development program was launched in 2016 that includes developing all aspects of economic, social, and cultural life under the name "Saudi's 2030 vision" [6]. This vision aims primarily to reduce dependence on oil resources and rely on the resources of all industrial, including tourism and other sectors of society. It also includes several development programs. This vision has received support from various groups of the Saudi population, especially the intellectual classes and the youth, and it is an approach that this study and those in charge of it see as worthy of support and encouragement [6].

5.2. Jeddah's Development Projects and Their Effects on the Historic Jeddah Area

As a result of the development pursued by Saudi Arabia, represented by the issuance of "Saudi's 2030 vision", the nation's political leadership adopted an ambitious program to modernize Jeddah as a tourist destination for the country and establish it as its second capital [44,45]. This includes removing and clearing many slum areas and the existing unplanned districts of the city and replacing them with new districts designed according to the latest global systems aimed at the well-being of citizens. This included the removal of some unplanned districts adjacent to the historical area in Jeddah, the case study area, starting in October 2021 [46].

Other sovereign decisions were also issued to expropriate the historical area, including the protection and conservation area, starting March 2022; a final masterplan for developing the area has not been announced to date, other than announcing that the removal of districts aimed to create surrounding areas to support the role of the historical area, and the expropriation of the historical area will contribute greatly to an easier and less re-strictive urban control and development process in the area [46,47] (Figure 10).

Figure 10. Removal and clearance of slum areas around the historical area and expropriation area within the scope of the historical area according to the recent development decisions in Jeddah made in 2021 and 2022, respectively [44].

It is expected that these modern development decisions for the city, which have begun to be implemented, will help confront the challenges facing the urban preservation process of historic Jeddah. Such challenges can be summed up by the need for more investment and funding to complete the preservation and restoration of the urban heritage of the area and then rehabilitate and repurpose it for the various proposed activities mentioned above. This is the actual purpose of those recent development decisions: they provided the necessary support for the expropriation of the area which will then allow for investment through a plan currently being prepared. Increased investment will contribute to creating an autonomous economy for the area that will enable it to continue to prosper as a historical tourist destination with diversified activities. In addition, starting the preservation and restoration work of such expansion, which is supervised by the highest sovereign authorities in Saudi Arabia, will create an opportunity for local cadres to train in conservation and restoration work, which is what the area in particular and Saudi Arabia in general needs [47,48].

The expropriation of the historical area will also contribute to confronting another of the most important challenges facing the area, which is represented by the existing imbalance in its demographic composition after its original people abandoned it and foreign expatriate workers moved in, which posed a threat to the area due their lack of belonging to it or association with its identity.

Other problems resulting from this demographic shift included the workers' misuse of the existing buildings by crowding in them (as a result of their low-level income), which resulted in many other dangers that the area suffered from for long periods, such as the outbreak of fires and a lack of attention being paid to periodic maintenance operations, ultimately causing the area to lose many of its historical and heritage buildings [46,47].

Furthermore, the removal and clearance of entire districts and huge areas surrounding the historic area has given the relevant administrative authorities a golden opportunity to replace them with new, more contemporary districts and activities that support the historical area culturally, artistically, touristically, economically, etc., and that contribute to its continuation and prosperity. This will be observed in the coming period once all master and detailed plans for that area are approved and their implementations begin [46] (Figure 11).

Figure 11. In the map to the left, the protection area (NP) within the scope of historic Jeddah as it appears before the demolition work. In the map to the right, historic Jeddah within the scope of the demolition of entire districts and huge areas that took place, which will give the relevant administrative authorities an opportunity to replace them with new and contemporary districts featuring activities that support the historical area and contribute to its continuity and prosperity [44].

6. Discussion

After reviewing urban preservation plans, programs, and strategies for the historic Jeddah area as a case study for historical areas within existing cities, this study concludes with the following results:

A. Historic Jeddah has a distinctive urban and architectural heritage, which qualified it to be registered on the UNESCO World Heritage List in 2014.

B. Interest in the urban and architectural heritage that Jeddah possesses did not begin until the last quarter of the twentieth century, and the historical area lost many of its historical components before that time, such as its historical walls, which were demolished in 1947. This long-term neglect also caused various forms of urban deterioration.

C. Despite the great interest in historic Jeddah, especially at the beginning of the current century, the successful attempt to register it on the UNESCO World Heritage List, and the attention it has received through "Saudi's 2030 vision", the area is still suffering from urban deterioration as a result of the entire period of neglect and the ac-cumulation of its causes, in addition to the large size of the historical area, both in terms of its area and in terms of the number of buildings within it which need restoration and ongoing maintenance.

D. Through the presentation and review of the various plans and strategies that have been developed and implemented, it can be noted that the urban preservation strategies in the historical area in Jeddah have been based on two main trends:

  - The first trend is dealing with the historical area itself and its urban fabric, where "developing" and "upgrading" strategies are followed through implementing some projects and following many procedures in the area, such as organizing vehicle traffic, allocating paths for pedestrians, establishing the fire-fighting network, and so on.
  - The second trend is dealing with historical and heritage buildings by following one of three strategies according to each individual case. Strategies are determined based on a special classification of buildings that was adopted from three basic classification levels that represent the feasibility, or lack thereof, of preservation and restoration operations. These strategies include implementing strategies of "restoration" for the first level of buildings, "replacement" for the second level of buildings, and the last strategy is for "new construction" for vacant lands or to employ a strategy of "clearance & renewal" for lands recently cleared of modern buildings that pollute or distort the urban visual image.

Moreover, the "Adaptation/Reuse" strategy is followed for historical and heritage buildings that are being restored, in accordance with Robert Matthew's proposal and classification in this regard.

E. Although the Jeddah municipality has adopted a number of strategies to deal with historical and heritage buildings depending on the conditions of each building individually, the number of restored buildings is considered very small, as the study observed that 82% of the total buildings in the "Nominated Property" (NP) need preservation and restoration operations. This may be due to the lack of funding necessary for such major work, which requires huge open budgets; the recent era of local cadres' involvement in preservation and restoration work; as well as the scarcity of specialists, which always requires the use of foreign expertise, thus increasing time, effort, and costs.

F. As a result of "Saudi's 2030 vision", the political leadership adopted a development program that includes removing many of the existing districts of Jeddah. This included the removal of some districts adjacent to historic Jeddah, and other sovereign decisions were also issued, such as expropriating the historical area of Jeddah, without announcing a final master plan to develop the area, aside from declaring that these removals aim to create surrounding areas that work to support the role of the historical area and that expropriation of the historical area will make the urban control and development process of the area easier and less restrictive. Perhaps as a result of this decision, we can expect that this will contribute to advancing the conservation process by providing all the financial, human, technical, and logistical capabilities necessary to implement the conservation plans and strategies being followed in the area.

7. Conclusions

This study dealt with plans, programs, and strategies for the urban preservation of the historic Jeddah area as a case study of historical areas within existing cities in the Middle East. In addition, the study covered the challenges that face the authorities responsible for urban preservation operations and which led to the recent development decisions to remove many districts around the area and to the complete expropriation of historic Jeddah. Therefore, the study concludes that, in order for the protection of historic Jeddah to be more effective, a masterplan must be put in place for the development of the area and the surrounding cleared areas that adopts integrated development, economic, urban, and social policies that take into account the preservation of what expresses the city's local character and identity, especially the urban fabric, the relationships between the buildings and the surrounding environment, and the construction methods and materials used.

This, of course, does not mean not constructing modern, contemporary buildings, but rather constructing them in a way that benefits the area functionally and visually.

In addition, the remaining original dwellers of the area should be involved in making decisions related to the preservation, development, or rehabilitation of the historical area.

All of this should also be presented to the public for discussion, especially since a state of satisfaction prevails among all Saudis as a result of the issuance of "Saudi's 2030 vision" and the resulting tangible openness in Saudi society. This opening trend includes all aspects of life, from interest in culture and arts to the formation of a "Ministry of Culture" for the first time in the history of Saudi Arabia and the establishment of an arts authority that includes the fine and musical arts and others. Therefore, this state of openness must be exploited to advance this praiseworthy trend by starting to consult public opinion when making decisions, especially when such decisions are related to their lives, heritage, and cities.

Based on the previous findings of the study, the authors make the following recommendations:

- *First—general recommendations:*

A. Attention should be paid to the rules and principles of shaping the urban and architectural fabric of historic Jeddah when following development policies, so that the area does not lose its continuity and maintains its population structure, with its unique social and economic characteristics.
B. Participation of the original dwellers is recommended in concerted efforts to confront existing challenges through popular, non-governmental institutions that enjoy official and public support.
C. Integrated studies of the value of the architectural and urban formation in the historical area should be conducted. The value of these historical formations should be documented and linked to the social values and principles that are rooted in the social fabric of the community of that area before starting on development processes that aim to preserve and consolidate historical formations, so that the science of preservation does not turn into a mere financial investment only.

- *Second—recommendations on the scope of the urban content of historic Jeddah:*

A. The division of the historic Jeddah area into several administrative areas should be studied in order to facilitate dealing with the urban contents therein within the framework of a comprehensive plan for the entire historical area, as one of the current challenges is how to deal with the enormity of the area as a whole.
B. A set of controls, requirements, guidelines, and technical standards should be established that would achieve the strategies of the development plan, ensuring that a balance is achieved between the general directions of development and individual desires for improvement and investment (facilitating communication and integration between investors, institutions, and public bodies), with the aim of preserving the historical, urban, and architectural heritage of historic Jeddah.
C. The process of restoring and reusing buildings, rehabilitating the area as a whole, rebuilding its dilapidated buildings with appropriate shaping of their original urban structure, and reviving the original urban character of the area as well as original activities or those compatible with the area's buildings should be accelerated through adherence to sustainable urban development standards.

D. The strategy of demolishing some modern buildings built in later periods should be stopped, and instead their appropriate reuse within the historic area should be studied. The history of a city must also be read through the passage of time and the scars that time leaves behind, so that the historic center is not reshaped in a new and false way.
E. The remaining original dwellers of the historical area should be encouraged to remain in the area by improving and restoring their residences while learning about their needs and providing them with various services, facilities, etc. They should al-so be given priority over any other projects, with the aim of maintaining their function in a large part of the historical houses in the area, which gives the historical area sustainability and opportunity for continuous development.
F. Infrastructure rehabilitation operations should be continued as a basis for development operations to ensure the possibility of the area's continued development according to the requirements of modernity and without prejudice to historical value within the framework of improving the urban environment.
G. Providing a strategic plan for the ongoing periodic maintenance of restored, repurposed, or reconstructed buildings within the framework of the sustainable development of the region is necessary.
H. A comprehensive plan must be developed to protect the historical area by establishing scenarios for how to deal with disasters, sources of pollution, potential dangers, and other threats.

- *Third—recommendations on the scope of the removed areas (the areas surrounding historic Jeddah):*

A. The current removed areas surrounding historic Jeddah must have an explicit primary mission to serve as an additional buffer zone to the historical area in Jeddah, and this option should be absolutely preferred over others when setting out to define uses for these areas and the choice of projects with them.
B. A clear strategy must be developed and the necessary legislation must be enacted to regulate the reconstruction and urbanization process in the areas surrounding historic Jeddah that have been removed, so that the architectural character of the area, its fabric, and its urban formation are preserved and integrated through all possible means. In addition, the definition of new uses must be compatible with the needs of the historical area.
C. Modern urbanism can achieve contemporaneity with local belonging by meeting the requirements of modern life, provided that the urban cultural dimension of Jeddah's historical heritage is preserved. This can be achieved by drawing inspiration from the spirit, philosophy, characteristics, and features of local architecture and by carrying out building and construction process using the latest contemporary materials and methods.

Author Contributions: Author Contributions: conceptualization, M.A., E.M. and K.M.H.; methodology, M.A.; software, E.M.; validation, I.E., M.G. and E.M.; formal analysis, K.M.H.; investigation, I.E.; resources, M.A.; data curation, E.M.; writing—original draft preparation, K.M.H.; writing—review and editing, I.E. and M.G.; visualization, E.M.; supervision, K.M.H.; project administration, M.A. and M.H.A.; funding acquisition, M.A. All authors have read and agreed to the published version of the manuscript.

Funding: This work was supported and funded by the Deanship of Scientific Research at Imam Mohammad Ibn Saud Islamic University (IMSIU) (grant number IMSIU-RG23116).

Data Availability Statement: The raw data supporting the conclusions of this article will be made available by the authors on request.

Conflicts of Interest: The authors declare no conflicts of interest.

References

1. Spahic, O. Historic Jeddah as a Unique Islamic City. *IIUM J. Relig. Civilizational Stud.* **2021**, *4*, 60–81.
2. Tawab, A.; El-Din Ahmed, A.G.; Al-Darwish, A.I.; Al-Mutahar, H.A.A.-Y. Achieving the Sustainable Preservation of Historical Urban Commercial Spaces–Case Study–Historic Jeddah. *Eng. Res. J.* **2021**, *5*, 239–240. Available online: https://digitalcommons.aaru.edu.jo/erjeng/vol5/iss4/1 (accessed on 30 October 2023).
3. Badawy, S.; Shehata, A.M. Sustainable Urban Heritage Conservation Strategies—Case study of Historic Jeddah Districts. In Proceedings of the International Conference on Cities' Identity through Architecture and Arts (CITAA 2017), Cairo, Egypt, 11–13 May 2017; Taylor & Francis Group: London, UK, 2018; pp. 83–97. Available online: https://www.taylorfrancis.com/chapters/oa-edit/10.1201/9781315166551-8/sustainable-urban-heritage-conservation-strategies%E2%80%94case-study-historic-jeddah-districts-samaa-badawy-ahmed-shehata (accessed on 26 January 2024).
4. Walid, S.; Nada, S. Policies for conservation of the urban heritage in historical centers historic Jeddah Case study. *J. Gulf Arab. Penins. Stud.* **2020**, *185*, 379–389.
5. UNESCO World Heritage Convention. Historic Jeddah, the Gate to Makkah. Date of Inscription: 2014. Available online: https://whc.unesco.org/en/list/1361/ (accessed on 27 January 2024).
6. Kingdom of Saudi Arabia "Saudi Vision 2030". 2016. Available online: https://www.vision2030.gov.sa/media/cofh1nmf/vision-2030-overview.pdf (accessed on 27 December 2023).
7. UNESCO. Operational Guidelines for the Implementation of the World Heritage Convention. Document-57-19, WORLD HERITAGE CENTRE, Paris, 2019. Available online: https://whc.unesco.org/en/compendium/154 (accessed on 30 October 2023).
8. Khaled Mahmoud, H. *Development and Employment of Archaeological Areas in Historic Cities: Examples of Historic Arab Cities*; Noor Publishing House: Rodgau, Germany, 2016.
9. Jeddah Municipality; General Authority for Tourism and Antiquities. *Historic Jeddah: A Journey of the Past, Present and Future*; General Authority for Tourism and Antiquities: Riyadh, Saudi Arabia, 2014.
10. UNESCO; The World Heritage Committee. Decisions Adopted by the World Heritage Committee at its 35th Session, Paris, UNESCO Headquarters 19–29 June 2011; p. 199. Available online: https://whc.unesco.org/archive/2011/whc11-35com-20e.pdf (accessed on 27 January 2024).
11. Rawiya, H. Developing Urban Spaces as an Entry Point to Preserving Historical Areas. In Proceedings of the Al-Azhar 7th International Engineering Conference, Cairo, Egypt, 7–10 April 2003; Faculty of Engineering, Al-Azhar University: Cairo, Egypt, 2000; pp. 11–12.
12. John, R. *The Lamp of Memory*; Nabu Press: Charleston, SC, USA, 2012.
13. Abdul Sattar, I. *Methods and Strategies for Heritage Conservation*; Research and Training Center for Monuments Conservation, City and Regional Planning Department, University of Engineering and Technology: Lahore, Pakistan, 2003.
14. Islam Hamdi, E.-G. The impact of implementing projects to preserve historical areas in improving the city's urban fabric, a case study: Developing the historical area in the city of Muharraq in the Kingdom of Bahrain. In Proceedings of the First National Architectural Heritage Forum; Saudi Commission for Tourism and Antiquities, Jeddah, Saudi Arabia, 11–17 November 2011; pp. 174–180.
15. Heba, K. Preserving architectural heritage in historical cities: Study of the architectural heritage with the end of the nineteenth century and the beginning of the twentieth century in Alexandria, Egypt. In Proceedings of the Structural Repairs and Maintenance of Heritage Architecture XI, (STREMAH 2011), Chianciano Terme, Italy, 4–7 September 2011; Volume 118, pp. 239–240. [CrossRef]
16. Khaled Mahmoud, H.; Islam Hamdi, E.-G. *Introduction to the Preservation and Development of Historical and Heritage Areas and Buildings*; Scientific Publishing and Translation, Qassim University Qassim: Buraydah, Saudi Arabia, 2018; pp. 16–19.
17. Abdul Baqi, I. *Rooting Cultural Values in Building the Contemporary Islamic City*; Center for Planning and Architectural Studies: Cairo, Egypt, 1986.
18. Dalia, A.-D.; Islam Hamdi, A.-G. *Historic and Heritage Areas: Preservation and Development, an Introduction to Improving the City*; Foundation for Cultural Development: Alexandria, Egypt, 2011.
19. Feilden, B. *Conservation of Historic Buildings*; Routledge: London, UK, 2003.
20. Pickard, R. *Conservation in the Built Environment*; Addison Wesley Longman Ltd.: Chelmsford, UK, 1996.
21. ICOMOS. International Charter for the Conservation and Restoration of Monuments and Sites. (THE VENICE CHARTER 1964), Adopted by ICOMOS in 1965. Available online: https://www.icomos.org/images/DOCUMENTS/Charters/venice_e.pdf (accessed on 31 October 2023).
22. UNESCO. Convention Concerning the Protection of the World Cultural and Natural Heritage. The General Conference of UNESCO Adopted on 16 November 1972. Available online: https://whc.unesco.org/en/conventiontext/ (accessed on 31 October 2023).
23. UNESCO. Recommendation Concerning the Safeguarding and Contemporary Role of Historic Areas. UNESCO. General Conference, 19th, Nairobi, Kenya. 1976. Available online: https://unesdoc.unesco.org/ark:/48223/pf0000114038.page=136 (accessed on 31 October 2023).
24. The Australia ICOMOS. Burra Charter 1979. Adopted by in 2013. Available online: https://australia.icomos.org/publications/burra-charter-practice-notes/ (accessed on 4 November 2023).

25. ICOMOS. Charter for the Conservation of Historic Towns and Urban Areas. (WASHINGTON CHARTER 1987), Adopted by ICOMOS General Assembly in Washington, DC, October 1987. 1987. Available online: https://www.icomos.org/images/DOCUMENTS/Charters/towns_e.pdf (accessed on 4 November 2023).
26. Available online: https://en.wikipedia.org/wiki/Jeddah (accessed on 8 November 2023).
27. Ibn al-Mujawar, A.; Jamal al-Din, A.-Y.; Mamdouh Hassan, M. *The Description of the Countries of Yemen, Mecca, and Some of the Hijaz, Called the History of the Clairvoyant*, 1st ed.; Library of Religious Culture: Cairo, Egypt, 1996; pp. 53–61.
28. Available online: https://en.wikipedia.org/wiki/Emirate_of_Nejd_and_Hasa (accessed on 9 November 2023).
29. Khaled Mahmoud, H. *Saudi National Heritage and UNESCO World Heritage List through the Saudi Vision (2030)*; The General Authority for Tourism and National Heritage: Qassim, Saudi Arabia, 2019. [CrossRef]
30. Available online: https://www.google.com/maps/@21.4992151,39.1820272,9331m/data=!3m1!1e3?hl=en&entry=ttu (accessed on 27 January 2024).
31. Available online: https://lostopportunitiesandwastedtimes.blogspot.com/2013/07/blog-post_28.html (accessed on 27 January 2024).
32. Available online: https://en.wikipedia.org/wiki/Jeddah#/media/File:Jeddah-1938.jpeg (accessed on 27 January 2024).
33. Available online: https://www.wafyapp.com/en/article/jeddahs-ancient-gate-bab-makkah (accessed on 28 January 2024).
34. Abdullah bin Zahir, A.-T. Architecture in the City of Jeddah in the Ottoman Era 923-1334 H/1517-1916 AD. Master's Thesis, King Abdulaziz University, Riyadh, Saudi Arabia, 2015.
35. Available online: https://al-turath.com/arabic/study-consolidation-and-restoration-of-jeddahs-high-risk-historic-buildings-umar-baeshen-house-historic-jeddah/?lang=en (accessed on 27 January 2024).
36. Available online: https://www.arabnews.com/node/1702936/saudi-arabi (accessed on 28 January 2024).
37. Mathew, R. *Study of the Historical District of Jeddah*; Ministry of Municipalities and Rural Affairs: Jeddah, Saudi Arabia, 1980.
38. Jeddah Municipality. Restoration Manual for Historical Buildings in Historical Jeddah. Saudi Arabia. Available online: https://al-turath.com/pdf/11.%20JEDDAH%20RESTORATION%20MANUAL.pdf (accessed on 28 January 2024).
39. Available online: https://www.arabnews.com/node/338618 (accessed on 28 January 2024).
40. Available online: https://news.iium.edu.my/?p=137354 (accessed on 28 January 2024).
41. Saudi Society for Urban Sciences and King Saud University. *Historic Jeddah: A World Heritage Site*; An Introductory Bulletin for Tourist Trails in the Historic City of Jeddah; Saudi Society for Urban Sciences: Riyadh, Saudi Arabia, 2014.
42. Alhojaly, R.A.; Alawad, A.A.; Ghabra, N.A. A Proposed Model of Assessing the Adaptive Reuse of Heritage Buildings in Historic Jeddah. *Buildings* **2022**, *12*, 406. [CrossRef]
43. Jeddah Municipality, Government Area. Historic Jeddah Building Requirements System, Jeddah Municipality, Saudi Arabia. 2014. Available online: https://www.jeddah.gov.sa/Business/LocalPlanning/HistoricalJeddah/pdf/Arabic/5.pdf (accessed on 7 November 2023).
44. King Abdulaziz City for Science and Technology (KACST) Institute of Earth and Space Sciences. Saudi Arabia. Available online: https://kacst.gov.sa/2022 (accessed on 28 October 2023).
45. Ministry of Municipal and Rural Affairs. The Comprehensive Urban Vision for Jeddah Governorate. Future Saudi Cities, Ministry of Municipal and Rural Affairs & the United Nations Human Settlements Programme-UN-Habitat, Riyadh, Saudi Arabia. 2019. Available online: https://unhabitat.org/sites/default/files/2020/04/jeddah.pdf (accessed on 6 November 2023).
46. Available online: https://en.wikipedia.org/wiki/Demolitions_in_Jeddah (accessed on 14 October 2023).
47. Ayman, I.; Mansour, H.; Amr, A.; Ibrahim, H. Exploring the Quality of Open Public Spaces in Historic Jeddah. *ACE Archit. City Environ.* **2023**, *18*, 52. [CrossRef]
48. Zaki, S.K.; Hegazy, I.R. Investigating the Challenges and Opportunities for Sustainable Waterfront Development in Jeddah City. *Int. J. Low Carbon Technol.* **2023**, *18*, 809–819. [CrossRef]

buildings

MDPI

Article

Between Laws and Trends: Unraveling the Dynamics of Vertical Housing Units' Development under Institutional Forces in the Brazilian Amazon

Frederico Guilherme Pamplona Moreira [1,*], Alexander Hierro Ferreira de Souza [2], José Maria Coelho Bassalo [2] and Milton Cordeiro Farias Filho [3]

1 Postgraduate Program in Civil Engineering, Federal University of Pará, Belém 66075-110, PA, Brazil

2 Faculty of Civil Engineering, Federal University of Pará, Belém 66075-110, PA, Brazil; alexander.souza@itec.ufpa.br (A.H.F.d.S.); jbassalo@ufpa.br (J.M.C.B.)

3 Multidisciplinary Postgraduate Program, University Center of Pará, Belém 66060-000, PA, Brazil; milton.filho@prof.cesupa.br

* Correspondence: fgpmoreira@ufpa.br

Abstract: This paper is based on the recognition of a phenomenon occurring in the real estate market of Belém, Pará, in the Brazilian Amazon. The phenomenon refers to the relative increase in the size of the balconies of residential apartments launched in the city between 2005 and 2017 by construction companies, with the aim of increasing the number of units built per tower. Given that balconies were originally areas of low frequency for the occupants of apartments in this region, the aim of this article was to identify and analyze the institutional factors that support this product development strategy, which is becoming increasingly consolidated among companies. To investigate this phenomenon, we collected data from official documents provided by the Municipal Urban Planning Department (Secretaria Municipal de Urbanismo—SEURB), specifically focusing on the launch of residential high-rise buildings during the specified period. Regression techniques with ordinary least squares methods were applied, as well as econometric models of spatial autocorrelation. The results of the hypothesis tests confirmed the relationship between the restrictive parameters imposed by the city's Urban Master Plan and the relative increase in apartment balconies. The spatial autocorrelation models confirmed the spatial spillover effect, showing that apartment projects with this characteristic tend to be concentrated in certain areas of the city, influencing each other. In order to discuss this phenomenon, the institutional theory was the protagonist of the analysis, showing how an organizational strategy can adapt to the authority that regulates the occupation of urban land in the city, meeting the two market forces and the demand and supply of apartments.

Keywords: real estate market; product development strategy; urban master plan; institutional theory; spatial spillover

1. Introduction

The product development strategies in the real estate market—residential and commercial units—represent a broad research field. As they establish the starting point of a production chain with strong sectoral interconnections, studies focused on this research area yield a variety of theoretical and empirical contributions. Macro- and micro-economic issues can be addressed, as well as those related to urban planning, health and safety management, the environment, and social well-being.

Despite its breadth, this research field has aspects that are not extensively explored by the scientific community. Between 2007 and 2021, the literature addressing product development for the real estate market has been dedicated to proposing models related to data and information management for business process development [1], models addressing subjective product attributes in purchasing decisions [2], and models addressing cost management [3]. Other studies associated real estate product development with the customer or market perspective [4–6] and project developer perspective [7].

In this context, the above-mentioned approaches have predominantly focused their efforts on understanding product development processes from the company's (builder's) perspective to the end of its value chain. In other words, these processes or strategies have been investigated based on the company's interpretation of consumer needs, tastes, and preferences, as well as the operational management of these processes, aiming to meet those tastes and preferences.

Therefore, the product development process that incorporates aspects related to the conception phase of the architectural project—from the company's perspective to the beginning of its value chain—remains underexplored. In this sense, the present study, although aligning with approaches emphasizing product attributes, as well as micro- and macro-economic aspects, primarily focuses on the impact that the environment, represented by institutional relations, has on the product development process. This process will be investigated in its preliminary phase when the architectural configuration of the property is defined by the builder and approved by regulatory authorities.

The institutional theory adopted as a theoretical framework in real estate market studies is not uncommon in academic circles. However, the authors of this research are unaware of those who have related it to the product development process. Hu et al. [8] made a special recommendation to adopt institutional theory as a theoretical foundation, seeking a better understanding of mega-project management, particularly in managing stakeholders, project planning, and procurement, as well as monitoring and controlling these projects. Prior to this contribution, Sawyer et al. [9] addressed changes and the increased complexity of the real estate market from its social structures and the importance of ICT.

Between 2019 and 2021, institutional theory has been employed to examine the role of institutions in shaping scenarios and decision-making structures within the housing industry, involving various actors such as NGOs, developers, and government [10]. Torreros et al. [11] conducted an evaluation of how institutions facilitate the emergence of new demand formats for fixed or seasonal residential units and the ensuing processes through which these units are booked via digital platforms. Additionally, Nappi and Eddial [12] delved into the disparities between the discourses of real estate market drivers regarding flexible office costs and the actual "hidden costs".

From our perspective, there is an important research gap to be filled. Obviously, cost, taste, and preference issues on the demand side remain important. However, this gap relates to the investigation that isolates the impact of institutions on product development processes in the real estate market, on the supply side.

In other words, at the moment when the construction company requests the architectural office to design residential unit projects in the building construction segment, important questions arise in this context: What is the influence of coercive institutions on these processes? How do companies react? Are business reactions homogeneous? What aspects do companies seek to reconcile, beyond institutions, when deciding what to build and how to sustain their choices? These questions will be better specified in the next section of this paper and will guide the results discussions.

Thus, our objective in this article is to identify and analyze the impact of institutions on the development of products—vertical residential units—in the city of Belém, Pará, which is situated in the Brazilian Amazon. The theoretical argument linked to institutional theory emphasizes its coercive aspect, represented mainly by the city's urban master plan, which restricts and provokes the strategic thinking of construction company managers, aiming for the maximization of economic performance. Therefore, starting from the conception of the architectural design commissioned by the builders, it must, from the established coercive constraints, encompass meeting the expectations of both the demand and supply sides.

This study uses a database composed of 122 residential buildings constructed in Belém between 2005 and 2017. Quantitative data were incorporated into inferential statistical models (regressions). Variables representing the physical characteristics of the apartment and the built land, as well as legal-coercive, socioeconomic, urbanistic, and temporal nature, were incorporated into the models as independent and control variables, aiming to identify their levels of correlation with the dependent variable.

This empirical strategy allowed for the identification of important results regarding the definition of the "balcony index" (dependent variable), which was measured based on the proportion of this architectural element—in square meters—relative to the total area of the apartment. Spatial models complemented and reinforced the analysis of the results, demonstrating the presence of "spatial spillover" among the observations (apartment buildings) in the sample.

The results achieved by this article represent theoretical and managerial contributions. The adoption of institutional theory in the discussion of results expands, diversifies, and supports the understanding of the strategic thinking of construction industry managers. The adoption of statistical models also broadens the types of possible empirical approaches, enriching this field of research. Managerially, the results found here can be useful for the development of public policies, especially those related to improving regulations governing urban space occupation.

The subsequent sections of this paper are organized as follows: Section 2 shows the contextual backdrop and outlines the research problems, along with stating this study's hypotheses. Section 3 delineates the data and empirical strategy, and the outcomes of the analysis are discussed in Section 4. The initial segment of Section 4 comprises descriptive analyses and sample characterization, while the latter part delves into explicating the institutional influences on the variation observed in the balcony index.

2. Context, Research Problem and Hypotheses

2.1. Study Area

This research evaluates the continental region of the municipality of Belém, the capital of the state of Pará. According to the Brazilian Institute of Geography and Statistics [13], Belém is the second most populous city in the Brazilian Amazon region, with around 1.3 million inhabitants. In addition, the municipality has a total area of 1059 km^2, of which its continental region covers 176.56 km^2.

The city has quite peculiar characteristics in terms of the occupation of urban land since its foundation in 1616. Its foundation was based on the Presépio Fort, built by the Portuguese in the Old Town to protect the city from invasions during the colonial period [14]. It is noteworthy that, since its foundation, the city has grown in a disorderly fashion [15], adapting to its restrictive characteristics: topographical and socioeconomic.

The mainland region of the municipality resembles a peninsula and is spatially divided into 44 neighborhoods, as shown in Figure 1. Approximately 40% of its territory is made up of areas known as "baixadas". These areas are influenced by the 14 watersheds that exist throughout the city's territory, and they are either flooded or subject to periodic flooding [16].

Regions with firm soil were the main targets of public and private investment, while areas with floodable soil were neglected during the beginning of the municipality's spatial development [17]. As a result, the waterfront of the municipality's continental region is underutilized by the real estate market.

Regional, environmental, and legal/historical factors also act as restrictive parameters for the exploitation of the municipality's territory by the real estate market. Around 18% of the territory consists of environmental conservation units, while the city's oldest neighborhoods—Reduto and Cidade Velha—have had the verticalization process contained due to legislation protecting historical heritage [16].

These restrictive factors have led to a severe shortage of land for residential buildings on the city's mainland. In recent years, this scenario has intensified the phenomenon of verticalization, which began in Belém in the 1930s, as a more modern way of living [18].

This urban context has created a challenging competitive environment for the companies that make up the oligopoly in the construction of vertical residential buildings. On the supply side, a relatively small number of companies are responsible for a significant portion of residential building launches, competing for the demand for apartments. In this context, the institutional role of Belém's Urban Master Plan (UMP) shares the leading role with these companies, regulating this market.

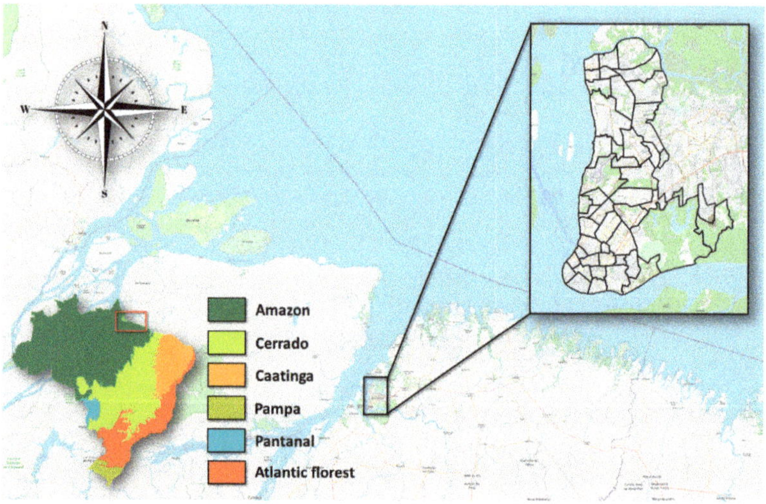

Figure 1. Belém, Pará—mainland region.

2.2. *Verticalization and the Phenomenon of Larger Balconies*

An attentive observer can easily see the increasing presence of residential apartment buildings in Belém, built since the beginning of the 2000s, with two strong characteristics: (a) buildings with a large number of floors; (b) apartment balconies that appear to be increasingly larger in relation to the total area of the apartment (balcony index).

Furthermore, these characteristics appear to be present in buildings concentrated in the city's noblest neighborhoods. Regardless of the total area of the apartment, as well as the construction company, the phenomenon of an increase in the balcony index (BI) seems to have been consolidated in the city. These findings seem to indicate that this is a product development strategy in Belém's real estate market, adopted by the aforementioned oligopoly in the pursuit of economic performance.

Given that the main objective of these companies is economic performance above the sector average, we would point out that the supply of residential apartments serves not only the interest of home buyers but also investors looking to receive long-term income through rentals. In this context, what would be the relationship between the proportional increase in balconies and the increase in the number of floors and, therefore, organizational performance?

We start with the simplest reasoning: in a competitive environment, the greater the number of vertical residential units built, the greater the company's overall sales value (OSV), and, therefore, the greater the business volume. This sequence of reasoning can lead to a reasonable number of research questions. The first research question is as follows: is there any relationship between an increase in the balcony index and an increase in units built and, therefore, an increase in organizational OSV?

If so, other questions arise: what is the involvement of the city's PDU in this context? Is it acting as a regulatory authority for the occupation of urban land? If there is a geographical concentration of buildings with apartments that have disproportionately large balconies, do they influence the launch of buildings with the same characteristics in neighboring districts (spatial spillover)?

Other questions might include the following: are there socioeconomic and urban infrastructure variables that could determine a greater geographical concentration of this phenomenon? Why does this strategy only seem to have been adopted since the early 2000s? Could the complementary competitive environment (complementary sectors) to the construction industry be supporting the occurrence of this phenomenon?

2.3. The Institutional Role of the Urban Master Plan

The intensity of verticalization in Belém has led to an institutional response from the Urban Master Plan to organize the growth and functioning of the city, seeking to guarantee the population a suitable place to live, work, and live with dignity [16]. As a result, construction companies are seeking competitive advantages, but not without first adapting to the UMP's restrictions.

In this study, we have simplified the investigation of the UMP's influence, exposing the restrictive issues of the plan, emphasizing only those that are clearly related to the research problems: the urban model (UM) and the utilization coefficient (UC), which are closely related to the size of the plot and its location.

According to Annex V of the UMP, which was made available by the city council, Belém's urban territory is segmented into seven urban environment zones (UEZ) (Zonas de Ambiente Urbano—ZAU) based on specific landscape and urban occupation patterns, urban problems, and potential and specific objectives [19]. UEZ3 and UEZ6 are subdivided into sectors (Sectors I to V). In line with the aim of this study and following Annex X of the UMP, the focus of this paper is on multi-family housing. Thus, distributed between UEZ1 and UEZ6, there are five types of urban models (UMs)—UM2 to UM6—as shown in Table 1.

Table 1. Urban models and main restrictions.

Urban Models	Lot's Area (m^2)—Min/Max	UC
UM2	360/—	1.4
UM3	400/—	2.0
UM4	450/—	2.5
UM5	600/—	3.3
UM6	750/—	3.5

"—" without restrictions.

The UMs are sets of more specific standards (parameters) that determine the construction restrictions linked to their respective sectors and UEZs. Thus, it is the UM that determines which parameters must be observed by the architectural firm when it is hired by the construction company to design the architectural project for a residential building in a given region of the city.

As an example, Table 1 shows that the UC in model UM2 is 1.4, while in model UM4, the UC is 2.5. Thus, the total area for an apartment building built on a 3000 m^2 lot could be a maximum of 4200 m^2 if it follows the UM2 model standard. In contrast, the UM4 standard allows this area to be up to 7500 m^2. Thus, the most promising residential apartment building projects for companies, in terms of OSV, are those that, firstly, are developed on large lots and, secondly, where the lot is located in a UEZ that allows the adoption of a UM with a high UC.

2.4. The Balcony Index and the Land's Legal Building Potential

According to the Urban Master Plan of the Municipality of Belém [19], the utilization coefficient "is the index that, multiplied by the area of the land, results in the maximum construction area allowed". Obviously, the higher the UC, the larger the total built area. In the case of family housing, only the private areas of the dwelling units are computed for the application of the UC, excluding, in some situations, balconies.

According to the Complementary Urban Control Law No. 02, of 19 July 1999 [20], in its Annex I, glossary of terms used, balconies are deemed non-computable when calculating the utilization coefficient, provided they do not exceed 5% of the area exclusively used by autonomous units up to 120.00 m^2. For larger units exceeding 120.00 m^2 and up to 180.00 m^2, the non-computable threshold is 10% of the exclusive use area. Additionally, for units surpassing 180.00 m^2, the non-computable threshold related to the exclusive use area is 15%.

Therefore, the areas that cannot be computed by the utilization coefficient are not part of the total legal building potential of the plot, even if they are part of the whole marketable area on the real estate market. From a legal point of view, the marketable area of each apartment or housing unit (UH) is made up of its fraction of the area established by the legal building potential, plus those that are not part of it (totally or partially), such as the areas of balconies.

It is, therefore, in the real estate market's financial interest to distribute the legal building potential over as many units as possible, and for this to happen, it is a fundamental strategy to minimize the use of the areas computable by the UC and maximize those that are not part of it. Figure 2 illustrates how local urban planning legislation allows the strategy of maximizing the marketable areas of the UHs, leaving balconies completely out of the calculation of a development's legal building potential.

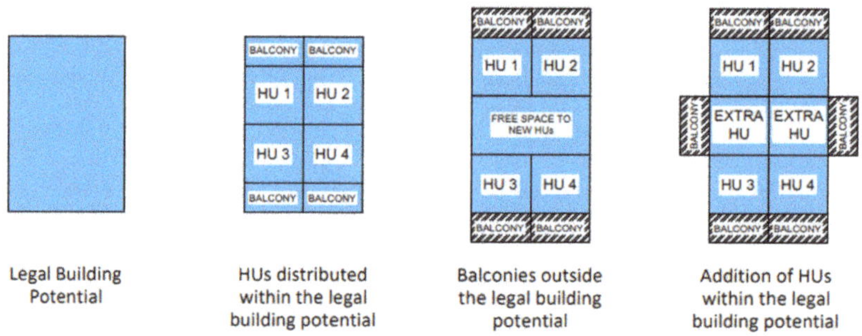

Figure 2. Balconies and the maximization of HUs within the legal building potential.

As we can see, increasing the area of balconies is a good strategy for making maximum use of the legal construction potential of each plot, within the limits of the law, and for increasing the financial results by increasing the number of floors in the building. Given the above, it is possible to formulate the first hypotheses of this research.

Hypothesis 1a. *The balcony index of apartments built between 2005 and 2017 in Belém, Pará, varies according to the urban model of the building.*

Hypothesis 1b. *The balcony index of these apartments is negatively related to the total area of the land on which the building was constructed.*

2.5. The Balcony Index and Spatial Spillover

In the real estate market, the social environment imposes standards and defines what is acceptable or permitted—through laws, such as the master plan, and regulations—restricting the heterogeneity of organizations [21]. In this way, and guided by these restrictive factors, the best-known type of organization—a company—seeks legitimacy in its operating environment.

This phenomenon, known in the institutional literature as isomorphism, consists of the tendency to adopt similar strategies, processes, and, consequently, products. Thus, institutional isomorphism forces the homogenization of companies facing similar market conditions and can occur through a combination of any of the following mechanisms: coercive, normative, and mimetic [22].

In the context of Belém, where the main restrictive factor for the construction of residential high-rise buildings is coercive—UMP—coercive isomorphism can have a direct influence on intensifying the process of adopting larger balconies. Coercive isomorphic pressures regulate the organizations' behaviors through the application of rules and their consequent monitoring, as well as the application of sanctions in the event of non-compliance [21].

In addition, construction companies, faced with the need to gain a competitive advantage and maximize production, tend to exploit the opportunities present between the lines of the market [23]. Therefore, when faced with various coercive pressures, these companies tend to develop similar strategies in order to avoid such penalties and gain the enhancement of the competitive position of their products and services within their industry [24].

The perception of organizations in relation to the successful strategies of others in the same industry leads to the mimetic aspect of isomorphism [25]. In the construction market, imitation occurs when companies adopt similar practices, both in technical and organizational aspects. Thus, institutional isomorphism—coercive and mimetic—may contribute to the phenomenon of spatial spillover.

Spatial spillover, induced by isomorphism in the construction industry, might be a determining factor in understanding how the strategies adopted by a construction company in a given location can influence neighboring areas. The imitation of successful practices not only reflects the search for operational efficiency and competitiveness [26] but also encourages the spread of construction standards on a geographical scale.

When a company incorporates new practices that have proved effective in its context, the spread of these practices to construction companies in nearby locations can result in a more comprehensive transformation of the regional construction scene. Thus, institutional isomorphism not only influences the individual decisions of organizations but also triggers a process of spatial convergence, where the adoption of similar practices contributes to greater homogeneity in strategies and operating methods throughout the production sector [22].

Thus, given the construction restrictions present in the territory of Belém, combined with the search by construction companies to maximize the production of units per building, the second hypothesis of this study can be presented:

Hypothesis 2. *The concentration of buildings with increasingly large balconies encourages the adoption of this strategy by neighboring apartment buildings.*

2.6. The Balcony Index and the Control Variables: IPC; INFRA; YEAR

Hypotheses 1a and 1b refer to the impacts that two restrictive parameters of the UMP may be having on the variation that the BI has shown in the vertical residential projects launched and built in Belém, Pará. Hypothesis 2 refers to the spatial influence of these projects. Both are the protagonists of this study.

However, it is not only these variables that can help explain the phenomenon of the increase in balconies in the period covered by this research. It is important to remember that this study investigates residential buildings distributed geographically across a city of more than 1 million inhabitants and 176 km^2 of territorial area.

Therefore, if the relationship between BI and the parameters of the UMP is confirmed, how does it behave when some variables associated with the urban region in which the focal residential building was built are incorporated into the model? These variables can also influence the occurrence of the phenomenon studied. To this end, this article proposes three more variables, called "control variables": (1) per capita income; (2) urban infrastructure; and (3) the year the residential building was built.

The per capita income of a given urban region relates the wealth of that location to its population. Therefore, if the per capita income of a region is above the average in a city, it can be said to have a high per capita income. Individual or family income determines financing capacity on the buyer's side, as well as influencing issues related to urban land occupation [27–29].

The amount of financing a family can obtain depending on their socioeconomic status is strongly correlated with the choice and decision to buy housing. Thus, the maximum amount a buyer can borrow to purchase a home is related not only to the value of the property but also to their income [27].

In a broader context, in terms of the urban real estate market, growing income inequality can have significant implications. At the upper extremes, an increase in the price-income ratio (HPIR) indicates a greater risk of a housing market bubble. Conversely, at the lower extremes, there can be an increase in the vacancy rate (HRV), resulting in "ghost towns" [28].

In cities with a severe shortage of land for buildings, such as Belém, another argument that highlights the issue of income in defining the locations of residential buildings is highlighted by Vergara-Perucich [29]. According to this author, higher income segments stimulate the supply of apartments by construction companies, with the aim of "financializing housing", i.e., UHs dedicated to investors looking to generate long-term income rather than housing.

Thus, in urban regions with higher per capita incomes—as in this study—a greater concentration of residential building launches is expected. Thus, we can say that the per capita income variable influences the definition of the locations of the buildings in our sample. In addition to influencing the locations of these buildings, we also found that, on average, they have an increasing balcony index. Therefore, we affirm that the relationship between land area and BI is controlled by the per capita income variable.

As for infrastructure, it is essential to understand the elements that contribute to sustainability and quality of life in urban areas and their consequent influence on the strategies employed in the Belém real estate market. To this end, in this study, the city's infrastructure is assessed through the proportions of coverage of public water supply services, energy supply, waste collection, and sewage treatment.

Water and energy are necessities for most, if not all, human activities, and the infrastructure needed to provide these features is essential for making urban spaces more habitable and productive. In addition, waste collection and sewage treatment are services that have a direct impact on the health and environment of cities, preventing the proliferation of diseases and the pollution of natural resources. Thus, urban infrastructure is a determining factor for the population's quality of life and for the economic development of regions [30,31].

In addition, infrastructure also influences the real estate market, as it affects the demand for and supply of real estate in urban areas. The demand for real estate depends, among other factors, on consumer preference, as consumers tend to look for places with better infrastructure and accessibility [32,33]. The supply of real estate, in turn, depends on the availability of land and the profitability of projects, which are affected by the level of existing infrastructure and the possibility of expansion or improvement [34].

Therefore, in regions with a higher percentage of urban infrastructure coverage, a higher concentration of vertical residential building launches of the standard covered by this research is expected. Therefore, similarly to the IPC variable, the infrastructure variable is expected to influence the definition of the locations of the buildings in our sample. In addition to influencing the locations of these buildings, we can also say that, on average, they have an increasing balcony index. Therefore, we affirm that the relationship between total land area and BI is controlled by the infrastructure variable.

With regard to the control variable "Year", we emphasize the upward trend in the balcony index of the UHs. From 2005 to 2017, the buildings in our sample showed an oscillation in the BI, with an average variation of seven percentage points. We, therefore, postulate that, like per capita income and infrastructure, this variable is statistically significant in our research context.

It is important to mention that regardless of the population's per capita income, the level of urban infrastructure coverage, and the period studied, the argument to be supported is that, on the supply side, these variables are related to the organizational strategy of increasing the proportion of balconies in apartments, especially as a response to institutional pressures.

However, on the demand side, how does the buyer receive a property with this particularity? How would this "extra area" be incorporated into the daily life of the occupants of the property, if this architectural element, by its very nature, is considered to be less frequented by the occupants of the property?

In this way, the opportunity arises to include another important aspect in the discussion of the results: is there an exogenous element to the building construction oligopoly that could contribute to the acceptance of this "novelty"? This question will not be the subject of hypotheses as it is outside the scope of this study. However, due to its relevance, it will be included in the discussion of the results of this research and could certainly be a future research agenda.

3. Research Methods

3.1. Data and Variables

In this paper, we analyze the balcony indexes of the UHs of high-rise buildings—comprising 5 or more floors [35]—present in the continental region of Belém, Pará. The initial database was assembled using data collected from the Municipal Urban Planning Department (Secretaria Municipal de Urbanismo—SEURB) between April 2018 and April 2019, based on access to three types of documentation relating to the buildings: (1) building permit/license; (2) habitation letter/housing certificate; and (3) architectural projects. As a result, the initial database had 194 observations, corresponding to high-rise buildings—residential and commercial—launched in Belém between 2005 and 2017.

In order to adjust the database to the research's scope, we excluded the observations relating to commercial buildings. In addition, to avoid specification problems in the spatial modeling, buildings with more than one UH were identified. For these buildings, the medians of the variables related to the structural characteristics of the apartments were taken as the representative value. This process resulted in a final database made up of 122 observations, corresponding to the representative UHs of each building. Notably, this dataset, representing 62.89% of the total high-rise buildings launched between 2005 and 2017, is considered a robust sample indicative of the study area.

For each observation, variables were collected relating to the structural characteristics of the housing units—APARTMENT_SIZE (total area, in square meters), BALCONY_SIZE (balcony area, in square meters), ROOMS (number of rooms), SUITES (number of suites), and YEAR (year of launch). Institutional aspects were captured through the variables MODEL (urban planning model), LAND_AREA (land area in square meters), and ADDRESS (complete location information). Furthermore, by utilizing the ADDRESS variable, we employed a geocoding service to obtain the geographical coordinates of each building and incorporated the LATITUDE and LONGITUDE variables into the database.

The dependent variable—BALCONY_INDEX (balcony index)—was computed by applying the ratio between the BALCONY_SIZE and APARTMENT_SIZE variables. In addition, to analyze the impact of the building's distribution on balcony index variation, we included the locational variable DOWNTOWN (distance to the historic city center in meters). This variable was calculated from the geographical coordinates of the buildings and Forte do Presépio, the founding site of Belém, using the Haversine formula, which measures the angular distance between two points on Earth (i.e., along a great circle).

Furthermore, we associated each building with its Human Development Unit (HDU) in order to compute the variables related to the socioeconomic and infrastructure conditions of the area in which the buildings are located. The HDUs, according to the United Nations

Development Program [36], are divisions of the municipality's urban fabric that have similar socioeconomic and infrastructure conditions. We emphasize that the boundaries of the HDUs do not necessarily coincide with the divisions of the neighborhoods and may cover multiple neighborhoods, or a single neighborhood may contain more than one HDU.

Socioeconomic influence was therefore incorporated by including the IPC variable (income per capita, in BRL/hab). Additionally, infrastructure was quantified by the INFRA variable, which is defined as the arithmetic mean of four indicators: (1) the percentage of households served by public water supply; (2) the percentage of households served by electricity supply; (3) the percentage of households served by an adequate sewage distribution system; and (4) the percentage of households served by waste collection and disposal services. Table 2 shows all the variables used in this study.

Table 2. Variables definition.

Variable	Description	Category
BALCONY_INDEX	Ratio between the balcony size and the total size of apartments	Structural
ROOMS	Number of rooms in the apartment unit	Structural
SUITES	Number of suites in the apartment unit	Structural
MODEL	Building urbanistic model (levels: M2; M4; M5; M6)	Structural
LAND_SIZE	Land size, in square meters (m^2)	Structural
YEAR	Building construction year	Structural
DOWNTOWN	Distance to downtown, in meters (m)	Locational
IPC	Income per capita, in the UDH where the building is situated	Socioeconomic
INFRA	Infrastructure coverage, in the UDH where the building is situated	Infrastructure

3.2. Empirical Strategy

To test the hypotheses of our study, various statistical techniques were employed. Among the most fundamental approaches, we used the ordinary least squares (OLS) regression model [37], as shown in Equation (1):

$$y = X\beta + \varepsilon \quad (1)$$

where y represents the vector of the dependent variable, X represents the matrix of independent variables, β is the vector containing the regression coefficients belonging to each independent variable, and ε is the vector of error terms.

Through this technique, our study aims to evaluate the influence of various characteristics—the structure of the UHs, location, and socioeconomic and urban infrastructure conditions—on the variation of the balcony indexes of housing units in Belém, Pará. Therefore, the natural logarithmic specification of the variable related to the HUs balcony index (LN_BALCONY) was used as the dependent variable of our study. This approach aims to examine relationships that can be represented in multiplicative terms, either on an exponential scale (level–log) or in the form of potentiation (log–log).

With regard to the independent variables, a natural logarithmic transformation was performed for the continuous variable LAND_SIZE in order to improve its distribution. It should be noted that this procedure resulted in a more significant association between the dependent variable and the transformed variable (LAND), compared to the original relationship as measured through Pearson's correlation. In addition, in order to address the heteroscedasticity, present in the OLS model, the robust standard errors technique was applied [38].

To assess the presence of spatial autocorrelation in the data, Moran's index was used [39], which ranges from −1 to 1, where a positive/negative value suggests positive/negative spatial autocorrelation or the presence of clusters in an observed area. The neighborhood matrix associated with the Moran index was computed using the k-nearest-neighbors approach [40]. This approach aims to avoid unbalanced connectivity throughout the neighborhood matrix since any observation will have the same number of neighbors, k.

The optimum value of k was determined using the Lagrange multipliers (LM) test applied to the LN_BALCONY variable, as well as to the OLS model residuals. The initial value of k was set to 1 and increased by 1 during each subsequent step. The selection process ended when the value of the LM test became insignificant, with k = 3 being the optimized number of neighbors, i.e., the maximum point of the Moran's index after its stabilization (LN_BALCONY Moran's I = 0.4359; p-value < 0.001 | OLS residual Moran's I = 0.0891; p-value < 0.05). Thus, positive spatial autocorrelation was established in both the dependent variable and the OLS residuals.

To incorporate spatial autocorrelation into the regression models used in this study, we adopted the classic spatial auto-regression (SAR) models. The SAR model takes two general forms: the spatial lag model (SLM) and the spatial error model (SEM). The spatial lag model assumes that the dependent variable is affected not only by the independent variables but also by the values of the dependent variable in nearby observations.

Thus, in the SLM a spatially lagged component—representing the spatial weighted average around the dependent variable—is added to the regression [41]. The specification of the spatial lag model is given by Equation (2):

$$y = \rho W y + X\beta + \varepsilon \quad (2)$$

where ρ is the spatial delay parameter and W is the spatial weight matrix that defines the neighborhood of each observation.

In contrast, the SEM assumes that spatial autocorrelation is due to unmodeled effects or other specification errors. Thus, spatial dependence, as well as its interaction, are treated as disturbances in the error term, rather than being incorporated as a spatially lagged component in the equation [41]. Equation (3) demonstrates the SEM specification.

$$y = X\beta + \mu \quad (3)$$

where $\mu = \lambda W\mu + \varepsilon$ is the vector of spatially correlated error terms, λ is the spatial error parameter, and ε is the vector of independent error terms.

Determining the most appropriate specification for spatial modeling was also guided by the LM test. The choice was based on the strength of the rejection of the test's null hypothesis, which postulates that the regression coefficient of the specified model is statistically equal to zero [42]. Thus, the spatial model chosen for use in this study was the SLM (LMSLM p-value < 0.05 | LMSEM p-value > 0.05), since it is the model that rejected the null hypothesis.

It is worth noting that the estimators of the spatial lag model, unlike those found in OLS, are not directly interpretable. In the spatial lag model (SLM), the presence of feedback effects from lags in the dependent variable causes changes in neighboring observations. Therefore, the parameters estimated by the SLM should be seen as representations of an equilibrium state in the modeling process, incorporating the effects of spatial diffusion [43].

In this scenario, the effects of each variable take the form of a matrix. This means that in order to fully understand the impact of each variable, we need to consider not only its direct effects but also how these effects spread and interact in surrounding areas. To this end, LeSage and Pace [44] recommend using scaling indicators to interpret the SLM model's estimators. These indicators include the following:

- Average direct effect, representing the effects caused by observations of an independent variable;
- Average indirect effect, which quantifies the diffusion effect between observations due to changes in an independent variable;
- Average total effect, which encompasses the total effect, direct and indirect, received by the dependent variable.

Finally, the comparison between the non-spatial and spatial models applied was performed using the Akaike Information Criterion (AIC). The AIC represents the relationship between the log-likelihood function—a statistical test used to compare the adequacy of two or more statistical models [45]—and the number of variables evaluated. The best model result is the one with the lowest AIC value.

4. Results and Discussion

4.1. Sample Characterization and Descriptive Analyses

The basic descriptive statistics of the variables used in this paper are shown in Table 3.

Table 3. Descriptive statistics.

Variable	Mean	SD	Min	Max
BALCONY_INDEX	0.163	0.0681	0.0283	0.376
SUITE	2.37	1.20	0.00	5.00
LAND_SIZE (m^2)	3565.00	9320.652	158.50	92,594.9
DOWNTOWN (m)	3899.85	2248.35	476.95	13,392.01
IPC (BRL/hab)	2573.49	1156.92	370.59	4342.04
INFRA (%)	98.23	1.39	87.03	99.56

The HUs evaluated have an average balcony index of 0.163, i.e., on average, around 16.3% of the area of these apartments is dedicated to balconies, ranging from lower values of 0.0283 (2.83%) to higher values of 0.376 (37.6%). Furthermore, the land size, representing one of the main limiting factors, has an average of 3565.00 m^2, with minimum values of 158.50 m^2.

Figure 3a shows that most of the buildings constructed on lots ranging from 158.50 m^2 to 1904.72 m^2 (59.83% of the sample) are concentrated in the south and southwest regions—one of the main centers of economic activity (commerce and services) [14]. This configuration results mainly from two aspects: greater financial attractiveness for companies and land scarcity due to the various factors that restrict exploitation by the real estate market. This condition also contributes to the prevalence of progressively smaller lots of urban land becoming available.

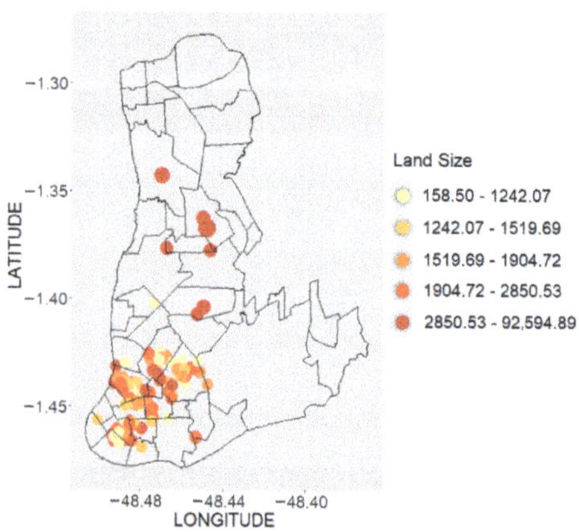

(a)

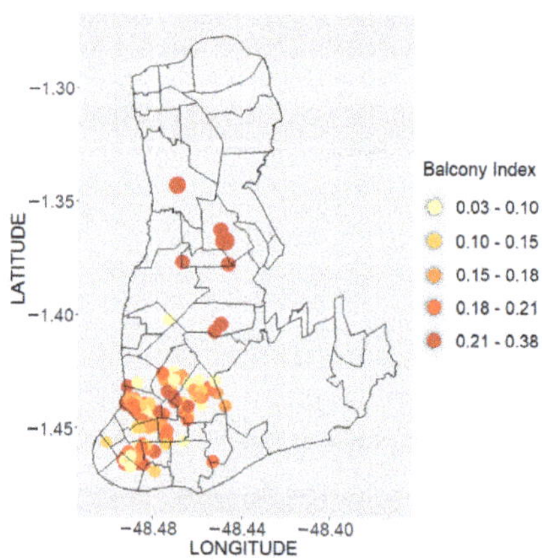

(b)

Figure 3. Distribution of buildings in mainland Belém. (**a**) Variation in land size; (**b**) variation in balcony index.

Figure 3 shows that the balcony index (Figure 3b) has an inverse spatial distribution to the land size (Figure 3a), tending to be higher in the south and southwest regions. In this way, the plot size and the balcony index appear to have a negative relationship, i.e., the smaller the land, the higher the balcony index. This relationship will definitively be tested by the inferential models (Hypothesis 1b).

In the context of socioeconomic aspects, Table 3 reveals a wide variability in income per capita, ranging from 370.59 BRL/hab to 4342.04 BRL/hab. This disparity, combined with the substantial standard deviation (SD) of the IPC variable—1156.92 BRL/hab—points to the heterogeneity present in this variable.

On the other hand, the significant average infrastructure coverage of the HDUs (98.23%) indicates that the residential buildings in this sample are well served by public services such as water supply, energy, garbage collection, and sewage. In addition, the low standard deviation (1.39) suggests that the HUs are concentrated in regions that are considerably homogeneous in terms of infrastructure coverage.

The variability in income per capita and infrastructure coverage in the neighborhoods of mainland Belém is shown in Figure 4.

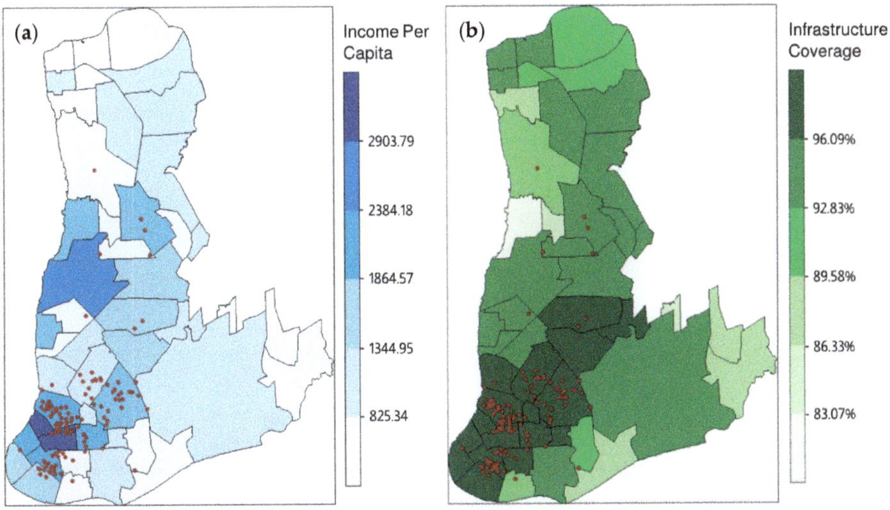

Figure 4. Spatial distribution of neighborhood characteristics in mainland Belém. (**a**) Per capita income; (**b**) infrastructure coverage.

Figure 4a shows the heterogeneity present in the distribution of income per capita and reveals that a large part of the sample (41.80%) is concentrated in neighborhoods with an IPC higher than 2384.18 BRL/hab: Batista Campos, Nazaré, São Brás, and Umarizal. Figure 4b shows that 96.72% of the UHs are located in neighborhoods with infrastructure coverage higher than 96.09%, reinforcing the homogeneity in the distribution of this variable.

Analysis of the correlation matrix, shown in Table 4, revealed significant relationships between the independent variables and the HUs balcony index. The variables LAND_SIZE and DOWNTOWN showed a negative relationship with BALCONY_INDEX, while SUITE, IPC, and INFRA showed a positive relationship. Furthermore, it can be seen in Table 4 that none of the independent variables show a substantially high correlation with each other, since none of the Pearson correlations exceed 0.7, showing that multicollinearity should not be a problem for the inferential models in this article.

The variation over time in the balcony index of housing units launched between 2005 and 2017 is illustrated in Figure 5. It can be seen that during the period evaluated, there was an increase of 7.73 percentage points in the average balcony index of apartments, rising from 0.1316 (13.16%) in 2005 to 0.2089 (20.89%) in 2017. The year 2015 saw the launch of units with the largest balcony areas in relation to the apartment total area, 3 units, with an average balcony index of 0.2622 (26.22%).

Table 4. Correlation matrix.

N	Variable	1	2	3	4	5	6
1	BALCONY_INDEX	1					
2	SUITE	0.47 ***	1				
3	LAND_SIZE	−0.23 *	−0.11	1			
4	DOWNTOWN	−0.38 ***	−0.35 ***	0.58 ***	1		
5	IPC	0.54 ***	0.45 ***	−0.29 **	−0.38 ***	1	
6	INFRA	0.25 **	0.15	−0.24 **	−0.60 ***	0.41 ***	1

* p-value < 0.05; ** p-value < 0.01; *** p-value < 0.001.

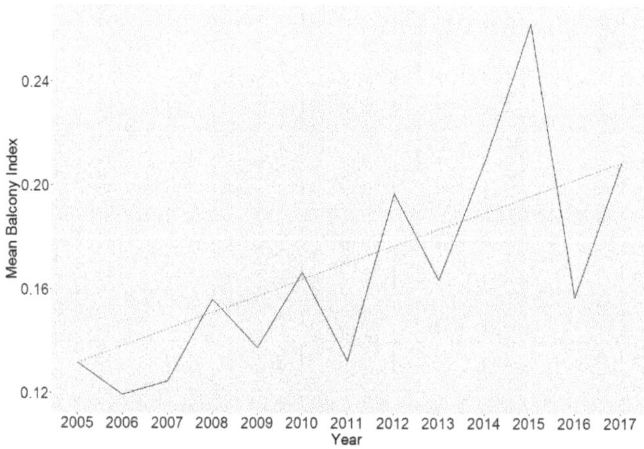

Figure 5. Temporal variation in the balcony index of HUs between 2005 and 2017.

Although the phenomenon of increasing balconies has varied over the years, with periods of abrupt decline, such as 2011, 2013, and 2016, in general, the adoption of this strategy seems well-established among construction companies during the period studied. Therefore, the research problem of this article becomes relevant given the need to understand the variables involved in this phenomenon in the context of the real estate market in Belém of Pará, considering its legal, economic, and urban implications.

Regarding the categorical variable related to the constructive models of the HUs (MODEL), it can be seen—based on Figure 6—that during the period studied, only five buildings were built, following the M2 standard.

Given that this UM is the most restrictive (refer to Table 1), these companies' behavior was already expected. Another characteristic common to these five observations is that they have a balcony index below 0.10 (10%), as shown in Table 5. This behavior may be reinforced by the fact that these buildings' temporal distribution shows that four observations were launched in the period from 2005 to 2006 and only one in 2009.

Table 5. Characteristics of the HUs adapted to the M2 urban model.

ID	BALCONY_INDEX	YEAR	NEIGHBORHOOD
3	0.0405	2005	Jurunas
7	0.0610	2009	Marco
10	0.0635	2006	Batista Campos
18	0.0863	2005	Batista Campos
24	0.0959	2006	Jurunas

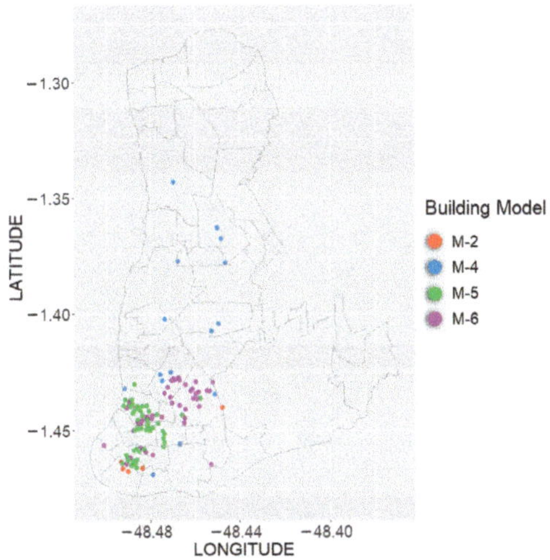

Figure 6. Spatial distribution of the HUs' building models.

Therefore, it can be seen that this specific pattern has been avoided throughout the period studied; this is possibly due to its low utilization coefficient (UC = 1.4). Furthermore, the low balcony index found in these buildings dates back to the early stages of the study period, when the strategy of adopting larger balconies might not have been established among construction companies.

4.2. *Hypotheses Testing*

The first question of our research asks the following: "Is there any relationship between the increase in the balcony index and the increase in the number of units (apartments) built and, consequently, the increase in the companies' OSV?" In order to answer this question, the first argument to consider is that the gradual increase in the balcony index in residential apartments has been a real phenomenon in the Belém real estate market since the early 2000s. This phenomenon is shown in Figure 5.

Although the number of launches stagnated in the last 4 years of the period studied—due to a serious economic and political crisis not covered in the scope of this article—the average number of launches was 10.5 apartment buildings per year [46]. Thus, despite the crisis, this performance seems to show that the gradual increase in balconies continues to serve both the supply side, construction companies, and the demand side, apartment buyers.

The second aspect refers to the legal argument of this context and incorporates the second question of our research, complementary to the first: "What is the involvement of the city's Urban Master Plan (UMP) in this context, as a regulatory authority?" To put it another way, in order to offer housing units with ever-larger balconies, construction companies need to adapt to the institutions that organize the growth and functioning of the city. In this way, the gradual increase in the balcony index must be legitimized by the city's UMP.

4.2.1. The Role of Institutional Theory

Answering the two questions, Hypotheses 1a and 1b relate the increase in the balcony index to the urban model (UM) and the utilization coefficient (UC), both determined by the UMP. The former determines all the parameters that restrict the construction of a building, and the latter is the parameter directly related to the size of the plot on which the apartment building is built. The OLS regression model in Table 6 confirms both hypotheses.

Table 6. OLS regression model.

Variable	Estimate	White Standard Error
(Intercept)	−0.8311 ***	20.94
SUITE	0.1057 ***	0.03013
MODEL: UM4	0.3854 °	0.2194
MODEL: UM5	0.4532 *	0.1973
MODEL: UM6	0.3741 °	0.1954
LAND	−0.1309 **	0.04887
YEAR	0.04011 ***	0.01035
DOWNTOWN	-2.669×10^{-5}	2.495×10^{-5}
IPC	8.773×10^{-5} *	4.069×10^{-5}
INFRA	8.202×10^{-3}	0.02747

Adjusted R-squared: 0.5067; Max. VIF: 3.6285; AIC: 96.2134 Breusch-Pagan Test: 17.302 *; White Test: 8.37 *

° *p*-value < 0.10; * *p*-value < 0.05; ** *p*-value < 0.01; *** *p*-value < 0.001.

In Hypothesis 1a, selecting UM2 as the reference, Table 6 shows that models UM4 (0.3854; *p*-value < 0.10), UM5 (0.4532; *p*-value < 0.05), and UM6 (0.3741; *p*-value < 0.10) have statistical significance and a positive sign, demonstrating that these three models are more associated with the increase in balconies than UM2. This result is consistent with the fact that Model 2 is the most restrictive among the four urban models and, probably for this reason, arbitrates the urban areas that received only 5 of the 122 buildings in our sample.

In Hypothesis 1b, Table 6 shows that the total land area (LAND) has statistical significance and a negative sign (−0.1309; *p*-value < 0.01), showing that the apartments with the highest balcony rates are found in buildings that were built on the smallest plots. Therefore, the influence of the urban model and the utilization coefficients linked to these plots of land are shown.

To deepen our discussions, we turn to institutional theory (IT). We begin by emphasizing the seminal contributions of the field, highlighting institutions as cognitive, normative, and regulatory structures and activities that provide stability and meaning to social behavior [47], reducing the uncertainty of the environment by establishing stable structures of interaction [48]. These structures are expressed by conventions, codes of conduct, norms of behavior, laws, and contracts,

When viewed from the perspective of human behavior, the application of IT is associated with two main focuses. The first covers individuals or groups guided by calculations that maximize their interests. On the other hand, the second focuses on actors who follow social commitments guided by habits, customs, and values, without the logic of optimization [49].

In this context, companies are organizations participating in an environment driven by relationships between well-defined parties, with some degree of clearly recognized order [50]. The company is a classic type of formal organization; it is the subject of various approaches to organizational analysis. These organizations seek material and social progress and are guided or mediated by technologies [51]. In this type of approach, the central role of institutions is highlighted, perceiving the competitive environment with a more ecological vision, focused on the role of selection and adaptation [52].

All this reasoning, adapted to our research problem, gives the UMP a central institutional role, establishing rules and norms that bring stability to the competitive environment among construction companies in Belém. In order to maximize their interests, these companies implement product development strategies that are legitimized by institutions. Entrepreneurial capacity determines the ability to adapt, which some companies possess more than others.

Thus, by confirming Hypotheses 1a and 1b, the first and the second research questions are answered: the UMP, through restrictions related to urban models, as well as the utilization coefficient, regulates the occupation of urban land and triggers the strategic thinking of construction industry managers. This strategic thinking establishes the proportional increase of balconies in relation to the area of the apartment, "freeing up" more space for the construction of a greater number of apartments/floors, up to the limit of the utilization coefficient. Consequently, an increase in the organization's OSV results in enhanced investment capacity in new technologies and management methods, thereby impacting the construction sector as a whole. Other significant impacts in this industry are linked to job creation and income generation.

4.2.2. The Spatial Spillover

The institutional theory can also be applied to answer the third research question of this paper: "If there is a geographical concentration of buildings with apartments that have disproportionately large balconies, do they influence the launch of buildings with the same characteristics in neighboring districts (spatial spillover)?" In order to answer this question, the presence of positive spatial autocorrelation found in the balcony indexes of the HUs evaluated during the empirical strategy section must be considered (LN_BALCONY Moran's I = 0.4359; *p*-value < 0.001).

Such phenomenon is strengthened by Figure 3b, which shows that apartments with the largest balcony areas in relation to their total area are located in nearby regions, mainly in the Batista Campos, Marco, Nazaré, and Umarizal neighborhoods. Answering this question, Hypothesis 2 states that the concentration of buildings with increasingly large balconies encourages the adoption of this strategy by neighboring buildings. The SLM regression model shown in Table 6 confirms this hypothesis.

In Hypothesis 2, Table 7 shows that the spatial lag parameter (ρ) is statistically significant (0.2308; *p*-value < 0.05) and also exhibits a positive sign, confirming the presence of spatial spillover in the strategy of maximizing the legal building potential by adopting larger balcony indexes. Additionally, the AIC test between the OLS (96.2134) and SLM (92.117) models was highly significant, indicating that the spatial lag model provides a better fit to the data due to its lower AIC, as shown in the Empirical Strategy subsection.

Table 7. Spatial specification—SLM.

Variable	Estimate	Standard Error
(Intercept)	−87.654 ***	19.516
SUITE	0.0958 ***	0.02801
MODEL: M4	0.3776 °	0.20462
MODEL: M5	0.423 *	0.18498
MODEL: M6	0.3414 °	0.18279
LAND	−0.1015 *	0.04546
YEAR	0.0423 ***	0.00964
DOWNTOWN	-2.156×10^{-5}	2.375×10^{-5}
IPC	6.055×10^{-5}	3.878×10^{-5}
INFRA	0.0136	0.02557
Rho (ρ): 0.2308 *; Adjusted R-squared: NA; AIC: 92.117		

° *p*-value < 0.10; * *p*-value < 0.05; *** *p*-value < 0.001.

In the case of our sample, there are a variety of companies that have incorporated the strategy of increasing the balcony index. This spatial pattern can be interpreted as a manifestation of institutional isomorphism, which is another approach to help explain the competition phenomenon among companies. This approach considers that the environment is dominated by institutional rules and constraints, resulting in the similarity of organizational models, based on the phenomenon of institutional isomorphism [22].

For our research problem, the type of institutional isomorphism seems to be coercive, since it results from strong external influences, both formal and informal, exerted on organizations by stakeholders [22], as well as the application of sanctions in the event of non-compliance [21].

Mimetic isomorphism also seems to occur from the perception of organizations in relation to the successful actions of others in the institutional field [25]. Thus, mimetic processes reveal the influential role of uncertainty, acting as a formidable force that drives imitation and the broader embrace of institutional rules and norms. In certain cases, entities face ambiguous problems or uncertain environments, making imitation a viable solution [26].

In the real estate market, mimicry or imitation occurs when companies adopt similar practices, both in technical and organizational aspects. This reaction is a response to uncertainty and a lack of knowledge about the most effective or market-accepted practices. In addition, due to the high costs required to solve specific problems, organizations tend to adopt similar successful organizations' strategies as a reference in order to adapt to the demands of the sector and seek competitive advantage [53].

Furthermore, institutional isomorphism plays a crucial role in shaping the dynamics of the real estate market in Belém, Pará. Companies operating in this context, driven by external pressures, particularly those arising from the Urban Master Plan, act strategically to formulate effective business strategies. As a result, they manage to enhance their competitive advantage through the exploration of opportunities within the UMP, consequently maximizing the production and sales of HUs.

Moreover, as companies observe the successful strategies employed by their counterparts in response to these external pressures, a pattern of imitation emerges. These observed practices become a viable approach to ensuring adaptability within the industry. This mimetic behavior, driven by the uncertainty inherent in the real estate market, further contributes to the homogenization of strategies among companies.

Thus, by confirming Hypothesis 2, the third research question is answered: the geographical concentration of buildings with apartments that have disproportionately large balconies reflects a spillover effect in neighboring buildings, influencing the launch of buildings with similar characteristics. This result is a convergence towards common practices, emphasizing the powerful influence of institutional isomorphism in fostering similarities and shared responses among market participants.

Regarding the coefficients of the SLM model, we calculated the average of the direct and indirect effects to obtain the semi-elasticity of the model. This enabled us to capture the effects caused by changes in the characteristics of neighboring buildings. As indicated in the Empirical Strategy subsection, the direct effects represent the impact caused by observations of an independent variable, while the indirect effects quantify the diffusion effect among observations due to changes in an independent variable—spatial spillover [44].

Table 8 shows the direct, indirect, and total effects, which are equal to the sum of both the direct and indirect effects, of the explanatory variables.

Table 8. SLM—direct, indirect, and total effects.

Variable	Direct	Indirect	Total
SUITE	0.0972 ***	0.0274 °	0.1246 ***
MODEL: M4	0.3828 °	0.1081	0.4909 °
MODEL: M5	0.4288 *	0.1211	0.5499 *
MODEL: M6	0.3461 °	0.0977	0.4438 °
LAND	−0.1029 *	−0.0291	−0.1319 *
YEAR	0.0428 ***	0.0121 °	0.0549 ***
DOWNTOWN	-2.185×10^{-5}	-6.173×10^{-6}	-2.803×10^{-5}
IPC	6.138×10^{-5} °	1.734×10^{-5}	7.872×10^{-5} °
INFRA	0.0138	3.906×10^{-3}	0.0177

° *p*-value < 0.10; * *p*-value < 0.05; *** *p*-value < 0.001.

Concerning the direct effects, as shown in Table 8, all the explanatory variables, except for DOWNTOWN and INFRA, are significant and exhibit the expected signs. The confirmation of Hypotheses 1a and 1b by the OLS model in Table 6 is corroborated by the

SLM direct effects. This is evident in all the urban models, UM4 (0.3828; p-value < 0.10), UM5 (0.4288; p-value < 0.05), and UM6 (0.3461; p-value < 0.10), and the LAND variable (−0.1029; p-value < 0.05) maintain their significance. These results reinforce the central institutional role of the UMP in bringing stability to the competitive environment among construction companies in Belém and serve as a robustness check for our previous findings.

As for the structural variable concerning the number of suites of each HU, it can be seen in Table 8 that both its direct (0.0972; p-value < 0.001) and indirect effects (0.0274; p-value) are positive and significant. In other words, it can be inferred that the direct effect associated with the variable SUITE indicates that for each extra suite, the balcony index of a housing unit will increase by approximately 9.72%, while the indirect effect is approximately 2.74%. This could be due to the fact that an apartment can be indirectly influenced by being located next to other apartments with better characteristics.

The significant spillover effect of SUITE can be attributed to the fact that the balcony index of an HU "i" is influenced not only by the balcony index of the neighboring housing units (as can be seen in Figure 3b) but also by the structural characteristics of housing units surrounding HU "i". Put simply, improvements in my neighbors' apartment features will likely "compel" me to enhance mine as well, given that they are my competitors. Consequently, apartments in close proximity are more likely to share similar structural characteristics rather than the other way around. Hence, the HUs with similar characteristics are clustered geographically [54].

4.2.3. Control Variables Implications

In relation to the control variables, when dealing with the IPC variable, the two models, OLS and SLM—direct effects—showed a positive relationship: (8.773×10^{-5}; p-value < 0.05) and (6.138×10^{-5}; p-value < 0.1), respectively. These results indicate that HDUs with higher per capita income are more likely to receive buildings with higher balcony indexes. This association can be attributed to the buyer's financing capacity [27–29] and its subsequent influence on the urban occupancy pattern.

The historic factor of Belém's urban development also plays a significant role in explaining this result. HDUs with lower per capita incomes tend to exhibit less verticalization and are often avoided by the middle and high-end real estate market [17]. As a result, the neighborhoods with a higher number of buildings, such as Batista Campos, Marco, Nazaré, and Umarizal, consequently exhibiting a greater mean balcony index, are often situated in HDUs with high per capita incomes.

Additionally, the absence of significance found for the variable related to infrastructure coverage (INFRA) in both OLS and SLM models shows that this result can be associated with the homogeneity present in the distribution of this variable, as shown in Figure 4b. Therefore, although urban infrastructure is a determining factor for the economic development [31,32] and the consequent exploitation of the region by the real estate market [34], its uniformity in the sample under analysis may explain the lack of a discernible impact on the dependent variable of the models.

Finally, regarding the control variable YEAR, both OLS and SLM—direct effects—models showed a positive relationship: (0.04011; p-value < 0.001) and (0.0428; p-value < 0.001), respectively. These findings imply that newer buildings tend to have higher balcony indexes. This association can primarily be attributed to the nature of the construction industry, where achieving the intended durability and complexity of the final product is capital-intensive, involves a large number of laborers, and takes a long time to complete [55].

These findings are further substantiated by Figure 5, illustrating a gradual growth in the balcony index of housing units at the onset of the analyzed period. Given that a majority of the buildings in our sample had construction times ranging from 4 to 5 years, it is clear that the integration of larger balconies has been a strategy that requires time for acknowledgment and legitimization within the social environment [56]—Belém real estate market.

Furthermore, the argument gains strength from the examination of the spillover effect associated with the YEAR variable in the SLM indirect effects (0.0121; p-value < 0.1)—Table 8. This aspect reveals that as buildings become newer, there is a corresponding increase in the balcony index of new housing units developed in neighboring areas. In essence, the newer the buildings, the higher the probability that the balcony index increases in the new HUs developed in adjacent regions, emphasizing a broader impact of the environment and temporal trends in the adoption of new characteristics by organizations [57]—in the context of this paper, architectural features.

4.2.4. The Demand Side

The arguments used in the discussion of the results so far have established that the strategy of increasing the draw rate is a real strategy and has also been tested and confirmed by inferential models. Thus, the phenomenon studied on the supply side (companies) was evidenced. However, in a market such as real estate, with products classified as an extreme case of differentiated goods [58], for a strategy to be successful, demand plays an important role.

Therefore, given that this particular product has been well received on the demand side, even though balconies are originally spaces that do not constitute part of the apartment's living area [20], why would a purchaser agree to buy a housing unit with increasingly larger "wasted" areas? Although the demand side has not been the protagonist in our paper, we consider it important to briefly incorporate this topic into the discussion, as it could be a viable future research agenda.

Our brief answer incorporates the theoretical framework of the organizational ecosystem, which consists of different "species" of companies, complementary or substitutes, acting in a competitive environment, sometimes collaboratively, sometimes as competitors [57]. In this specific study, we highlight the "balcony glazing" companies. These companies act as a complement to building firms, by incorporating this type of technology. In fact, the glazing technology has been around for 25 years but has only been established in Belém for a little more than 10 years.

This consolidation has led to the point where building companies have developed balconies in advance that are fully compatible with this technology, allowing them to be perfectly protected from adverse weather conditions and aesthetically aligned with the building's facades. In other words, balconies, which were not originally part of the living area of the apartments, ended up being attached to living rooms, bedrooms, and suites, thereby becoming fully usable areas for the apartment's occupants.

5. Conclusions

The aim of this paper was to identify and analyze the impact of institutions on the development of real estate products—vertical residential units—in the city of Belém, Pará, which is situated in the Brazilian Amazon. To this end, we investigated the increasingly consolidated occurrence of balconies that are proportionally larger when compared to the total area of the apartment (balcony index). This is a curious fact since, regionally, balconies have always been considered less frequented areas by the occupants of these apartments, mainly for climatic reasons.

Considering economic performance issues as primary conditions for companies in the sector, we first showed that the relative increase in balconies actually allows the company to launch and build more apartments per tower built. Thus, we established and confirmed the main hypotheses by relating the increase in the balcony index to restrictive factors present in the city's Urban Master Plan. In this way, by increasing the balcony index, companies meet the requirements of the law, obeying the urban planning models to which the plots belong, as well as the utilization coefficient linked to these plots and, consequently, their dimensions.

In addition, we also confirmed the hypothesis that towers with relatively large balconies influence their neighborhoods. Put another way, there are more and more launches with apartments with the same characteristics in certain areas of the city, characterizing the concentration of these towers, which the literature calls "spatial spillover".

In this sense, it was possible to consistently understand the strong level of influence of institutions in Belém's real estate market, legitimizing their strategies. The central lessons of this research can be understood from at least two aspects: the first concerns the very nature of the relationship between the regulatory authority and the companies, which is a relationship that at first would be simply and exclusively one of limitation and coercion and can, in fact, provoke a re-signification of the concept or way of living, "bringing into" the apartment areas that were previously less used. The second aspect reinforces the understanding of the residential real estate market product as a differentiated good, establishing a competitive environment of high investments in which companies imitate each other to an important degree, seeking to reduce uncertainties.

With regard to the impacts on the city, we can consider that the proportional increase in the balcony index, although well accepted by apartment buyers, can certainly bring up issues that should be discussed by the people drafting the Urban Master Plan. As the glazing of balconies spreads, issues related to the availability of views or natural ventilation for immediate neighbors may be questioned, as the apartment building takes on an opaque format. In a city located in the Amazon region, the free and strong circulation of wind can be very important for general well-being. Therefore, these concerns highlight the potential necessity for a revision of the city's master plan.

In reality, regardless of the region in which a city is located, verticalization is a common phenomenon in Brazilian big cities. In this context, even though the institutions that regulate this environment differ significantly among the states, another essential conclusion is that this type of debate can be faced by the entire Brazilian territory. Finally, our research shows the importance of engineering and architecture in mediating the dilemma of a business–client city.

One of the limitations of our study concerns the period studied. The number of years and, consequently, the sample size could have been greater. However, the data were collected from an institution through manual extraction from hard-copy records, which were not available in digital format. The institution retains these documents, with complete access granted solely from the year 2005 onwards. Moreover, it is important to note that data predating 2005 were subject to poor conservation and had many missing values. Prior to 2005, the institution utilized a different data collection method, resulting in a database with incomplete values and an inadequate format.

Concerning the acquisition of data beyond 2017, limitations arise due to delays in negotiating the renewal of permission to collect the data, which remains in a non-digital format and is not accessible through open access. Upon the conclusion of these negotiations, it will be crucial to extend this period. Future research agendas could explore the relationships between complementary and substitute companies in Belém's real estate market, with the potential to identify innovation in the construction sector. Additionally, incorporating the perceptions of apartment buyers through questionnaires could provide valuable insights, helping to identify new trends in architectural configurations.

Author Contributions: Conceptualization, F.G.P.M. and J.M.C.B.; methodology, F.G.P.M. and A.H.F.d.S.; software, F.G.P.M. and A.H.F.d.S.; validation, F.G.P.M. and A.H.F.d.S.; formal analysis, F.G.P.M. and A.H.F.d.S.; investigation, A.H.F.d.S.; resources, F.G.P.M. and A.H.F.d.S.; data curation, F.G.P.M. and A.H.F.d.S.; writing—original draft preparation, F.G.P.M., A.H.F.d.S., J.M.C.B., and M.C.F.F.; writing—review and editing, F.G.P.M., A.H.F.d.S., J.M.C.B. and M.C.F.F.; visualization, F.G.P.M., A.H.F.d.S., J.M.C.B. and M.C.F.F.; supervision, F.G.P.M.; project administration, F.G.P.M.; funding acquisition, F.G.P.M. All authors have read and agreed to the published version of the manuscript.

Funding: The authors received no financial support for the research, but received financial support for the publication of this article from the Qualified Publication Support Program of the Department of Research and Postgraduate Studies at the Federal University of Pará (PROPESP/PAPQ/UFPA).

Data Availability Statement: Restrictions apply to the availability of these data. The data was obtained from SEURB Department and not are available due to a term of commitment signed by the institutions and the authors.

Conflicts of Interest: The authors declare no conflicts of interest.

References

1. Yunita, Y.; Galinium, M.; Lukas, L. Integrating data mining technique and AHP in market analysis to propose new product development in real estate. *IOP Conf. Ser. Mater. Sci. Eng.* **2017**, *166*, 012030. [CrossRef]
2. Llinares, C.; Page, A.F. Kano's model in Kansei Engineering to evaluate subjective real estate consumer preferences. *Int. J. Ind. Ergon.* **2011**, *41*, 233–246. [CrossRef]
3. de Melo, R.S.S.; Granja, A.D. Guidelines for target costing adoption in the development of products for the residential real estate market. *Ambiente Construído* **2017**, *17*, 153–165. [CrossRef]
4. Christensen, P.; Robinson, S.; Simons, R. The Application of Mixed Methods: Using a Crossover Analysis Strategy for Product Development in Real Estate. *J. Real Estate Lit.* **2016**, *24*, 429–451. [CrossRef]
5. Oliva, C.A.; Granja, A.D. Proposta para adoção do Target Value Design (TVD) na gestão do processo de projeto de empreendimentos imobiliários. *Ambiente Construído* **2015**, *15*, 131–147. [CrossRef]
6. Steegmans, J.; de Bruin, J. Online housing search dataset: Information flows of real estate platform users. *Data Brief* **2021**, *38*, 107327. [CrossRef] [PubMed]
7. Kron, C. Application of Product Development Processes in the Early Phases of Real Estate Development: A Feasibility Study. In Proceedings of the IGLC 23, Perth, WA, Australia, 29–31 July 2015; p. 589.
8. Hu, Y.; Chan, A.P.; Le, Y.; Jin, R. From construction megaproject management to complex project management: Bibliographic analysis. *J. Manag. Eng.* **2015**, *31*, 04014052. [CrossRef]
9. Sawyer, S.; Wigand, R.T.; Crowston, K. Redefining access: Uses and roles of information and communication technologies in the US residential real estate industry from 1995 to 2005. *J. Inf. Technol.* **2005**, *20*, 213–223. [CrossRef]
10. Cheah, C.W. The social-political roles of NGOs: A study on a triadic business network. *J. Bus. Ind. Mark.* **2019**, *34*, 994–1004. [CrossRef]
11. Torreros, L.G.; Medina, J.L.S.; Uvario, K.M.L.; Castro, R.C. Airbnb en Ajijic, Jalisco. Una nueva forma de turismo residencial e impulso al desarrollo inmobiliario. Un análisis desde la Nueva Teoría Institucional. *Pasos Rev. Tur. Patrim. Cult.* **2020**, *18*, 143–158.
12. Nappi, I.; Eddial, H. Real cost of flex-offices: Discourse and reality. *J. Corp. Real Estate* **2021**, *23*, 225–242. [CrossRef]
13. IBGE. "Censo 2022 | IBGE", Instituto Brasileiro de Geografia e Estatística—IBGE. Available online: https://www.ibge.gov.br/estatisticas/sociais/trabalho/22827-censo-demografico-2022.html (accessed on 29 October 2023).
14. Chaves, C.; Gonçalves, A.P.C. *O Mercado Público em Belém: Arquitetura e Inserção Urbanística. IV Colóquio Comércio E Cid. Uma Relaç. Origem Uberlândia Laboratório Comércio E Cid. Fac. Arquitetura E Urban*; Universidade São Paulo: São Paulo, Brazil, 2013.
15. Arruda, M.E.I.; Lucas, A.T.; Doebeli, L.S. Integração global e dissociação local: Uma análise temporal do planejamento de GPDUs por meio dos projetos Estação das Docas e Belém Porto Futuro. *Rev. Bras. Estud. Urbanos E Reg.* **2021**, *23*, e202113. [CrossRef]
16. SEGEP. "Aspectos do Município de Belém", Secretaria Municipal de Coordenação Geral do Planejamento e Gestão. Available online: https://anuario.belem.pa.gov.br/aspectos-do-municipio-de-belem/ (accessed on 14 March 2023).
17. Mendes, L.A.S. A produção do urbano e do imobiliário na região metropolitana de Belém. *Confins Rev. Fr. Brés. Géographie* **2020**, *147*. [CrossRef]
18. Chaves, C.; Miranda, L. Avenida Presidente Vargas: Onde Belém foi mais moderna. Um estudo sobre a verticalização da Avenida Presidente Vargas. *Semin. Arquitetura Mod. NA Amaz.* **2016**, *1*. Available online: https://arquiteturamodernanaamazonia.weebly.com/uploados/7/0/0/22/70024539/artigo_avenida_presidente_vargas_onde_bel%C3%89m_foi_moderna__2_.pdf (accessed on 15 May 2023).
19. SEGEP. "Lei de Diretrizes Orçamentárias para 2008", Secretaria Municipal de Coordenação Geral do Planejamento e Gestão. Available online: http://www.belem.pa.gov.br/segep/ldo_2008/ldo_2008.htm (accessed on 7 June 2023).
20. SEGEP. "Lei Complementar de Controle Urbanístico de 1999", Câmara Municipal de Belém. Available online: http://www.belem.pa.gov.br/planodiretor/pdfs_legislacao/lccu.pdf?id_lei=724 (accessed on 11 June 2023).
21. AbouAssi, K.; Bies, A. Relationships and resources: The isomorphism of nonprofit organizations' (NPO) self-regulation. *Public Manag. Rev.* **2018**, *20*, 1581–1601. [CrossRef]
22. DiMaggio, P.J.; Powell, W.W. The Iron Cage Revisited: Institutional Isomorphism and Collective Rationality in Organizational Fields. *Am. Sociol. Rev.* **1983**, *48*, 147–160. [CrossRef]
23. Sigalas, C.; Economou, V.P.; Georgopoulos, N.B. Developing a measure of competitive advantage. *J. Strategy Manag.* **2013**, *6*, 320–342. [CrossRef]
24. Farida, I.; Setiawan, D. Business Strategies and Competitive Advantage: The Role of Performance and Innovation. *J. Open Innov. Technol. Mark. Complex.* **2022**, *8*, 163. [CrossRef]

25. Hwang, K.; Choi, M. Effects of innovation-supportive culture and organizational citizenship behavior on e-government information system security stemming from mimetic isomorphism. *Gov. Inf. Q.* **2017**, *34*, 183–198. [CrossRef]
26. Assis, L.B.; Andrade, J.O.; Carvalho Neto, A.; Tanure, B.; Carrieri, A. O isomorfismo entre executivos nas maiores empresas brasileiras. *Gerais Rev. Interinst. Psicol.* **2010**, *3*, 95–107.
27. Tiwari, P.; Moriizumi, Y. Efficiency in housing finance: A comparative study of mortgage instruments in Japan. *Eur. J. Hous. Policy* **2003**, *3*, 267–288. [CrossRef]
28. Zhang, C.; Jia, S.; Yang, R. Housing affordability and housing vacancy in China: The role of income inequality. *J. Hous. Econ.* **2016**, *33*, 4–14. [CrossRef]
29. Vergara-Perucich, J.-F. *Typological Study of Financialized Housing in Chile: Verticalization in Estación Central*; Horizon Research Publishing: San Jose, CA, USA, 2021.
30. Ouyang, Y.; Cai, H.; Yu, X.; Li, Z. Capitalization of social infrastructure into China's urban and rural housing values: Empirical evidence from Bayesian Model Averaging. *Econ. Model.* **2022**, *107*, 105706. [CrossRef]
31. Sun, C.; Meng, X.; Ouyang, X.; Xu, M. Social cost of waste-to-energy (WTE) incineration siting: From the perspective of risk perception. *Environ. Impact Assess. Rev.* **2023**, *102*, 107204. [CrossRef]
32. Cui, N.; Gu, H.; Shen, T.; Feng, C. The impact of micro-level influencing factors on home value: A housing price-rent comparison. *Sustainability* **2018**, *10*, 4343. [CrossRef]
33. Mulley, C. Accessibility and Residential Land Value Uplift: Identifying Spatial Variations in the Accessibility Impacts of a Bus Transitway. *Urban Stud. Edinb. Scotl.* **2014**, *51*, 1707–1724. [CrossRef]
34. Dalberto, C.R.; Staduto, J.A.R. Uma análise das economias de aglomeração e seus efeitos sobre os salários industriais brasileiros. *Rev. Econ. Contemp.* **2013**, *17*, 539–569. [CrossRef]
35. Trindade, T.J., Jr. *Formação Metropolitana de Belém (1960–1997)*; Editora Paka-Tatu: Belém, Brazil, 2016.
36. UNDP. "Ipea—Instituto de Pesquisa Econômica Aplicada", Atlas do Desenvolvimento Humano no Brasil. Available online: https://www.ipea.gov.br/portal/categoria-projetos-e-estatisticas/9941-atlas-do-desenvolvimento-humano-no-brasil (accessed on 12 November 2023).
37. Greene, W.H. *Econometric Analysis*, 4th ed.; Prentice Hall: Upper Saddle River, NJ, USA, 2000; pp. 201–215.
38. White, H. A heteroskedasticity-consistent covariance matrix estimator and a direct test for heteroskedasticity. *Econom. J. Econom. Soc.* **1980**, *48*, 817–838. [CrossRef]
39. Moran, P.A. The interpretation of statistical maps. *J. R. Stat. Soc. Ser. B Methodol.* **1948**, *10*, 243–251. [CrossRef]
40. Bivand, R.S.; Pebesma, E.J.; Gómez-Rubio, V.; Pebesma, E.J. *Applied Spatial Data Analysis with R*; Springer: Berlin/Heidelberg, Germany, 2008; Volume 747248717.
41. Anselin, L. Under the hood issues in the specification and interpretation of spatial regression models. *Agric. Econ.* **2002**, *27*, 247–267. [CrossRef]
42. Anselin, L. *Spatial Econometrics: Methods and Models*; Springer Science & Business Media: Berlin/Heidelberg, Germany, 1988; Volume 4.
43. Ward, M.D.; Gleditsch, K.S. *Spatial Regression Models*; Sage Publications: New York, NY, USA, 2018; Volume 155.
44. LeSage, J.; Pace, R.K. *Introduction to Spatial Econometrics*; Chapman and Hall: London, UK; CRC: Boca Raton, FL, USA, 2009.
45. Akaike, H. A new look at the statistical model identification. *IEEE Trans. Autom. Control* **1974**, *19*, 716–723. [CrossRef]
46. Lima, G.V.B.d.A.; de Carvalho, A.C.G.; Moreira, F.G.P.; Bassalo, G.H.M. Real Estate Tendencies in High-Rise Residential Buildings: Case Study in Belém, Amazonia, Brazil. *J. Urban Plan. Dev.* **2021**, *147*, 05021033. [CrossRef]
47. Scott, W.R. W. Richard Scott (1995), Institutions and Organizations. Ideas, Interests and Identities: Paperback: 360 pages Publisher: Sage (1995) Language: English ISBN: 978-142242224. *M@n@gement* **2014**, *17*, 136. [CrossRef]
48. Meyer, J.W.; Rowan, B. Institutionalized Organizations: Formal Structure as Myth and Ceremony. *Am. J. Sociol.* **1977**, *83*, 340–363. [CrossRef]
49. David, R.J.; Tolbert, P.S.; Boghossian, J. Institutional Theory in Organization Studies. In *Oxford Research Encyclopedia of Business and Management*; Oxford University Press: Oxford, UK, 2019. [CrossRef]
50. Clegg, S.R.; Lawrence, T.B.; Hardy, C. *The Sage Handbook of Organization Studies*; Sage: London, UK, 2006; Available online: http://digital.casalini.it/9781446206898 (accessed on 24 October 2023).
51. Reed, M. Organizational Theorizing: A Historically Contested Terrain. In *The Sage Handbook of Organization Studies*; Sage: London, UK, 1996; Chapter 1.
52. Hannan, M.T.; Freeman, J. *Organizational Ecology*; Harvard University Press: Cambridge, MA, USA, 1989.
53. Currie, W.L. Institutional Isomorphism and Change: The National Programme for IT—10 Years on. *J. Inf. Technol.* **2012**, *27*, 236–248. [CrossRef]
54. Crone, T.M. House prices and the quality of public schools: What are we buying? *Bus. Rev.* **1998**, *9*, 3–14.
55. Pheng, L.S.; Hou, L.S. The Economy and the Construction Industry. In *Construction Quality and the Economy: A Study at the Firm Level*; Pheng, L.S., Hou, L.S., Eds.; Springer: Singapore, 2019; pp. 21–54. [CrossRef]
56. Odou, P.; Schill, M.; Chaney, D.; Roznowicz, C. Store support for local producers as a driver of legitimacy and purchase intentions: A moderated mediation model. *J. Clean. Prod.* **2023**, *394*, 136361. [CrossRef]

57. Morgan, G. *Imagens da Organização: Edição Executiva*; ATLAS: São Paulo, Brazil, 2007; p. 380.
58. Hill, R.J. Hedonic price indexes for residential housing: A survey, evaluation and taxonomy. *J. Econ. Surv.* **2013**, *27*, 879–914. [CrossRef]

buildings

MDPI

Article

Power Dynamics in Collaborative Governance Processes: A Case Study of a Disadvantaged Neighbourhood in Southern Spain

Antonio Delgado-Baena and Antonio Sianes *

Research Institute on Policies for Social Transformation, Universidad Loyola Andalucía, 14004 Cordoba, Spain; adelgadob@uloyola.es

* Correspondence: asianes@uloyola.es; Tel.: +34-957-222-100 (ext. 2220)

Abstract: This study focuses on assessing collaborative governance from the perspective of power dynamics in a disadvantaged neighbourhood in southern Spain: Las Palmeras. Increasing global urbanisation in recent decades has exacerbated problems of segregation within cities, posing a major challenge to achieving healthy urban environments for harmonious coexistence. Public policies have struggled to address this challenge, particularly following the adoption of the United Nations' 2030 Agenda for Sustainable Development and its Sustainable Development Goals (SDGs), which have underlined the importance of fostering inclusiveness and sustainability in urban environments. In this context, collaborative governance has emerged as a prominent strategy for democratising the implementation of public policies in urban contexts. However, the efficacy of collaborative governance is hampered by the dynamics of power relations, which limits its effectiveness. The aim of this study is understanding how these dynamics mediate collaborative processes in urban regeneration contexts, an area little addressed so far. Through a case study and the systematisation of experiences, the study proposes strategic recommendations for managing these dynamics within collaborative governance. The insights drawn from this study can serve as a basis for better addressing urban challenges, thus fostering greater inclusiveness and sustainability in urban areas as well as offering valid lines for future research in the field.

Keywords: collaborative governance; deliberative democracy; power; participation; 2030 agenda

1. Introduction

This study focuses on exploring power dynamics within collaborative governance processes in urban regeneration contexts. To contribute to this body of literature, this study focuses on the experience of Las Palmeras, a disadvantaged neighbourhood in Córdoba (Spain). For the last eight years, this neighbourhood has been developing a collaborative governance process promoted by a grassroots initiative: the "Comprehensive Plan" of Las Palmeras. The relevance of this study lies in its ability to shed light on the power dynamics shaping the implementation of this "Comprehensive Plan", highlighting their effects on the efficacy of the process. In addition, the uniqueness of the case is demonstrated by its status as a bottom-up experience.

Comprehensive planning in urban settings poses crucial challenges, as citizen collaboration and governance are intertwined in complex ways. To reach consensus on decisions, public bodies and stakeholders are articulated in a deliberative process called collaborative governance [1]. However, recent studies have noted the paradoxical and conflicting nature of collaborative governance [2–4]. Moreover, many authors have pointed out how such governance is mediated by power dynamics [5–11]. In this sense, the most recent trends argue for the importance of analysing collaborative governance processes with a logic of power, applying it to practical experiences at the micro level [3,5,9].

1.1. Contextual Framework: Collaborative Governance for Urban Regeneration

The increase and consolidation of segregated territories in urban spaces is a reality shared throughout the world. As Carlos [12] points out, cities reproduce social stratification through historical, economic, cultural, and political factors. Thus, structural and inequality issues shape certain suburban neighbourhoods, sometimes referred to as disadvantaged neighbourhoods [13,14]. These are usually formed from the breakdown of working-class neighbourhoods, marked by the gradual dismantling of the welfare state. Today, they host multiple problems ranging from precariousness to delinquency; hence, these neighbourhoods have been converted into territories of residential segregation.

A complex of circumstances characterises these as territories with high levels of conflict and violence, which causes a social stigma to their inhabitants. This structural issue, together with the complex nature of the reality of these neighbourhoods (multidimensional, multiscale, and multifactor), questions the urban regeneration models proposed to date [14]. This questioning has fostered a relevant debate on power dynamics [5,15–17].

The 2030 Agenda and the Sustainable Development Goals (SDGs) have echoed this debate, encouraging a participatory approach (SDG 16, Peace, Justice, and Strong Institutions) when promoting more inclusive and sustainable cities (SDG 11, Sustainable Cities and Communities). In this context, collaborative governance is understood as a deliberative process in which public entities and non-state actors work together to solve territorial or sectoral problems, democratising the implementation of public policies [18].

In line with it, numerous studies about collaborative governance have been conducted in recent years [13,19–22], highlighting their democratic and power-sharing nature [9,22]. Urban studies, such as those developed by Kamruzaman and others [23] and Amini Pishro and colleagues [24], have highlighted how mobility impacts communities and how approaches such as Transit-Oriented Development (TOD) offer strategies to overcome risks such as the displacement of communities [25], addressing governance challenges [26], influencing urban design for inclusion [27], and promoting the use of sustainable transport [28]. These approaches can balance environmental benefits with their social implications, highlighting the importance of collaborative governance that incorporates diverse voices to address power dynamics and ensure inclusive and equitable urban development. The social processes analysed by such a body of literature have in common the inclusion of stakeholders in the governance process. However, the characteristics of each experience depend on contextual factors. For example, in Spain, collaborative governance processes normally have revolved around the so-called "Comprehensive Plans" [29–31].

A "Comprehensive Plan" is a collaborative governance process that includes a multifaceted, comprehensive, and coordinated approach to the problems and challenges of an area or territory. Applied in disadvantaged urban contexts, it aims to improve quality of life, reduce inequality, and promote development. Such plans are called comprehensive as they address the complexity of the problems in areas such as housing, health, or education; imply the coordination of all levels of public administration, whether state, regional, or local; and assume the collaborative participation of the different types of actors present in the focal territory, whether political, technical, or civic. Thus, each "Comprehensive Plan" is unique because it is adapted to the particular circumstances of its territory [29]. The object of this study is, precisely, of one of those plans, as proposed by the citizens in a particular neighbourhood: Las Palmeras.

1.2. Conceptual Framework: Power Dynamics in Collaborative Governance Networks

Power dynamics, understood as the unequal relationships and subsequent tensions between individuals and groups in terms of influence, resources and authority, play a crucial role in decision-making about common interests [32]. Thus, power imbalances affect collaboration and decision-making in governance contexts. The most recent research has highlighted the importance of expanding the understanding of power dynamics to improve collaborative governance [5,16,17]. Understanding power relations is deemed essential for improving stakeholder engagement and motivation in collaborative governance [17].

Traditionally, power has been defined as the ability to influence, control, or direct the behaviours and actions of other people or groups through coercion, compensation, or persuasion [33]. The degree, scope, or even historicity of power has been studied, whereby it has been noted that power is mediated by the social and cultural structures affecting the structuring of organisations [34–36]. Therefore, applying this view to collaborative governance processes is crucial.

Collaborative governance presupposes that any deliberative process with different stakeholders jointly reaches consensus and makes collective decisions [9]. However, this implies planning and management [1]. Thus, collaborative governance processes must be analysed as a network of interacting actors, mediated by the power tensions of their roles. To navigate such tensions, collaborative governance adopts strategies to achieve the objectives of the network in a framework of democratic consensus. However, recent authors [7,15] have identified that the power relations among actors are projected in the collaborative network, generating situations of power asymmetry. In this way, collaborative governance becomes a dialectical tension between the benefit of sharing power and the asymmetry of power [9].

Such asymmetry of power shows that not all stakeholders present themselves on equal terms or with the same amount of power. Therefore, for the correct analysis of collaborative governance networks [9,11,37], an evaluation of the sources of power of every stakeholder is needed. This analysis can define their ability to influence their network. Accordingly, the literature has begun to clarify the different sources of power, such as formal authority, resource control, or discursive legitimacy, involved in participatory processes. These sources of power determine the roles of participants in a dynamic way, as they fluctuate depending on the arenas and flows of power [5].

The most refined theoretical framework for analysing the exercise of power in collaborative governance networks is the one proposed by Ran and Qi [9], a framework of contingent factors and a series of analytical hypotheses. In their proposal, the authors propose a specific hypothesis, i.e., how the more widespread the sources of power are, the more beneficial for the effectiveness of collaborative governance it is to share them. Their analysis also points to a dual categorisation of collaborative governance networks: mandatory or voluntary networks. The former are those subject to legal-based regulations, supported by hierarchical and bureaucratic mechanisms, and sustained through medium- or long-term planning. The latter are based on voluntary collaboration characterised by reciprocity, consensus, and mutual trust and linked to urgency and emergency.

Other authors have focused on the participants who collaborate in the network [38], who are commonly classified by the legal nature of their organisational structure as political, technical, or civic actors [39,40]. Other authors incorporate into this distinction actors' practices, discourses, or rationalities [41–43]. Thus, actors with political rationality, technical rationality, neighbourhood rationality, profit-making rationality, or bureaucratic rationality emerge. What is relevant in this view is that each stakeholder is traversed by various rationalities subject to spatial–temporal contexts [38]. The rationalities of the actors at each moment in the process condition their objectives in the network [41] and articulate the processes of collaboration in it, as rationality maintains a dialectical relationship with praxis [44].

Summarizing, collaborative governance networks and their effectiveness appear to be mediated by the type of network that is constituted; the nature and rationality of the actors that make it up; and contingent factors specific to each context. All these elements are traversed by the logics of power.

The objective of this study is to deepen the understanding of these power dynamics through the presentation of a localised case study, a methodology that has been proven effective for this purpose. To this end, this study examines how these power dynamics cut across collaborative governance practices in the urban regeneration of the Las Palmeras neighbourhood in Cordoba, offering new perspectives on the implementation of these practices in disadvantaged urban contexts.

As noted above, the case of Las Palmeras represents a differential experience of collaborative governance in a deprived urban area, as it is based on a grassroots initiative.

A combined approach of the systematisation of experiences [45] and case studies [46] has been used. This approach allowed for the collection and analysis of qualitative information from multiple sources, including interviews, participant observations, and documents of the process itself. This approach was selected for its ability to systematise practice and simultaneously provide a detailed and contextual understanding of governance processes, social relations, and their symbolic value.

The research reveals how collaborative governance practices in the context under study, while improving the implementation of public policies through collaboration between stakeholders, are traversed by power dynamics that affect governance, as Ansell and Gash [1] argue. Consequently, differences in access to resources and influence can affect the effectiveness of collaborative governance, as Ran and Qi [9] point out. Thus, the study suggests the need to take power dynamics into account in the practice of collaborative governance in order to influence its effectiveness.

This study offers significant practical insights, highlighting the importance of citizen autonomy, self-regulation, agenda co-management, and accountability. These aspects, together with an analysis of power dynamics, provide useful tools for those involved in urban governance and community development.

To this goal, the study is structured as follows: Following the above presentation of the contextual and conceptual frameworks, the adopted methods and materials are elaborated to build the narrative of the focal case study. The systematisation of the results obtained is then presented according to the categorisation established based on the participation levels, methodologies, and actors. Finally, the findings are discussed with respect to the conceptual framework. We close the study by making specific action proposals before appropriate conclusions are drawn.

2. Materials and Methods

To carry out the study, we apply a systematisation of experiences approach [45,47] to a case study [46,48]. The former connects with the tradition of Participatory Action Research, an approach that emphasises participation in the research process, oriented towards action and social change [49–51] and parallels storytelling, a method that uses personal and collective narratives as a research tool to capture experiences and perceptions, facilitating the qualitative analysis of social phenomena [52,53]. These approaches are essential to investigate the case in its complex circumstances, allowing for generalisable lessons to be drawn. It reconstructs the objective and subjective elements of an event from an analytical and critical reflection together with participants [47]. The latter investigates such event by considering its complex circumstances with the intention of understanding the focal phenomenon, which allows for the extraction of generalisable lessons [54].

This combination of approaches was selected given the capacity of systematisation to valorise the experience by generating knowledge for the participants themselves [45] and the case study given the uniqueness of the experience as a bottom-up initiative governance process [48].

As Jara suggests, the reconstruction of processes must incorporate the voice of those who have played a leading role in them. Various factors must be scrutinised: the political, economic, and social context of the experience; the specificity of the experience; the subjectivity of those who have lived the experience; the salient results or effects; and the relevant relationships [55]. In parallel, the case study method involves an empirical question related to the focal event and is relevant when it is difficult to distinguish between the context and the phenomenon [48,54,55].

Both approaches therefore involve the collection of a large amount of information from various sources. Thus, different sources of information are used to integrate the scientific and social discourse [54]. Although mixed methods is the most common methodology used in case studies [46], systematisation is typically based on qualitative sources of evidence such as interviews, documents, and observations [48], particularly in the dialogical approach [56]. In this way, the two approaches have been integrated, starting with the reconstruction of the process through narrative and dialogic techniques (systematisation) and moving on to a contextualised exploration of collaborative governance in a real-life setting (case study). This methodological combination facilitates a comprehensive understanding of power dynamics, linking the reflexive analysis of systematisation with the contextualisation offered by the case study.

In this study, this methodological approach has been applied to a collaborative governance process called the "Comprehensive Plan", which began in 2015 in the Las Palmeras neighbourhood (Córdoba, Spain). The objective of the research is to reconstruct this process from its genesis to the present by addressing the following research question: how are power dynamics configured in a collaborative governance process arising from a grassroots initiative?

The relevance of the case is illustrated by its clear representation of a bottom-up process that starts with a neighbourhood organisation and is configured to promote comprehensive policies of social inclusion and urban regeneration. Bottom-up processes are not a common focus in the literature, where top-down cases tend to prevail [1].

In the compilation of the study materials, the proposal of Koch [57] has been followed in gathering the following sources of information: academic papers on the case; documents generated in the process; the protocols of each phase; participant observations; notes from the main actors; and 5 interviews with key informants.

Interviews were selected following the criteria of direct involvement in the governance of the Integrated Plan ensuring diversity including community leaders, civil servants, and academics. Sources of information were selected following the criteria of relevance to the collaborative governance process or its stakeholders.

The information collected was managed with the qualitative research software NVIVO 12 [58], which was used to code and organise the data collected through the labelling of text segments, allowing for the identification of emerging patterns that make up the categories or themes of analysis [59,60]. This has facilitated the application of thematic analysis during the examination of the data, which allows for the identification, organisation, description, and analysis of the issues identified in the dataset [61,62]. This methodology is commonly used in case studies [20,21,63].

Braun and Clark's six-step model was used for coding [64]. This model allows researchers to identify patterns in the texts studied by co-differentiating them through the themes around which their analysis is structured [61]. To do this, a deep dive into the data was conducted, and then initial codes were generated by labelling relevant data (e.g., Comprehensive Plan, initiative, political perception, etc.). Subsequently, topics were identified by grouping the generated codes (e.g., participation, power, etc.). Finally, topics were reviewed to finally define and categorise the themes (levels of participation, methodologies, and actors) that open the door for systematisation.

To address potential biases and limitations inherent to the methodological approach adopted, such as those related to case selection or interpretative subjectivity, data triangulation was implemented using different sources (interviews, documentary analysis, observations, etc.), and criteria were applied to select the case, such as the fact that it was a bottom-up initiative, that it was alive, that it was a neighbourhood of exclusion, and that the promoters were still active. In addition, validation was sought from the participants by returning the study. These measures seek to ensure the robustness and accuracy of the findings, contributing to the rigour of the analysis.

Thus, to answer the research question, three themes of analysis were selected: the level of participation, observed through who holds the initiative in the process; the methodologies of participation, identified through the mechanisms of participation, the spaces, and the times; and the actors, determined through their identification and the rationalities that they activate. The conceptual model is introduced as Table 1.

Table 1. Themes and sub-themes in the thematic analysis.

Themes	Sub-Themes
Levels of participation	Initiatives
Methodologies	Mechanisms, spaces, and times
Actors	Participants, rationalities, and motivations

Source: author's own elaboration.

Given that the case is an ongoing process, once the analysis had been carried out, several implications and proposals were identified to address the limitations of this process, which can also be applied to other analogous processes affected by power dynamics.

3. Results

3.1. Presentation of the Case Study

The Andalusian city of Cordoba in southern Spain is a historic city. With a population of just 300,000, the city has four UNESCO World Heritage declarations, signalling the weight of the importance of the city's past. However, it now faces contemporary challenges typical of a southern European urban environment, such as social and economic inequality, and is home to four of the fifteen poorest neighbourhoods in Spain [65]. One of them is the Las Palmeras neighbourhood. With a population of around 3000 inhabitants and a housing stock of 719 homes, Las Palmeras is located at the extreme west of the city. It originated as a temporary shelter for hundreds of Cordovan families who had to be relocated due to the floods in 1963. The development of such neighbourhoods was part of the urban policy at the time, dominated by urban renewal and urban redevelopment initiatives [66].

The city's population dynamics, lack of resources and spatial segregation have shaped this neighbourhood with increasingly high indicators of social exclusion [67,68]. It has a per capita income of less than EUR 5000 per year and an unemployment rate of more than 60%, resulting in the strong presence of an underground economy, insecurity, and urban deterioration [69,70]. The persistence of these exclusion standards has motivated the implementation of urban and social regeneration programmes, as the neighbourhood is highly affected by public administration. However, such interventions have not reversed any situation of exclusion.

In 2015, this situation led residents in the neighbourhood to self-organise, following the tradition of social struggle in Las Palmeras, and the Neighbourhood Association "Unión y Esperanza de Las Palmeras" was established. It began its journey with the intention of developing a "Comprehensive Plan" for the development of the neighbourhood and to demand public administrations to develop more effective policies for social inequality in the neighbourhood. After a period of neighbourhood consultation, the "Comprehensive Plan" of Las Palmeras was approved in January 2018. According to the criterion of control or management of the Plan, four different relevant periods can be observed: 2015–2018, dominated by citizen action; 2018–2020, controlled by the public administration; 2020–2022, characterised by outsourcing; and 2022 to date, in which citizen relocation takes place.

3.1.1. 2015–2018: Genesis of the "Comprehensive Plan"

This period began in 2015 with the constitution of the Neighbourhood Association "Unión y Esperanza de Las Palmeras", which initiated the bottom-up consultation process. This period culminated in 2018 when the City Council, meeting in plenary session, unanimously approved the document provided by the Neighbourhood Association, and, thus, the "Comprehensive Plan" was born.

In this period, the process was managed by the Neighbourhood Association, which takes the initiative to open a participatory process to build a text with which to commit the public administration to implement a "Comprehensive Plan" in the neighbourhood. In the words of a neighbourhood leader,

"... that defence that this Neighbourhood Association is doing and that it will continue to do, is going to be the defence of the representation of this Comprehensive Plan of its neighbours"

(Neighbourhood leader 2)

The Neighbourhood Association implements listening mechanisms such as individual door-to-door interviews and neighbourhood meetings:

"So, the idea that occurred to me at that time, or occurred to us, was a bit of doing a neighbourhood interview. Block by block, that is, door to door, right? With what little I knew how to write that my daughter taught me. I know very little. I took my notebook and asked them, Hey, what would you like? How would you like things to be done in the neighbourhood? What do you think we should do?"

(Neighbourhood leader 1)

In this period, deliberative mechanisms were applied. In 2015, e.g., tables were set up in each square to discuss the needs of the neighbourhood. They also developed strategic mechanisms, such as the functional reinforcement of an existent roundtable of associations in the neighbourhood. Although in this phase protagonism was sustained by the Neighbourhood Association, the entities in the association's roundtable and neighbours without affiliation participate actively.

The spaces where the different milestones were developed are multiple: the squares, where informal meetings took place through word of mouth; public devices as the one known as "La Foggara", located on the edge of the neighbourhood; facilities of Loyola Andalucía University, with which the Neighbourhood Association signed a collaboration agreement in 2017; and the City Council itself, in which the Plan was ultimately approved in 2018.

This period was thus dominated by a neighbourhood strategy whose rational logic was to value the needs expressed and felt by the citizens. On the other hand, it also activated mechanisms of a political nature to position its process and its product in the "Comprehensive Plan". Meanwhile, the agreement with Loyola University maintained a technical-citizen rationale, shaping the technical orientation of the endogenous process to improve its political impact.

3.1.2. 2018–2020: The "Comprehensive Plan" as Public Policy

This period began with the approval of the "Comprehensive Plan" in January 2018 and ended in mid-2020 amid changes in political signals after the municipal elections which changed the conceptualisation of the plan and the model of dialogue with the neighbourhood. Also, mobility limitations imposed by the COVID-19 pandemic made face-to-face participatory actions difficult, thus defying the participative nature of the process. It was dominated by milestones related to the Public Administration, particularly the beginning of the ERACIS, a regional policy for combatting social exclusion heavily financed by the European Union.

Once the "Comprehensive Plan" was approved in the municipal plenary session, the initiative of the process passed into the hands of the City Council, while the Neighbourhood Association moved to the background while maintaining an active role in different community processes.

The City Council put up 5 of its own roundtables for the development of the "Comprehensive Plan". However, it ultimately only convened the education roundtable and the employment roundtable. In parallel, all the proposals included in the "Comprehensive Plan" were formally included in the *ERACIS* policy but without any consultative or participatory process. This mechanism guided the "Comprehensive Plan" towards community social services.

Regarding spaces, the City Council activated consultative/informative mechanisms, such as the aforementioned education and employment roundtables. However, the space established through the *ERACIS* prevailed: another roundtable at District level. This one gathered many social entities that provide services at the local, regional, or even state level but not the Neighbourhood Association or other neighbourhood-based entities. In this space, technical rationality and unidirectional information prevailed. A summary by a neighbourhood community radio station of one of these meetings is as follows:

> *"During the meeting, there was an exhibition of the actions carried out in the area and the situation of the labour itineraries carried out so far. Then, in the round of questions and answers, the attendees explained the challenges and difficulties that the ERACIS strategy was facing"* [71]

In parallel, the Neighbourhood Association and Loyola University initiated a socioeconomic diagnosis of the situation in the neighbourhood, establishing training and consultative mechanisms. While technical organisations participated in the roundtables promoted by the City Council, the Neighbourhood Association and youth of the neighbourhood collaborated in the socioeconomic diagnosis.

Beyond this initiative, the space par excellence became "La Foggara", the public building located on the edge of the neighbourhood run by community social services. The strategies implemented were based on a political–technical logic in which variables such as time and resources have important weight. In this sense, the impact of the COVID-19 pandemic and municipal elections delayed the activation of the mechanisms provided in the "Comprehensive Plan". The election delayed the process, while the pandemic limited its mechanisms with a participatory nature, affecting the implementation of the Plan very negatively.

3.1.3. 2020–2022: Outsourcing

In 2020, there were two outstanding events that conditioned the development of the process: the concession of a Horizon 2020 project of the European Union to the University of Córdoba in alliance with the City Council of Córdoba and the Association of Neighbours, called In-Habit, and the signing of an agreement between the Córdoba City Council and Loyola University to intervene in the vulnerable neighbourhoods of the city, including Las Palmeras. Both projects had high external financing.

Although, in a certain way, neighbourhood movements were integrated into both plans, the "Comprehensive Plan" was diluted by the City Council, which seemed to constitute a process of outsourcing. Specifically, the process outsourced the community action in the neighbourhood via these two interventions that were highly financed and implemented by the two universities in the city. Despite the fact that each university had its own participatory strategy, a certain shift in the role of the neighbourhood entities can be observed. The Neighbourhood Association was, however, a partner in both projects.

The main actors in this period were the City Council, the University of Córdoba, the Loyola Andalucía University, and the Neighbourhood Association, in addition to the various entities and organisations linked to the processes in each project, mainly service providers.

Once again, the space par excellence became "La Foggara", and other institutional spaces such as Loyola Andalucía University where training was carried out. However, in the neighbourhood, intervention actions were developed mainly in the parochial halls by the entities located in the neighbourhood.

In this period, the "Comprehensive Plan" was diluted by the usual efforts of the City Council without any participatory structure being articulated. Municipal actions tend to cover some proposals concerning actions in a territory based on a large, outsourced project. For this reason, this phase was dominated by an eminently technical rationality.

3.1.4. 2022 to Date: Neighbourhood Reconnection

This last period began in March 2022, when, at the initiative of the Neighbours Association and Loyola Andalucía University, a local team was established that promotes the process in the neighbourhood: the "Driving team". It is a local platform of neighbourhood entities and residents without affiliation, with the aim of promoting the "Comprehensive

Plan" with a bottom-up logic. It serves as a space for deliberations on interventions in the neighbourhood. The team was organised into working groups and began to promote some proposals included in the "Comprehensive Plan", such as a community kitchen or urban gardens.

In 2023, amid the call for new municipal elections, the "Driving team" organised a discussion in the neighbourhood with the main political forces in the city. Additionally, in 2023, Loyola University trained another group of young people from the neighbourhood in social communication, aimed to support the civic processes that take place in their neighbourhood.

In this same year, there was a conflict with the regional Public Administration concerning the schooling of the minors in the neighbourhood. There is an important mobilisation linked to the education sector including the "Driving team", neighbourhood families, and municipal technicians.

In this period, the City Council maintained its proposal to outsource the intervention and thus the dilution of the "Comprehensive Plan". However, in the face of this strategy, there was a neighbourhood protest, which took up the initiative of the "Plan" with the driving force to operationalise it and to put it again on the political agenda. This period is thus dominated by neighbourhood and social action.

Various participatory mechanisms were introduced: consultative mechanisms, such as surveys; deliberative mechanisms, such as assemblies; and strategic mechanisms, such as working groups. Although the prominence and initiative of the Neighbourhood Association are notable, the social entities that provide services, neighbourhood entities, neighbours with a private capacity, teams of educational centres, and universities, participate in the "Driving team". In this way, the dominant strategy becomes the articulation of civil society in the territory by neighbourhood leadership. Thus, the articulation of territorialised social entities and connections with the administration is the aim, whereby constant tension is maintained between technical and civic rationality. In this period, the spaces of reference are the parochial halls located at the extreme west of the neighbourhood.

4. Discussion

4.1. Levels of Participation

By levels of participation, we consider how power is redistributed in the participation process [72]. To make a process operational, the type of link that occurs between institutions and social organisations must be analysed, establishing a gradation of participation [38,43]. Specifically, the approach of Ortiz de Zárate is followed [73] while adjusting the participation ladder [72] to include "nonparticipation", i.e., a lack of citizen involvement in decision-making processes; "symbolic participation", involving citizens in certain activities but with no real impact on final decisions; and "citizen power", where citizens have an active and meaningful role in decision-making. The starting point for this analysis is to assess who takes the initiative.

In general terms, the collaborative governance process was based on an initiative led by organised neighbours. Even though the initiative has fallen throughout the process into technical and political hands, the grassroots initiative has kept the "Comprehensive Plan" active throughout the process:

> *"So, it is a Comprehensive Plan that is made by the residents that the neighbourhood association has done; it is defending it, it is carrying it out, it is leading it"*
>
> (Neighbourhood leader 2)

The specific initiative has varied in the four periods discussed above, based in the neighbourhood at the beginning and at the present but passing through political and technical initiatives. This illustrates what Ran and Qi propose [9] when they argue that the roles in collaborative governance vary depending on the sources of power. In the two central periods, when the initiative was run by the Public Administration with political and technical control, the source of power was formal authority and the control of resources. Formal authority emerged as soon as the plan was approved and became managed by the municipality, and the resources were controlled when it intervened in hyper-financed programmes such as *ERACIS*, *In-Habit*, or the agreement with Loyola. On the other hand, when the initiative remained in the neighbourhood, the source of power was discursive legitimacy:

> *"One so simple and so clear, like going with the heart in the hand everywhere with the truth ahead, Eh? And humility. Because when you tell a person your reality... your truth, huh? There will be those who accept it and there will be those who do not accept it, but at least they consider that you are a person who goes honestly, that you go with your face and that you go with your truth in your hands"*
>
> (Neighbourhood leader 1)

In the periods in which the initiative was in the hands of the institutions, the process of the "Comprehensive Plan" became lethargic, exemplifying the postulate of Grey and Purdy [74] suggesting that power differences create major obstacles in the successful development of collaborative governance. Basically, those who dominate the sources of power become central actors and control the process. Thus, a dialectical relationship is maintained between the agents of the collaborative network for the control of the sources of power within their capacities, which generates asymmetries of power that articulate and mediate the process.

The evolution of the levels of participation in the "Comprehensive Plan" highlights how the power dynamics between stakeholders have affected the management of the process due to power asymmetries marked by the agenda-setting capacity or control of resources of some stakeholders vis-à-vis others (the administration or highly funded entities).

In short, although the collaborative governance network that emerges from the "Comprehensive Plan" is part of a grassroots initiative, coinciding with the last steps of the participation ladder [72] or what Zárate [73] calls citizen power, the dynamics of power allow for the loss of control over the process among promoter actors and the co-optation of it by actors who dominate the sources of power, such as the City Council, certain public devices, or universities.

4.2. Methodologies

Methodologies refer to the characteristics of the strategies and mechanisms activated by some agents and others to develop participation within the collaborative network. First, they include the participation mechanisms, which differ according to the objectives of the moment in such a way that they complement each other [22]. In turn, they are mediated by space and time. The former allows researchers to analyse the formal or informal character, as well as the symbolic value, of the places of participation. The latter makes it possible to determine whether they are permanent processes or circumscribed to conjunctural periods or mechanisms of process or synthesis, reflecting the degree of institutionalisation of the participation instruments [38]. The characteristics of each of these three elements are explained below.

First, regarding the participation mechanisms, their implementation has varied according to the moment in which the process has been developed. While during the initial stage, the mechanisms applied were listening, consulting, and strategizing, in the following stages, these mechanisms diminished or were diluted via the small strategies that the different large projects implemented in the neighbourhood.

The participation mechanisms that allow for managing the control of power are mediated by the type of collaborative governance network. In the first period and in the current period, in which the network has been voluntary in nature, the strategies and mechanisms used have allowed the social organisations to increase the strength of their source of power: discursive legitimacy. On the other hand, in the periods of political–technical control, the networks became mandatory [9]. These central periods have been characterised by the application of hierarchical and bureaucratic mechanisms, as well as by a decrease in participation and a loss of neighbourhood control, coinciding with the following hypotheses of these authors: "the more voluntary the type is, the more beneficial collaborative governance is for sharing power" [9].

Power dynamics have been reflected in the mechanisms of participation by mediating agenda-setting capacity, the control of resources, or access to information depending on the type of network that was constituted, where more horizontal mechanisms reduce the asymmetry of power compared to more hierarchical mechanisms.

The loss of oversight of the "Comprehensive Plan" by the social actors has coincided with the decrease in the participatory mechanisms. In this way, both the nature of the network (voluntary or mandatory) and the flexibility and adaptability of the implemented mechanisms have affected the relationship between shared power and the asymmetry of power, aligning with the literature [9].

Second, regarding the spaces for participation, these form a physical and symbolic context [75] that affects the collaborative network. The first period of the "Comprehensive Plan" is identified within the space of the neighbourhood; thus, its journey is carried out in the squares or even door to door. In times of technical–political control, the space par excellence became the publicly owned building "La Foggara". However, in the last period, the spaces that dominated were again in the neighbourhood, but formal, such as the parochial halls.

Different physical spaces symbolise the contexts where the power takes place. Thus, these spaces became related to the actors that dominated each period, their sources of power, or the natures of the mechanisms that were applied. The hypothesis of Ran and Qi [9] maintains that the stronger the institutional context is, the more beneficial it is to sharing power. In contrast, in the case analysed, the more institutional the context became, the more distrust it generated among the actors.

Third, the times of the "Comprehensive Plan" have been mediated by exogenous and endogenous issues: changes in the government bodies; the COVID-19 pandemic; the dynamics of groups; or emerging events in the neighbourhood. Thus, time has configured the identity and genuine character of the process [76].

The four periods analysed above have been differentiated according to who had the initiative, articulating each period as a source of power. Thus, when the process was managed by the Public Administration, either directly or by outsourcing, the asymmetry of power that had been activated allowed for the lethargy of the participatory mechanisms of the "Comprehensive Plan". This highlights the risk of co-optation of the process by the most powerful actors.

The literature recognises that the neighbourhood-dominated processes are marked by urgency and emergency, while the administrative periods are dominated by exogenous factors such as elections or budget cycles. Given that the "Comprehensive Plan" involves a long-term planning and implementation process, the investment in time remains very high, now approaching 10 years, entailing the establishment of a horizontal structure with a permanent vocation: the "Driving team". However, the urgent and emerging processes that have motivated the neighbourhood articulations require a temporary hierarchical structure [9,77] to be effective. This has led to a loss of citizen control in the process, in line with Ran and Qi [9], who note that the less demanding the mission fulfilled by collaborative governance is, the more beneficial it is to share power for collaborative governance.

The methodology is traversed by power dynamics, sustaining power asymmetries by maintaining dialogical tensions such as those between the mechanisms of temporary and permanent networks or between voluntary and mandatory networks. The control of resources, access to information, interdependence, or adaptability are aspects that reflect the methodologies implemented that project existing power asymmetries, where tight deadlines and inaccessible locations can limit participation. However, it is observed that when mechanisms address these issues, they minimise the impact of these asymmetries, enabling interest in participation, as has been the case in the initial and current stages of the process.

4.3. The Actors

The actors are those individuals, groups, organisations, or institutions participating in the network, and they are identified according to their particular position and their specific interests in the process [38]. Authors such as Alberich [78] and Villasante [40] have classified these actors into three broad categories: technical, political, and civic. All of them contextualise their performances through judgements of relevance or rationalities [38,42,43], which is the element taken into account in this analysis.

The profile of the actors involved in the network has been shaped as the process has developed. First, local, or territorial organisations and individual neighbours who participated from the beginning stand out. A preexisting culture of networking was already established in the neighbourhood, the association's roundtable, which facilitated the constitution of a collaborative governance network of a voluntary nature [9]. This has favoured the emergence of consent, reciprocity, and trust:

> *"And the neighbours are trusting. [. . .] They are having enough patience to trust, [. . .] after so much suffering, they are already living a "Comprehensive Plan" that was drawn up by the neighbours"*
>
> (Neighbourhood leader 2)

This is how the "Comprehensive Plan" is configured as a bottom-up process [79]. This is a characteristic that in some way persists, standing out in the constitution of the "Driving team" in the last period, where the collaboration network was once again more civic:

> *"We wanted many more entities to be added, many more people. In addition to the associations, roundtable of internal and external groups that we already had in it and so it was decided. So, I don't know, well, the associations roundtable as such stops working, but the representatives of the associations table are still in the Driving team"*
>
> (Neighbourhood leader 3)

The Plan was born, therefore, with a civic rationality as it was activated by a coalition of social entities linked to the territory and neighbours. They did so because it was urgent, which is characteristic of civic rationality [38]. And it managed to be approved in the municipal plenary session because this urgency was translated into a highly technical document: the "Comprehensive Plan".

This approval supposed the emergence of a political rationality, reflected in the agenda of the political authorities, their capacity for dialogue, and the collaborative associations that they establish, marked by a logic of accumulation of power [38]. However, due to power asymmetries, political rationalities have fluctuated among technical, political, and citizen agents. This was observed, e.g., when the Neighbourhood Association made use of the media or when they activated alliances with other entities, such as universities. In any case, from its consolidation into a technical–political document, technical rationality became present throughout the process, fluctuating across all the categories of actors, to the point that the Neighbourhood Association became a formal partner in one of the projects managed by the University of Córdoba.

The dense networks among actors, based on positive experiences and power sharing, explain the survival of the "Comprehensive Plan" after 10 years of ups and downs, a survival consistent with the literature [80]. The process and the governance network of the "Comprehensive Plan" are sustained because they legitimise the acceptance of the process, the perception of its opportunities, and the desire to share power. This is possible because it maintains a diverse and complex network of actors, preventing constant tension between the mandatory network and the voluntary network via the small achievements of the process. These coincide with the three major types of legitimacy of political systems (input, throughput, and output) identified by Martí-Costa and others [81].

Ultimately, the "Comprehensive Plan" has involved a diverse network of actors (technical, political, and civic), whose interactions and rationalities evolved throughout the process, reflecting power dynamics. The fluctuation of power among these actors is seen in how the power asymmetries, agenda-setting capacity, and control of resources have impacted the plan. The variation in actors, and their rationalities, has brought unequal legitimacy and access to information mediating the collaborative web of governance. On the other hand, these tensions lead to practices such as the formation of alliances and the use of media that actors such as the Neighbourhood Association apply to balance these asymmetries.

4.4. Implications and Proposals

The analysis of power dynamics in this study highlights the paradoxical and dialectical nature of collaborative governance. This question implies limitations in practice, which calls for the activation of contingent solutions [3]. Therefore, within the framework of a strategy for addressing these paradoxes and the power dynamics that they imply, in light of the preceding analysis, four strategic proposals are proposed: citizen autonomy, self-regulation, comanaged agenda, and accountability. These proposals address the limitations of the focal case by addressing the dialectical nature of collaborative governance.

First, we propose mechanisms for guaranteeing citizenship quotas regarding autonomous participation within an agenda outside the network. The establishment of such participatory structures will allow civic sectors to have their own agendas. This could contribute to reducing the danger of public co-optation of the process, as well as the game of power asymmetry. In the case analysed, progress has been made with the establishment of the "Driving Team" and their working groups. The former allows for the management of the process in the neighbourhood, and the latter allows for the establishment of networks of proximity and collaboration among the different actors.

Second, to minimise power asymmetries and limit the potential co-optation of the process, we propose to develop practices such as self-regulation, already proved in similar processes as a participatory budget [19,82–84]. This entails developing a participatory document that organises the rules of the process in all its steps and actions [84]. Some authors suggest that a process of self-regulation should be based on the principle of parity under the criteria of social justice, focused on participatory synergies [19,83]. Through a participatory and open methodology, self-regulation can address the limitations resulting from the asymmetries of power and the paradoxical nature of collaborative governance, promoting transparency, deliberation, and decision-making in a framework of power distribution. In the case studied, this issue is a pending task.

Third, it is known that whoever controls the agenda controls the outcome, either by excluding certain issues or problems, by excluding important lines of action, or by making use of the structure itself [85]. The establishment of an agenda is mediated by the dynamics of power and has important consequences for deliberative processes [85,86]. Thus, we propose the co-management of the deliberative space, which could enable the balancing of power dynamics through shared control and the adjustment of the agenda. The active collaboration of the actors in this line is facilitated through balanced and structured information, as well as the incorporation of didactic materials, considering the contextual factors [87]. In the case analysed, internal, external, formal, and informal mechanisms are beginning to be configured, allowing for movement between the public agenda and the political agenda, from deliberative processes to decision-making. However, there is still room for advancement.

Fourth and last, numerous academics have already noted the importance of accountability in collaborative governance [10,88,89]. Accountability is defined in terms of the relationship between an actor and a forum, where the former has the obligation to justify his or her actions and face the consequences of the forum's deliberation [90]. This implies a relationship of coresponsibility between the members of the network who are accountable (account-givers) and those who are rendered account-holders [89]. The creation of horizontal and informal accountability mechanisms can strengthen the deliberative capacity of the network, mutual trust, and legitimacy [10,91] by ensuring transparency and fairness in decision-making [92]. However, as it is an essentially communicative process and information is a source of power, the system must be carefully designed to recognise any imbalance of power and minimise it [89]. In the case analysed, the group of young people trained in communication is called upon to do so, but their empowerment is still a pending task.

5. Conclusions

The results indicate that collaborative governance essentially maintains a paradoxical and dialectical nature whereby power dynamics have significant weight and are present across all times, levels, and procedures. This study has illustrated how, even when the participatory process starts from citizen empowerment, power dynamics mediate the process, either limiting or expanding it but always making it more complex. It also illustrates how the nonrecognition of this paradoxical and essentially conflicting nature leads to power dynamics that affect the benefits of power sharing for collaborative governance. In addition, it has highlighted the need for carefully designed strategies to balance these forces and strengthen the legitimacy of the process, and some strategies have been proposed: citizen autonomy, self-regulation, a comanaged agenda, and accountability.

Furthermore, this study contributes significantly to the literature on co-operative governance by offering a case study exploring power dynamics, in line with what Ran and Qi [9] suggest. This analysis provides valuable insights into how these dynamics affect stakeholder interaction and participation mechanisms, deepening the understanding between power asymmetries and participation. We also present a novel methodological approach to the analysis of power dynamics in specific contexts, thus contributing to the expansion of analytical tools available for future studies in collaborative governance.

Despite these findings, this study also has important limitations. First, the qualitative and participatory nature of this research, where the researchers are agents involved in the process, could lead to bias. On the other hand, the uniqueness of the focal case and its specific history call for prudence when extending the abovementioned results, especially in the application of the strategic proposals in other contexts. Furthermore, the temporality and complexity of the case do not allow for capturing the full multiplicity of factors; collaborative governance is inherently fluid and contingent, subject to unexpected changes.

Nevertheless, this study contributes significantly to the understanding of collaborative governance and citizen participation in a framework of power relations and offers a deep understanding of the challenges and complexities inherent to these relations. Likewise, it offers practical proposals for strengthening citizen participation and legitimacy in similar processes, highlighting the importance of preserving the autonomy of the community amid the co-optation of the most powerful. The combination of limitations and insights open future lines of research, including additional case studies that address the dynamics of power in collaborative governance. They could shed more light on the proposals indicated above, especially in terms of the autonomy of citizens and the danger of co-optation.

Possible Directions for Future Studies

This study of power dynamics in collaborative governance in the Las Palmeras neighbourhood opens up several avenues for future research. One promising direction is the application and refinement of the concepts and methods developed in this study in diverse geographical and cultural contexts. This would allow researchers to explore how power dynamics and participatory mechanisms designed to balance these asymmetries work in contexts with different socio-economic structures.

In addition, further research is suggested on the design and implementation of participatory mechanisms that foster inclusion, assessing their effectiveness in empowering marginalised stakeholders. A longitudinal study examining the evolution of collaborative governance in Las Palmeras, or in similar contexts, would provide valuable insights into the long-term impact of the strategies implemented and how power dynamics and citizen participation change over time.

Exploring new strategies that promote effective self-regulation and citizen empowerment in the face of the constraints and challenges encountered in this study also represents a direction for future work. Furthermore, addressing the limitations of our study, such as the focus on a single urban context and the limited temporal scope of the analysis, comparative studies that expand the geographical and temporal focus and the inclusion of a wider range of voices and perspectives would enrich the understanding of collaborative governance and the power dynamics involved.

Author Contributions: All authors contributed equally. All authors have read and agreed to the published version of the manuscript.

Funding: This research received no external funding.

Institutional Review Board Statement: The study was conducted in accordance with the Declaration of Helsinki, in the framework of the project "Proceso de investigación participativa para la transformación social de zonas desfavorecidas en Andalucía", which was approved by the Ethics Committee of Universidad Loyola Andalucía in the 28 April 2021.

Informed Consent Statement: Informed consent was obtained from all subjects involved in the study.

Data Availability Statement: The data presented in this study are available on request from the corresponding author. The data are not publicly available due to ongoing research activities.

Conflicts of Interest: The authors declare no conflicts of interest.

References

1. Ansell, C.; Gash, A. Collaborative Governance in Theory and Practice. *J. Public Adm. Res. Theory* **2008**, *18*, 543–571. [CrossRef]
2. Ospina, S.M.; Esteve, M.; Lee, S. Assessing Qualitative Studies in Public Administration Research. *Public Adm. Rev.* **2018**, *78*, 593–605. [CrossRef]
3. Qi, H.; Ran, B. Paradoxes in Collaborative Governance. *Public Manag. Rev.* **2023**, *2023*, 2196290. [CrossRef]
4. Waardenburg, M.; Groenleer, M.; de Jong, J.; Keijser, B. Paradoxes of Collaborative Governance: Investigating the Real-Life Dynamics of Multi-Agency Collaborations Using a Quasi-Experimental Action-Research Approach. *Public Manag. Rev.* **2020**, *22*, 386–407. [CrossRef]
5. Molinengo, G. Flows of Power: An Analytical Framework for the Study of Collaboration. *Crit. Policy Stud.* **2023**, *17*, 101–122. [CrossRef]
6. Janes, J.E. Democratic Encounters? Epistemic Privilege, Power, and Community-Based Participatory Action Research. *Action Res.* **2015**, *14*, 72–87. [CrossRef]
7. Booher, D.E.; Innes, J.E. Network Power in Collaborative Planning. *J. Plan. Educ. Res.* **2002**, *21*, 221–236. [CrossRef]
8. Holdo, M. Power Games: Elites, Movements, and Strategic Cooperation. *Political Stud. Rev.* **2019**, *18*, 189–203. [CrossRef]
9. Ran, B.; Qi, H. Contingencies of Power Sharing in Collaborative Governance. *Am. Rev. Public Adm.* **2018**, *48*, 836–851. [CrossRef]
10. Triantafillou, P.; Hansen, M.P. Introduction to the PMR Special Issue on Accountability and Legitimacy under Collaborative Governance. *Public Manag. Rev.* **2022**, *24*, 655–663. [CrossRef]
11. Purdy, J.M. A Framework for Assessing Power in Collaborative Governance Processes. *Public Adm. Rev.* **2012**, *72*, 409–417. [CrossRef]
12. Carlos, A.F.A. El Concepto de "Producción Del Espacio" y La Dinámica Urbana Contemporánea Bajo El Dominio Del Capital Financiero. *Rev. Geogr. Norte Gd.* **2022**, *2022*, 89–107. [CrossRef]
13. Bellet Sanfeliu, C. La Producción de Un Barrio Desfavorecido En Los Márgenes de Una Ciudad Media: La Mariola (Lleida). *Scr. Nova Rev. Electrónica Geogr. Y Cienc. Soc.* **2021**, *25*, 75–103. [CrossRef]
14. Gallego Gamazo, C. El Enfoque Integrado En Los Programas de Regeneración Urbana de Barrios Desfavorecidos: Una Visión Relacional. Ph.D. Thesis, E.T.S. Arquitectura (UPM), Madrid, Spain, 2015. [CrossRef]
15. Ran, B.; Qi, H. The Entangled Twins: Power and Trust in Collaborative Governance. *Adm. Soc.* **2018**, *51*, 607–636. [CrossRef]
16. Cao, J.; Prior, J.; Giurco, D.; Gu, D. Power Relations Are Central to Shaping Collaborative Governance of the Urban Sharing Economy. *Humanit. Soc. Sci. Commun.* **2023**, *10*, 85. [CrossRef]
17. Pollard, S.; Chikunda, C.; Mohale, T.; Goredema, L.; Kotschy, K. Power Taken, Power Given: Lessons for Collaborative Governance Praxis and Co-Learning from Land Reform and Co-Management in the Lekgalameetse Nature Reserve (South Africa). In *Sustainable Development in Africa*; Springer: Cham, Switzerland, 2021; pp. 709–731. [CrossRef]
18. Healey, P. *Collaborative Planning: Shaping Places in Fragmented Societies*, 2nd ed.; Palgrave Macmillan: London, UK, 2006; ISBN 9781137086006.
19. Barragán, V.; Sanz, J.M.; Romero, R. Midiendo Las Experiencias Participativas. El Caso de Los Presupuestos Participativos de Sevilla. *Obs. Iberoam. Desarro. Local Y Econ. Soc.* **2014**, *8*, 1–26.

20. Chan, E.T.H.; Li, T.E. The Effects of Neighbourhood Attachment and Built Environment on Walking and Life Satisfaction: A Case Study of Shenzhen. *Cities* **2022**, *130*, 103940. [CrossRef]
21. Dekker, K.; Haw, A. Understanding Conviviality in Australian Suburbs with High Muslim Concentrations: A Qualitative Case Study in Melbourne. *Aust. J. Soc. Issues* **2023**, 1–14. [CrossRef]
22. Franklin, A.L.; Ho, A.T.; Ebdon, C. Participatory Budgeting in Midwestern States: Democratic Connection or Citizen Disconnection? *Public Budg. Financ.* **2009**, *29*, 52–73. [CrossRef]
23. Kamruzzaman, M.; Wood, L.; Hine, J.; Currie, G.; Giles-Corti, B.; Turrell, G. Patterns of Social Capital Associated with Transit Oriented Development. *J. Transp. Geogr.* **2014**, *35*, 144–155. [CrossRef]
24. Amini Pishro, A.; Yang, Q.; Zhang, S.; Amini Pishro, M.; Zhang, Z.; Zhao, Y.; Postel, V.; Huang, D.; Li, W.Y. Node, Place, Ridership, and Time Model for Rail-Transit Stations: A Case Study. *Sci. Rep.* **2022**, *12*, 16120. [CrossRef]
25. Padeiro, M.; Louro, A.; da Costa, N.M. Transit-Oriented Development and Gentrification: A Systematic Review. *Transp. Rev.* **2019**, *39*, 733–754. [CrossRef]
26. Papa, E. Implementing Transit Oriented Development in Greater London. In *Transit Oriented Development and Sustainable Cities*; Edward Elgar Publishing: Cheltenham, UK, 2019; pp. 186–197. [CrossRef]
27. Jacobson, J.; Forsyth, A. Seven American TODs: Good Practices for Urban Design in Transit-Oriented Development Projects. *J. Transp. Land Use* **2008**, *1*, 51–88. [CrossRef]
28. Nasri, A.; Zhang, L. The Analysis of Transit-Oriented Development (TOD) in Washington, D.C. and Baltimore Metropolitan Areas. *Transp. Policy* **2014**, *32*, 172–179. [CrossRef]
29. Villasante, T. Las Ciudades Hablan y Hacen Programas Integrales. *Boletín CF+S* **1998**, 1–13.
30. Ortega Fernández, J. *Evaluación y Posibles Alternativas al Plan Integral Zona Norte de Alicante: Efectos y Consecuencias En El Desarrollo Local Del Barrio Juan XXIII 2º Sector*; Publicacions de la Universitat d'Alacant: Sant Vicent del Raspeig, Spain, 2016; pp. 173–186. [CrossRef]
31. Torres Gutiérrez, F.J. Polígono Sur En Sevilla. Historia de Una Marginación Urbana y Social. *Scr. Nova Rev. Electrónica Geogr. Y Cienc. Soc.* **2021**, *25*, 105–129. [CrossRef]
32. Robert, D.A. The dilemma of pluralist democracy. In *Classes and Elites in Democracy and Democratization*; Eva Etzioni Halvey: New York, NY, USA, 2020; pp. 267–274. [CrossRef]
33. Galbraith, J.K. *La Anatomía Del Poder*; Plaza & Janés: Barcelona, Spain, 1985; ISBN 84-01-33253-2.
34. Raffestin, C. ¿Hacia Dónde va La Geografía Política? Reflexiones Críticas Sobre El Ejercicio Práctico Del Poder En El Espacio. *Geopolítica(s) Rev. Estud. Sobre Espac. Y Poder* **2020**, *11*, 29–37. [CrossRef]
35. Mann, M. *Las Fuentes Del Poder Social*; Alianza: Madrid, Spain, 1986; ISBN 842062666X.
36. Lukes, S. *Power: A Radical View*, 2nd ed.; Macmillan, P., Ed.; Palgrave Macmillan: New York, NY, USA, 2005; ISBN 9780333420911.
37. Offutt, A.J. Overcoming the Hypolimnion: Stakeholder Influence in Transboundary Water Quality Governance. Master's Thesis, Oregon State University, Corvallis, OR, USA, 2020.
38. Pagani, M.L. Las Piezas de La Participación: Dimensiones de Análisis, Un Rompecabezas Para Armar. In *Debates Teóricos y Metodológicos Actuales Sobre las Políticas Públicas*; University of La Plata: La Plata, Argentina, 2017; pp. 119–147. ISBN 978-950-34-1594-8.
39. Gutiérrez, P.M. El Sociograma Como Instrumento Que Desvela La Complejidad. *Empiria Rev. Metodol. Cienc. Soc.* **1999**, 129–152. [CrossRef]
40. Villasante, T.; Gutiérrez, P.M. Redes y Conjuntos de Acción: Para Aplicaciones Estratégicas En Los Tiempos de La Complejidad Social. *Redes Rev. Hisp. Para El Análisis Redes Soc.* **2006**, *11*, 1–22. [CrossRef]
41. Osvaldo Alonso, P. La Lógica de Los Actores y El Desarrollo Local. *Rev. Pilquen* **2004**.
42. Lazega, E. Racionalidad, Disciplina Social y Estructura. *Redes Rev. Hisp. Para El Análisis Redes Soc.* **2004**, *5*. [CrossRef]
43. Lapalma, A.I. El Escenario de La Intervención Comunitaria. *Rev. Psicol.* **2001**, *10*, 61–70. [CrossRef]
44. Freire, P. *La Naturaleza Política de La Educación. Cultura, Poder y Liberación*; Paidós: Barcelona, Spain, 1990.
45. Jara, O. Systematisation of Experiences: New Paths to Academic Work in Universities. *Int. J. Action Res.* **2020**, *16*, 62–74. [CrossRef]
46. Creswell, J.W. *Qualitative Inquiry and Research Design: Choosing among Five Approaches*, 3rd ed.; Knight, V., Ed.; Sage: Los Angeles, CA, USA, 2013; ISBN 9781412995306.
47. Jara, O. *Para Sitematizar Experiencias*; Alforja: San José, Costa Rica, 1998.
48. Yin, R.K. *Qualitative Research from Start to Finish*; Guilford Press: New York, NY, USA, 2011; ISBN 9781606239773.
49. Delgado-Baena, A.; Serrano, L.; Vela-Jiménez, R.; López-Montero, R.; Sianes, A. Epistemic Injustice and Dissidence: A Bibliometric Analysis of the Literature on Participatory Action Research Hosted on the Web of Science. *Action Res.* **2022**, *20*, 318–342. [CrossRef]
50. Hensler, L.; Alatorre Frenk, G.; Merçon, J. *Participatory Action Research*; Transcript: New Rockford, ND, USA, 2023.
51. McIntyre, A. *Participatory Action Research*; Sage Publications: Los Angeles, CA, USA, 2008; ISBN 9781412953665.
52. López-Montero, R.; García-Navarro, C.; Delgado-Baena, A.; Vela-Jiménez, R.; Sianes, A. Life Stories: Unraveling the Academic Configuration of a Multifaceted and Multidisciplinary Field of Knowledge. *Front. Psychol.* **2022**, *13*, 960666. [CrossRef]
53. McCall, B.; Shallcross, L.; Wilson, M.; Fuller, C.; Hayward, A. Storytelling as a Research Tool and Intervention around Public Health Perceptions and Behaviour: A Protocol for a Systematic Narrative Review. *BMJ Open* **2019**, *9*, e030597. [CrossRef]
54. Meyer, S.; Mader, M.; Weiser, A. Case Study. In *Participatory Action Research*; Thorsten Philipp, T.S., Ed.; Transcript: New Rockford, ND, USA, 2023; pp. 31–40. ISBN 978-3-8376-6347-1.

55. Coral-Díaz, A.M.; Espejo Fandino, F. Sistematización de Experiencias y Rastreo de Procesos Como Métodos de Investigación En Ciencias Sociales: Reflexiones Para Su Aplicación. *Opción RevCienc. Humanas Y Soc.* **2019**, *25*, 449–481.
56. Marková, I. Dialogicality as Epistemology of Daily Life and of Professional Practices. In *Dialogical Mind: Common Sense and Ethics*; WE-Book Citation Index–Social Sciences; Univ Stirling, Psychol: Stirling, UK, 2016; pp. 91–95. ISBN 978-1-107-00255-5.
57. Koch, P. Bringing Power Back in: Collective and Distributive Forms of Power in Public Participation. *Urban Stud.* **2013**, *50*, 2976–2992. [CrossRef]
58. QSR International Pty Ltd. NVivo (Version 12) [Software]. 2018. Available online: https://www.qsrinternational.com/nvivo-qualitative-data-analysis-software/home (accessed on 15 November 2023).
59. Fatmawati, F.; Mustari, N.; Haerana, H.; Niswaty, R.; Abdillah, A. Waste Bank Policy Implementation through Collaborative Approach: Comparative Study—Makassar and Bantaeng, Indonesia. *Sustainability* **2022**, *14*, 7974. [CrossRef]
60. Conti, D.d.M.; Guevara, A.J.d.H.; Heinrichs, H.; da Silva, L.F.; Quaresma, C.C.; Beté, T.d.S. Collaborative Governance towards Cities Sustainability Transition. *Urbe Rev. Bras. Gestão Urbana* **2019**, *11*, e20190046. [CrossRef]
61. Lochmiller, C.R. Conducting Thematic Analysis with Qualitative Data. *Qual. Rep.* **2021**, *26*, 2029–2044. [CrossRef]
62. Nowell, L.S.; Norris, J.M.; White, D.E.; Moules, N.J. Thematic Analysis: Striving to Meet the Trustworthiness Criteria. *Int. J. Qual. Methods* **2017**, *16*, 1–13. [CrossRef]
63. van Pinxteren, M.; Colvin, C.J.; Cooper, S. Using Health Information for Community Activism: A Case Study of the Movement for Change and Social Justice in South Africa. *PLoS Glob. Public Health* **2022**, *2*, e0000664. [CrossRef]
64. Cernasev, A.; Axon, D.R. Research and Scholarly Methods: Thematic Analysis. *J. Am. Coll. Clin. Pharm.* **2023**, *6*, 751–755. [CrossRef]
65. INE Instituto Nacional de Estadística. *Indicadores Urbanos Edición 2022*; INE Instituto Nacional de Estadística: Madrid, Spain, 2022.
66. Fernández García, M. Regeneración Urbana En Perspectiva Andaluza. Los Programas ARB (1996–2012). *Anduli* **2018**, *17*, 67–88. [CrossRef]
67. Sianes, A.; Vela-Jiménez, R. Can Differing Opinions Hinder Partnerships for the Localization of the Sustainable Development Goals? Evidence from Marginalized Urban Areas in Andalusia. *Sustainability* **2020**, *12*, 5797. [CrossRef]
68. García-Araque, J. Inclusive Capacity of Public Space in Disadvantaged Urban Areas: Case in the City of Valladolid. *OBETS* **2022**, *17*, 101–120. [CrossRef]
69. Ayuntamiento de Córdoba—Estadísticas 2021. Available online: https://www.cordoba.es/la-ciudad/cifras-estadisticas/estadisticas-de-poblacion/estadisticas-2021 (accessed on 10 April 2023).
70. Ministerio de Transportes, Movilidad y Agenda Urbana. Catálogo de Barrios Vulnerables. n.d. Available online: https://portalweb.mitma.es/aplicaciones/portalweb/BarriosVulnerables (accessed on 15 November 2023).
71. Palmeras, O. Reunión de La Mesa ERACIS de Distrito de Las Palmeras | ONDA PALMERAS. Available online: http://ondapalmeras.org/reunion-de-la-mesa-eracis-de-distrito-de-las-palmeras/ (accessed on 21 October 2023).
72. Arnstein, S.R. A Ladder of Citizen Participation. *J. Am. Inst. Plan.* **1969**, *35*, 216–224. [CrossRef]
73. Ortiz de Zárate, A. ¿Por Qué Esta Obsesión Con La Participación Ciudadana? *Open Gov. Gob. Abierto* **2010**, 29–49.
74. Gray, B.; Purdy, J. *Collaborating for Our Future: Multistakeholder Partnerships for Solving Complex Problems*; Oxford University Press: Oxford, UK, 2018; pp. 1–251. [CrossRef]
75. Licona Valencia, E. Construcción Simbólica Del Espacio Urbano. *Rev. Antropol. Exp.* **2012**, *12*, 61–75.
76. Hinton, L. Temporality and the Torments of Time. *J. Anal. Psychol.* **2015**, *60*, 353–370. [CrossRef]
77. Moynihan, D.P. The Network Governance of Crisis Response: Case Studies of Incident Command Systems. *J. Public Adm. Res. Theory* **2009**, *19*, 895–915. [CrossRef]
78. Alberich Nistal, T. Iap, Redes y Mapas Sociales: Desde La Investigación a La Intervención Social. *Portularia Rev. De Trab. Soc.* **2008**, *8*, 131–151.
79. Provan, K.G.; Kenis, P. Modes of Network Governance: Structure, Management, and Effectiveness. *J. Public Adm. Res. Theory* **2008**, *18*, 229–252. [CrossRef]
80. Zimmerman, M.A.; Zeitz, G.J. Beyond Survival: Achieving New Venture Growth by Building Legitimacy. *Acad. Manag. Rev.* **2002**, *27*, 414. [CrossRef]
81. Martí-Costa, M.; Blanco, I.; Parés, M.; Subirats, J. Regeneración Urbana y Gobernanza: ¿Cómo Evaluar La Participación En Una Red de Gobernanza? Tres Perspectivas Teóricas y Un Estudio de Caso. In *Participación, políticas públicas y territorio: Aportes a la construcción de una perspectiva integral*; Rofman, A., Ed.; Universidad Nacional de General Sarmiento: Buenos Aires, Argentina, 2016; pp. 27–52.
82. Martinez-Sanchez, W. Participatory Budgeting: Characteristics and Participatory Quality in Institucional Desings. *OBETS Rev. De Cienc. Soc.* **2023**, *18*, 133–154. [CrossRef]
83. Navascues, J. Vas, Propones y Votas... ¿así de Fácil?: Un Año de Presupuestos Participativos En Sevilla. In *La Pedagogía de la Decisión: Aportaciones Teóricas y Prácticas a la Construcción de las Democracias Participativas*; CIMAS—Observatorio Internacional de Ciudadania y Medio Ambiente Sostenible; Delegación dle Participación Ciudadana—Ayuntamiento de Sevilla: Sevilla, Spain, 2005; pp. 77–86. ISBN 846112443X.
84. Vérgez, J.C. Principales Cuestiones Derivadas de La Elaboración Del Autorreglamento. In *La Experiencia de los Presupuestos Participativos en los Entes Locales*; Dykinson: Madrid, Spain, 2011; pp. 139–184. ISBN 978-84-9031-841-6.

85. Lang, A. Agenda-Setting in Deliberative Forums: Expert Influence and Citizen Autonomy in the British Columbia Citizens' Assembly. *Des. Deliberative Democr. Br. Columbia Citiz. Assem.* **2008**, 85–105. [CrossRef]
86. Elstub, S.; Carrick, J.; Farrell, D.M.; Mockler, P. The Scope of Climate Assemblies: Lessons from the Climate Assembly UK. *Sustainability* **2021**, *13*, 11272. [CrossRef]
87. Harvey, J. How to Make America Democratically Deliberate Again: A Proposal for Widespread Deliberative Mini-Publics. Ph.D. Thesis, University of British Columbia, Kelowna, BC, Canada, 2021. [CrossRef]
88. Lægreid, P.; Rykkja, L.H. Accountability and Inter-Organizational Collaboration within the State. *Public Manag. Rev.* **2022**, *24*, 674–694. [CrossRef]
89. Wachhaus, T.A. Anarchy as a Model for Network Governance. *Public Adm. Rev.* **2012**, *72*, 33–42. [CrossRef]
90. Bovens, M. Analysing and Assessing Accountability: A Conceptual Framework1. *Eur. Law J.* **2007**, *13*, 447–468. [CrossRef]
91. Lee, S. When Tensions Become Opportunities: Managing Accountability Demands in Collaborative Governance. *J. Public Adm. Res. Theory* **2022**, *32*, 641–655. [CrossRef]
92. Page, S.B.; Stone, M.M.; Bryson, J.M.; Crosby, B.C. Public Value Creation by Cross-Sector Collaborations: A Framework and Challenges of Assessment. *Public Adm.* **2015**, *93*, 715–732. [CrossRef]

buildings

MDPI

Article

The Contextuality of Dongdaemun Design Plaza and Park, an Iconic Building in Seoul—A Study Based on Fuzzy Statistical Analysis

Zeheng Cao and Jae-Eun Yoon *

Culture Design Lab, Graduate School of Techno Design, Kookmin University, Seoul 02707, Republic of Korea; caozeheng@kookmin.ac.kr
* Correspondence: dreamask@hanmail.net

Abstract: For many cities around the world, the construction of iconic buildings is a major strategy to improve their urban space and enhance recognition. Yet, newly built iconic buildings usually spark controversy owing to their relationship with the urban context of the city. In this study, revolving around Dongdaemun Design Plaza and Park, we analyze and evaluate the functional influences of its reconstruction on the urban context of Seoul. We found that though the DPP has a positive effect on the city image and cultural exchange of Seoul, it is barely satisfactory for maintaining and extending the urban context. Through the combination of a random sampling questionnaire and triangular fuzzy number analysis, we transfer previous subjective comments about the DPP into an objective evaluation, analyze its influence on the dimension of culture, and call for consideration of the continuity of the urban context in future urban planning. The results of this study provide a new perspective on quantifying the urban contextual influence of iconic buildings, reminding city planners of the balance between the development of business and the sustainability of the urban context.

Keywords: regional study; iconic building; fuzzy statistical analysis; urban context

1. Introduction

In the current global construction of cities, as one of the resources of renewable energy for improving city image, iconic buildings have gained significant importance in the construction of cities [1]. Shaping the skylines of cities with their unique characteristics, iconic buildings imbue cities with distinct historical and aesthetic value that indulges locals and tourists [2]. Along with the fast urbanization that has occurred since the beginning of the 21st century, and as a result of city governors' speculative strategies that have prioritized the construction of iconic buildings, a multitude of iconic buildings have sprung up around the world to attract tourists and global investors, affecting themselves, their surroundings, and their communities with their peculiar features [3]. In many cities, iconic buildings have been built regardless of their urban context, in that they have been rejected by their surroundings as an "empty monument" rather than seen as an inclusive place of cultural exchange [4]. Especially in East Asia, new iconic buildings are arousing controversy in a more frequent manner due to their contrast with the environment [5]. A city is an archive center that records the old-time trails of its dwellers and its own environment for different sections of its history. Over time, they constitute the unique culture and identity of their citizens, which may be lost due to arbitrary planning and construction [6]. Seoul is one of the representatives of this dispute.

Entering a new era, Seoul shifted from the pure development of its economy to cultural regeneration, a more nature- and culture-friendly strategy and pattern of city update, to sustain its urban context [7,8]. Under such a policy, a series of programs for relic renovation have been undertaken, such as the restoration of Gyeongbokgung Palace, the reconstruction of

Gwanghwamun Square, and the replanning of Insadong to become a national tourist zone [9]. Meanwhile, new iconic buildings, like Lotte World Tower and Mall, the 63 Building, and others, have been constructed. Yet, an iconic building is different from a monument in that not only is it a symbolization of the scale of its city or country, but it is also of historical and cultural value [10]. The Design Plaza and Park (DPP), a piece by world-renowned designer Zaha Hadid, has caused protests due to its impact on the urban context of Seoul, on the basis of its appearance, the protection of relics, the regionality of South Korea, cultural memory, etc. Although the main construction of the DPP and its historical park has played an important role in revitalizing the economy of the district, attracting tourists through the visual miracle itself [11], during the process of its construction, there were conflicts between the construction party and local citizens, resulting in protests by various groups, including citizens, sports people, and culturalists [12], against the work of a foreign designer that sabotaged the urban context of Seoul. The iconic building of the DPP was completed, for the government met the protestors with concessions that the memories of the site for various groups would be replicated, with considerations of relic protection. However, there are still ongoing discussions about the urban contextual influence of the DPP.

Revolving around the DPP, in this study, we aim to discuss the urban contextual influence of newly constructed iconic buildings, through which we give suggestions for policies and practices for city renewal. Previous studies have centered, by and large, on the urban contextual influence of the DPP, interviews with people related to the project [13,14], and judgements and analyses based on policy papers and earlier studies [15]; in this study, meanwhile, by analyzing previous studies, we summarize the indexes for evaluating the urban contextual influences of iconic buildings and update the models for depicting the relationship between iconic buildings and their urban contexts. For an objective evaluation of the urban contextual influences of iconic buildings, we collected data using random sampling questionnaires. Meanwhile, triangular fuzzy statistics were employed to analyze the data, whose triangular fuzzy values were then obtained using triangular fuzzy statistics [16]. The abstract concept of context was objectively quantified via barycentric value; thus, we reached our conclusions from the final sequence of indexes that we chose. With the combination of the above methods, we could handle indeterminacy in the subjective evaluation, in turn evaluating the urban contextual influence of an iconic building in a more accurate manner. Through an investigation of the complexity and inner contradictions of the integration of iconic buildings into their urban contexts, we evaluated the urban contextual influence of the DPP to attempt to achieve a relatively optimal solution that is compatible with the local situation, which could shed light on other programs of iconic buildings and their cities.

2. Materials and Methods

2.1. The Definition of Iconic Buildings

Iconic buildings are those that raise worldwide attention due to their uniqueness and attractiveness, reinforcing urban regeneration [17,18]. In *The Iconic Building*, the artificial marvels of the modern age are referred to, by Charles Jencks, as "buildings", playing roles that symbolize power and interest, rather than "architectures", for "it is an age that everything can be an icon" [19].

One of the fundamental features of an iconic building is to astonish its visitors, and this means being overwhelmed by its height, shape, or unique location, which means that the original intention when designing an iconic building is to raise attention visually by outracing its surroundings [20]. They are of importance in the construction of cities [1], in turn reinforcing their cities with additional economic, political, and cultural effects in the competition among cities [21]. On account of this, more cities have grasped and utilized the uniqueness granted by newly constructed iconic buildings to attract attention and tourists [22]. Various cultural and business campaigns are held in which citizens are routinizing the field of the iconic buildings as the place of their everyday communications [23]. The sections of interconnections between the day-to-day lives of citizens and their groundings into events with the iconic buildings included are compiled by the memory of the community, yielding episodes of consecutive space–time fields [3], which in turn benefit the community and shape the iconic building into a symbol with incentive and moralization function, endowing the city with a unique cultural field.

However, since iconic buildings are mostly funded by establishments, they symbolize not only the scale of the cities or countries but also their cultural or historical value, namely mighty powers in politics, religion, or business that feature the establishments behind them [10]. Unlike the latter, they do not play the cultural function of turning the collective memory of a certain era into shared values of all times; thus, disjunction is seen between iconic buildings and the city they are located in [4].

2.2. The Development of Context

"Context" is a concept that originated in linguistics, underlying the relationship between a part and the whole [24], and was introduced into architecture in the 1960s [5]. Against the background of postmodernism, context has been endowed with different qualities in culture, with it being understood as interdependent communication between different cultural venations and the consequences of change and mergence. Context is a concept including both perceptions of the future and related experiences in the past, while the cultural turn of the concept triggers qualitative integrations of concepts in deeper layers [25]. Venturi blended the concept of "context" and "environment". In the 20th century, he emphasized the relationship between buildings and the environment and that between buildings [26]. Generally, people define themselves by means of their own past and reference objects from their lives in the past; thus, the self-identities of people are constituted partly by the physical environments of cities [27], which in some situations trigger intense talks as tokens of people's commemoration, resulting in the formation of a field of memory [28]. Entering the 1960s, Colin Rowe was among the figures initiating public censure against the failure of modernism urban planning and its damage to civic cultural inheritance. He criticized the fact that modern urban planning and building design gave rise to the school of contextualism [29], which emphasizes the inner connection of cities and their cultural backgrounds as an interconnected system. Contextualism advocates consideration of history and social environment in urban planning to maintain the uniqueness and sustainability of history, meeting the needs in function on the one hand and reinforcing the inner connection between cities and their urban context on the other. Contemporary contextualism underlies protections of constructions and their intangible culture [30].

2.3. The Relationship between Iconic Buildings and Urban Context

Iconic buildings serve not only as visual landmarks but also as carriers of urban context and identity. They interact with their surroundings by means of unique design language, attracting attention and triggering the urban effect depicted by the Bilbao effect [31], so that in many cities, iconic buildings are built strategically to raise their international profiles. Iconic buildings interrelate with the urban context in many facets [Table 1], and the latter can be expounded by important pairs of relationships including people and buildings and buildings and cities, as well as people and cities, which profoundly impact the image, recognition, and cultural inheritance of a city. The relationship between iconic buildings and the urban context is featured in the following facets:

- Identifiability: The basic property of iconic buildings. Iconic buildings enhance the identifiability of their cities by means of their uniqueness.
- Publicity: Being large in scale, iconic buildings provide public space for citizens, which indirectly activates public life in the city.
- Cultural field: By blending into their surroundings and the local intangible culture as part of integrations, iconic buildings serve as fields of cultural exchange and social interaction, promoting the cultural vitality of cities, which in turn internalizes as a component of the foundation of urban culture that intertwined with the image and shape of cities, constituting a compact network of urban contexts and cultural fields [5].

- Event: By staging various campaigns covering business, culture, and entertainment, iconic buildings connect citizens [23].
- Collective memory: While iconic buildings witness the history of the region that they bear, they deepen the connection as well between the history of cities and their citizens who are meanwhile strengthened in both their confidence in and the loyalty to the culture of their cities [25,30]. Citizens tend to preserve such connections, for long-term memories of surrounding urban regions make up the overall image of their cities [32].
- Locality: Contextualists advocate coalescing iconic buildings into their surrounding intangible culture and the aggregation of history and tradition, which constitute the integrity to underlie the uniqueness of the region [33].
- Heritage and sustainability: Iconic buildings carry the responsibility of relic protection and promote the sustainability of cultures and communities because not a single race, group, or individual survives and breeds in a new environment without historical continuity [34].

Table 1. Evaluation indexes of the relationship between iconic buildings and the urban context.

Property	Factor	The Relationship between Iconic Buildings and the Urban Context	Bibliography
Property of Constructions	Identifiability	Uniqueness and attractiveness are the fundamental properties of iconic buildings	Jencks C, (2006) [17] Raevskikh E, (2018) [18] Charles J, (2005) [19]
	Publicity	The indoor part of iconic buildings serves its users, while the outdoor part serves as the background of public lives	Gospodini A, (2004) [1] Klaus R, (1999) [31]
Property of Urban Context	Cultural field	A place to breed, spread, and inherit culture	Marcus and Francis, (1997) [23] Liu yun, (2010) [5]
	Event	Large-scale activities such as festivals, parades, exhibitions, etc.	Marcus, C.C. and C. Francis, (1997) [23]
	Public memory	Bearing public memories and bringing out resonance	Krzyżanowska N, (2016) [28]
	Regionality	Heterogenization	Robert V, (2002) [26] Stern; R.A.M, (1983) [32] Rong. Z, (2015) [27] Hubbard, P., (1996) [33]
	Inheritance and sustainability	Protection of cultural relics and the sustainability of communities	Pepper, S.C, (1942) [25] Zhuyuan, (2012) [34] Liu, X, (2008) [30]

Not only do iconic buildings decorate the city space or act as tools for economic growth, but they are also vital carriers of history and cultural heritage. By visual means, they work and are experienced as other architectures do/are in basic and cultural resources. As motivations of change to the city, iconic buildings endow the city with unique identifiability, while the identifiability of the city is constituted by people's memories and recognition of different places, according to Kevin Lynch. Thus, the evaluation of iconic buildings in urban renewal should be expanded from a mere discussion of the direct effects of iconic buildings to that of their long-term contribution to the urban context.

According to the summary above, our concept is based on a hypothesis to deconstruct the model in which an iconic building improves the image of a city through various urban contextual variables and the properties of the building itself [Figure 1]. In the framework of the model, an iconic building is considered a multi-dimensional socio-cultural phenomenon and influences five aspects of urban context (marked as H4-1, H4-2, H4-3, H4-4, and H4-5) through five variables (marked as H3-1, H3-2, H3-3, H3-4, and H3-5). H4-1 represents the combination of an iconic building with its surroundings, forming a platform of communication and interaction in culture, which enhances the impetus of urban culture, tightening the urban context and cultural field. H4-2 refers to an iconic building's functioning as a stage of various activities, promoting urban cultural vitality and stimulating the city and its social communication. H4-3 symbolizes an iconic building's function, as a witness of the past, in binding the connection between citizens and the history of the city, boosting the sense of belonging for its citizens. H4-4 signifies the integration of an iconic building into geography, culture, history, and tradition, emphasizing its regionality and the uniqueness of the city. H4-5 indicates the role of an iconic building in the promotion of the inheritance of culture, the protection of relics, and support for the sustainable development of the community.

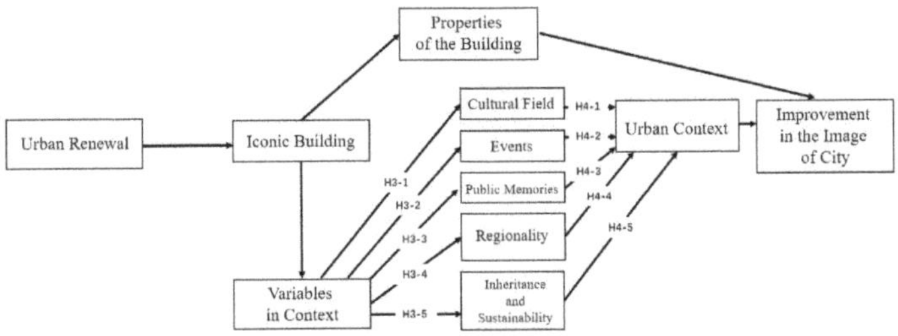

Figure 1. Research hypothesis model (source: one of the author's drawings).

With what we have shown above, the influence of the relationship between iconic buildings and the urban context is multidimensional, ranging from the visual identifiability of the city to improvements in cultural activeness, the creation of public space, the fortification of public memory, and emphasis on local culture. All of the factors above constitute the overall image of a city, promoting the continuity of culture and sustaining the community. After thorough discussion, we gained further awareness that when illustrating the role of an iconic building in urban renewal, attention should not be paid to merely its immediate effects, such as the attraction of tourists and financial profits. Rather, we should note its long-term contribution to the urban context. For this, we built a model of evaluation, aimed at the quantification of such influences. In the following sections, a specific case was chosen in order to verify the model and examine how to integrate the newly constructed iconic buildings into the historical and cultural context of the city they are located in.

3. Object of the Study

3.1. Introduction of the Design Plaza and Park

Located in the Dongdaemun district of Seoul (Figure 2), the Design Plaza and Park was a training ground for the army of the Yi dynasty and a facility for the training and accommodation of troops to protect the capital [6]. In the 18th century, with the booming of business, the DPP became one of the three main market squares in Seoul. With the signing of the Japan–Korea Treaty in 1905, the Gwangjang Limited company was founded by the Koreans, as a countermeasure against businessmen from Japan. This is the first permanent market to protect the native capitals of the Koreans and their interests and is a typical Korean marketplace. When the Japanese occupied the Korean peninsula in 1924, the first modern stadium, Gyeongseong Stadium, was built to celebrate a wedding held by the imperial house of Japan. Since the stadium witnessed several victories by Korean athletes in confrontation with the Japanese, it is thus regarded as "a place for Korea to fight Japan" [35].

In 1945, following the defeat of Japan, Gyeongseong Stadium was thereafter renamed as Seoul Stadium, along with a series of changes within society in Korea. With several new modern stadiums constructed, Seoul Stadium was used less frequently, and then once again renamed as Dongdaemun Stadium. Being named after a district rather than a city marked the decline of its influence [36]. As for Gwangjang Market, the neighbor of the stadium, the Korean War made it the biggest clothing wholesale market in Korea.

(a) (b)

Figure 2. (**a**) Dongdaemun Stadium aerial view (source: online); (**b**) map of the Design Plaza and Park. The red part is the main building of the DDP (source: one of the author's drawings).

Entering the 1980s, Seoul upgraded gradually from a city of heavy industry to a city with information technology-intensive industry [12]. The development of the economy boosted the self-confidence of the Korean nation. During Mayor Oh Se-hoon's term of office, Dongdaemun district was chosen as the core region of Seoul to improve "the brand value of the culture, art, design and environment of Seoul" [37], featuring the city with its design-oriented path of urban renewal. In the 1990s, a batch of shopping malls and duty-free shops belonging to fashion brands were built around the stadium. In 2008, the stadium was demolished, and, following the designs of Zaha Hadid, a famous designer from Britain, it was rebuilt into a versatile iconic building of Seoul, in which a series of cultural and art activities have been hosted.

3.2. Reason for Studying the DPP

In this study, we chose the Design Plaza and Park as a test case and analyzed its urban contextual influence. As an iconic building in Seoul, the DPP is an integration of historical relics, cultural memories, wounds of the nation, and modern design. From the accommodation of the army to one of Seoul's three main marketplaces [6], the role of the DPP was shifted by the emerging modern culture in Korea, which is a "local" change, driven by internal power. In the following historical period, the land received mandatory upgrades by the Japanese, which is an external factor that came into play. Up to now, the district contains the famous iconic building that represents Seoul. Observing the course of the development of the land, we see the transition from historical relics and a stadium bearing public memories to a project of urban renewal centered on design and culture [14]. Yet, the project aroused various disputes [Table 2]. Citizen groups headed by "Cultural Solidarity" protested against the tearing down of the old stadium, with an alternative solution given to reconstruct it into public sports facilities and amusement parks [12]. Historians and archeologists proposed to protect and excavate the site since it had been part of the Hangyangdoseong Capital City Wall [38], and along with its excavation, the Seoul Metropolitan Government (SMG) built Dongdaemon Museum to exhibit cultural relics found in the archeological site. In the sports circle, the Korean baseball league spoke against the proposal on behalf of sports people, in that there is no proper baseball field in Seoul. The SMG met with the baseball league over 150 times, reaching an agreement that the new construction was to be identified as a field of sport by keeping the lighting tower and the flame-holder as symbols of a stadium. The baseball league let the matter drop after a sports museum and seven other baseball fields were put forward in the plans. After coordination with different groups of interests, Dongdaemon Stadium was deconstructed in 2008, including criticism from historians, the worries of sports people, and protests from local businessmen [Table 2].

Table 2. General outline and design element characteristics of the study site.

Discommender of the DPP	Reason	Appeal
Historians	Protection of cultural relics	Protection of archaeological remains
Sports people	Continue cultural memories	Rebuild the stadium into a sports museum
Citizens	Maintain the place bearing emotions and memories	Register it as a heritage site
Businessmen	Survival pressure	Retain the marketplace

Source: Seoul Metropolitan Government [37].

Our results show, however, some of the citizens' awareness of the positive effect of the DPP in raising the reputation of Seoul and staging cultural and art activities. This controversy illustrates the dual role identity of the DPP that it is both the motivation that drives the modernization of the city and a potential threat to the sustainability of the urban culture (Table 2). Below is a general outline and the design element characteristics of the study sites.

Above all, we chose the DPP as the case of the study for it displays the diversity and complexity of the relationship between iconic buildings and the urban context. By exemplifying the DPP, we may investigate its urban contextual influence on Seoul, in turn shedding light on the urban contextual development of iconic buildings [Table 3].

Table 3. General outline and design element characteristics of study site.

Division		Attributes Included in the DPP
Cultural Characteristics	Historical cultural facilities	Remains of the capital in history and commercial plot in history
	Cultural field	Art gallery and Dongdaemun History Park
	Event	Large-scale activities including festivals, parades, art exhibitions, etc.
	Regionality	Newly developed constructions and foreign designers
	Public memories	Memory of the stadium
	Sustainability in culture	Iconic building in Seoul
Cultural Characteristics	Shopping	DPP Design Store
	Business activities	Business campaigns and exhibitions
	Dining	Restaurants
	Entertainment	Exhibitions and light shows
	Hotels	/

4. Results

4.1. Investigation Method

4.1.1. Examination of the Object under Investigation

In total, 220 copies of the questionnaire were distributed and 220 of them were recollected. For diversity in the data, we distributed the questionnaire in various places, with large-scale activities avoided over time. Respondents were stratified into groups of different ages, avoiding members of large tourist groups and visitors taking part in large-scale activities. Although the samples are mediocre in quantity, the effectiveness of them was ensured.

Before the questionnaire survey, an extensive search of the literature was carried out using both Web of Science and Google Scholar to ensure the feasibility and reliability of the study. The retrieval range includes SCI, SSCI, AHCI journals, and other highly cited papers. By providing a summary of the properties of the urban context, we specified several dimensions of the urban context in which iconic buildings have an influence and drew progressive conclusions. The questionnaire had in total of 23 options, in which part 1 records the basic information of the respondent, such as age and education, and part 2 records the type of respondent and the purpose of their visit. Among all of the sections, part 3 is the core part of the questionnaire, with it listing all of the relative properties of iconic buildings, including properties in business and urban contexts, based on the literature review in the previous part of the study. This was carried out to verify the model we proposed. In the evaluation of all the properties, a five-point Likert scale (Table 4) was employed. To exclude invalid questionnaires, question 4 about local identity and question 10 about the property of cultural memories were limited to respondents with a local identity relating to Seoul to establish mutually exclusive conditions. Since the questions were not appropriate for explicit analysis, accurate points in the evaluation were replaced by five fuzzy options, such as asking the respondent if he/she is satisfied with the new construction's reproduction of the urban contextual elements. In the final analysis, options are assigned with different values, which are then fuzzified and defuzzified with a triangular fuzzy formula, among which part three touches on the evaluations analyzed using a five-point Likert scale [Table 4]. Since the questions were not appropriate for explicit analysis, accurate points were replaced by fuzzy options.

Table 4. Survey items.

Division	Subsection
Demographic information	Age and education
Type	Citizen or tourist and type of tourist
Purpose of the visit	Visiting a historical relic, visiting an exhibition, shopping, or dining
Context elements	Identifiability, publicity, cultural field, event, public memory, regionality, inheritance and sustainability, and protection of the cultural relic
Commerce characteristics	Commercial exhibition, entertainment, art exhibition, attraction for international tourists, and improvement in national profile
User satisfaction	I am very satisfied with the element of the building (1.0) I am satisfied with the element of the building (0.75) I am just okay with the element of the building (0.5) I am not quite satisfied with the element of the building (0.25) I am not satisfied with the element of the building (0.0)

4.1.2. The Transformation of the Qualitative Data during Fuzzy Statistical Analysis

Regarding the urban contextual influence of iconic buildings, researchers are prone to studying cases using qualitative methods. Conclusions are usually drawn from interviews with relative people, previous studies, and policy documents. In disposing of evaluation of respondents and capturing the fuzziness of the concept, researchers are often limited by subjective factors, while the abstractness of the concept of urban context is usually intangible. Thus, in this study, we employed fuzzy statistical analysis, quantifying the qualitative evaluations we collected by means of triangular fuzzy numbers to reach more accurate and reliable conclusions.

Fuzzy theory is a theory that quantifies fuzzy, implicit quantitative information and qualitative data by means of mathematical concepts [39]. Since people's evaluation of complex emotions and abstract concepts is usually blurry and implicit, to counter this, we must reduce errors that are caused by language and recognition [40]. Since traditional evaluations of elements related to the urban context, such as the cultural field, public memories, and locality, hinder the respondents from accurate scoring; in this study, we employed triangular fuzzy statistical analysis.

Moreover, the results of the study can be calculated and analyzed by means of a combination of fuzzy variables and triangular fuzzy indexes, for qualitative evaluations can be effectively quantified by triangular fuzzy numbers, a mathematical tool that is also capable of the coding and analysis of indeterminacy and subjectivity in the data of the survey. The method employed in this study based on an integration of the three data points in the triangular fuzzy index and quinquepartite method was feasible, according to another study using the same method [16]. After assigning value to the options in Table 4, fuzzy sets and fuzzy logic are built and then projected to concrete fuzzy numbers, in which each selected result A can be categorized into a triple (l,m,u), in which l_A is the lower limiting value, m_A is the possible value, and u_A is the upper limiting value (Formula (1)).

These fuzzy numbers are expressed by means of triangular fuzzy numbers, with which we construct a triangular fuzzy matrix for each urban contextual element so that we can delimit the range of indeterminacy for each option (Table 5). Next, we defuzzify each result of the option via the center of mass method (Formula (2)) for further analysis and relatively appropriate and scientific results. After we verified the rationality of the end results with SPSS, we obtained our final fuzzy rate of the citizens' and visitors' satisfaction with the DPP's urban contextual influence on Seoul.

$$\mu_A(x) = \begin{cases} \frac{x-l_A}{m_A-l_A}, & for\ l_A \le x < m_A \\ 1, & for\ x = m_A \\ \frac{u_A-x}{u_A-m_A}, & for\ m_A < x \le u_A \\ 0, & otherwise \end{cases} \tag{1}$$

Table 5. Analysis of the visit purpose of visitors.

Options	Triangular Fuzzy Number
A. Very satisfied	(0.75, 1, 1)
B. Satisfied	(0.5, 0.75, 1)
C. Okay	(0.25, 0.5, 0.75)
D. Not quite satisfied	(0, 0.25, 0.5)
E. Not satisfied	(0, 0, 0.25)

According to the results of the questionnaire and a verification of rationality using SPSS, we obtained the results of the degree of satisfaction of citizens and tourists regarding the urban contextual influence of the iconic building. A relatively objective result was reached after defuzzification of the data (Formula (2)) and further comparative analysis.

$$x^* = \frac{\int_{l_A}^{u_A} x \cdot \mu_A(x)dx}{\int_{l_A}^{u_A} \mu_A(x)dx} \tag{2}$$

4.2. Result of the Survey

Among the 220 questionnaires retrieved, we excluded 3 that were invalid due to answers in the mutually exclusive questions. In the 217 valid questionnaires, the respondents were distributed the most in the age range of 30–39 (n = 93.4%) and the least in the under 18 age range (n = 0.9%). We saw a large number of respondents who received college education (n = 66.7%), while there were only a few with a doctor degree (n = 2.7%) (Figure 3).

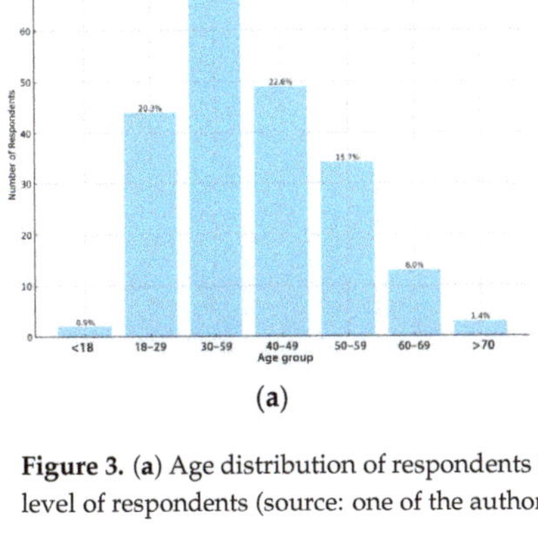

(a)

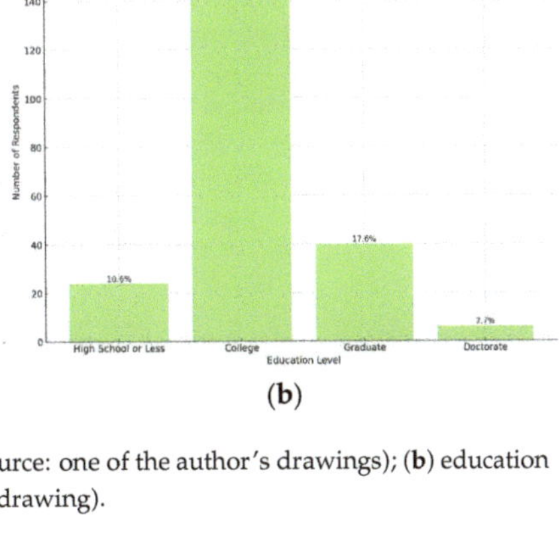

(b)

Figure 3. (**a**) Age distribution of respondents (source: one of the author's drawings); (**b**) education level of respondents (source: one of the author's drawing).

Of the visitors to the DPP involved in the survey, 85% ($n = 186$) are Seoul locals, while foreign tourists made up a percentage of 14.3% ($n = 31$). Among all tourists, only 3.68% of tourists ($n = 8$) were history enthusiasts, and 1.5% ($n = 7$) of tourists visited the DPP for historical relics, while most of them visited the site for exhibitions (34%, $n = 164$) and cultural activities (29%, $n = 140$) (Table 6). This result demonstrates to some extent both the DPP's attractiveness as an exhibition and cultural activity hall and its deficiency in the protection and sustainability of history and context (Table 7).

Table 6. Analysis of types of tourists.

	Options					
Types	**History Enthusiasts**	**Tourists in Family**	**Tourists for Leisure**	**Tourists in Couple**	**Visitors for Business**	**Visitors for Academic Purpose**
Date	8 (3.68%)	32 (14.7%)	78 (35.94%)	32 (14.7%)	45 (20.7%)	22 (10.1%)
Seoul locals	186 (85.7%)				Foreign tourists	31 (14.3%)

Table 7. Analysis of the purpose of visits.

Types	Visiting Historical Relics	Visiting Exhibitions	Visiting the Building	Shopping	Participating Activities of Culture	Casual Tour
Date	7 (1.5%)	164 (34.0%)	80 (16.5%)	31 (11.2%)	140 (29.0%)	37 (7.7%)

4.3. Regression Analysis

Regression analysis was employed in this study for a thorough comprehension of the quantitative relation of the DPP's urban contextual influence on Seoul. We categorized the dimensions of evaluation in the questionnaire as a series of independent variables, including public memories, events, cultural fields, regionality, and sustainability and the protection of cultural relics; meanwhile, we categorized the DPP's urban contextual influences on Seoul as the dependent variable. In the regression analysis using SPSS [Table 8], we saw that the modified R2 of each variable of the DPP's urban contextual influence on Seoul is greater than 0.3 ($n = 0.515$), which means that the effectiveness of the explanation of the model reached a percentage of 51.5%, while the value of P was less than 0.05, representing profound influences of each value on the urban context. The result of the regression analysis generally verified the objectivity and effectiveness of the evaluation of the urban contextual influence of an iconic building on the city.

Table 8. Regression analysis results for Context of Design Plaza and Park.

Mode	Standardized Coefficients (β)	T	Sig.	Collinearity Statistics	
				Tolerance	VIF
Public memories	0.155	2.266	0.013	0.685	1.303
Events	0.086	3.776	0.035	0.630	1.587
Cultural field	0.187	3.840	0.029	0.473	1.677
Regionality	0.241	3.184	0.019	0.684	1.463
Cultural sustainability	0.22	2.847	0.049	0.749	1.553
Protection of cultural relics	0.204	3.153	0.010	0.853	1.172

R2 = 0.539, Adjusted R2 = 0.515, F = 22.750, and $p < 0.05$

The result of the regression analysis showed that the most significant variable is the regionality of the building, whose standardized coefficient value is 0.246, the value of t is 3.184, and the value of the significance level is 0.019. This is the variable that makes the greatest contribution to the DPP's urban contextual influence on Seoul. Moreover, publicity, public memories, and the property of the cultural field represented by the building as variables affect the evaluation of the urban context as well. In the results of the multicollinearity statistics of the regression analysis, the value of tolerance of each of the independent variables was above 0.1, with each of the VIF values lower than 5, so that there was no multicollinearity between variables. The model in this study is therefore feasible and can be expanded to the evaluation of other iconic buildings.

According to the regression analysis, cultural status, publicity, and cultural coordination are the most important three factors that influence the experience of people visiting the DPP and their recognition of the function of it. Its value in terms of cultural status and cultural coordination highlights the importance of the DPP's influence on developing urban culture and shaping the image of the city. Yet these results expose the tension between cultural relics and modernization, which are to be properly balanced in future projects of urban renewal. These results provide specific guidance for future projects for urban planning and the construction of iconic buildings, emphasizing the importance of the protection and development of a continuing urban culture.

4.4. Analysis of the Result of Triangular Fuzzy Statistics

In this section, we carry out further analysis of the result of the questionnaire with triangular fuzzy statistics to establish results for each analyzed index on the DPP given by the respondents. The index "event" is subdivided into the protection of cultural relics, business activities, and art exhibitions, making our analysis more accurate. In the diagram below (Table 9), indexes ranked in first and second place are art exhibitions ($n = 0.6706$) and business activities ($n = 0.6555$), respectively, illustrating the positive effect of the DPP in these aspects. The two indexes at the bottom of the rank are attraction of international tourists ($n = 0.2377$) and representation of the public memories ($n = 0.2970$), respectively, which means that the DPP leaves much to be desired in these aspects. It thus appears that the DPP was not considered as an urban contextual token of public memories that connects the citizens with its history.

Table 9. Analysis of types of tourists and the purpose of their visit.

Division		Triangular Fuzzy Number	Center of Gravity	Rank
Factors	**Subsection**			
Iconic building	Identifiability	(0.3324, 0.53135, 0.7233)	0.529	5
	Publicity	(0.43175, 0.677, 0.85085)	0.6532	3
Cultural characteristics	Cultural field	(0.266, 0.4964, 0.4964)	0.4948	6
	Public memories	(0.0981, 0.2728, 0.5202)	0.2970	11
	Regionality	(0.13675, 0.314, 0.55575)	0.3355	10
	Protection of cultural relics	(0.15385, 0.33555, 0.5815)	0.3569	9
	Inheritance and sustainability	(0.1607, 0.35492, 0.59417)	0.3699	8
	Business activities	(0.42925, 0.67425, 0.6742)	0.6555	2
	Art exhibitions	(0.449, 0.694, 0.86875)	0.6706	1
	Improvement of international image	(0.3295, 0.54325, 0.75125)	0.5413	4
	Attraction of international tourists	(0.1325, 0.2344, 0.3463)	0.2377	12
	Entertainment	(0.32402, 0.51167, 0.6393)	0.4916	7
The urban contextual influence of the DPP on Seoul		(0.1659, 0.3871, 0.6325)	0.3952	

When we focused on the DPP's property as a building, we saw that most respondents scored highly on the identifiability ($n = 0.529$) and publicity ($n = 0.6532$) of the site, meaning that most respondents approve of the design of the DPP in terms of its function in the urban landscape and public space. The DPP affects the public by means of its visual impact and symbolization.

However, when it comes to the contribution of the DPP to urban history and cultural inheritance, the score is relatively low, with no values over 0.5 and lower positions in rank. The indexes of context are obviously lower than that of commerce characteristics and iconic buildings, which shows that although the DPP is of great significance in commerce and modern art, there is no overt public realization of the DPP's function in strengthening urban cultural memories (n = 0.2970), regionality (n = 0.3355), the protection of cultural relics (n = 0.3569), and connection with and inheritance of urban culture (n = 0.3699). Among all of the functions of the DPP, people are not satisfied with it in sustaining the urban context. The DPP's urban contextual influences on Seoul trend from negative medium to negative maximum (n = 0.3952), according to the diagram. Although the DPP is a great site for commercial and artistic use, playing a positive role in improving the international image of Seoul, It does not resonate with the public in terms of connecting the past and the present and reinforcing urban cultural memories. For city planners, this result calls for their consideration, in future projects for iconic buildings, of the influence of different demands from visitors and local citizens, as they are two different groups asking for different functions for the iconic building. The strategy of constructing iconic buildings should be made from the business frame for both economic development and cultural heritance.

The project of the DPP is the embodiment of Seoul's global strategic city planning. It was built to erase the mark of the colonization of the Korean people, emphasizing their new identity as an independent society [41]. Through a deep analysis of the project of the DPP, local citizens' dissatisfaction regarding the ignorance shown toward the sustainability of the urban context through the project is exposed. The project serves well as a reflection of how to sustain the unique identity of a nation and the continuity of its culture in rapid development and globalization and as a good reference for future projects for iconic buildings.

4.5. Summary

Through data collection and literature summarization, we know that the DPP has a close relationship with the planning of the image of Seoul. Within the strategy of building Seoul into a global city, it is necessary to reproduce the locality and identity of Seoul, whilst also creating new cultural spaces and fields. The traditional design art of Korea was supposed to be displayed with its modern variant by means of the integration of history, the park, and the design plaza, but looking at whether the evaluation by researchers and the local citizens or the result of the analysis of data, the finished construction seems counterproductive in those aspects. Indeed, the building and history park of the DPP has played an important role in revitalizing the Dongdaemun district and creating visual wonder, promoting the image of the city, and communicating cultures. Yet it is more of a building out of nowhere that separates itself from its surroundings and history, covering the historical aspects of the land. This discovery is of great importance in city planning at present, shedding light on future design projects for urban districts.

One of the prime targets of urban renewal is to sustain local history and culture and restore and protect districts with historical sites, which plays a central role in utilizing resources and maintaining existing socio-cultural structures. Through the case of the DPP, we know that the function of an iconic building is not merely to fill in the space in the district that it is assigned to. The purpose of an iconic building is to be a narrator, a representative, of the revolution of the local society around and the relics beneath it, especially when built in an Asian city featuring multi-layer cultures. This is conducive in reinterpreting "Asian-ness" by some people [42]. Otherwise, these cities, along with their newly constructed iconic buildings, are to be consumed by the tide of globalization, which is not only challenge in Seoul but also in East Asia.

5. Conclusions

In this study, by employing fuzzy statistical analysis, we provide a method to transfer the qualitative evaluation of the urban contextual influence of iconic buildings, which improves the accuracy and objectivity of the evaluation data collected. According to the study, fuzzy statistical analysis has a significant advantage in the quantification of the evaluation of abstract notions, such as urban context. By using a method such as this, we provide a concrete evaluation of the influence of the DPP from the dimensions of the economy and the urban context, providing new quantitative perspectives of the influence of a new iconic building in city planning and supporting overall research on the urban contextual influence of new iconic buildings. This study also has limitations for specific regions and a relatively smaller number of samples, without considering a broader international environment. In future research, the number of studies can be expanded to different countries and cities, so as to exploit the application of fuzzy statistical analysis in evaluating iconic buildings and their urban context with more samples and to improve the feasibility of the method.

Author Contributions: Conceptualization, Z.C. and J.-E.Y.; methodology, Z.C. and J.-E.Y.; formal analysis, Z.C. and J.-E.Y.; investigation, Z.C.; resources, Z.C.; data curation, Z.C. and J.-E.Y.; writing—original draft preparation, Z.C.; writing—review and editing, Z.C.; visualization, Z.C.; supervision, J.-E.Y. All authors have read and agreed to the published version of the manuscript.

Funding: This research received no external funding.

Data Availability Statement: Data is contained within the article.

Conflicts of Interest: The authors declare no conflicts of interest.

References

1. Gospodini, A. Urban morphology and place identity in European cities: Built heritage and innovative design. *J. Urban Des.* **2004**, *9*, 225–248. [CrossRef]
2. Jones, C.; Svejenova, S. The architecture of city identities: A multimodal study of Barcelona and Boston. *Res. Sociol. Organ.* **2018**, *54*, 203–234.
3. Elhagla, K.; Nassar, D.M.; Ragheb, M.A. Iconic buildings' contribution toward urbanism. *Alex. Eng. J.* **2020**, *59*, 803–813. [CrossRef]
4. Zukin, S. *Naked City, The Death and Life of Authentic Urban Places*; Oxford University Press: Oxford, UK, 2010.
5. Liu, Y. Urban Landmark Preservation and the Continuity of Urban Literature—An Example of Landmark Preservation in the United States. Ph.D. Thesis, Central Academy of Fine Arts, Beijing, China, 2015.
6. Chung, H. A Critical Review on Regenerating a Place's Economic Value through Landscape Restructuring: The Case of Dongdaemun Stadium. *J. Korean Geogr. Soc.* **2009**, *44*, 161–175.
7. Hwang, Y.-S. Korean Urban Regeneration:Design Emphasizing Historical, Regional, and culture values. *J. Inter. Des.* **2021**, *46*, 3–9.
8. Jeong, S. *Seoul's Experience in Cultural Heritage, Sustainable Tourism and Urban Regeneration*; Seoul Solution: Seoul, Republic of Korea, 2018.
9. Kim, W.B. The viability of cultural districts in Seoul. *City Cult. Soc.* **2011**, *2*, 141–150. [CrossRef]
10. Wang, S. *A History of Modern Architecture*; China Architecture & Building Press, China Building Industry: Beijing, China, 2012; p. 425. ISBN 10 7112141419.
11. Yeonghwa, G. Dongdaemun undongjang, dijain gongwon euro saeropget'aeeonada [Dongdaemun Stadium, reborn as a design park]. *Maru* **2007**, *66*, 160–163.
12. SDI (Seoul Development Institute); Nomura Research Institute. *Can Seoul Become a World City? Comparison & Analysis of Northeast Asia's Six Cities: Seoul, Tokyo, Hong Kong, Beijing, Shanghai, and Singapore*; SDI: Seoul, Republic of Korea, 2003.
13. Hwang, J.-T. Territorialized urban mega-projects beyond glob convergence: The case of Dongdaemun Design Plaza & Park Project, Seoul. *Cities* **2014**, *40 Pt A*, 82–89.
14. Yoon, K. Recreating Dongdaemun Stadium in South Korea: Beyond Japanese Colonial Memories and towards a Global City. *Seoul J. Korean Stud.* **2018**, *31*, 99–128.
15. Yun, J. A copy is (not a simple) copy: Role of urban landmarks in branding Seoul as a global city. *Front. Archit. Res.* **2019**, *8*, 44–54. [CrossRef]
16. Choi, Y.E.; Chon, J.; Yoon, J.A. An Analysis of Design Elementsand Satisfaction on the Usability of City Squares—Focused on Gwanghwamun Square and Geumbit Square. *J. Korean Inst. Landsc. Archit.* **2014**, *42*, 111–123. [CrossRef]
17. Jencks, C. The iconic building is here to stay. *City* **2006**, *10*, 3–20. [CrossRef]

18. Raevskikh, E. Anticipating the "Bilbao effect": Transformations of the city of Arles before the opening of the Luma Foundation. *Cities* **2018**, *83*, 92–107. [CrossRef]
19. Charles, J. *The Iconic Building*; Rizzoli Press: New York, NY, USA, 2005; p. 20.
20. David, H. *Models of Democracy*; Stanford University Press: Redwood City, CA, USA, 1996.
21. Gottmann, J. Why Skyscraper? *Geogr. Rev.* **1966**, *56*, 190–212. [CrossRef]
22. Sklair, L. Iconic Architecture and Capitalist Globalization. *City* **2010**, *10*, 21–47. [CrossRef]
23. Marcus, C.C.; Francis, C. *People Places: Design Guidelines for Urban Open Space*, 2nd ed.; John Wiley and Sons: New York, NY, USA, 1997.
24. Bognar, B.; Isozaki, A. *Contemporary Japanese Architecture: Its Development and Challenge*; Van Nostrand Reinhold: New York, NY, USA, 1983; p. 125.
25. Pepper, S.C. *World Hypotheses, a Study in Evidence*; University of California Press: Berkeley, CA, USA; Los Angeles, CA, USA, 1942; p. 244.
26. Robert, V. *Complexity and Contradiction in Architecture*; The Museum of Modern Art: New York, NY, USA, 2002.
27. Rong, Z. *Collective Memory of City and Urban Form—Research of History and Culture Continuity in the Perspective of Psychology and Sociology*; Southeast University, PHD: Nanjing, China, 2015.
28. Krzyżanowska, N. The discourse of counter-monuments: Semiotics of material commemoration in contemporary urban spaces. *Soc. Semiot.* **2016**, *26*, 465–485. [CrossRef]
29. Jencks, C. *The New Paradigm in Architecture: The Language of Post-Modernism*, 7th ed.; Yale University Press: New Haven, CT, USA, 2002; pp. 78–79.
30. Liu, X. *Theories of Modern Architecture*; China Architecture & Building Press: Beijing, China, 2008; p. 41.
31. Reichold, K. *Buildings That Changed the World*; Prestel Publishing Press: Munich, Germany, 1999.
32. Stern, R.A.M. After modernist movement. *Architect* **1983**, *15*, 201–205.
33. Hubbard, P. Urban design and city regeneration: Social representations of entrepreneurial landscapes. *Urban Stud.* **1996**, *33*, 1441–1661. [CrossRef]
34. Zhu, Y. *The Utopia of the Present*; Southeast University Press: Nanjing, China, 2012; p. 122.
35. Kim, M.; Pak, K. Hanguk geunhyeondae seupocheu ui sansil, Dongdaemun undongjang [The cradle of modern Koreansports, Dongdaemun stadium]. *Seupocheu Illyuhak Yeongu* **2010**, *5*, 47–69.
36. Youn, S.-H. The Impact of the Colonial Architectural Heritage on South Korean's National Identity. Ph.D. Thesis, University of Surrey, Guildford, UK, 2014.
37. Seoul Metropolitan Government (SMG). *A White Paper of Dongdaemun*; Design Plaza and Park; Business; SMG: Seoul, Republic of Korea, 2013; Volume 1.
38. Ji, H. A culture city is one that contains human life: A critique of Dongdaemun Stadium Park Project. *Munhwagwahak* **2008**, *53*, 538–547.
39. Zadeh, L.A. Outline of a new approach to the analysis of complex systems and decision processes. *IEEE Trans. Syst. Man Cybern.* **1973**, *SMC-3*, 28–44. [CrossRef]
40. Yoon, S.H.; Yun, D.K. Using triangle fuzzy numbers and semantic differential scale to evaluate service quality. *J. Korean Soc. Qual. Manag.* **2004**, *32*, 182–197.
41. Kong, L. Cultural Icons and Urban Development in Asia: Economic Imperative, National Identity, and Global City Status. *Political Geogr.* **2007**, *26*, 383–404. [CrossRef]
42. Logan, W.S. The Cultural Role of Capital Cities: Hanoi and Hue, Vietnam. *Pac. Aff.* **2005**, *78*, 559–575. [CrossRef]

buildings

MDPI

Article

Design Strategies to Improve Metro Transit Station Walking Environments: Five Stations in Chongqing, China

Chungui Yao, Gaoyuan Li and Shuiyu Yan *

School of Architecture and Urban Planning, Chongqing University, Chongqing 400044, China; yaochungui@stu.cqu.edu.cn (C.Y.); 27120504@alu.cqu.edu.cn (G.L.)
* Correspondence: yanshuiyu@cqu.edu.cn; Tel.: +86-136-484-26928

Abstract: While transit-oriented development (TOD) has been widely adopted in urban design alongside the expansion of urban metro transit, the creation of pedestrian-friendly environments has often been overlooked during implementation. This has resulted in a lower walking advantage around metro transit stations. To address this issue and encourage walking and public transport use in metro transit station areas, this study undertook a quantitative comparative analysis of the pedestrian environment in five Chongqing metro transit station areas. The analysis focused on three key dimensions: "comprehensive evaluation", "basic scale", and "structural quality". The comprehensive evaluation considered factors such as the pedestrian catchment area ratio, POI kernel density distribution, and crowd agglomeration. The basic scale dimension comprised floor area ratio, building density, pedestrian road density, and the quantity of station entrances and exits. Finally, structural quality factors included land use type mixing degree, POI function mixing degree, intersection connectivity, median street length, pedestrian route directness, and green view index. Based on these analyses, this study proposes a series of pedestrian environment design strategies including land use and transportation. The strategies for land use advocate for "developing compact and diverse land use", "strengthening attraction of station center", "positioning large projects on the edge", "restricting private transportation capabilities". The strategies for transportation consist of "increasing pedestrian road density", "traffic calming organization", "subdivision of road types", and "three-dimensional pedestrian traffic system". These strategies aim to create a more humanized and environmentally friendly pedestrian environment, proactively rise to the challenge of climate change, thereby cultivating sustainable urban development.

Keywords: metro transit station area; walking environment; TOD mode; pedestrian catchment area; walkability; walking environment evaluation; station optimization strategies

1. Introduction

Since the Industrial Revolution, the swift expansion of mechanized forms of transportation, including cars, trains, and subways, has led to urban challenges such as traffic jams, air pollution, and energy shortages. With environmental concerns becoming more pressing, the urban development model heavily reliant on private vehicles is proving to be unsustainable. The consensus for advancing sustainable urban transportation comprises controlling the expansion of private vehicular traffic, enhancing public transportation systems, and encouraging green commuting practices among residents. Urban metro transit has become a crucial solution to mitigate traffic congestion in high-density urban areas. Walking, as a transportation option, plays a critical role not only as a primary means of interchange for metro transit but also in connecting various service offerings in metro transit station zones [1]. However, the reality in China demonstrates that metro transit construction often stalls at the TOD planning phase, failing to fully capitalize on the "walking advantages" in station areas. The *2020 Annual Statistical and Analysis Report on Urban Metro Transit* notes that China's urban metro network extended to 7969.7 km in 2020, with the network's size increasing annually. However, data from 2012 to 2020 reveal a decline in passenger density alongside the growth of the metro network's reach. This indicates that urban metro systems should not merely focus on expansion but should prioritize enhancing the functionality and environment of existing stations through specific design improvements or urban rejuvenation projects, aiming for sustainable operation at each location [2]. The current limitations in designing pedestrian-friendly spaces around metro transit stations are primarily evident in several key areas: (1) a lack of coordination between transportation planning and land use, leading to areas around stations that lack appeal; (2) a prioritization of expanding the urban transportation network without equal emphasis on optimizing land use in the vicinity of metro transit stations, indicating potential for better land use efficiency; (3) an oversight in developing a conducive built environment and managing pedestrian flow in metro transit station areas, which reduces residents' inclination to walk; and (4) a lack of attention to how urban landscapes are visually perceived and the implementation of visual navigation systems for transfers at stations, which curtails the benefits of pedestrian movement in these areas. Therefore, it is of significant importance to develop pedestrian-friendly environments in metro transit station areas for enhancing the willingness of pedestrians to walk to public stations, increasing the utilization rate of metro transit, and responding to climate change.

TOD is an integrated approach to urban planning that combines transportation infrastructure with land use, enhancing urban efficiency and promoting sustainable urban development [3–6]. The primary goals of TOD include encouraging walking, boosting public transit ridership, reducing congestion, and yielding environmental benefits [7–11]. This development strategy is widely implemented in urban planning and strategic development efforts around the globe [12–15]. Schlossberg highlights that understanding opportunities for pedestrian movement is a key component in assessing TOD projects [16]. Estupiñán emphasizes the importance of defining pedestrian spaces around transit stations to support the use of public transportation [17]. Jacobson [18] and Vale [19] argue that the walkability of communities surrounding transit stations is crucial to their success, as it facilitates convenient, safe, and direct access to transit nodes. This, in turn, encourages a greater use of transit stations and ensures that TOD residents have access to local amenities to meet their needs. King explores how the walkability of communities around transit stops can inform future TOD developments [20]. Morency, after conducting scenario simulations for four different plans in the Montreal metropolitan area, found that TOD areas could reduce walking travel distances by 28% [21]. Nasri studies the impact of the built environment on mode of transportation choice within the context of TOD, using discrete choice models and focusing on the built environment at the origin of trips [22]. Given that trips in rail transit begin and end with walking, walkability is a significant concern in TOD communities.

Regarding the assessment studies of urban walkability or pedestrian environments, the field of research can be classified into three categories: the objective spatial environment, the subjective spatial experience, and comprehensive evaluations. From the objective environment perspective, Dill [23], Marshall [24], and Jones [25] have explored the correlation between residents' modes of transportation (i.e., cycling and walking) and the urban form. They emphasize the significant influence that the structural properties of the street network exert on road connectivity. Chinese researcher Lu has introduced the Walk Score system adopted in the United States, acknowledging its utility in demonstrating how the layout of everyday facilities can affect the walkability score of urban areas [26]. Yin has connected the walk score with urban design spatial variables across streets through the application of computer vision, presenting a 3D analytical perspective on urban walkability evaluation [27]. Peiravian [28] and Mayne [29] have simplified the evaluation process through developing a pedestrian environment index that incorporates indicators such as land use diversity, population density, commercial density, and intersection density. On the subjective evaluation front, Millstein [30] and Kurka [31] have shifted the focus from a broad analysis of street connectivity to the specific elements that may affect residents' willingness to walk in the built environment. They have carried out a Microscale Audit of Pedestrian Streetscapes (MAPS) scoring evaluation from the pedestrian's perspective. Gou, utilizing semantic differential (SD) analysis, has performed landscape perception evaluations of street spaces in Nanjing and Shanghai [32]. Similar studies include those by Shatu, who, in the absence of direct data, referenced existing factors influencing pedestrian path choices and employed computer virtual environments to facilitate the assessment of street walkability [33]. Evaluations of the objective spatial environment predominantly rely on quantitative analysis of urban spatial walkability, spatial morphological characteristics, and specific functional layouts. Subjective spatial experiences are derived from the pedestrian's perspective, focusing on surveying and understanding residents' preferences and satisfaction levels with the walking space, thereby summarizing relevant patterns to further inform spatial design. This approach represents a type of qualitative research; comprehensive evaluations bridge the gap between objective and subjective viewpoints, finalizing a weighted scoring following a holistic analysis.

Research in various regions, both in our own borders and globally, on the subject of pedestrian environments near metro transit stations, reveals a consensus: creating a walk-friendly atmosphere extends beyond merely ensuring accessible road networks to encourage walking. It also involves curating a rich variety of destinations in the spatial layout to maintain residents' interest in walking. Scholars, including Yan [34], Guo [35], and Zhao [36], are of the opinion that managing the pedestrian space around metro transit stations should prioritize aspects such as connectivity, safety, comfort, and the holistic design of public spaces. Through the lens of the Ningbo case, Bai illustrated how communities around urban metro transit stations can be forged with thoughtful pedestrian environment planning [37]. Olaru's research has validated that enhancing public transport access on a city-wide scale can broaden the scope of residents' activities and lessen their dependence on personal vehicles. This strategy necessitates the aggregation of public amenities near stations in reasonable walking distances as part of land development initiatives [38]. Jeffrey and other scholars assessed the potential for transit-oriented development (TOD) at various stations along the Melbourne train loop, employing 14 pedestrian-centric criteria throughout their study [39]. Lamour emphasized the importance of considering public space quality (safety, comfort) in the walking experiences of those living near station areas for the successful implementation of TOD models [40]. Strategies to encourage walking among residents include diversifying land use types, introducing street-level commercial ventures, and interspersing small open areas (such as plazas and green spaces) along walkways. The inclusion of leisure amenities also plays a crucial role in making walking an attractive option for residents.

A comprehensive review of the existing literature identifies a limited scope of research focused specifically on evaluating pedestrian environments in these areas. Existing studies either lack granularity by focusing on the broader impact of TOD or utilize limited evaluation indicators that fail to reflect the full complexity of pedestrian experiences. Moreover, there appear to be insufficient systematic summaries of improvement strategies and empirical research on walking environments in Chinese metro transit systems. To address this gap, this study undertakes a quantitative comparative analysis of five metro transit station areas in Chongqing, China. The analysis focuses on three key dimensions: "comprehensive evaluation", "basic scale", and "structural quality". The comprehensive evaluation considered factors such as the pedestrian catchment area ratio, POI kernel density distribution, and crowd agglomeration. The basic scale dimension comprised floor area ratio, building density, pedestrian road density, and the quantity of station entrances and exits. Finally, structural quality factors included land use type mixing degree, POI function mixing degree, intersection connectivity, median street length, pedestrian route directness, and green view index. By analyzing the shared spatial characteristics of these station areas and uncovering the linear relationships between relevant environmental variables, this study proposes optimization strategies for pedestrian-focused design in such situations. These findings can inform future planning efforts aimed at creating more walkable and accessible environments around metro transit stations, thus promoting sustainable transportation practices.

2. Research Object and Scope

2.1. Selection of Research Subjects

The central urban area of Chongqing is characterized by its cluster-style layout, heavily influenced by the surrounding mountainous geography. Metro transit emerges as a vital connector among these urban clusters. Beyond metro transit, Chongqing's transportation network comprises private vehicles, public buses, and pedestrian pathways, while the topography renders non-motorized modes such as cycling impractical. The city is also outfitted with an extensive array of pedestrian infrastructures, including pathways, staircases, elevators, and escalators, all of which facilitate ease of movement for residents and boost both the efficiency and the appeal of walking as a commuting option. Walking, as a transfer method to metro transit, stands at the heart of TOD and represents a crucial strategy for achieving sustainable development in both land use and transportation planning in Chongqing.

The cluster-based urban structure in Chongqing creates a significant opportunity for the development of transfer stations, with various-sized regional commercial hubs emerging around the central metro transfer stations in each cluster. This setup draws on a unique combination of "urban characteristics + transportation features" associated with metro transit stations. For the purpose of this study, five stations have been chosen for in-depth analysis: Guanyinqiao Station in the Jiangbei District, Ranjiaba Station in the Yubei District, Daping Station in the Yuzhong District, Shapingba Station in the Shapingba District, and Nanping Station in the Nan'an District (Figure 1).

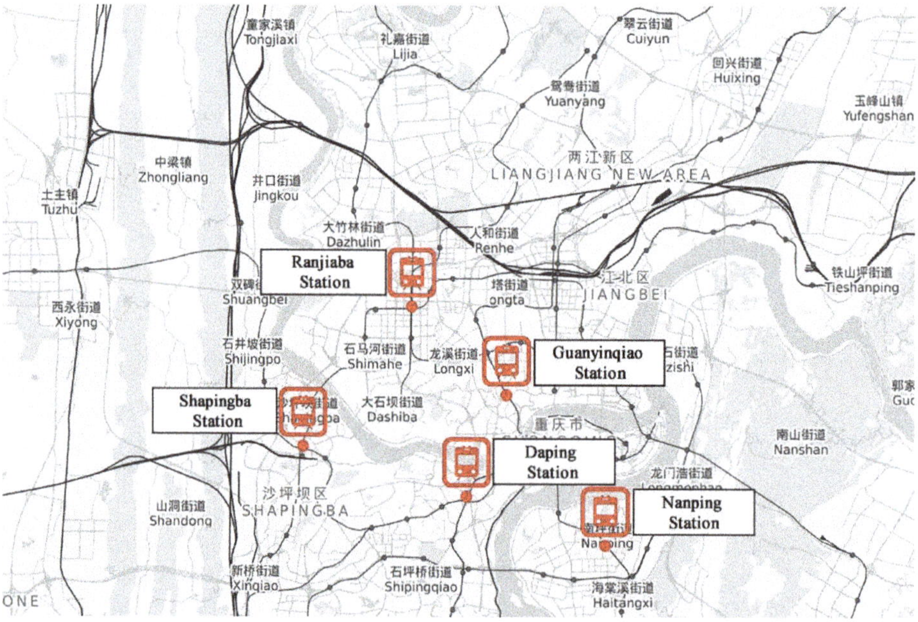

Figure 1. Distribution map of research object locations.

2.2. Research Data

The data on land use for this research were acquired through the color-translating in detailed planning maps from prior project disclosures. Information on the road network was sourced from Baidu Maps and the Open Street Map (OSM) website, with additional data on metro transit, building vectors, and Points of Interest (POI) sourced from the Baidu Maps website. Street view imagery and demographic scale data were also compiled from Baidu Maps. The preliminary data processing included converting the data to the WGS-84 coordinate standard and applying corrections through the ArcMap platform. This phase also involved the cleaning and supplementation of road data and the selective filtering of POI data.

2.3. Research Scope

TOD theory suggests that the vicinity of metro stations should be organized with these stations as the central area, forming a layered structure in space that is within walking distance of 400~800 m, or walking time of 5~10 min [3,41–44]. However, changes in related environmental factors suggest the necessity to extend the analysis beyond the immediate scope of study. Accordingly, this research extends the area of influence for metro stations to 1500 m, while confining the discussion to a maximum acceptable distance of 1000 m. With respect to individual factors, the focus is confined to within 400 m of the central area (Figure 2).

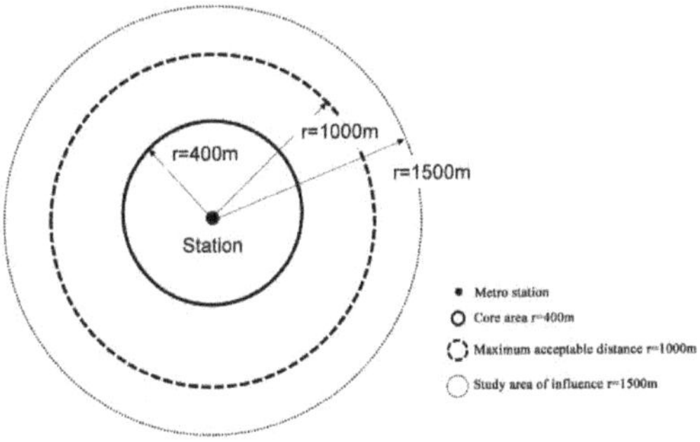

Figure 2. Division of research area and influence area.

In this study, we refer to "station domain" as the ring buffer area centered on the station, including the pedestrian catchment area (PCA) derived from simulating actual walking paths [45–47]. The former is an ideal model based on the TOD theory, suggesting a core area with a specific radius, while the latter is determined by mapping out the accessible area on foot, taking into account the locations of station entrances and exits over a uniform distance. During the PCA's calculation, it is crucial to define certain service area attributes, notably interruption values, setting a core area of 400 m and extending up to a maximum radius of 1000 m. To maintain the integrity and accuracy of road data, it is important to consider barriers such as campuses and gated communities. The approach incorporates "multiple facility point options" into the service areas, guided by the defined interruption values. Road data for this study are derived from the transformation of road centerlines. Additionally, a "trimming area" of 50 m is introduced to streamline the analysis of road width and the impact of building setbacks on the ease of travel for residents, reflecting the differences at the boundaries of pedestrian catchment areas (Figure 3).

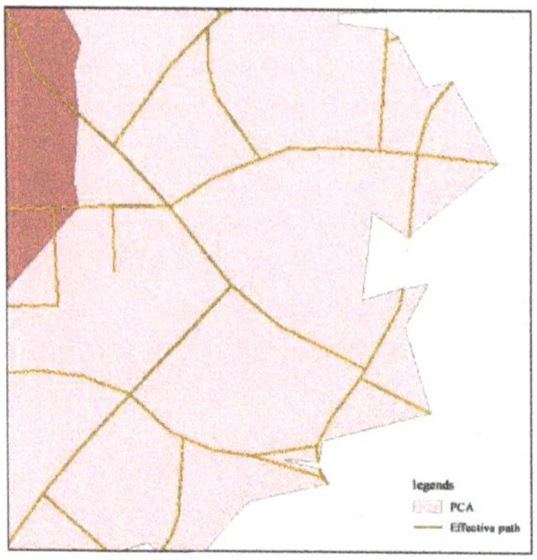

(a) No trimming area

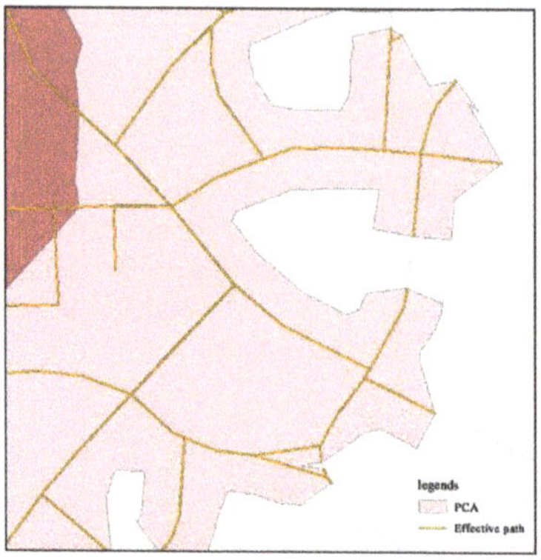

(b) Setting the trimming area

Figure 3. Comparison of PCA before and after trimming area setting.

3. Factors and Quantitative Methods for Evaluating the Impact of Pedestrian Environment

The focus of this article is on exploring the factors that shape the pedestrian environment near metro transit stations, specifically through the lens of the physical layout encountered by residents during their commute to metro transit hubs or central areas. The objective is to identify ways to enhance travel conditions and motivate residents to opt for walking by adjusting key planning and design parameters. The key issue of our discussion is the "3D" principle of TOD, encapsulating Density, Diversity, and Design [48–50]. This study further evaluates identifying factors that affect the pedestrian surroundings near metro transit stations, aiming to support pedestrian activities. These factors include walking opportunities, purposes, walkability, and the overall walking experience, grouped under comprehensive evaluation, basic scale, and structural quality indicators [51,52] (Table 1). Through comprehensive evaluation, we seek to reflect the differences in phenomena caused by the development of land and traffic organization in different station areas. The basic scale indicator serves to measure the development prospects of a metro station's surrounding area, while the structural quality indicator seeks to unravel potential challenges.

Table 1. Factors influencing the pedestrian environment around metro transit stations.

Categories	Encodings	Indicators	Calculation Method
Comprehensive evaluation category	A1	Pedestrian catchment area ratio	$PCR = S_{PCA} * S_{buffer}$
	A2	POI kernel density distribution	ArcMap kernel density calculation tool
	A3	Crowd gathering	Baidu heat map vectorization
Basic scale category	B1	Floor area ratio	$FAR = (\sum_{i=1}^{n} S_{arc} * Floor) / S_{buffer}$
	B2	Building density	$D_{arc} = S_{arc} / S_{buffer}$
	B3	Pedestrian road density	$D_{walk} = L_{walk} / S_{buffer}$
	B4	The quantity of entrances and exits at metro transit stations	information access and field statistics
Structural quality category	C1	Land use type mixing degree	$D_{land\ mix} = -\sum_{n=1}^{N} S_i * \log S_i$
	C2	POI function mixing degree	$D_{poi\ mix} = -\sum_{p=1}^{P} N_p * \log N_p$
	C3	Intersection connectivity	$L_{junction} = 2m/N$
	C4	Median street length	$M_{stree} = L_{(m+1)/2}$ $M_{stree} = \left(L_{\frac{m}{2}} + L_{\frac{m}{2}+1}\right)/2$
	C5	Pedestrian route directness	$PRD = M_{walk} / L_{buffer}$
	C6	Green view rate	OpenCV tools

Comprehensive evaluation indicators offer a preliminary evaluation of the effects on development of metro transit station areas, thereby reflecting the holistic efficacy of the pedestrian environment and identifying areas with suboptimal pedestrian flow related to the walking environment. Comprehensive evaluation factors are operationalized through three indicators: pedestrian catchment area ratio, POI kernel density distribution, and crowd gathering.

On a fundamental level, basic scale indicators offer a preliminary judgment of the relationship between land utilization and transportation systems in the vicinity of metro transit stations. The presence of a robust population and comprehensive transportation services is key to optimizing pedestrian flow. Through an analysis of these basic scale indicators, it is possible to conduct a comparative analysis across different metro transit stations. Included in these indicators are the floor area ratio, building density, pedestrian road density, and the quantity of entrances and exits at metro transit stations.

Comprehensive evaluation factors offer a broad analysis of various differential phenomena, while indicators of structural quality shed light on the specific differences arising from land development and the organization of transportation systems. These structural quality indicators are instrumental in identifying limitations in the design of pedestrian environments around metro transit station areas. They consist of six key indicators: land use type mixing degree, POI function mixing degree, intersection connectivity, median street length, pedestrian straightness coefficient, and green view rate. The pedestrian environmental impact factors and calculations for the metro station area are shown in Table 1.

4. Research Results and Analysis of Pedestrian Environment Factors

4.1. Research Results of Pedestrian Environment Factors

4.1.1. Comprehensive Evaluation Comparison

(1) Pedestrian Catchment Area Ratio (A1)

With respect to the pedestrian catchment area ratio, we observe a spectrum ranging from the lowest to the largest, the stations are Daping Station (48.27%), Shapingba Station (53.28%), Nanping Station (61.34%), Ranjiaba Station (66.10%), and Guanyinqiao Station (66.32%).

Figure 4 offers insights into the spatial distribution of each station's pedestrian catchment area (PCA). Specifically, Shapingba Station's PCA is constricted in the northwest due to the closed management of nearby university campuses. Ranjiaba Station exhibits a relatively uniform, near-ideal rhombus-shaped PCA, with minor compression in the 400 m PCA range to the north and a more significant compression in the 1000 m PCA range due to the presence of Panxi River Park on its eastern flank. Guanyinqiao Station's PCA takes on a leaf-like structure extending from northwest to southeast, largely influenced by the dense pedestrian network in this direction and the location of exit No. 8 on the south side. At Daping Station, natural and human-made boundaries such as Fotu Mountain Park, Hutouyan Tunnel, and the Xizhong metro line on the south and north sides force pedestrian into detours, resulting in a compressed 400 m PCA. Nanping Station's actual pedestrian network and PCA stretch along both sides of the metro line, mirroring the shape observed at Guanyinqiao Station. However, a higher road density on the east and west sides contributes to a more elliptical form.

In general, the size and morphology of each metro transit station's PCA are shaped by a confluence of factors, including road network density and structure, land use characteristics, management and ownership, block size, and topographical features.

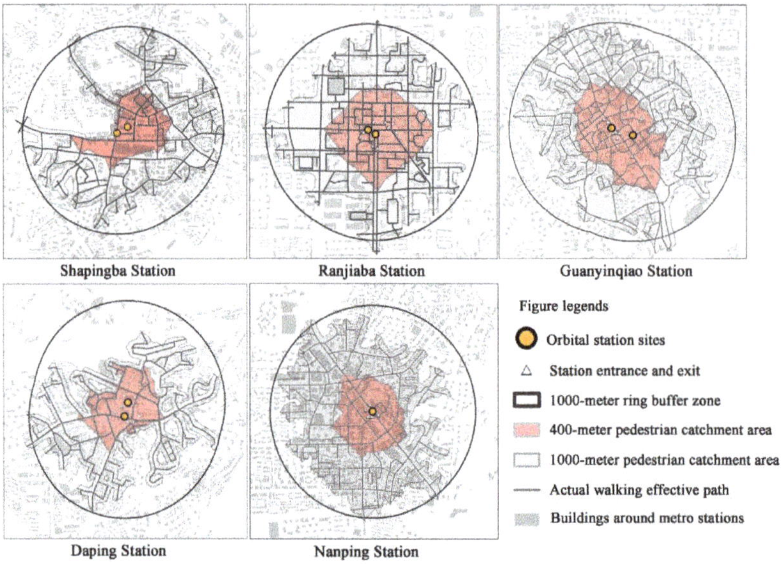

Figure 4. Illustration of pedestrian catchment areas (PCAs) for various metro stations.

(2) POI Kernel Density (A2)

Analyzing the POI kernel density, we observe a spectrum of values from lowest to highest as follows: Ranjiaba Station (0.1896), Shapingba Station (0.4258), Nanping Station (0.4427), Daping Station (0.4646), and Guanyinqiao Station (0.6011).

The proximity of high-density POI clusters to metro transit station and the overall distribution of POI kernel density can reflect the degree to which station functionality correlates with pedestrian pathways.

Figure 5 offers a detailed visualization of the POI kernel density distribution for each metro transit station. Specifically, Shapingba Station's POI density concentrates around the centrally located Sanxia Square, with a smaller cluster toward the southeast. The presence of the university campus in the northwest creates a dividing line, causing POIs to spread along its periphery and cultivating a vibrant street life atmosphere. Ranjiaba Station's highest POI density, conversely, is situated in the Longhu Yuanzhu commercial district on the western edge, at a considerable distance from the station itself. This phenomenon is potentially attributable to the grid-like road network, which encourages a more dispersed distribution of POIs. Guanyinqiao Station presents a unique linear pattern in its POI kernel density distribution. This is due to the presence of numerous large-scale commercial facilities, such as Guanyinqiao Pedestrian Street, Longhu North City Tianjie, and Maoye Tiandi, which are situated contiguously along the east side of Jianxin North Road. Daping Station, facing competition from other commercial districts and transportation hubs in the southwest, no longer enjoys the advantage of spatial agglomeration. This is reflected in the shift of its highest POI kernel density toward the southwestern edge, suggesting that the station's function as a transportation hub surpasses its commercial function. In other words, people are more likely to visit Daping Station for transfer purposes than for commercial activities. Finally, Nanping Station also displays a linear trend in its POI kernel density distribution, with the concentration primarily located in close proximity to the station itself.

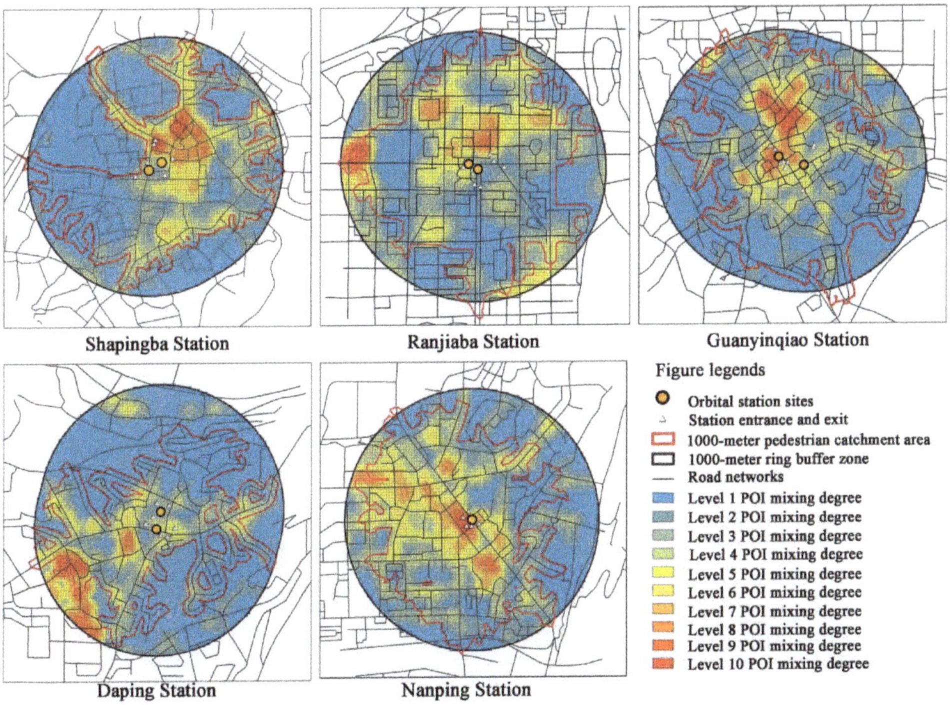

Figure 5. Distribution of POI kernel density at each metro transit station.

(3) Crowd Gathering Situation (A3)

Considering the emphasis on pedestrian behavior in metro transit stations areas, understanding the motivations driving pedestrian traffic—particularly commercial and transfer-related activities—is crucial for maximizing the economic and environmental benefits of station area development. Comparing pedestrian congregation patterns on weekends and weekdays offers valuable insights into a station's ability to attract crowds and the underlying motivations for pedestrian visits (Figure 6).

Shapingba Station experiences consistent pedestrian congregation in the central Sanxia Square, irrespective of weekends or weekdays, while surrounding campuses exhibit lower pedestrian density. Ranjiaba Station displays the least pedestrian gathering among all stations. Specifically, during weekends, the focal point of pedestrian gathering at Ranjiaba is not the nearby Airong Hui commercial center, but rather the residential area surrounding the Red Star Macalline furniture market to the northwest. Even during peak hours, no significant pedestrian congregation appears, suggesting Ranjiaba Station's limited appeal to pedestrian traffic. Guanyinqiao Station exhibits consistently high pedestrian density throughout the week, indicating its strong attraction to pedestrians. At Daping Station, the need for commuting and transfers during weekdays results in significantly higher pedestrian activity compared to weekends. Conversely, Nanping Station experiences greater pedestrian traffic on weekends, with crowds concentrated around commercial outlets along Jiangnan Avenue.

Among the five stations, except for Ranjiaba Station, the remaining stations exhibited significant activity. According to the general observation, Guanyinqiao Station attracted substantial crowds on both weekdays and weekends, indicating its potential as a central hub. Daping Station, on the other hand, served primarily as a transfer point for commuters, while Nanping Station attracted crowds primarily for commercial and leisure activities.

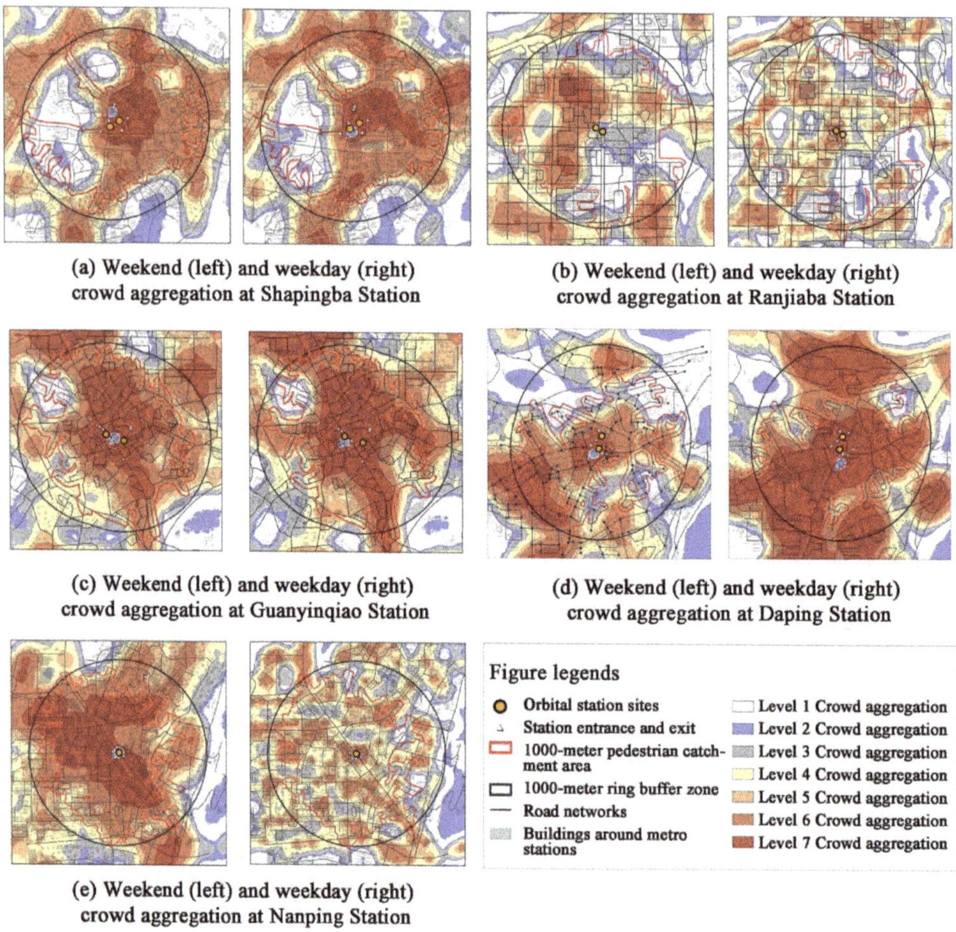

Figure 6. Crowd gathering at various metro transit stations on weekends and weekdays.

Considering the three key evaluation indicators, namely the dimensions and form of the PCR, the POI density distribution at focal points, and the patterns of crowds gathering across various times in the vicinity of metro transit stations, this study posits that Guanyinqiao Station exemplifies an optimal equilibrium in its overall efficacy, standing out as the top performer among the five research subjects. Shapingba Station is second, followed by Nanping Station. Ranjiaba Station and Daping Station occupy the final two positions, in that order.

4.1.2. Basic Scale Comparison

(1) Floor Area Ratio (B1)

Analysis of floor area ratio indicates that Guanyinqiao Station possesses the highest value (2.5318) in its 1000 m ring buffer zone. In contrast, Shapingba Station exhibits a higher floor area ratio (3.0022) when considering the 1000 m PCA range. Conversely, Ranjiaba Station demonstrates the lowest floor area ratio in both the 1000 m ring buffer and PCA range. Specifically, the discrepancies in FAR between the two spatial scales for the five stations are relatively insignificant (Table 2).

Table 2. Calculation results of floor area ratio (B1) for each metro station area.

Station	FAR—in 1000 m Ring Buffer Area	FAR—in 1000 m Pedestrian Catchment Area	Magnitude of Change
Shapingba Station	2.1876	3.0022	0.8146 ↑
Ranjiaba Station	1.6908	1.8186	0.1278 ↑
Guanyinqiao Station	2.5318	2.8306	0.2988 ↑
Daping Station	1.7881	2.4058	0.6177 ↑
Nanping Station	2.2051	2.7032	0.4981 ↑
			↑ indicates rising

(2) Building Density (B2)

An analysis of building density indicates that Nanping Station exhibits the highest concentration (26.06%) in a 1000 m ring buffer zone, while Shapingba Station has the highest density (29.57%) in a 1000 m PCA range. Conversely, Ranjiaba Station displays the lowest building density in both the 1000 m ring buffer and 1000 m PCA range. Specifically, the five stations demonstrate significant discrepancies in building density between the 1000 m ring buffer and 1000 m PCA range, suggesting potential changes in spatial distribution patterns (Table 3).

Table 3. Calculation results of building density (B2) for each metro station area.

Station	BD—In 1000 M Ring Buffer Area	BD—in 1000 m Pedestrian Catchment Area	Magnitude of Change
Shapingba Station	25.36%	29.57%	4.21% ↑
Ranjiaba Station	16.85%	17.25%	0.4% ↑
Guanyinqiao Station	25.81%	28.26%	2.45% ↑
Daping Station	21.32%	28.17%	6.85% ↑
Nanping Station	26.06%	29.54%	3.48% ↑
			↑ indicates rising

(3) Pedestrian Road Density (B3)

A study of pedestrian road density indicates that Ranjiaba Station consistently exhibits the highest values in both the 1000 m ring buffer and 1000 m PCA range. Conversely, Daping Station displays the lowest pedestrian road density across both indicators. Specifically, the ranking of pedestrian road density for all five metro transit stations remains consistent across both the ring buffer and PCA, suggesting minimal differences in pedestrian road density (Table 4).

Table 4. Calculation results of pedestrian road density (B3) for each metro station area.

Station	Pedestrian Road Density—in 1000 m Ring Buffer Area (km/km^2)	Pedestrian Road Density—in 1000 m Pedestrian Catchment Area (km/km^2)	Magnitude of Change
Shapingba Station	9.7834	12.5154	2.732 ↑
Ranjiaba Station	13.5049	15.2173	1.7124 ↑
Guanyinqiao Station	11.5436	13.8259	2.2823 ↑
Daping Station	8.2592	11.5939	3.3347 ↑
Nanping Station	11.0591	12.9663	1.9072 ↑
			↑ indicates rising

(4) Quantity of Station Entrances and Exits (B4)

With regard to metro transit station access points, Guanyinqiao and Daping Stations have the highest number at nine each, followed by Shapingba and Ranjiaba Stations with eight, while Nanping Station possesses the lowest count of six. The number of station entrances and exits will influence the extraction and comparison of PCA-related indicators.

Measurements of the basic scale of the metro transit station areas indicate that four of the five research subjects exhibit characteristics of high-density urban development. Ranjiaba Station, however, presents an exception, with its lower basic scale score suggesting a less compact land development model. This is reflected in the failure of building structures to effectively enclose street spaces, resulting in a scattering of open spaces in individual plots.

4.1.3. Structural Quality Comparison

(1) Land Use Type Mixing Degree (C1)

Analysis of land use mixing degree indicates that Shapingba and Daping stations demonstrate a high degree of diversity in their immediate surroundings. While Shapingba exhibits a strong presence of educational and commercial land uses, the remaining four stations are predominantly characterized by residential land use, both in the 1000 m buffer zone and the 1000 m PCA (Table 5).

Table 5. Calculation results of land use type mixing degree (C1) for each metro station area.

Station	1000 m Ring Buffer Area			1000 m Pedestrian Catchment Area		
	Land Use Type Mixing Degree	Dominant Type	Percentage	Land Use Type Mixing Degree	Dominant Type	Percentage
Shapingba Station	0.9132	Educational land	32.32%	0.9154	Business land	30.78%
Ranjiaba Station	0.8386	Residential land	41.48%	0.8235	Residential land	38.31%
Guanyinqiao Station	0.80002	Residential land	42.04%	0.8217	Residential land	38.15%
Daping Station	0.9491	Residential land	29.64%	0.9121	Residential land	32.28%
Nanping Station	0.6812	Residential land	55.95%	0.6447	Residential land	51.15%

(2) POI Function Mixing Degree (C2)

Analysis of POI data indicates a consistent pattern across all five metro transit stations, with shopping services constituting the dominant function in both the 1000 m buffer and PCA. While specific business formats remain largely similar across stations, a difference exist in the overall quantity of POIs present (Table 6).

Table 6. Calculation results of POI functional mixing degree (C2) for each metro station area.

Station	1000 m Ring Buffer Area			1000 M Pedestrian Catchment Area		
	POI Mixing Degree	Dominant Type	Percentage	POI Mixing Degree	Dominant Type	Percentage
Shapingba Station	0.9394	Shopping Services	25.30%	0.9242	Shopping Services	26.68%
Ranjiaba Station	0.9750	Shopping Services	24.16%	0.9597	Shopping Services	25.82%
Guanyinqiao Station	0.9217	Shopping Services	26.74%	0.9075	Shopping Services	27.80%
Daping Station	0.9587	Shopping Services	23.24%	0.9423	Shopping Services	26.22%
Nanping Station	0.9622	Shopping Services	23.67%	0.9452	Shopping Services	23.31%

(3) Intersection Connectivity (C3)

Daping Station's 1000 m PCA exhibits lower intersection connectivity compared to its 1000 m ring buffer, suggesting a higher prevalence of dead-end roads in its pedestrian catchment area, Guanyinqiao Station, despite having a relatively high number of intersections, which does not reach the expected connectivity levels. Conversely, Ranjiaba and Nanping Stations, despite having fewer intersections, demonstrate superior connectivity. This suggests that a grid-like road network layout cultivates greater connectivity between intersections, minimizes dead-end roads, and offers pedestrians a wider range of route choices. This study utilizes the OD Cost Matrix Analysis from the Network Analyst toolset and the Inverse Distance Weighted (IDW) interpolation method to comprehensively analyze and evaluate intersection connectivity. As illustrated in Figure 7, the intersection connectivity of each metro transit station displays a concentric distribution pattern, with values decreasing outward from the center. Specifically, all stations are located in the innermost zone characterized by the highest connectivity, indicating that their spatial layout has already established them as core areas for TOD development (Table 7).

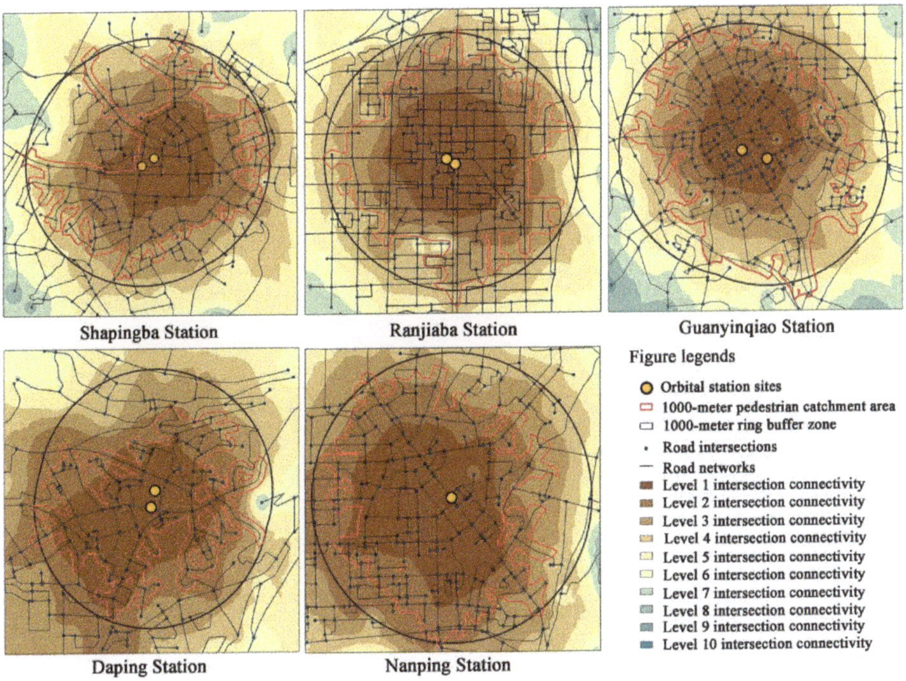

Figure 7. Intersection connectivity of each metro transit station area.

Table 7. Calculation results of intersection connectivity (C3) for each metro station area.

Station	1000 m Ring Buffer Area			1000 m Pedestrian Catchment Area		
	Number of Sections	Number of Intersections	Intersection Connectivity	Number of Sections	Number of Intersections	Intersection Connectivity
Shapingba Station	324	206	3.1456	244	144	3.3889
Ranjiaba Station	420	253	3.3202	337	197	3.4213
Guanyinqiao Station	463	306	3.0261	400	256	3.1250
Daping Station	198	133	2.9774	154	105	2.9333
Nanping Station	354	228	3.1053	282	175	3.2229

(4) Median Street Length (C4)

By comparing the average, median, and maximum values of street lengths, we can determine the distribution trends of street lengths in the ring buffer areas and pedestrian catchment areas of each station. The results indicate that, except for Nanping Station, all five research sites exhibit a characteristic where the median street length is lower than the average street length. This holds true for both the ring buffer area and the pedestrian catchment area (Table 8). This suggests that the street length values at the respective stations tend to concentrate in intervals higher than the median. Considering the specific averages and medians of the five stations, the average street length in metro transit station areas should be between 90 and 120 m, with room for individual variations.

Table 8. Calculation results of median street length (C4) for each metro station area.

Station	1000 m Ring Buffer Area			1000 M Pedestrian Catchment Area		
	Mean (M)	Med (M)	Max (M)	Mean (M)	Med (M)	Max (M)
Shapingba Station	135.6924	117.7710	730.4128	122.8114	107.6503	730.4128
Ranjiaba Station	141.0615	118.9701	549.0844	130.9317	112.9775	543.0548
Guanyinqiao Station	118.3296	95.2120	513.6606	108.7982	89.6119	522.7639
Daping Station	190.3246	164.2972	1241.0906	164.9648	153.0923	647.6561
Nanping Station	130.0667	150.7599	669.9187	117.4318	151.0518	345.9430

(5) Pedestrian Route Directness (PRD) (C5)

China's guidelines provided by the *Code for transport planning on urban road* (GB 502220-95) [53] assert that the coefficient reflecting the indirectness of public transport routes should ideally not surpass 1.4. A PRD exceeding this threshold suggests a more complex network of pedestrian pathways in areas surrounding metro transit stations. Notably, Guanyinqiao Station exhibits the lowest PRD for both the 400 m and 1000 m ring buffer areas. This suggests an extensive array of navigational choices available at Guanyinqiao Station. Conversely, the 1000 m ring buffer area around Daping Station and the 400 m ring buffer area around Nanping Station both present PRDs that exceed 1.5 (Table 9, Figure 8). Further analysis of spatial data revealed that the increased PRDs at Daping Station, particularly near the northern Huacun Interchange and Fotuguan Park, as well as near Nanping Station's eastern Guohui Mountain Park, can be attributed to the detours necessitated by the mountainous terrain.

Table 9. Calculated pedestrian route directness (C5) for each subway station domain.

Station	PRD-400 m	PRD-1000 m
Shapingba Station	1.2917	1.2003
Ranjiaba Station	1.2347	1.2437
Guanyinqiao Station	1.2169	1.2275
Daping Station	1.3487	1.5264
Nanping Station	1.5183	1.3047

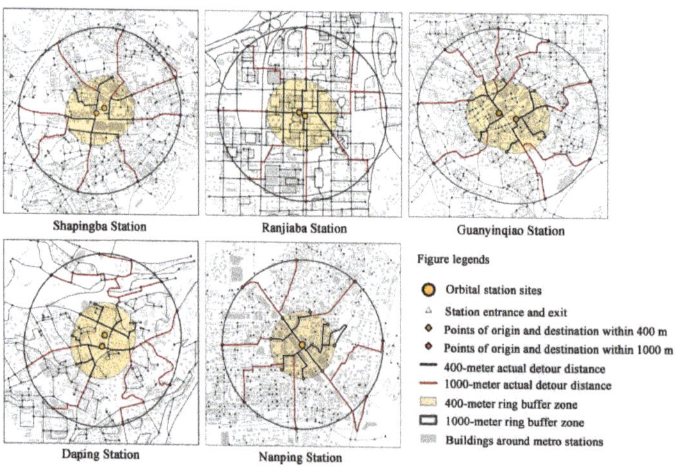

Figure 8. Illustration of PRD (pedestrian route directness) for each metro transit station area.

(6) Street Green View Index (C6)

In terms of Street Green View Index (C6), the stations are ranked from the least to the most green view as follows: Guanyinqiao Station (19.93%), Shapingba Station (20.97%), Ranjiaba Station (23.35%), Nanping Station (29.07%), and Daping Station (34.57%). Relevant studies indicate an inverted U-shaped correlation between the street green view index and the alleviation of pedestrian congestion. A green view perception between 15% and 25% is regarded as excellent, while a percentage beyond 25% is considered superior. However, excessive green view can also affect interface transparency [54–56]. Despite this, the greenery levels at all five stations are commendable, with Daping Station's particularly green environment offering opportunities for enhancement through the strategic thinning of plant density.

4.2. *Explanation and Analysis of Research Results*

4.2.1. Evaluation Criteria and Comprehensive Results

According to the previous study on various environmental factors, the correlation between the trend of each environmental factor and the quality of the pedestrian environment can be defined (Table 10). It was discovered that the length of the median street (C5) and the pedestrian directness coefficient (C6) share an inverse relationship, whereas the number of station entrances/exits (B4) and the street green view index (C6) are neutral. The remaining environmental factors display a positive relationship with the quality of the environment. A greater median street length (C5) suggests a reduced number of intersections in the vicinity of the station, limiting pedestrian pathway choices and reducing the road network's service area. This scarcity of incidental opportunities for pedestrians to pause and linger renders the area less appealing for relaxed, aimless strolls, potentially leading to quicker cognitive exhaustion and loss of interest in walking. A higher pedestrian directness coefficient (PRD) (C5) signifies a more circuitous route for pedestrians around the metro station area. A significant detour in a specific quadrant could reduce the PCA in that direction. The quantity of station entrances/exits (B4) and the street green view index (VGI) (C6), while not directly dictating the pedestrian environment's quality, offer valuable insights for urban planning.

Table 10. Correlation between environmental factors and pedestrian environment quality.

Categories	Encodings	Indicators	Correlation
Comprehensive evaluation category	A1	Pedestrian catchment area ratio	Positive correlation
	A2	POI kernel density distribution	Positive correlation
	A3	Crowd gathering	Positive correlation
Basic scale category	B1	Floor area ratio	Positive correlation
	B2	Building density	Positive correlation
	B3	Pedestrian road density	Positive correlation
	B4	The quantity of entrances and exits at metro transit stations	--
Structural quality category	C1	Land use type mixing degree	Positive correlation
	C2	POI function mixing degree	Positive correlation
	C3	Intersection connectivity	Positive correlation
	C4	Median street length	Negative correlation
	C5	Pedestrian route directness	Negative correlation
	C6	Green view rate	--

4.2.2. Descriptive Statistics of Pedestrian Environment Factors

This study also evaluates the PCA's internal variability, demonstrated the descriptive statistics—minimum, maximum, average values, standard deviation, and skewness—for various factors across the five subjects, as presented in Table 11.

Table 11. Descriptive statistics of environmental factors in PCA of each metro station area.

Environmental Factors	Minimum Value	Maximum Values	Average Value	Standard Deviation	Variance	Skewness
PCA-1000 m	2.1912	3.1477	2.637360	0.3854551	0.149	0.343
PCA-400 m	0.5135	0.8062	0.634960	0.1540198	0.024	0.600
PCR-1000 m	0.4827	0.6632	0.590620	0.0802029	0.006	−0.571
POI kernel density maximum	0.1896	0.6011	0.424760	0.1485818	0.022	−0.947
Number of POI	3853	12,896	8098.80	3570.97574	12,751,867.7	0.190
Floor area ratio	1.8186	3.0022	2.552080	0.4643300	0.216	−1.172
Building density	0.1725	0.2957	0.265580	0.0524640	0.003	−2.145
Pedestrian road density	11.5939	15.2173	13.223760	1.3748889	1.890	0.551
land use type mixing degree	0.6447	0.9154	0.823484	0.1098610	0.012	−1.332
POI function mixing degree	0.9075	0.9597	0.935780	0.0202316	0.000	−0.471
Intersection connectivity	2.9333	3.4213	3.218280	0.2001652	0.040	−0.552
Median street length	89.6119	153.0923	122.876760	28.0321382	785.801	0.163
PRD-1000 m	1.2003	1.5264	1.300520	0.1319505	0.017	1.804
PRD-400 m	1.2169	1.5183	1.322074	0.1212430	0.015	1.336

The differences in the PCA scale are most pronounced in the 400 m core area. The positive skew of the numerical statistics indicates an asymmetrical distribution leaning toward the right and the minimum influencing value observed at Shapingba Station, whereas the comparative results of PCR are negatively skewed for all five stations.

The specific changes in POIs relate to land development and the spatial draw of the metro station area. Irrespective of Kernel_Max or the number of POIs, the minimum value for both is found at Ranjiaba Station.

Land development-related factors, such as the plot ratio and building density, tend to lean toward lower values, influenced by the baseline set by Ranjiaba Station. However, on average, the correlation factors at the other stations still maintain relatively higher values.

The diversity in the pedestrian environment is affected by the mix of land use and the variety of POI functions, which play a crucial role in fulfilling the needs of non-transfer travel. The statistical results demonstrate a consistent negative skewness across all continuous factors, though the standard deviation of POI function mixing degree is not significant. In contrast, Nanping Station stands out for its high proportion of residential land, indicating opportunities for transformation.

The minimum values for pedestrian road density and intersection connectivity are both observed at Daping Station. This suggests that Daping Station needs further improvements to the pedestrian road capacity.

The median street length and PRD, despite their inverse relationship, both exhibit a tendency toward higher values, suggesting that optimization efforts should target the station marking the lowest values. Specifically, Guanyinqiao Station has the lowest one for a 400 m radius, and also records the shortest median street length. Evaluating the performance of each station, the enhancement of the transportation network could take cues from the spatial features unique to Guanyinqiao Station.

In summary, this study performs an analysis of the relationship and asymmetry among different variables. It is found that Guanyinqiao Station and Shapingba Station exhibit superior performance in comparison to the other three stations under review. It is suggested that these remaining stations could benefit from tailored design improvements, taking into account their unique geographical and transportational attributes.

4.2.3. Pearson Correlation Coefficients of Transportation Factors

The previous study identified strong correlations between various environmental factors in the TOD transport framework. Therefore, a Pearson correlation analysis was applied to appraise the environmental factors surrounding each metro station area (Table 12).

Table 12. Pearson correlation coefficients of various metro station area transportation system factors.

	PCA 1000 m	Building Density in 1000 m Pedestrian Catchment Area	Pedestrian Road Density in 1000 m Pedestrian Catchment Area	Intersection Connectivity in 1000 m Pedestrian Catchment Area	Median Street Length in 1000 m Pedestrian Catchment Area	PRD 1000 m
PCA 1000 m	1	−0.042	0.826	0.295	−0.733	−0.634
Building density in 1000 m pedestrian catchment area	−0.042	1	−0.47	−0.193	−0.074	−0.218
Pedestrian road density in 1000 m pedestrian catchment area	0.826	−0.47	1	0.628	−0.554	−0.628
Intersection connectivity in 1000 m pedestrian catchment area	0.295	−0.193	0.628	1	−0.445	−0.816
Median street length in 1000 m pedestrian catchment area	−0.733	−0.074	−0.554	−0.445	1	0.775
PRD 1000 m	−0.634	−0.218	−0.628	−0.816	0.775	1

The analysis results demonstrate that the correlation coefficient between the size of the PCA and the density of pedestrian pathways is notably high at 0.826, signifying a strong and positive correlation. There appears to be a significant negative correlation between PCA size and both the median length of streets and PRD. Meanwhile, the relationship between building density and the degree of intersection interconnectivity appears to be minor. Overall, it is evident that factors associated with the transport framework influence both the dimensions and configuration of the PCA, with the variance in these factors collectively shaping the land use patterns around metro station zones. To boost land use efficiency and enhance walkability around metro station areas, priority should be given to improving the quantity and quality of roadway infrastructure in their spatial layout. Moreover, it is crucial to implement design adjustments in response to land system feedback promptly. Promoting spatial diversity is key to accommodating a range of travel needs in the PCA to the fullest extent.

5. Chongqing Metro Transit Station Walking Environment Optimization Study

5.1. Design Strategies for Metro Transit Station Walking Environment

This study argues that the determinants of the pedestrian environment in metro station zones can be categorized into three categories: comprehensive evaluation, basic scale, and structural quality. Nonetheless, each factor is integral to the broader frameworks of land use and transportation. Metro station areas optimized for pedestrian access should exhibit certain spatial organizational traits, and identifying these characteristics can lead to effective design approaches for enhancing the pedestrian experience in these areas.

In terms of the land system: (1) Implement compact development in metro station areas to ensure functional mix and diversity. This approach, characterized by high density and intensity, shortens distances between functions and increases the number of potential metro transit users and pedestrians in walking distance. In addition, each metro station should emphasize land-use compatibility. (2) In station core areas, attractive public centers should be created. Such centers highlight the advantages of pedestrian traffic and strengthen the connection between pedestrian behavior and urban space. This convergence of activity and transportation hubs offers mutually beneficial growth potential. (3) To ensure accessibility and maintain the integrity of pedestrian catchment areas and potential paths, large-scale functions and projects should be positioned with specific requirements on the outskirts of the stations, but in acceptable walking distances. The nature of these developments—including large land areas, closed management, and potentially poor land compatibility—can disrupt pedestrian pathways. (4) The advantages of walking by restricting private motorized traffic in core station areas should be emphasized. Metro station TOD should prioritize pedestrian traffic. Limiting private vehicles highlights the benefits of walking, reinforces positive pedestrian attitudes, and encourages potential pedestrians to embrace walking.

In terms of the transportation system: (1) Increase the density of pedestrian walkways to enhance the network of pedestrian paths, thereby reducing their dependency on roads designated for motor vehicles. Pedestrian routes offer the advantage of flexibility, allowing them to exist independently of vehicular lanes. By combing urban streets, alleys, and adjacent structures to incorporate pedestrian facilities, transform building sites, and carry out other design measures, it is possible to significantly increase the network of walkways near metro stations. This approach will lead to a comprehensive pedestrian path network, improving the connectivity of the road system. (2) In the core areas of stations, where the intermingling of pedestrian and vehicle traffic is especially significant, introducing traffic calming strategies is necessary. This can be achieved by directing traffic flow away from these congested zones using permeable transportation options, realigning roads to naturally reduce vehicle speed, creating separate paths for pedestrians and vehicles through underpasses, and designing traffic plans that favor pedestrian movement. Such measures aim to balance the needs of pedestrian and private vehicle traffic around metro stations, without entirely excluding the latter. (3) Promote pedestrianization in metro station areas, categorize road types and align them in a manner that favors pedestrian access in metro station vicinities. This involves a detailed analysis of urban road designations, construction methodologies, and the utilization of land adjacent to these roads. The objective is to distinguish between main roads utilized for traffic and local roads that serve the daily needs of residents, assigning them appropriate importance. Main roads should be relocated to the periphery, while local roads should be situated in the core station area, either parallel to each other or in a circular layout. This strategy focuses on road design with a priority on pedestrian movements in and around metro station zones. (4) Development of a multilevel pedestrian transit system: Enhancing the efficiency of pedestrian movement necessitates the creation of a three-dimensional transportation network. The traditional, flat-plane approach to traffic movement can be inefficient and disorienting for pedestrians. A multilevel system offers varied path options across different planes and elevations, integrating with surrounding buildings to minimize travel distances.

5.2. Strategies for Optimizing Pedestrian Environments at Various Orbital Station Sites

The areas to the north and south of Shapingba Station face spatial constraints due to the campus's enclosed management, while the extensive centralized greenery of Shaping Park, located on the station's southwest side, adversely affects the station area's overall land use efficiency. Moreover, the high volume of traffic and volume of motor vehicles on nearby roads create human-made barriers, which restrict pedestrian movement. To address these challenges, we recommend the following optimization strategies: (1) To alleviate spatial compression, it is advisable to remove walls to extend the campus boundaries, thereby facilitating temporary pathways. This approach aims to mitigate the tension between land usage and transportation demands. By directing the flow of passengers to utilize internal roads during peak commuting times and holidays, we can offer a greater variety of routes and thereby broaden the accessible area around the metro station. (2) By reimagining centralized green areas, transitioning them from broad expanses to linear shapes, and eventually to focal points, we can boost land utilization. It is essential to promptly integrate underutilized parcels in the station's vicinity into urban regeneration initiatives. This includes the creation, renovation, and enlargement of urban green spaces, such as street-side greenery, mini-parks, and green corridors, enhancing the living environment. Such measures should also aim to coordinate the spatial relationships between various land uses and the station. (3) To overcome the limitations imposed by artificial barriers, introducing vertical spatial integration and establishing access points that cross these barriers, is recommended. This involves separating pedestrian and vehicle traffic vertically in the traffic management design and constructing station entrances and exits to improve pedestrian transfer convenience.

The volume rate and the construction density at Ranjiaba Station rank as the lowest when compared to four other stations, and its geometric layout combined with the organization of factors does not favor the establishment of a TOD area in the vicinity of the metro station. The variety of POI functions at Ranjiaba Station is considered moderate, yet it has the smallest number of POIs among the stations in question. The areas surrounding the station predominantly consist of residential zones and government buildings, featuring uniform primary functions and enclosed spatial development patterns for each block, which restricts pedestrian access to city services. The layout of the streets and pathways at Ranjiaba Station results in a scattered arrangement of POIs, exhibiting less concentrated POI clusters compared to other nearby regions. In light of these challenges, we propose the following strategies for enhancement: (1) Adopt a compact development approach to address the issues related to the low density and intensity of development. This can be achieved by setting minimum thresholds for building density and volume rate through rigorous planning controls, thereby fostering a balance between environmental quality and the efficient use of space. Adjusting the share of land designated for roads and encouraging the creation of open spaces in blocks will support the desired high-density and vibrant spatial characteristics specific to the metro station area. (2) Introduce a variety of factors to cultivate appealing public areas near the station. (3) Restructure the distribution of road access to facilitate the development of pedestrian zones in the metro station area. This involves prioritizing pedestrian mobility services and imposing limitations on private vehicle traffic, with the aim of enhancing and protecting the community's preference for walking as a mode of transport.

The quantified results from Guanyinqiao Station exhibit that, while several aspects are excellent, the connection between intersection buffering zones and PCA is notably weak. The landscape in the station's vicinity features varying elevations, with roads that are both straight and winding, alongside the presence of cul-de-sacs in certain sections. Among the five stations analyzed, Guanyinqiao Station scored the lowest on the VGI, highlighting a less than ideal pedestrian experience. To enhance the walkability around the station, it is crucial to address the issue of cul-de-sacs by integrating them into a more comprehensive pedestrian network. Improving the design of street landscapes could also play a significant role in rendering longer walks more appealing, thereby encouraging greater participation in walking as a viable mode of transport.

With respect to Daping Station, it is observed that its connectivity as a transit hub surpasses its commercial appeal, and it displays a rather dense development pattern. However, in its vicinity, there exist underdeveloped plots such as aged residential blocks, governmental bodies, and public enterprises. Moreover, the 1000 m PRD assessment points out that Daping Station experiences the most significant detours for pedestrians among the stations studied. To foster a better walking environment around the station, it is necessary to promptly redevelop underutilized lands to boost land use efficiency, albeit in a constrained scope. Embracing high-density urban layouts where feasible will elevate the overall density of the station area. Concurrently, enhancing the pedestrian network by increasing the density of walkways and broadening their reach is essential for rendering the area more accessible and walk-friendly.

The residential land use of Nanping Station is excessively concentrated, paired with a significant lack of diversity in both land utilization and the functions of POIs. Constructed in the mountainous terrain, Nanping Station contends with significant elevation differences, complicating the creation of efficient spatial connections with the nearby buildings. This situation adversely affects the walkability and convenience for pedestrians, particularly in navigating transfers. Moreover, an analysis with the 400 m PRD metric reveals a pronounced tendency for circuitous routes around Nanping Station. To address these challenges, we propose several optimization strategies aimed at enhancing the area's functionality and accessibility: (1) promoting land use adjustment and improving the diversity of metro station domain. (2) In consideration of the significant elevation difference, a three-dimensional walking system to be established for walking efficiency enhancement in different directions and spatial levels is required. (3) Optimize the 400 m core circle layer road network to improve detours and highlight the advantages of walking. In addition to increasing pedestrian road density, attention should also be paid to the layout of surrounding land use. Encouraging non-residential functions in the core vicinity as well as establishing compatible land use policies can ensure that the area remains open and accessible, free from constraints of spatial ownership. Moreover, managing the dimensions of streets and land parcels can help in developing a compact, densely built environment that supports a pedestrian-oriented metro station area, through joint land and transport planning efforts. (4) The connection between the station entrances and exits as well as the walking system can be strengthened to facilitate residents' walking transfers.

6. Conclusions

While metro transit demonstrably alleviates traffic congestion and improves the climate in densely populated urban areas, the pedestrian environment surrounding stations is often neglected during construction, hindering the full potential of walkability. This study addresses this critical gap by proposing pedestrian-focused improvement strategies, not only to enhance the walking experience in metro transit station areas but also to cultivate more human-centric and sustainable urban spaces. Optimizing the pedestrian environment in the station domain can increase the appeal of public transportation, encouraging a modal shift toward metro transit and promoting low-carbon walking habits. This transformation is expected to effectively reduce greenhouse gas emissions from the transportation sector, thereby addressing the challenge of climate change, and contributing meaningfully to the realization of sustainable urban development.

This article evaluates five metro transit stations in Chongqing as case studies and categorizes the factors influencing the environmental aspects of metro station zones into three groups: "comprehensive assessment", "basic scale", and "structural quality". It employs a mix of quantitative and qualitative analyses to explore the relationship between various factors, facilitating a comparative review across different stations. Based on these analyses, this study proposes a series of pedestrian environment design strategies including land use and transportation. The strategies for land use advocate for "developing compact and diverse land use", "strengthening attraction of station center", "positioning large projects on the edge", "restricting private transportation capabilities". The strategies for transportation consist of "increasing pedestrian road density", "traffic calming organization", "subdivision of road types", and "three-dimensional pedestrian traffic system". This article concludes by identifying the challenges faced by the five stations and suggests improvement measures for pedestrian spaces at each location.

This paper carries out an objective evaluation of the metro transit station environment through quantitative analysis of key performance indicators and by informing optimization strategies for pedestrian flow and experience. Nonetheless, two limitations remain in this research: (1) The methodological approach lacks qualitative insights collected from pedestrian interviews. Future research should integrate quantitative and qualitative methodologies for a holistic evaluation. (2) The current indicators neglect the effect of weather conditions on pedestrian behavior. Further analysis is required to optimize and assess pedestrian-centric design solutions in extreme temperatures.

Author Contributions: Conceptualization, C.Y. and G.L.; methodology, G.L. and S.Y.; software, C.Y. and G.L.; validation, C.Y. and G.L.; formal analysis, C.Y. and G.L.; investigation, G.L.; data curation, G.L.; writing—original draft preparation, C.Y. and G.L.; writing—review and editing, C.Y.; visualization, C.Y. and G.L.; supervision, S.Y.; project administration, S.Y.; funding acquisition, S.Y. All authors have read and agreed to the published version of the manuscript.

Funding: This research is supported by National key R&D plan "Joint Research and Demonstration for Carbon Reduction Key Technologies in Urban areas and Neighborhoods" (NO.2022YFE0208700).

Data Availability Statement: Data are contained within the article.

Conflicts of Interest: The authors declare no conflicts of interest.

References

1. Liu, Q. Walking scale of tod area along rail transit line. *City Plan. Rev.* **2019**, *43*, 88–95. (In Chinese)
2. Pan, H. Urban rail transit and sustainable development. *Urban Transp. China* **2008**, *4*, 35–39. (In Chinese)
3. Calthorpe, P. *The Next American Metropolis: Ecology, Community, and the American Dream*; Princeton Architectural Press: New York, NY, USA, 1993.
4. Wey, W.-M.; Zhang, H.; Chang, Y.-J. Alternative transit-oriented development evaluation in sustainable built environment planning. *Habitat Int.* **2016**, *55*, 109–123. [CrossRef]
5. Papa, E.; Bertolini, L. Accessibility and Transit-Oriented Development in European metropolitan areas. *J. Transp. Geogr.* **2015**, *47*, 70–83. [CrossRef]
6. Cucuzzella, C.; Owen, J.; Goubran, S.; Walker, T. A TOD index integrating development potential, economic vibrancy, and socio-economic factors for encouraging polycentric cities. *Cities* **2022**, *131*, 103980. [CrossRef]
7. Sung, H.; Oh, J.-T. Transit-oriented development in a high-density city: Identifying its association with transit ridership in Seoul, Korea. *Cities* **2011**, *28*, 70–82. [CrossRef]
8. Kamruzzaman, M.; Baker, D.; Washington, S.; Turrell, G. Residential dissonance and mode choice. *J. Transp. Geogr.* **2013**, *33*, 12–28. [CrossRef]
9. Higgins, C.D.; Kanaroglou, P.S. A latent class method for classifying and evaluating the performance of station area transit-oriented development in the Toronto region. *J. Transp. Geogr.* **2016**, *52*, 61–72. [CrossRef]
10. Griffiths, B.; Curtis, C. Effectiveness of Transit Oriented Development in Reducing Car Use: Case Study of Subiaco, Western Australia. *Urban Policy Res.* **2017**, *35*, 391–408. [CrossRef]
11. Ibraeva, A.; Correia, G.H.d.A.; Silva, C.; Antunes, A.P. Transit-oriented development: A review of research achievements and challenges. *Transp. Res. Part A Policy Pract.* **2020**, *132*, 110–130. [CrossRef]
12. Feldman, S.; Lewis, P.; Schiff, R. Transit-oriented Development in the Montreal Metropolitan Region Developer's Perceptions of Supply Barriers. *Can. J. Urban Res.* **2012**, *21*, 25–44.
13. Topalovic, P.; Carter, J.; Topalovic, M.; Krantzberg, G. Light Rail Transit in Hamilton: Health, Environmental and Economic Impact Analysis. *Soc. Indic. Res.* **2012**, *108*, 329–350. [CrossRef]
14. Guthrie, A.; Fan, Y. Developers' perspectives on transit-oriented development. *Transp. Policy* **2016**, *51*, 103–114. [CrossRef]
15. Paulhiac Scherrer, F. Assessing Transit-Oriented Development Implementation in Canadian Cities: An Urban Project Approach. *J. Plan. Educ. Res.* **2019**, *39*, 469–481. [CrossRef]
16. Schlossberg, M.; Brown, N. Comparing Transit-Oriented Development Sites by Walkability Indicators. *Transp. Res. Rec.* **2004**, *1887*, 34–42. [CrossRef]
17. Estupiñán, N.; Rodríguez, D.A. The relationship between urban form and station boardings for Bogotá's BRT. *Transp. Res. Part A Policy Pract.* **2008**, *42*, 296–306. [CrossRef]
18. Jacobson, J.; Forsyth, A. Seven American TODs: Good practices for urban design in transit-oriented development projects. *J. Transp. Land Use* **2008**, *1*, 51–88. [CrossRef]
19. Vale, D.S. Transit-oriented development, integration of land use and transport, and pedestrian accessibility: Combining node-place model with pedestrian shed ratio to evaluate and classify station areas in Lisbon. *J. Transp. Geogr.* **2015**, *45*, 70–80. [CrossRef]
20. King, T.L.; Bentley, R.J.; Thornton, L.E.; Kavanagh, A.M. Does the presence and mix of destinations influence walking and physical activity? *Int. J. Behav. Nutr. Phys. Act.* **2015**, *12*, 115. [CrossRef] [PubMed]
21. Morency, P.; Plante, C.; Dubé, A.-S.; Goudreau, S.; Morency, C.; Bourbonnais, P.-L.; Eluru, N.; Tétreault, L.-F.; Hatzopoulou, M.; Iraganaboina, N.C.; et al. The Potential Impacts of Urban and Transit Planning Scenarios for 2031 on Car Use and Active Transportation in a Metropolitan Area. *Int. J. Environ. Res. Public Health* **2020**, *17*, 5061. [CrossRef] [PubMed]
22. Nasri, A.; Zhang, L. How Urban Form Characteristics at Both Trip Ends Influence Mode Choice: Evidence from TOD vs. Non-TOD Zones of the Washington, D.C. Metropolitan Area. *Sustainability* **2019**, *11*, 3403. [CrossRef]
23. Dill, J. Measuring network connectivity for bicycling and walking. In Proceedings of the Proceedings of the 83rd Annual Meeting of the Transportation Research Board, Washington, DC, USA, 11–15 January 2004; pp. 11–15.
24. Marshall, W.E.; Garrick, N.W. Effect of Street Network Design on Walking and Biking. *Transp. Res. Rec. J. Transp. Res. Board* **2010**, *2198*, 103–115. [CrossRef]
25. Jones, T.; Horton, D.; Mullen, C.; Pooley, C.; Scheldeman, G.; Strano, E.; Tight, M. The role of street network connectivity and access to everyday facilities in shaping everyday walking and cycling in English cities. In *Strategies for Sustainable Mobilities*; Routledge: London, UK, 2016; pp. 15–38.
26. Lu, Y.; Wang, D. Walkability measuring in america and lts enlightenment. *Urban Plan. Int.* **2012**, *27*, 10–15. (In Chinese)

27. Yin, L. Street level urban design qualities for walkability: Combining 2D and 3D GIS measures. *Comput. Environ. Urban Syst.* **2017**, *64*, 288–296. [CrossRef]
28. Peiravian, F.; Derrible, S.; Ijaz, F. Development and application of the Pedestrian Environment Index (PEI). *J. Transp. Geogr.* **2014**, *39*, 73–84. [CrossRef]
29. Mayne, D.; Morgan, G.; Wilmore, A.; Bauman, A.; Jalaludin, B.; Bambrick, H.; Rose, N.; Rodgers, B.; Bennett, C. An objective index of walkability for the Sydney metropolitan region. *J. Sci. Med. Sport* **2012**, *15*, S32–S33. [CrossRef]
30. Millstein, R.A.; Cain, K.L.; Sallis, J.F.; Conway, T.L.; Geremia, C.; Frank, L.D.; Chapman, J.; Van Dyck, D.; Dipzinski, L.R.; Kerr, J.; et al. Development, scoring, and reliability of the Microscale Audit of Pedestrian Streetscapes (MAPS). *BMC Public Health* **2013**, *13*, 403. [CrossRef]
31. Kurka, J.M.; Adams, M.A.; Geremia, C.; Zhu, W.; Cain, K.L.; Conway, T.L.; Sallis, J.F. Comparison of field and online observations for measuring land uses using the Microscale Audit of Pedestrian Streetscapes (MAPS). *J. Transp. Health* **2016**, *3*, 278–286. [CrossRef]
32. Gou, A.; Bo, W. SD method based street space vitality evaluation. *Planners* **2011**, *27*, 102–106. (In Chinese)
33. Shatu, F.; Yigitcanlar, T. Development and validity of a virtual street walkability audit tool for pedestrian route choice analysis—SWATCH. *J. Transp. Geogr.* **2018**, *70*, 148–160. [CrossRef]
34. Yan, T.; Ren, L.; Gao, H. Integrative urban design for urban railway station vicinity: Beiing M15 line shunyi section case. *Planners* **2012**, *28*, 42–45. (In Chinese)
35. Guo, W.; Hou, X. Desian strategies for urban pedestrian space under the tod mode. *Landsc. Archit.* **2015**, *5*, 100–104. (In Chinese) [CrossRef]
36. Zhao, P.; Wang, Y. The empirical study on pedestrian-friendly tod design-A case of beijing municipality. *J. Hum. Settl. West China* **2015**, *30*, 36–40. (In Chinese) [CrossRef]
37. Bai, W. Community Development of Rail Transit Terminal Area: Ningbo TOD Community Case. *Planners* **2015**, *31*, 63–68. (In Chinese)
38. Olaru, D.; Curtis, C. Designing TOD precincts: Accessibility and travel patterns. *Eur. J. Transp. Infrastruct. Res.* **2015**, *15*, 6–26. [CrossRef]
39. Jeffrey, D.; Boulangé, C.; Giles-Corti, B.; Washington, S.; Gunn, L. Using walkability measures to identify train stations with the potential to become transit oriented developments located in walkable neighbourhoods. *J. Transp. Geogr.* **2019**, *76*, 221–231. [CrossRef]
40. Lamour, Q.; Morelli, A.M.; Marins, K.R.d.C. Improving walkability in a TOD context: Spatial strategies that enhance walking in the Belém neighbourhood, in São Paulo, Brazil. *Case Stud. Transp. Policy* **2019**, *7*, 280–292. [CrossRef]
41. O'Sullivan, S. Walking Distances to and from Light Rail Transit Stations in Calgary. Master's Thesis, University of Calgary, Calgary, AB, Canada, 1996.
42. Nasri, A.; Zhang, L. The analysis of transit-oriented development (TOD) in Washington, DC and Baltimore metropolitan areas. *Transp. Policy* **2014**, *32*, 172–179. [CrossRef]
43. Liu, Q. Factors affecting circle structure of tod concentric models. *Urban Plan. Int.* **2017**, *32*, 72–79. (In Chinese) [CrossRef]
44. Jaafar Sidek, M.F.; Bakri, F.A.; Kadar Hamsa, A.A.; Aziemah Nik Othman, N.N.; Noor, N.M.; Ibrahim, M. Socio-economic and Travel Characteristics of transit users at Transit-oriented Development (TOD) Stations. *Transp. Res. Procedia* **2020**, *48*, 1931–1955. [CrossRef]
45. Andersen, J.L.E.; Landex, A. Catchment areas for public transport. *WIT Trans. Built Environ.* **2008**, *101*, 175–184.
46. Jun, M.-J.; Choi, K.; Jeong, J.-E.; Kwon, K.-H.; Kim, H.-J. Land use characteristics of subway catchment areas and their influence on subway ridership in Seoul. *J. Transp. Geogr.* **2015**, *48*, 30–40. [CrossRef]
47. Pafka, E.; Dovey, K. Permeability and interface catchment: Measuring and mapping walkable access. *J. Urban. Int. Res. Placemaking Urban Sustain.* **2017**, *10*, 150–162. [CrossRef]
48. Cervero, R.; Kockelman, K. Travel demand and the 3Ds: Density, diversity, and design. *Transp. Res. Part D Transp. Environ.* **1997**, *2*, 199–219. [CrossRef]
49. Cervero, R. Built environments and mode choice: Toward a normative framework. *Transp. Res. Part D Transp. Environ.* **2002**, *7*, 265–284. [CrossRef]
50. Lee, C.; Moudon, A.V. The 3Ds+R: Quantifying land use and urban form correlates of walking. *Transp. Res. Part D Transp. Environ.* **2006**, *11*, 204–215. [CrossRef]
51. Kamruzzaman, M.; Shatu, F.M.; Hine, J.; Turrell, G. Commuting mode choice in transit oriented development: Disentangling the effects of competitive neighbourhoods, travel attitudes, and self-selection. *Transp. Policy* **2015**, *42*, 187–196. [CrossRef]
52. Li, G. Research on the Influence Factors of Pedestrian Environment in Chongqing Subway Station Area—Take 5 Typical Stations for Example. Master's Thesis, Chongqing University, Chongqing, China, 2022. Available online: https://kns.cnki.net/kcms2/article/abstract?v=Skeo7MzZydZCr7gTm8SdgB2ESFbbhCSc8gGUQtH03Kl7qpEAnQLQ_hqXSvBViDOrywq5oyq0tJPiPsYa80iSLt7a-3f_JsmHOkLfz6sRBilOMsxYJbwIErDEA5DY4GLpu3XAjp4IuvlC27NMaexxnw==&uniplatform=NZKPT&language=CHS (accessed on 2 April 2024). (In Chinese)
53. *GB 50220-95*; Code for Transport Planning on Urban Road. Planning Press of China: Beijing, China, 1995.
54. Aikoh, T.; Homma, R.; Abe, Y. Comparing conventional manual measurement of the green view index with modern automatic methods using google street view and semantic segmentation. *Urban For. Urban Green.* **2023**, *80*, 127845. [CrossRef]

55. Zhang, W.; Zeng, H. Spatial differentiation characteristics and influencing factors of the green view index in urban areas based on street view images: A case study of Futian District, Shenzhen, China. *Urban For. Urban Green.* **2024**, *93*, 128219. [CrossRef]
56. Huang, D.; Jiang, B.; Yuan, L. Analyzing the effects of nature exposure on perceived satisfaction with running routes: An activity path-based measure approach. *Urban For. Urban Green.* **2022**, *68*, 127480. [CrossRef]

buildings

MDPI

Article

Research on the Classification of Rail Transit Stations and Passenger Flow Patterns—A Case from Xi'an, China

Li Chen [1,2], Yuan Chen [2,*], Yupeng Wang [2] and Ying Li [1]

1 China Railway First Survey and Design Institute Group Co., Ltd., Xi'an 710043, China; chenli.yy@crcc.cn (L.C.); liying.yy@crcc.cn (Y.L.)
2 School of Human Settlements and Civil Engineering, Xi'an Jiaotong University, Xi'an 710049, China; wang-yupeng@xjtu.edu.cn
* Correspondence: y.chen@stu.xjtu.edu.cn

Abstract: Transit-oriented development (TOD) has been promoted and implemented worldwide through the efficient integration of rail transit stations and land use. However, the interactions between stations and the surrounding catchment areas (CAs) are characterized by different features of the built environment and regional development. Therefore, it is necessary to develop a quantitative classification method for rail transit stations based on the built environment within a CA and to identify the passenger flow characteristics of different types of stations to develop targeted planning and design policies. In this study, Line 1 and Line 2 within the third ring road of Xi'an City were taken as the objects, and a station classification system was constructed by taking station traffic levels and different building functions within the CA as the classification factors. Secondly, indicators of the built environment, such as six different types of functions, were calculated through refined vector modeling, and 30 typical stations were typologically analyzed. Furthermore, 10 typical types—totaling 11 stations—were selected for passenger flow monitoring, and the passenger flow characteristics of the different types of stations were summarized in terms of the dimensions of stations and exits. Finally, the correlations between the indicators of the built environment and passenger traffic for different functions were quantified. This study provides a basis for the future optimization of stations and the built environment, as well as station design and management.

Keywords: transit-oriented development (TOD); urban built environment; station classification; land use; passenger flow monitoring

1. Introduction

In 2050, 68% of the total global population of 9.7 billion is expected to live in cities [1]. A massive population concentration leads to conflicts related to urban land use and transportation, resulting in severe traffic congestion [2]. In this regard, countries have developed rail transit on a large scale to alleviate urban transportation problems. By the end of 2022, 545 cities in 78 countries and regions operated urban rail transit systems, with a total mileage of more than 41,386 km [3]. In order to better promote the integration of rail transit and urban development, rail transit combined with a transit-oriented development (TOD) strategy has been recognized and advocated worldwide in recent decades [4]. TOD refers to the high-density mixed-use development of urban land within a certain range of the area surrounding a rail transit station, thus compactly and efficiently utilizing the surrounding land resources [5]. One important purpose of TOD is the study of the heterogeneity of stations' traffic characteristics and the development patterns of the surrounding land, as well as the formulation of strategies for integrating stations and the surrounding land based on different types of TOD. This includes strategies such as planning the site layout, optimizing the site design, and optimizing the surrounding land use and the built environment. Despite the simplicity of the basic principles of TOD, there are significant differences in the implementation of TOD in different countries and cities in North America, Europe, South America, and Asia due to the differences in population densities and the built environments of cities [6].

Typical TOD includes two features, i.e., providing efficient transportation services and diversified multifunctional services, such as work and recreation for the residents around the station [7]. However, in rapid TOD, there are problems with incompatibility between stations and the surrounding urban functions. Certain stations are unable to fully meet the transportation demand generated by the surrounding land, and the functional modes around the stations are unable to meet the diversified needs of the population. Previous studies have shown that different land-use functions and built environments around a station reflect heterogeneity in station traffic characteristics [8,9]. Therefore, determining station classifications based on different built environments is important for stations themselves and the surrounding areas [10]. For the transportation function of a station, it helps to explain and predict the heterogeneity of the station's passenger flow characteristics [8]. For the station's area function, it is helpful to formulate different strategies for different types of TOD that can be conducive to achieving the integrated development of the station and the surrounding land use [11,12]. Therefore, developing station classification methods and clarifying the passenger flow patterns of different types of stations can help in the understanding of stations' traffic demand, guide the integrated development of stations and land use, and, at the same time, provide data support and a basis for station TOD planning.

1.1. Literature Review

To establish quantitative classification methods for stations in built environments with different characteristics, academics have developed different methods based on the node–place model theory [13], and they have done so around different classification factors using clustering methods such as time-series or K-means clustering. Meanwhile, based on different types of station classifications, the spatiotemporal characteristics of passenger flow were further explored. According to the different classification factors, the classification of stations in established studies can mainly be categorized into the following three types.

(1) Classification by station passenger flow characteristics: A related study in Beijing classified 195 stations into eight different types through the time-series clustering of metro stations' passenger flow data [14]. A study in Shanghai classified the time series of stations based on card operation data from 288 stations, combining passenger flow and shopping district integration [15]. A related study in Chongqing utilized operational data to classify 25 stations into five classes using passenger traffic, station operations, and other transportation characteristics as classification factors [16].

(2) Classification by land use and building function around the station: Related studies determined the station type by analyzing the dominant urban functional types within a certain catchment area (CA) around a station. In Shanghai, for example, a study took 23 stations in new towns as an example, used a 600 m radius as the CA, characterized the functions of the land around the stations through point-of-interest (POI) data, and classified the stations into five types [17]. In Hong Kong, POI data within the 500 m CAs around stations were extracted to categorize stations according to seven indicators of the built environment, including the land-use diversity, commercial intensity, and population. The results found that commercial, residential, tourist, and industrial land uses had the most significant influence on the station type [7].

(3) Classification based on multiple indicators, such as passenger flow characteristics, surrounding urban functions, and accessibility: a study in Lisbon evaluated 50 stations based on their transportation accessibility, land use, and urban design [18]. Another study in Shanghai used POIs to classify 347 stations into five categories based on three factors: traffic characteristics, pedestrian accessibility, and comprehensive urban development [19]. A study in Ningbo combined traffic volume and land use to evaluate urban stations using metro operation data and cell phone signaling data

after developing a grading system [20]. A study in Xi'an used cell phone signaling and station traffic data to classify 66 stations into six categories based on traffic characteristics and the job–housing balance, and the traffic characteristics of different types of stations were analyzed [21].

Previous studies have achieved rich results related to station classification methods, but the following gaps still exist: 1. In studies that used a classification based on stations' land-use functions, POI and urban planning data were most commonly used. POI data cannot reflect the impacts of land and building areas on passenger flow [22], and land-use planning data are different from the actual built environment. 2. In studies that used a classification based on passenger flow characteristics, the passenger data were usually obtained using station operation data and cell phone signaling data. When using such methods, only the vertical passenger flow in the station can be analyzed, but not enough consideration is given to the more refined passenger flow at different exits. Therefore, it is necessary to explore station classification methods based on a refined view of the built environment and to analyze and validate the passenger flow characteristics of different types in station and exit dimensions.

Based on the above results of theoretical research, guidelines related to station classification have also been issued at the governmental level in various countries. In Japan, cities such as Tokyo and Tsukuba classify stations into three categories based on the station location, traffic intensity, and interchange coefficients based on traffic characteristics [23]. In Singapore, stations are classified into four classes based on their land structure and the layout characteristics around each station; the classifications are center stations, regional center stations, town center stations, and neighborhood center stations [24].

China is the country with the greatest increase in urbanization in recent decades, and it is currently building rail transit on a large scale. By the end of 2022, a cumulative total of 61 cities in China had opened and operated rail transit, with an operating mileage of 10,857 km [3]. In China, governmental departments have issued two guidelines to instruct the classification of rail transit stations in recent years. China's Ministry of Housing and Urban–Rural Development issued the Guidelines for the Planning and Design of Areas Along Urban Railways in 2015, in which stations were classified into six types, namely, hub stations, center stations, cluster stations, special control stations, end stations, and general stations, using traffic positioning as a classification factor [25]. This guideline only considered the transportation characteristics of the stations and did not consider the surrounding urban functions. In addition, the Urban Planning Society of China issued the Guidelines for the Spatial Planning and Design of Facilities in the Area Surrounding Urban Rail Transit Stations in 2021. The guideline divided stations into three simpler types based on stations' traffic characteristics. Meanwhile, stations were divided into six categories, namely, residential, commercial, industrial, transportation, special, and integrated stations, based on the land-use characteristics around them [26]. However, this specification does not clearly define the specific scope of the surrounding CAs, and the simple division of the three transportation types cannot reflect the complex situation of megacities. These guidelines provide instructions for the pre-construction phase of rail transit. However, these simple and fixed guidelines cannot meet the needs of developing countries and regions in which urbanization is rapidly advancing and the built environment is drastically changing because they do not specify a comprehensive quantitative method. More importantly, these classification criteria are only used as a reference for the preliminary stage, and the passenger flow characteristics of different types of stations after their completion have not been analyzed and verified in depth.

1.2. Motivation and Objective

This study attempted to address the above research gaps by taking Xi'an, a plains-type megacity in China, as an example. As shown in Figure 1, Xi'an takes the form of a checkerboard and ring road network. It had a population of nearly 13 million in 2022, a built-up area of 806 km^2, and an urbanization level of nearly 80% [27]. As of January 2024, the length of rail transit operation was 312 km, with a total of 214 stations on 10 lines [28], of which 6 lines operated within the city's third ring road. In 2023, the average daily passenger flow intensity was able to reach 3.64 million passengers/day. It is expected that by 2029, nearly 200 km of new lines will be added on top of this [29].

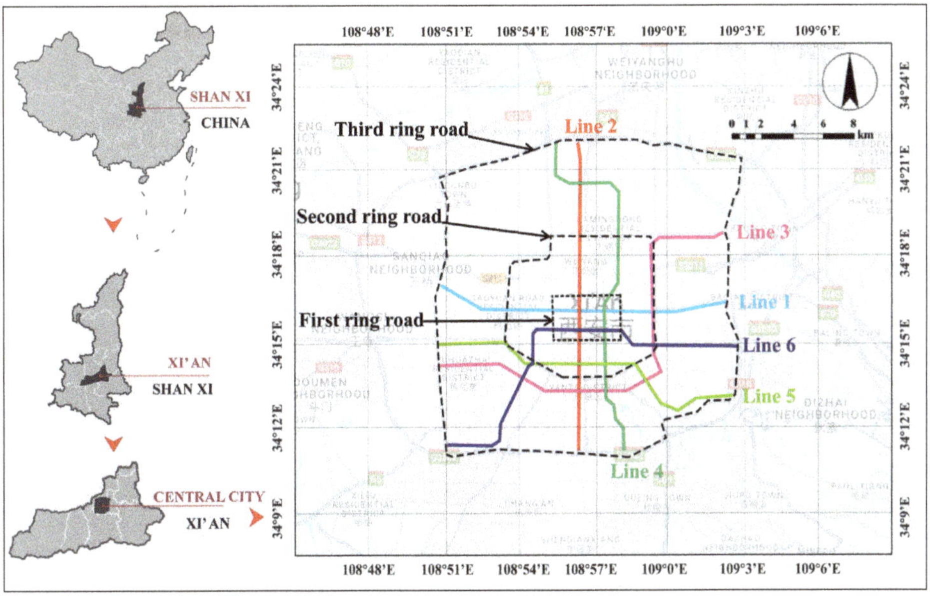

Figure 1. Location of the study area.

This study took 30 typical stations of Line 1 and Line 2 within the third ring road as specific research objects. Figure 2 shows the analytic flow of the study. Firstly, it was considered that the guidelines and standards issued by the government are mandatory in actual construction [30]. Based on the two relevant guidelines in China, this study extracted the four traffic levels and six types of building functions to form a new classification system. At the same time, a built environment dataset was constructed through fine vector modeling of the stations and the surrounding built environment, and 10 typical stations were typologically analyzed based on this classification system. Secondly, 11 typical stations of different types were selected, field monitoring was carried out, and the passenger flow characteristics of different types of stations were analyzed. Finally, the correlation of different factors of the built environment with passenger flow was verified. This study provides guidance and reference for similar megacities with ancient city centers in terms of the optimization of the urban spatial structure, the renewal of stations and the surrounding built environment, and the management of station operation.

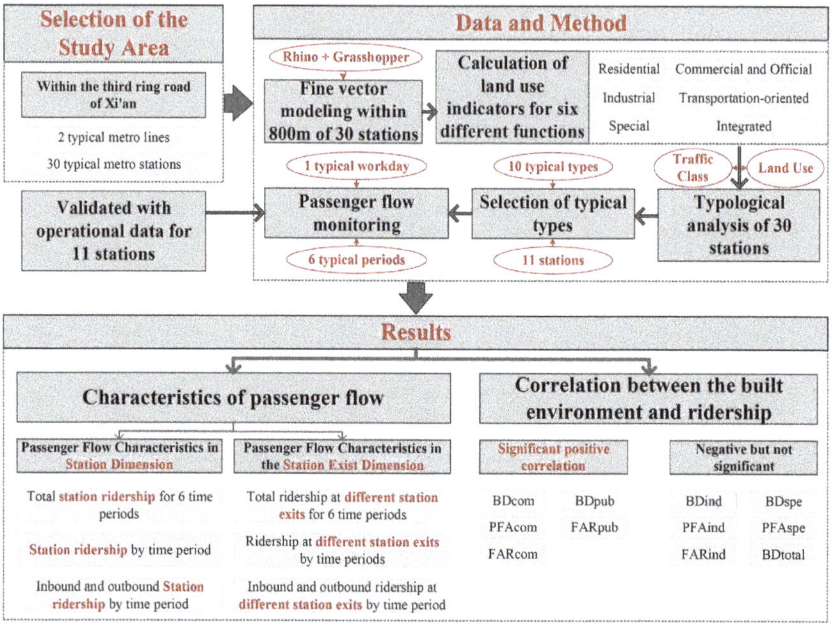

Figure 2. The analytic flow of this study.

2. Data and Methodology

2.1. Selection of the Study Area and Typical Metro Lines

In this study, the core urban area within the third ring road of Xi'an was selected as the study area, as it has the highest population density and development intensity [31]. At the same time, this area includes the old city within the first ring road, which is controlled through planning protection, as well as the new city outside the first ring road; thus, the different types of built environments in Xi'an are represented.

As typical lines, we selected Line 2 (opened in 2011) and Line 1 (opened in 2013), which were the first to be built in Xi'an. As shown in Figure 1, Line 1 and Line 2 are the north–south express lines in the Xi'an rail transit network. Among all lines, Line 2 has the highest passenger volume and the second highest traffic intensity, while Line 1 has the highest traffic intensity and the third-highest passenger volume [32]. Therefore, we took the 30 stations of these two lines within the third ring road as the study objects.

Previous studies have shown that people within the 500–800 m (i.e., those with a life circle accessible within 15 min of walking) CA around a metro station are the main targets of station services [10]. Considering the large scale of neighborhoods in China's first-tier cities, this study takes an 800 m radius as the CA. Figure 3a shows a vector model of the built environment of a station area, taking the North Street station as an example.

2.2. Urban Rail Transit Station Classification System

The traffic level and the surrounding urban functional characteristics are the core factors for station classification. In China, the classification methods issued by different government departments focus on a single factor and lack the integrated combination and quantification of these two factors. Therefore, in this study, the two core factors from two government guidelines were extracted to form a new classification system. As shown in Table 1, the station traffic level was based on the Guidelines for the Planning and Design of Areas Along Urban Railways, and according to the built environment of Xi'an, the classifications of special control stations and end stations were removed. The classification based on surrounding urban functions referred to the Guidelines for the Space Planning and Design of Facilities around Urban Rail Transit Stations and was based on the six types of functions in the guidelines; the types of public services were added, and the specific proportions were also determined with reference to the guidelines. Then, the combination of the two factors allowed a division into 28 different types of stations to finally obtain a new classification system for rail transit stations. Traffic levels and building function types have been widely used in existing station classification studies and are suitable for the actual built environments of quickly developing countries.

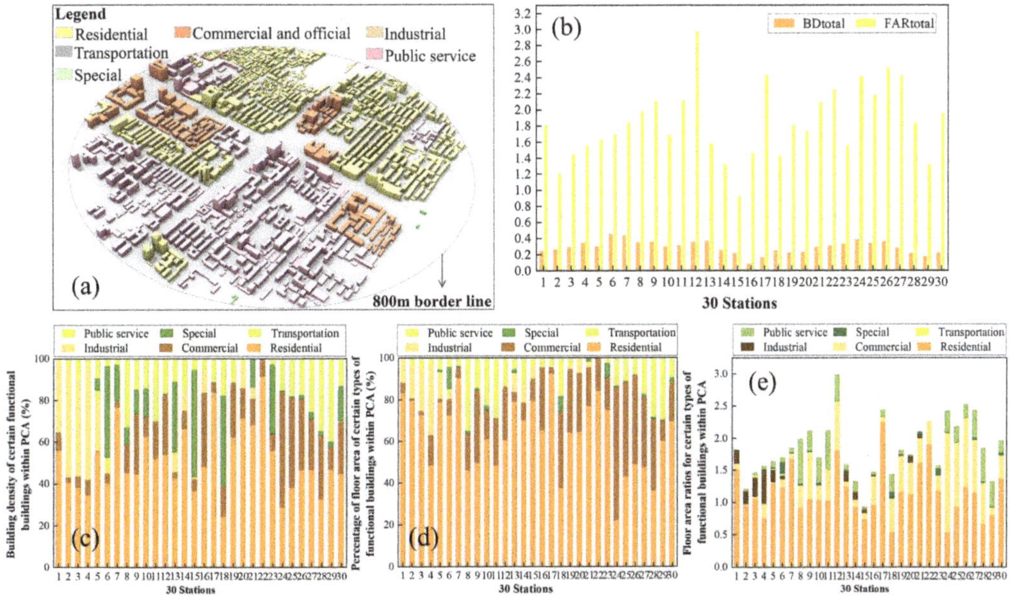

Figure 3. (**a**) Fine-scale vector modeling of the built environment in a station's CA using North Street as an example; indicators of the built environment within the CAs of 30 stations: (**b**) total building density and total floor area ratio; (**c**) building densities of certain types of functional buildings; (**d**) percentage of floor area of certain types of functional buildings; and (**e**) floor area ratios for certain types of functional buildings.

Table 1. Urban rail transit station classification system.

Types	Residential (A)	Commercial and Official (B)	Industrial (C)	Transportation (D)	Special (E)	Public Service (F)	Integrated (G)
Hub (I)	I A	I B	I C	I D	I E	I F	I G
Central (II)	II A	II B	II C	II D	II E	II F	II G
Cluster (III)	III A	III B	III C	III D	III E	III F	III G
Normal (IV)	IV A	IV B	IV C	IV D	IV E	IV F	IV G

2.3. Calculation of Building Functional Indices within Station CAs

Refined vector modeling of the built environment contained in the 800 m CAs of the stations was conducted using Rhino 7 with reference to Google Maps and Street View maps, and the Grasshopper plug-in was used to parameterize the calculation of six different indicators of building functions. The definition and calculation methodology for each indicator are shown in Table 2. Figure 3b–e show the results of calculating the built environment indicators for the 30 stations.

Table 2. Descriptions and calculations of the indicators for six different building functions.

(Unit)/Built-Up Indices (Abbreviations)	Description or Calculation Method
(%)/Building density of certain types of functional buildings within the CA ($BD_{certain\ type}$ *1)	Sum of the base areas of certain types of functional buildings within the CA/the area of the CA
(%)/Percentage of floor area of certain types of functional buildings within the CA ($PFA_{certain\ type}$ *1)	Gross floor area of certain types of functional buildings within the CA/total floor area within the CA
(null)/Floor area ratios for certain types of functional buildings within the CA ($FAR_{certain\ type}$ *1)	Gross floor area of certain types of functional buildings within the CA/the area of the CA
(%)/Total building density of the CA (BD_{total} *2)	Total base areas of all buildings within the CA/the area of the CA
(null)/Total floor area ratio of the CA (FAR_{total} *2)	Total floor area within the CA/the area of the CA

*1 Certain type in the subscript: certain functional buildings within the 800 m CA, including six types of buildings: residential, commercial, industrial, transportation, special, and public services. *2 Total in the subscript: all buildings within the 800 m buffer zone.

2.4. Analysis of the Station Typology

Thirty typical stations were analyzed and classified based on Table 1. The traffic levels of the stations were determined according to the positioning of the city center, sub-center, and cluster in the city master plan, as well as the locations of large-scale external transportation facilities, such as railway stations, high-speed rail stations, and airports. The dominant urban functions of the station CAs were calculated based on Figure 3c. Figure 4 shows the distribution of different types of stations. The results showed that Line 1 and Line 2 mainly contained nine types of stations. In terms of the traffic level, there were more central stations. In terms of urban function, there were many industrial and special stations on Line 1 and more commercial stations on Line 2. Line 1 passed through the east–west traffic arteries of Xi'an and connected many industrial parks and large green areas in the city. Line 2 passed through the north–south axis of Xi'an and connected many commercial districts.

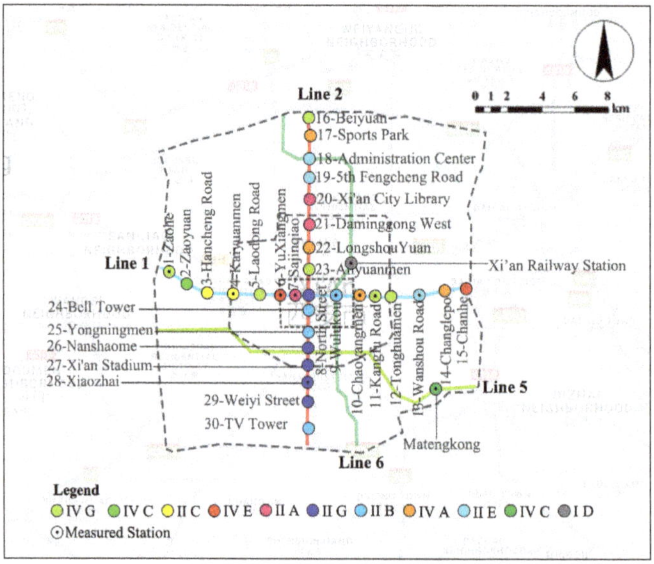

Figure 4. Results of the typology analysis of the 30 stations of Line 1 and Line 2 within the third ring road and the distribution of the measured stations.

2.5. Passenger Flow Data

We used measured passenger flow data to analyze the characteristics of passenger flow on a typical weekday and validated them using metro operation data.

2.5.1. Field Monitoring of Passenger Flow at Typical Stations

For different types of stations, six different typical time periods on a typical weekday were selected to monitor the flow of passengers through different exits of each station to obtain a typical weekday passenger flow dataset.

- Selection of Typical Stations for Different Types

Based on the results of the typology analysis described in Section 2.4, we found that the 15 stations of Line 1 covered nine types, while the 15 stations of Line 2 covered only six types. Therefore, we selected Line 1 as the monitoring object. Notably, in order not to confuse the passenger flow on North Street (of the II G type), where Line 1 and Line 2 intersect, we replaced this station with Xiaozhai (of the II G type), which had the highest passenger flow on Line 2. Meanwhile, Xi'an had four train stations, and we added the hub transportation type (I D), i.e., Xi'an Railway Station, as a supplement. In addition, considering that Line 1 and Line 2 did not have departure stations, we selected Matengkong, the eastern departure station of Line 5 near the boundary of the third ring road, as a supplement for this type. In the end, 11 stations were chosen for monitoring, and the distribution of the measured stations is shown in Figure 4. Figure 5 shows the differences in the built environments of the 11 different types of stations.

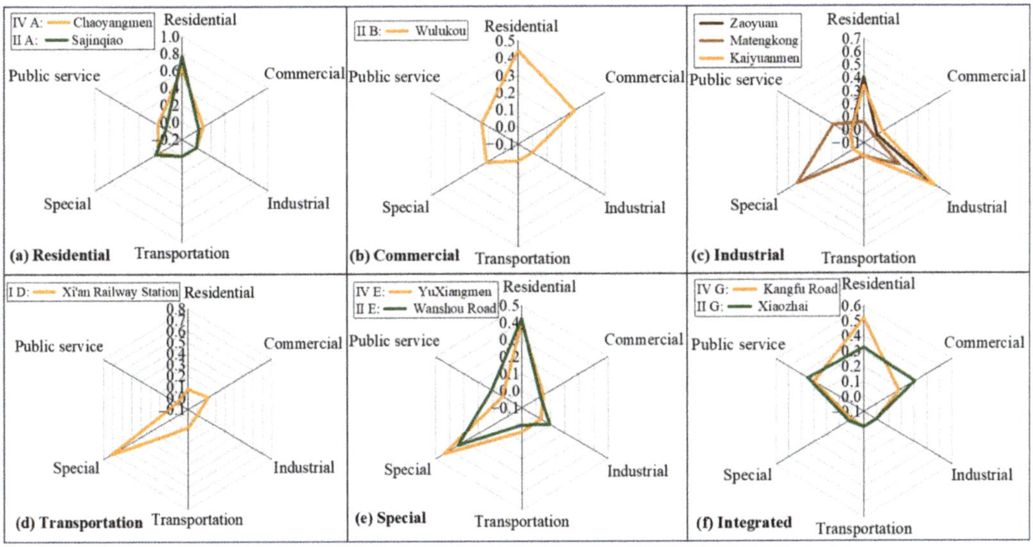

Figure 5. Radar map of the characteristics of the built environments of the 11 stations: (**a**) residential; (**b**) commercial and official; (**c**) industrial; (**d**) transportation-oriented; (**e**) special; and (**f**) integrated.

- Manual Monitoring of the Passenger Flow

To obtain the passenger flow data at different exits of the stations, manual counting was performed using push counters, as shown in Figure 6. On a typical weekday (Wednesday, 15 February 2023), the number of people entering and exiting the station at each exit was counted at six moments (8:00, 10:00, 12:00, 15:00, 18:00, and 20:00). Each exit was recorded for 5 min. We assumed that the passenger flow in 5 min represented the level of passenger flow in 1 h and normalized the 5 min data to the 1 h passenger flow. One person was positioned at each station during the measurement process, and the measurements were carried out at all stations simultaneously.

Figure 6. (**a**) Manual counters; (**b**) monitoring scenarios; and (**c**) inbound and outbound numbers were monitored and recorded simultaneously.

- Calculation of the Passenger Flow Indices

The passenger flow indices were divided into two dimensions—stations and exits—and each dimension included three indicators, as shown in Table 3.

Table 3. Definitions and calculation of the passenger flow indices.

(Unit)/Passenger Flow Indices	Definition and Calculation
(person trips)/Total station ridership for 6 periods	The sum of the measured number of passengers at a station for 6 typical periods
(person trips)/Station ridership in each different period	Measured number of passengers at a station in a different typical period
(person trips)/Inbound/outbound stations in each different period	Measured number of passengers entering/exiting a station in a different typical period
(person trips)/Total ridership at an exit in 6 periods	The sum of the measured number of passengers at an exit of a station for 6 typical periods
(person trips)/Total ridership at an exit in each different period	Measured number of passengers at an exit of a station in a different typical period
(person trips)/Inbound/outbound flow at an exit in each different period	Measured number of passengers entering/exiting an exit of a station in a different typical period

2.5.2. Validation of the Measured Data against Metro Operational Data

The operational ridership data from 11 stations on a typical weekday (13 December 2023) provided by the official Xi'an Metro were used for verification. The operation data included the daily inbound, outbound, and total passenger flow of the stations, and they excluded the dimensional data of the exits. As shown in Figure 7, the validation results showed that the measured and operational data of the 11 stations were significantly correlated with each other, with an R^2 of 0.96. This confirmed the feasibility of the monitoring method and the accuracy of the measured passenger flow data.

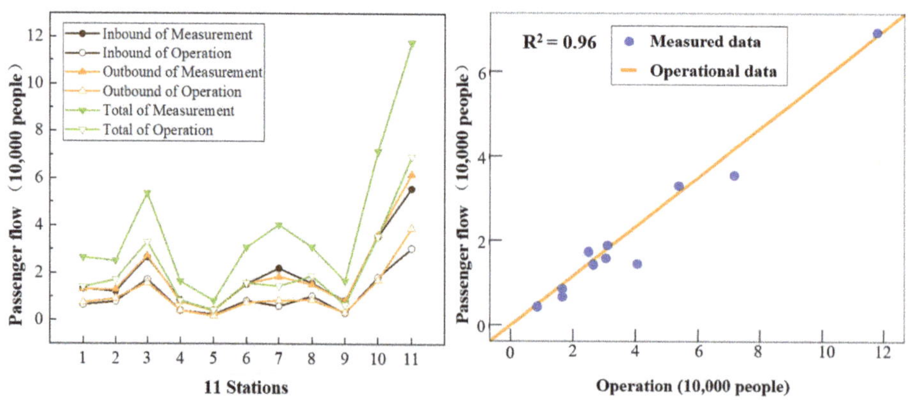

Figure 7. Validation of the measured data against metro operational data.

3. Results

3.1. Passenger Flow Characteristics of Different Types of Stations

3.1.1. Passenger Flow Characteristics in the Station Dimension

The passenger flow characteristics of the stations were analyzed from three perspectives: total station ridership for six periods, station ridership in each different period, and inbound/outbound flow in the station in each different period.

- Total Station Ridership for Six Periods

Based on the different traffic levels and building functions, the total station ridership for six periods was analyzed. Figure 8a shows the results with the traffic level as the classification factor. The average total station ridership for six periods for the four stations that were classified as central stations was the highest (31,551); those of general stations (16,176) and hub stations (14,496) were close to each other. Xi'an Railway Station, as an urban railroad hub station, did not have a high passenger flow. This suggested that a high-level hub connecting with the city's external transportation did not necessarily generate significant intra-city transfer traffic. Figure 8b shows the results when using building functions as the classification factor. The results showed that integrated (52,422) and commercial stations (33,000) had significantly greater values than those of other types; those of residential (15,660) and transportation-oriented stations (14,496) were normal, and those of special (12,750) and industrial stations (9532) were lower than those of the other types. This indicated that integrated functions, such as commercial and office functions, are the core functions that attract passengers.

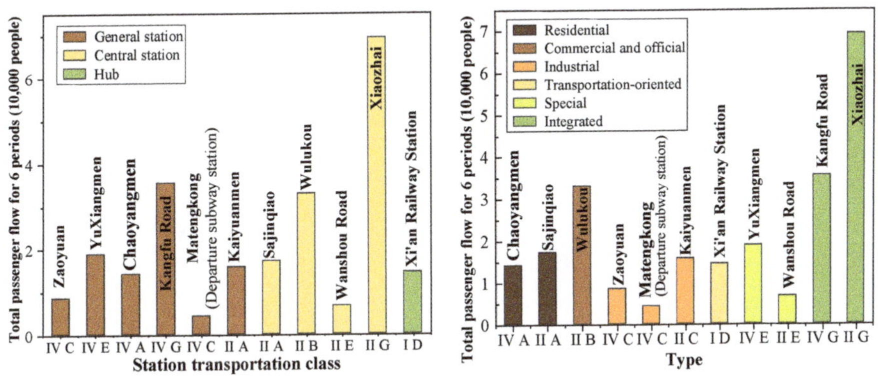

(a) (b)

Figure 8. The total station ridership for six periods with different classification factors: (**a**) traffic levels and (**b**) building functions.

- Station Ridership in Each Different Period

Figure 9a shows that there were significant differences in both the station ridership and the morning and evening peak characteristics for the different functional types. Figure 9b demonstrates the morning and evening peak coefficients of each station, with only evening peaks and no morning peaks for the integrated type; there was no bimodal phenomenon for the I D type, and double peaks were found at 8:00 and 18:00 for all the other types of stations. Overall, the morning peak was stronger than the evening peak, with the range of the morning peak coefficients being 0.14–0.42 and the range of the evening peak coefficients being 0.13–0.37. The specific characteristics of the different types of stations were as follows.

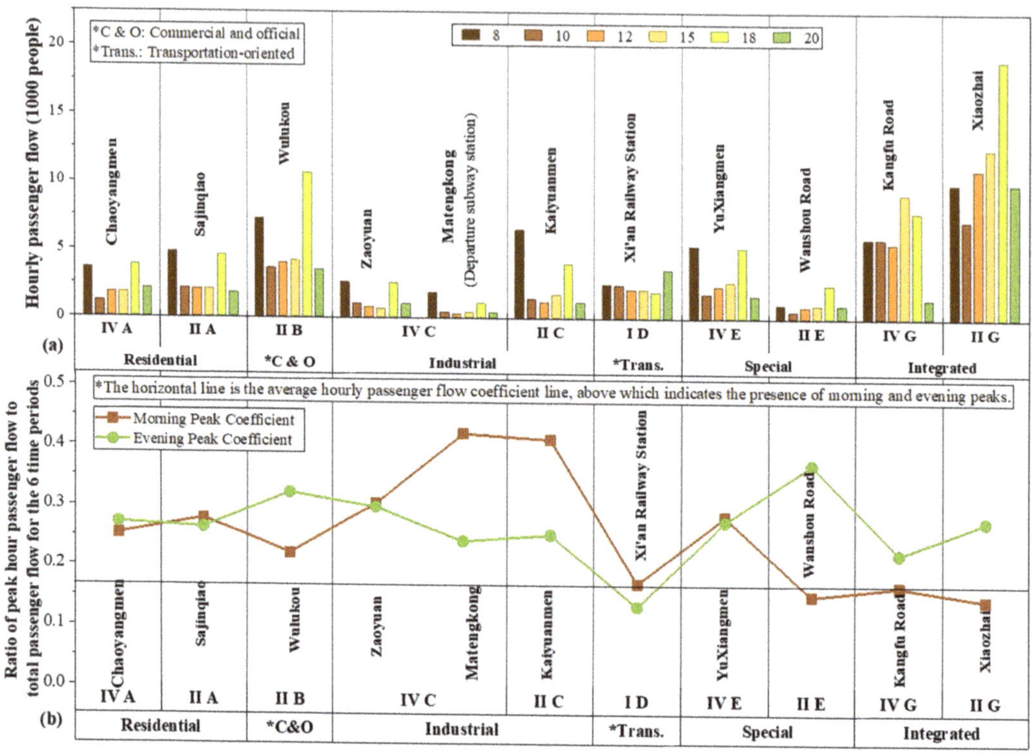

Figure 9. (**a**) Station ridership in each different period for the 11 stations; (**b**) morning and evening peak coefficients for the 11 stations.

(1) Residential: Both types of residential stations had a moderate passenger flow and bimodal peak intensity. There were significant morning and evening peaks, and the bimodal intensity was close, with an average peak coefficient of 0.27.

(2) Commercial and official: The commercial type had a large flow of passengers; there were morning and evening peaks, and the evening peak was significantly stronger than the morning peak; the evening peak coefficient was 0.32, and it was ranked second among those of all the types.

(3) Industrial: The industrial type had the lowest passenger flow, but the morning and evening peaks were the most significant. For the two stations of type IV C, the terminal Matengkong embodied a significant difference; the terminal's morning peak was significantly higher than the evening peak, and the morning peak coefficient was the highest among all of the types at 0.42. In addition, the characteristics of the changes in different time periods for the industrial type were similar to those for the residential type, but the off-peak passenger flow was significantly lower than that of the residential types.

(4) Transportation-oriented: The transportation-oriented type did not have morning and evening peaks, but the passenger flow was significantly higher at 20:00, which may have been influenced by the fact that external transportation, such as urban railroads, has more arrivals in the evening.

(5) Special: The two special types reflected uncertainty in both passenger flow and morning and evening peak intensities, with IV E having a large passenger flow and the morning and evening peak intensities being close to one another; IIE had a small passenger flow with only significant evening peaks; and the highest evening

peak coefficient of all types of stations was 0.37. The rest of the hours had strong heterogeneity in passenger flow.

(6) Integrated: The integrated type had the highest passenger flow, with only significant evening peaks and a low evening peak intensity, and the average coefficient was 0.25. In addition, the passenger flow showed a gradual increase from 8:00 to 18:00, and it significantly decreased at 20:00.

- Inbound/Outbound Flows of Stations in Each Different Period

Figure 10 shows the spatial and temporal distribution characteristics of inbound and outbound flows at each station during the six time periods.

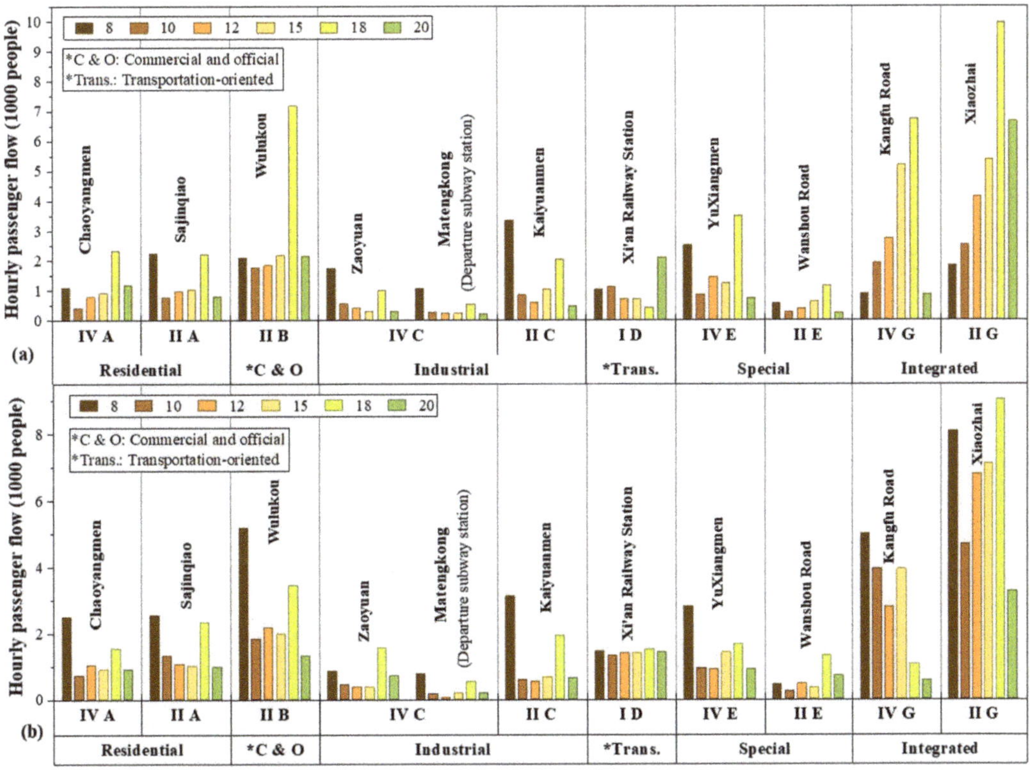

Figure 10. Distribution of flows in each different period for the 11 stations: (**a**) inbound and (**b**) outbound.

(1) Residential: There was a bimodal phenomenon for both inbound and outbound flows, but the intensity was different. IV A had a morning peak for outbound flows and an evening peak for inbound flows. This was not a typical characteristic of residential stations. This may have been because the employed population did not live on the residential land in the area of station IV A, thus resulting in the phenomenon of an imbalance between occupations and residences. II A had similar intensities for the inbound and outbound peaks in the morning and the evening, which also reflected a certain degree of imbalance between jobs and housing.

(2) Commercial and official: The inbound flow only had an evening peak, and the outbound flow's morning peak was stronger than the evening peak, which is a typical characteristic of employment-oriented access. This indicated that there were more employment flows out of the station during the morning peak, and attention should

be paid to controlling the outbound flows during the morning peak and the inbound flows during the evening peak.

(3) Industrial: Although the three stations had significant differences in passenger flow, they all had significant morning and evening peaks. For IV C, the terminal station Matengkong had the lowest passenger flow, and the intensity of the morning and evening peaks was close to that of IV C. II C had the highest passenger flow, and the inbound and outbound peaks were significantly stronger than the evening peaks.

(4) Transportation-oriented: The outbound flow was significantly larger than the inbound flow throughout the day, and the outbound flow was close to the same at all times. There was a night peak for inbound flows at 20:00, which may have been due to the external traffic brought by the railroad being more concentrated in the evening; this part of the passenger flow through the station to transfer to the city's internal traffic was larger. Therefore, for hub-type stations, the transfer passenger flow from external traffic to internal traffic should be controlled at 20:00, which is the peak hour in the evening.

(5) Special type: For inbound flows, the evening peak was more significant than the morning peak. For outbound flows, the morning peak was more significant than the evening peak. This was a significant characteristic of employment-oriented traffic. This was caused by the greater number of land uses attracting employment in this type of area.

(6) Integrated: The outbound flows had a double morning and evening peak, while the inbound flows had only an evening peak, and the evening peak occurred earlier. This was characterized by employment-oriented passenger flows. In comparison with the special and commercial and official types, this type had no morning peaks in the inbound flows, and the double-peak phenomenon occurred in the outbound flows. This indicated that this type was characterized by a high proportion of commercial and other public service functions, with the morning peak attracting employment and the evening peak attracting consumers.

3.1.2. Passenger Flow Characteristics in the Exit Dimension

The passenger flow characteristics of the exits were analyzed from three perspectives: total ridership at exits in six periods, total ridership at exits in each different period, and inbound/outbound flow at exits in each different period.

- Total Ridership at Different Exits in Six Periods

Figure 11 illustrates the total ridership at the different exits in six periods. Except for type I D, there was a significant difference in the passenger flows of the different exits of the other nine types. II A and II G both had one exit with significantly higher passenger flows than the other exits of the station, which was due to the fact that exit B of Sajinqiao was directly connected to a pedestrian commercial area and exit D of Xiaozhai was directly connected to the largest commercial complex in Xi'an; thus, the phenomenon of significantly higher passenger flows occurred at these two exits.

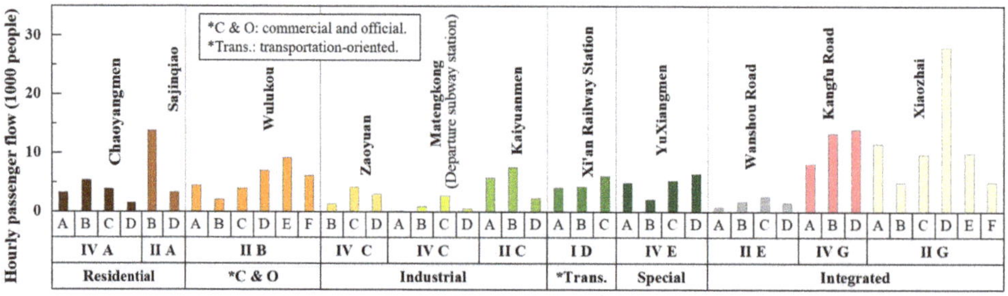

Figure 11. Total ridership at different exits in six periods.

- Total Ridership at Exits in Each Different Period

Figure 12 shows that the spatiotemporal characteristics of the passenger flows at different exits of different types of stations showed significant differences.

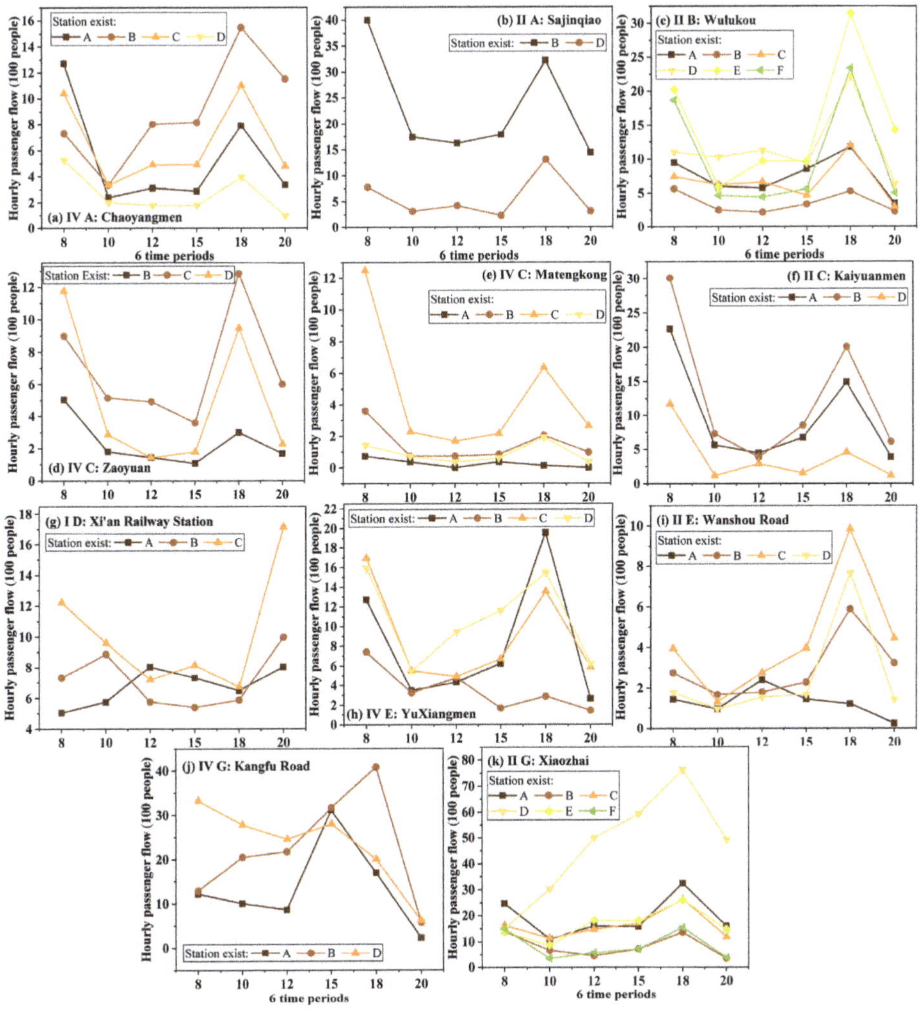

Figure 12. Total ridership at different exits in each different period for the 11 stations.

(1) Residential: Although there were significant differences in the passenger flows at each exit of the two types, the spatial and temporal characteristics were close to each other, and there were significant morning and evening peaks at each exit.

(2) Commercial and official: There were significant differences in the total passenger flows and the spatiotemporal characteristics of the exits. There was significant bimodality in the morning and evening at only two exits, non-significant bimodal characteristics at three exits, and significant evening peaks at only one exit. This was due to the different land-use functions of the parcels of land connected to the exits. In addition, although Figure 9 shows that the evening peak was stronger than the morning peak in the station dimension, there were still certain exits (B) where the morning peak was stronger than the evening peak.

(3) Industrial: There were significant differences in the total passenger flows at each exit, but the spatiotemporal characteristics were close to each other, with significant bimodal phenomena at all exists. This was similar to the findings for the residential type, but the passenger flows in off-peak hours were significantly lower than those of the residential type. Notably, the departure station Matengkong had uneven passenger flows at all exits, with one exit having significantly high passenger flows and a great intensity of morning and evening peaks. This can be managed operationally by guiding passenger flows to relieve congested exits. The morning and evening peaks at IV C (Zaoyuan) were close, and the intensity of the morning peak was stronger than that of the evening peak at all exits in II C (Kaiyuanmen).

(4) Transportation-oriented: Although the station ridership was close in each period (Figure 9), there was a significant difference in the spatiotemporal characteristics of the morning and evening peaks at each exit.

(5) Special: The overall spatiotemporal characteristics of the two types of passenger flows were close to each other, and there were morning and evening peaks at each exit, but the evening peak of the II E type was stronger than the morning peak, and the intensities of the morning and evening peaks of the IV E type were close to each other. Both types had individual exits with significantly lower passenger flows and insignificant bimodal phenomena.

(6) Integrated: The passenger flow characteristics of the IV G and II G types were close in the station dimension, but there were significant differences in the exit dimension. There were significant differences in the spatiotemporal characteristics of all the exits of the IV G type, and the morning and evening peak characteristics were different. The passenger flows and spatiotemporal characteristics of the II G type were close to each other at all exits except for exit D, which was directly connected to a commercial complex in Xiaozhai. This suggests that for integrated stations, the focus should be on controlling exits that connect to large commercial facilities.

- Inbound/Outbound Flows at Exits in Each Different Period

Figure 13 demonstrates the spatiotemporal characteristics of the inbound and outbound flows for different types of stations in the exit dimension in the six time periods.

(1) Residential: There were significant differences between the two types. The inbound flows in the IV A type were more significant at the evening peak at all exits, which was consistent with the station dimension shown in Figure 10. However, there was uncertainty in the outbound flow, and the evening peak was stronger than the morning peak at exit B for the outbound flow. The remaining three exits were consistent with the station dimension shown in Figure 10, with the morning peak being stronger than the evening peak. For II A, there were significant differences in the two exits, with the morning peak being stronger than the evening peak at exit B and the opposite at exit D, which was different from the station dimension shown in Figure 10.

(2) Commercial and official: All exits had stronger evening peaks than morning peaks for inbound flows, and there were complexities and uncertainties in the outbound flow. Exits A, B, and C had lower traffic and stronger morning peaks than evening peaks. Exit D had close outbound traffic in all time periods. Exit E had close morning and evening peaks, and F had only morning peaks. This indicated that the building functions connected to different exits had a significant impact on passenger flows, and management needs to pay attention to the key exits.

(3) Industrial: Three stations were close to each other in terms of the morning and evening peaks at all exits. II C was the most stable, with the morning peaks being stronger than the evening peaks at all exits, and there was uncertainty in the morning and evening peaks at all exits in IV C.

(4) Transportation-oriented: Compared with the other types, this type had the most significant difference in passenger flow characteristics among the exits. Only one exit had an inbound evening peak at 20:00. The inbound and outbound flows of each

exit were close to each other, and management needs to implement control at 20:00 in the evening.

(5) Special: The inbound peaks of IV E were uncertain, but the outbound peaks were stronger than the evening peaks. The outbound peaks of IIE had the same spatiotemporal characteristics, and there was only a significant evening peak. There was uncertainty in the morning and evening peak characteristics at all inbound exits.

(6) Integrated: The spatiotemporal characteristics of the inbound flows of the two types were close, and only significant evening peaks were manifested for inbound flows. However, for the outbound flows, there were significant morning peaks at two exits and only evening peaks at one exit at IV G. There was a significant evening peak at the exit that directly connected to a commercial complex at II G, and the remaining five exits exhibited stronger morning than evening peaks, which was the opposite of what is shown in Figure 12. This suggested that the multiple connections of integrated stations to commercial complexes have a significant impact on the overall passenger flow at these stations.

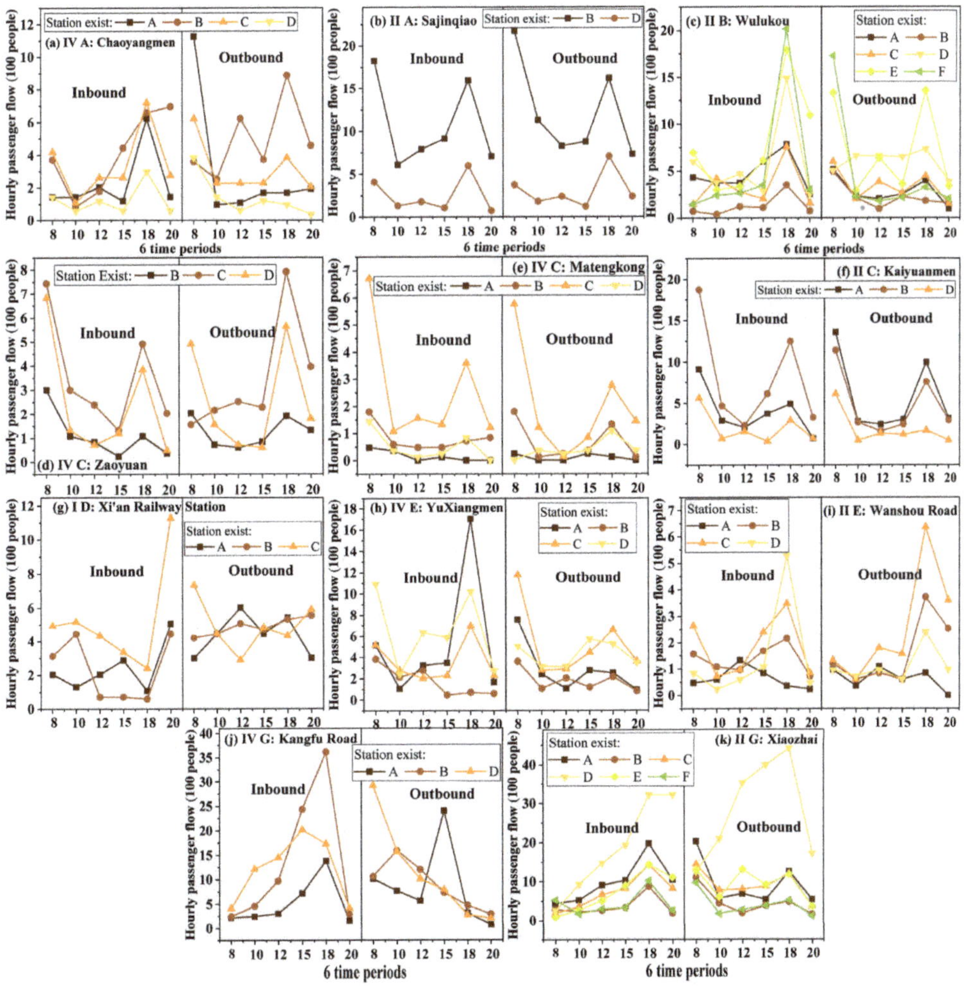

Figure 13. Inbound/outbound flows at different exits in each different period.

3.2. *Analysis of the Correlation between Passenger Flows and the Characteristics of the Built Environment in CAs*

To further analyze the relationship between different urban functions and intensities within the CAs and passenger flows, bivariate correlation analyses were conducted between the three operational passenger flow indices and seven types of indicators of the built environment for 30 stations. Figure 14 shows the details of the indices and the results.

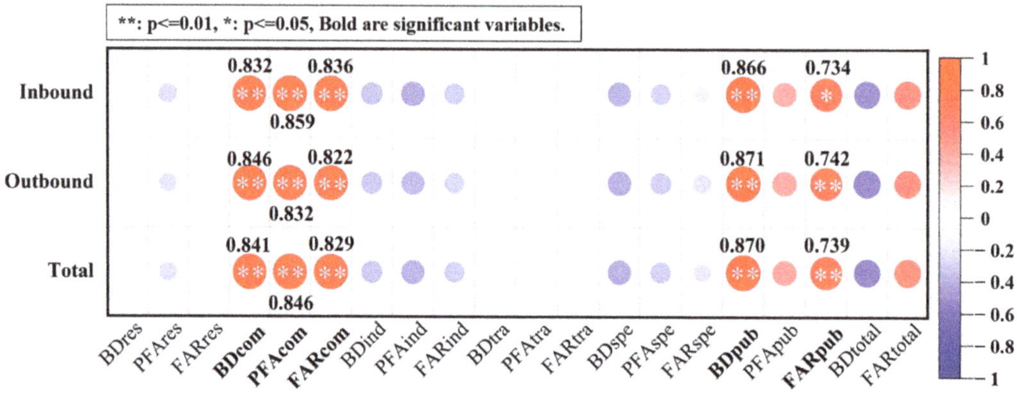

Figure 14. Analysis of the correlation between the built environment and passenger flows at the 11 stations.

The correlation of the commercial type was the most significant, and the three built environment indicators were significantly and positively correlated with the three passenger traffic indicators at the level of 0.01. The correlations of building density as a station classification indicator with inbound, outbound, and total passenger traffic were 0.832, 0.846, and 0.841, respectively. Next was the public service function; the building density was significantly and positively correlated with the inbound, outbound, and total flows at the level of 0.01, with coefficients of 0.866, 0.871, and 0.879, respectively. The floor area ratio was significantly and positively correlated with the inbound flows at the level of 0.05 and with the outbound and total flows at the level of 0.01. This indicated that commercial functions and public service facilities are the most significant variables affecting passenger intensity. The two indices BD_{total} and FAR_{total} did not manifest significant relations. This indicated that unilaterally considering the development intensity within a CA will not have a significant impact on passenger traffic.

In addition, although the industrial and special types did not manifest significant correlations, the three built-up indices were all negatively correlated with passenger traffic. This indicated that commercial functions and public services are the core functions that attract passenger traffic, while the industrial and special types are significant functions that decrease passenger traffic.

4. Discussion

In this study, based on the guidelines issued by the Chinese government, the station traffic level and different building functions within the CAs that were 800 m around the stations were extracted as classification factors. A new classification system was developed, and it is highly applicable to all cities in China. In addition, field measurements were performed to analyze the passenger flow patterns at different types of stations, demonstrating that the different functional characteristics of a station's close vicinity result in different station passenger flow characteristics.

The correlation between station ridership and the different functions of the station area in terms of land and building capacity shows that commercial and public service functions are positively correlated with ridership, while residential functions do not reflect significance. The results from Shanghai also showed that for all-day ridership, residential land use was the least significant function, while commercial and entertainment were the most significant variables [22]. In addition, similar studies mostly used station operation data, which do reflect what occurs at the exits. In this study, field measurements of the passenger flow at each station exit were conducted, and the variance of the measured data was verified using the operational data; the R^2 was as high as 0.96, which indicated the feasibility and accuracy of the monitoring method and provides a reference for similar studies.

This study provides guidance on the management of traffic demand at stations and the optimization of the built environment of CAs. (1) Guidance for station traffic demand and operation: for different types of stations, their traffic characteristics show different features due to the dominant surrounding building functions and their development intensity. Therefore, different operational strategies are adopted for different types of stations. Residential, industrial, and special stations should pay attention to controlling the morning and evening peaks for entering and leaving the station. Commercial and office stations should focus on controlling the morning peak for leaving the station and the evening peak for entering the station. Transportation-oriented stations should focus on controlling the inbound and outbound peak at 20:00. Integrated stations should control the inbound evening peak, outbound morning peak, and evening double peak. In addition, within the exit dimension, except for the transportation-oriented types, there were significant differences in passenger traffic through exits. Guidance is needed to relieve congested exits during the morning and evening peak periods and to try to equalize the passenger traffic flows at each exit. (2) Guidance for optimizing the built environment around the station area: most of the 11 typical stations for this study reflect passenger flow characteristics that are in accordance with the dominant building functions in the surrounding area, but neither of the two residential stations reflect the typical residential-oriented traffic characteristics. Similar studies have shown that this is due to the imbalance between station occupancy and residence [21], i.e., the residential and commercial office functions around the station are not coordinated. Therefore, for such stations, the station occupancy/residence balance can be optimized by adjusting the proportion of residential construction within the CA.

This study also has some limitations. First, only 10 types of stations on Line 1 and Line 2 were selected for this study, and for other types of stations, samples were not discussed. In addition, for the hub type, weekdays may not best reflect the stations' passenger flow, and holidays may better characterize the changes in flow. Furthermore, this study only analyzes the correlation between different building functions and ridership, and the impact of the built environment on station ridership needs to be further quantified, which will be the content of our future work as well.

5. Conclusions

This study took Line 1 and Line 2 within the third ring road of Xi'an as the research objects and constructed a station classification system by taking the station traffic level and the functions of different buildings within CAs as the classification factors. At the same time, the passenger flow characteristics of 10 typical types of stations in the station dimension and the exit dimension were summarized. Finally, the correlations between indicators of different functions of the built environment within CAs and passenger traffic were quantified. The results show that, with the influence of the dominant function within the 800 m CA of a station, there is a significant difference in the passenger flow characteristics of different types of stations. The ridership is significantly higher in the integrated type than in the other types of stations and lowest in the industrial and special types. In addition, the traffic characteristics of each type of station are significantly different at different times of the day and at different station exits. Therefore, the traffic demand management of stations should be based on the dominant functions around different types of stations, and the time periods and exits with higher ridership should be guided and controlled to realize the equal distribution of ridership at each station exit.

The results of the correlation analysis show that the commercial and public service functions in the station area are significantly positively correlated with the total station ridership for the six periods at the 0.01 level. For commercial functions, the highest correlation coefficient between PFA_{com} and total ridership is 0.841, and for public service functions, the highest correlation coefficient between BD_{pub} and total ridership is 0.870. This study provides a reference for the future operation and management of ridership at different types of stations and helps to promote the coordinated development of transportation demand and the built environment around the stations.

Author Contributions: Conceptualization, Y.W. and Y.L.; project administration, Y.L.; resources, Y.L.; methodology, L.C. and Y.W.; formal analysis, L.C. and Y.C.; investigation, L.C. and Y.C.; writing—review and editing, Y.C. and Y.W.; visualization, Y.C.; supervision, Y.W. and Y.L.; funding acquisition, Y.W. All authors have read and agreed to the published version of the manuscript.

Funding: This research was funded by China Railway First Survey and Design Institute Group Co., Ltd. (grant number 2022RJ22-01).

Data Availability Statement: The original contributions presented in the study are included in the article, further inquiries can be directed to the corresponding author.

Conflicts of Interest: Authors Li Chen and Ying Li were employed by the company China Railway First Survey and Design Institute Group Co., Ltd. The remaining authors declare that the research was conducted in the absence of any commercial or financial relationships that could be construed as a potential conflict of interest.

References

1. Sun, L.; Chen, J.; Li, Q.; Huang, D. Dramatic uneven urbanization of large cities throughout the world in recent decades. *Nat. Commun.* **2020**, *11*, 5366. [CrossRef] [PubMed]
2. Pang, L.; Jiang, Y.; Wang, J.; Qiu, N.; Xu, X.; Ren, L.; Han, X. Research of Metro Stations with Varying Patterns of Ridership and Their Relationship with Built Environment, on the Example of Tianjin, China. *Sustainability* **2023**, *15*, 9533. [CrossRef]
3. Han, B.M.; Xi, Z.; Sun, Y.J.; Lu, F.; Niu, C.Y.; Wang, C.X.; Xu, K.L.; Yao, Y.F. Statistical analysis of urban rail transit operation in the world in 2022: A review. *Urban Rapid Rail Transit* **2023**, *36*, 5–11.
4. Su, S.L.; Zhao, C.; Zhou, H.; Li, B.Z.; Kang, M.J. Unraveling the relative contribution of TOD structural factors to metro ridership: A novel localized modeling approach with implications on spatial planning. *J. Transp. Geogr.* **2022**, *100*, 103308. [CrossRef]
5. Hu, J.; Chen, J.; Li, P.; Yan, J.; Wang, H. Systematic Review of Socially Sustainable and Community Regeneration: Research Traits, Focal Points, and Future Trajectories. *Buildings* **2024**, *14*, 881. [CrossRef]
6. Lyu, G.; Bertolini, L.; Pfeffer, K. Developing a TOD typology for Beijing metro station areas. *J. Transp. Geogr.* **2016**, *55*, 40–50. [CrossRef]
7. Yu, Z.; Zhu, X.; Liu, X. Characterizing metro stations via urban function: Thematic evidence from transit-oriented development (TOD) in Hong Kong. *J. Transp. Geogr.* **2022**, *99*, 103299. [CrossRef]
8. Sung, H.; Oh, J.T. Transit-oriented development in a high-density city: Identifying its association with transit ridership in Seoul, Korea. *Cities* **2011**, *28*, 70–82. [CrossRef]
9. Jiang, Y.S.; Yu, G.S.; Hu, L.; Li, Y. Refined classification of urban rail transit stations based on clustered station's passenger traffic flow features. *J. Transp. Syst. Eng. Inf. Technol.* **2022**, *22*, 106–112. [CrossRef]
10. Lu, Y.; Wen, X.R.; Liu, M.; Yu, H. A Review on the functional orientation of rail station regional planning based on typology. *Urban Dev. Stud.* **2023**, *30*, 28–31.
11. Chen, J.; Pellegrini, P.; Yang, Z.; Wang, H. Strategies for Sustainable Urban Renewal: Community-Scale GIS-Based Analysis for Densification Decision Making. *Sustainability* **2023**, *15*, 7901. [CrossRef]
12. Singh, Y.J.; Fard, P.; Zuidgeest, M.; Brussel, M.; Maarseveen, M.v. Measuring transit oriented development: A spatial multi criteria assessment approach for the City Region Arnhem and Nijmegen. *J. Transp. Geogr.* **2014**, *35*, 130–143. [CrossRef]
13. Bertolini, L. Nodes and places: Complexities of railway station redevelopment. *Eur. Plan. Stud.* **1996**, *4*, 331–345. [CrossRef]
14. Yin, Q.; Meng, B.; Zhang, L.Y.; Bin, M.; Liying, Z. Classification of subway stations in Beijing based on passenger flow characteristics. *Prog. Geogr.* **2016**, *35*, 126–134. [CrossRef]
15. Zhang, J.W.; Li, K.L.; Guo, Z.C.; Wang, C.B. Grading research of subway stations based on traffic data analysis. *J. Donghua Univ. (Nat. Sci.)* **2018**, *44*, 652–657. [CrossRef]
16. Dong, Y.Z. Grade Evaluation of Urban Rail Transit Stations and Method for Grading of Regional Stations. *Urban Rapid Rail Transit* **2023**, *36*, 146–151.
17. Zhang, L.; Hou, P.; Qiang, D. Transit-Oriented Development in New Towns: Identifying Its Association with Urban Function in Shanghai, China. *Buildings* **2022**, *12*, 1394. [CrossRef]

18. Vale, D.S.; Viana, C.M.; Pereira, M. The extended node-place model at the local scale: Evaluating the integration of land use and transport for Lisbon's subway network. *J. Transp. Geogr.* **2018**, *69*, 282–293. [CrossRef]
19. Qiang, D.; Zhang, L.; Huang, X. Quantitative Evaluation of TOD Performance Based on Multi-Source Data: A Case Study of Shanghai. *Front. Public Health* **2022**, *10*, 820694. [CrossRef]
20. Yang, L.; Song, X.Y. TOD Typology Based on Urban Renewal: A Classification of Metro Stations for Ningbo City. *Urban Rail Transit* **2021**, *7*, 240–255. [CrossRef]
21. Feng, H.; Chen, Y.; Wu, J.; Zhao, Z.; Wang, Y.; Wang, Z. Urban Rail Transit Station Type Identification Based on "Passenger Flow—Land Use—Job-Housing". *Sustainability* **2023**, *15*, 15103. [CrossRef]
22. An, D.; Tong, X.; Liu, K.; Chan, E.H.W. Understanding the impact of built environment on metro ridership using open source in Shanghai. *Cities* **2019**, *93*, 177–187. [CrossRef]
23. Preis, T.; Zhong, C.; Batty, M.; Manley, E.; Wang, J.; Wang, Z.; Chen, F.; Schmitt, G. Variability in Regularity: Mining Temporal Mobility Patterns in London, Singapore and Beijing Using Smart-Card Data. *PLoS ONE* **2016**, *11*, e0149222. [CrossRef]
24. Ren, L.J.; Yun, Y.X.; Quan, H.Y. Study on Classification and Characteristics of Urban Rail Transit Station Based on Node-Place Model: Empirical Analysis and Experience Enlightenment of Singapore. *Urban Plan. Int.* **2016**, *31*, 109–116.
25. Ministry of Housing and Urban-Rural Development of the People's Republic of China. *Guidelines for Planning and Design of Areas along Urban Railways*; Ministry of Housing and Urban-Rural Development of the People's Republic of China: Beijing, China, 2015; p. 7.
26. Urban Planning Society of China. *Guidelines for Space Planning and Design of Facilities around Urban Rail Transit Station*; Urban Planning Society of China: Beijing, China, 2021; p. 16.
27. Xi'an 2022 Statistical Yearbook. Available online: http://tjj.xa.gov.cn/tjnj/2022/zk/indexch.htm (accessed on 24 January 2024).
28. Xi'an Metro Official Homepage. Available online: https://www.xianrail.com/#/index (accessed on 24 January 2024).
29. Xi'an Metro Statistics. Available online: https://metrodb.org/index/xian.html (accessed on 24 January 2024).
30. Zhang, X.; Sun, Y.; Luan, X. Regulation flexibility and legitimacy building in governing intercity railways: The polymorphous role of the Chinese provincial government. *J. Urban Aff.* **2023**, 1–16. [CrossRef]
31. Xi'an Bureau of Statistics. *Bulletin of Main Data of the Seventh National Population Census of Xi'an City*; Xi'an Bureau of Statistics: Xi'an, China, 2021; p. 1.
32. Xi'an City Planning Design and Research Institute. *Annual Report on Urban Transportation Development in Xi'an, 2022*; Xi'an Municipal Bureau of Natural Resources and Planning: Xi'an, China, 2023; pp. 11–12.

buildings

MDPI

Article

Spatial Gradient Differences in the Cooling Island Effect and Influencing Factors of Urban Park Green Spaces in Beijing

Chao Xu [1], Wenjing Wang [2] and He Zhu [1,*]

[1] Beijing Advanced Innovation Center for Future Urban Design, School of Architecture and Urban Planning, Beijing University of Civil Engineering and Architecture, Beijing 100044, China; xuchao@bucea.edu.cn
[2] China Urban Construction Design & Research Institute Co., Ltd., Beijing 100120, China; wangwenjingyuanlin@cucd.cn
* Correspondence: zhuhe@bucea.edu.cn; Tel.: +1-86-0021-1231

Abstract: Urban park green spaces exhibit significant cool island effects, which can effectively mitigate the urban heat environment. Clarifying the characteristics and differences in the cool island effects of urban parks across different spatial gradients within cities is instrumental in identifying potential issues and optimizing the structure and resource allocation of park green spaces in a scientifically rational manner. This study focuses on parks within the central urban area of Beijing, utilizing remote sensing image interpretation and land surface temperature (LST) inversion to extract relevant characteristics of park green spaces and the park cool island intensity (PCI) index. Various mathematical and statistical methods including correlation analysis, regression analysis, and cluster analysis are employed to conduct comparative studies across three gradients: within the 3rd ring road, between the 3rd and 5th ring roads, and outside the 5th ring road. The analysis reveals that both park green space characteristics and urban heat island effects exhibit spatial gradient differences, collectively influencing the cool island effects of urban park green spaces. PCI gradually decreases across the three spatial gradients. Cluster analysis identifies four distinct types of parks with different cool island effect characteristics, highlighting the need for optimization and improvement in over half of the parks. Various indicators of park green space characteristics show different correlations with PCI, with variations in correlation strength and thresholds across gradients. The fitting effects of regression equations for each characteristic indicator and PCI gradually worsen from within the 3rd ring road to outside the 5th ring road, with different factors playing important roles across gradients. This study enhances our understanding of the cool island effects of urban park green spaces and facilitates the proposition of differentiated optimization management strategies for urban park green space planning and system construction in different regions.

Keywords: landscape architecture; urban park; cold island effect; influencing factors; gradient differences

1. Introduction

Urbanization has altered surface structures [1], leading to increased absorption of solar radiation by urban surfaces. Simultaneously, it has brought about significant human activities such as high-density building developments, the expansion of transportation systems, and large-scale land use changes, which contribute to the emission of anthropogenic heat sources. These factors collectively result in the urban heat island (UHI) effect, characterized by a noticeable increase in urban temperatures [2]. Climate change is one of the major challenges facing the world today, and the UHI, as a significant phenomenon in the process of urbanization, is closely related to it [3]. With the intensification of global climate change, the impact of the UHI effect on the urban thermal environment and ecosystem has become more complex and serious [4]. This interaction not only has a serious negative impact on the living environment, ecosystem, and residents' health [5,6], but also causes a series of problems such as energy waste and environmental pollution. Consequently, it has become a crucial factor constraining the sustainable development of cities [7].

Urban park green space has an obvious cold island effect, which can effectively optimize the urban thermal environment [8]. It is a cost-effective and low-carbon thermal mitigation method in cities with scarce land [9]. Previous research has indicated that green vegetation in urban parks converts a large amount of heat into water vapor through transpiration, releasing it into the atmosphere. Additionally, vegetation can shield and reflect a significant portion of solar radiation, thereby maintaining lower surface temperatures [10]. Natural or artificial water bodies such as lakes, rivers, or ponds can absorb and release considerable heat, exerting strong temperature regulation effects [11]. The microclimate environment formed by green spaces can guide heat exchange through convective effects, consequently reducing surrounding surface temperatures [12]. Factors such as the scale, morphology, landscape composition, and spatial arrangement of urban parks directly influence the formation and intensity of their cold island effects [13–16]. They are also influenced by various factors such as the surrounding built environment and land use patterns [17,18]. Clarifying the underlying mechanisms of the cold island effect in urban parks not only helps enhance their cooling effects but also provides a scientific basis for urban park planning and construction [19]. At present, many studies take several parks in the urban area as samples to analyze the correlation between their characteristics and the intensity of cold islands and then reveal the internal objective laws, which can provide reference for park site selection [20], land cover optimization [21], and vegetation composition renewal [22].

The pattern of UHI effect across the entire urban area exhibits certain spatial gradient differences [23,24], primarily due to the spatial heterogeneity of urban spatial structure, including land use types, land cover, and construction intensity [25]. Existing studies have confirmed these differences. For example, the central heat island intensity of dense urban areas in cities is the strongest [26], the new urban areas in metropolitan areas are more vulnerable to the heat island effect [27], and highly continuous dense and sprawling urban expansion will enhance the heat island effect [28]. As an integral component of urban green space systems, urban parks are directly influenced by factors such as the positioning of urban functional zones [29], socioeconomic factors [30], and differences in heat island patterns [31], which affect their intrinsic characteristics and functions. Additionally, the cold island effect of urban vegetation varies across different gradients [32]. For example, urban parks close to the city center have higher cold island intensity [33], and the different surrounding land coverage will also affect the cold island intensity of the park [34]. These multi-dimensional differences are the objective facts that can not be ignored to promote urban parks to alleviate the UHI effect, and the planning and construction of urban parks should be fully combined with these differences for more targeted and comprehensive consideration. However, at present, the multi-sample comparative study between different gradient regions needs to be improved, and the relevant potential laws are not clear, which makes it difficult to provide differentiated references for the stock renewal and fine management of urban green spaces in the reduction and quality improvement stage.

This study focuses on parks within the central urban area of Beijing, utilizing remote sensing image interpretation and surface temperature inversion techniques. The park cool island intensity (PCI) is selected as the indicator of the cold island effect, and comparative analyses are conducted across three different spatial gradients using mathematical and statistical methods. The aim is to identify the distinct characteristics of urban park green space cold island effects within each gradient, reveal inherent differences, and clarify relevant influencing factors. Such insights are expected to facilitate the proposal of differentiated optimization and management strategies for urban park green space planning and system construction in various regions.

2. Materials and Methods

2.1. Study Area

According to the Beijing City Master Plan (2016–2035), the central urban area comprises six districts: Haidian, Chaoyang, Xicheng, Dongcheng, Shijingshan, and Fengtai, totaling approximately 1378 km^2. Due to factors such as high urbanization and population density, the central urban area has long been confronted with severe urban heat island issues [35]. Phenomena such as frequent hot weather, increasing heatwave occurrences, and persistently high nighttime temperatures negatively affect the energy consumption and the quality of life and health conditions of urban residents [36]. The construction of park green spaces serves as an effective means to alleviate the heat island effect in the central urban area. Presently, influenced by various factors such as urban development patterns, land use policies, and planning management, the construction status of parks across different spatial gradients within the region exhibits diversity. In the core old urban areas, park construction faces significant spatial and economic pressures due to high urban density, land scarcity, and relatively high costs. In the central areas adjoining the old urban areas and the suburbs, more available land with relatively lower density allows for the completion of park construction of a certain scale, in alignment with urban development needs. In the outermost suburban areas, although there is ample space for construction, the actual demand for use and the driving force for construction are relatively low. The current conditions of parks within the study area provide a solid foundation for researching the differences in their cold island effects (Figure 1).

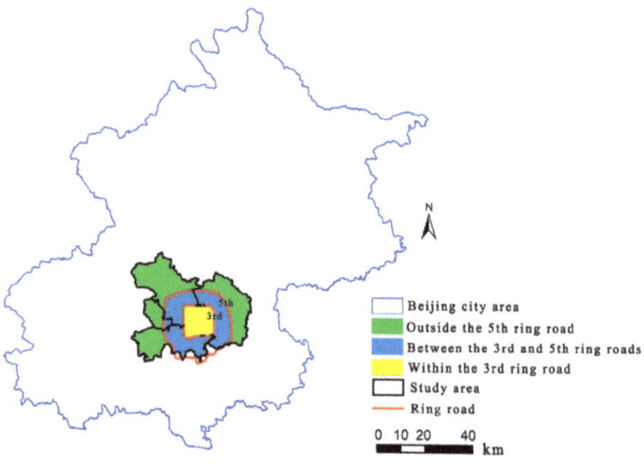

Figure 1. The location of the study area.

In this study, the study area is divided into three spatial gradient regions: within the 3rd ring road (3), between the 3rd and 5th ring roads (3–5), and outside the 5th ring road (5). Based on the list of parks published by the Beijing Municipal Forestry and Parks Bureau, a total of 493 parks are selected as research objects, including 101 parks within the 3rd ring road, 276 parks between the 3rd and 5th ring roads, and 116 parks outside the 5th ring road.

2.2. Data Sources

The main datasets used in this study include Landsat satellite imagery, remote sensing imagery, land use data, park boundary data, land surface temperature (LST) data, Normalized Difference Vegetation Index (NDVI), and ring road data. Landsat 8 OLI_TIRS satellite imagery was downloaded from the United States Geological Survey (USGS) Earth Explorer website for surface temperature inversion. The selected satellite imagery exhibits good quality with minimal cloud cover over the study area. Radiometric calibration and atmospheric correction were conducted to mitigate the effects of sensor and atmospheric factors. Land use data with a spatial resolution of 4 m were derived from 2021 data of the High-Resolution Satellite GaoFen-2 (GF-2). The data were obtained mainly through radiometric calibration, atmospheric correction, image mosaic, geometric correction, and region cropping. Then, it was classified into forestland, grassland, water bodies, farmland, buildings, artificial surfaces, roads, and bare land, and accuracy validation was performed using the support vector machine method with an accuracy of 92.3%. The categories of artificial surfaces and roads were merged as impervious surfaces. Park boundary data were obtained from both the park AOI (Area of Interest) in Baidu Map and data publicly released by the Beijing Municipal Forestry and Parks Bureau. These datasets collectively provide comprehensive information for the analysis and evaluation of park green spaces and their cold island effects within the study area (Table 1).

Table 1. Research data and sources.

Research Data	Types	Sources
Landsat 8 OLI_TIRS satellite imagery	Raster data (30 m)	USGS http://earthexplorer.usgs.gov/ (accessed on 14 June 2021)
Remote sensing imagery	Raster data	GF-2
Land use data	Raster data (4 m)	Derived from GF-2 satellite imagery
Park boundary data	Shapefile	AOI in Baidu Map, Beijing Municipal Forestry and Parks Bureau
LST data	Raster data (30 m)	Derived from Landsat satellite imagery
NDVI	Raster data (30 m)	Derived from Landsat satellite imagery
Ring road data	Shapefile	Open Street Map (OSM)

2.3. Calculation of Park Cool Island Intensity

The Split-Window algorithm [37,38] was employed to compute the LST data for the study area. This method relies on the differential atmospheric absorption between two adjacent thermal infrared channels. By geometrically manipulating the brightness temperatures of these two channels, the atmospheric effects are mitigated, thus allowing for the correction of atmospheric influences. Subsequently, aided by the surface emissivity, the remote sensing inversion of LST is achieved. The Split-Window algorithm is widely recognized as one of the most extensively used and accurate methods for thermal infrared remote sensing inversion of LST.

PCI was calculated by determining the average difference in LST between the park area and the surrounding buffer zone, with

$$\Delta T = LST_B - LST_P \quad (1)$$

where ΔT represents the PCI, measured in degrees Celsius (°C); LST_B represents the average LST within the buffer zone, measured in degrees Celsius (°C); and LST_P represents the average LST within the park area, measured in degrees Celsius (°C).

The equidistant radius method [39] was employed to calculate the buffer zone radius. This radius is defined as the radius of a circle with an area equal to that of the park area.

$$R_B = \sqrt{(A/\pi)} \quad (2)$$

where R_B represents the radius of the park buffer zone, measured in meters (m); and A represents the park area, measured in square meters (m^2).

2.4. Calculation of Relevant Characteristic Indicators of Park Green Spaces

In order to fully explore the potential influencing factors of the cold island effect of park green spaces, based on the effectiveness and feasibility of the research, as well as the value and feasibility of the application of the research results, we selected characteristic indicators from both the internal and external aspects of the parks. Internally, indicators such as area, perimeter, and perimeter–area ratio were chosen to represent the size and shape characteristics of the parks [40]. Additionally, the NDVI was utilized to assess the vegetation condition within the parks [41]. Furthermore, forest coverage, grassland coverage, water coverage, and impervious surface area coverage were selected to represent the basic land use composition within the parks, and most of the existing studies have been carried out around these aspects, and have achieved effective results [42,43]. As the most important vegetation type in the park, forestland is also an element that is easy to optimize the management inside the park [39]. Combined with relevant research [44] and considering the implementation of subsequent green space optimization regulation, the average patch area of forestland is selected to represent its spatial configuration characteristics. Externally, indicators including forest coverage, grassland coverage, water coverage, impervious surface area coverage, and building area coverage within the buffer zone were selected to represent the land use composition surrounding the parks. Previous studies have shown that these land characteristics around the park have a significant impact on its cold island effect [45,46]. The detailed calculation methods for each indicator are presented in Table 2.

Table 2. Relevant characteristic indicators of park green spaces.

Indicator (Abbreviation)	Description	Unit	Calculation Method
Area (A)	Area of a park	ha	Geometric statistics calculation in ArcGIS 10.6
Perimeter (P)	Total perimeter of a park	m	Geometric statistics calculation in ArcGIS 10.6
Perimeter–area ratio (PA)	Overall compactness of a park	Dimensionless	$\frac{P}{A}$
NDVI	Vegetation condition within a park	Dimensionless	$\frac{V_{NIR}-V_R}{V_{NIR}+V_R}$ V_{NIR} and V_R represent the pixel values of the near-infrared band and the red band, respectively
Forest coverage (FC)	Percentage of forest cover in a park	%	$\frac{\sum_{i=1}^{n} F_i}{A} \times 100$
Grassland coverage (GC)	Percentage of grassland cover in a park	%	$\frac{\sum_{i=1}^{n} G_i}{A} \times 100$
Water coverage (WC)	Percentage of water cover in a park	%	$\frac{\sum_{i=1}^{n} W_i}{A} \times 100$
Impervious surface area coverage (ISC)	Percentage of impervious surface area in a park	%	$\frac{\sum_{i=1}^{n} IS_i}{A} \times 100$
Average patch size of forest (F_MPS)	Average size of forest patches in a park	ha	Calculation using Fragstats 4.2
Forest coverage in a buffer (B_FC)	Percentage of forest cover in a buffer	%	$\frac{\sum_{i=1}^{n} B\_F_i}{B} \times 100$
Grassland coverage in a buffer (B_GC)	Percentage of grassland cover in a buffer	%	$\frac{\sum_{i=1}^{n} B\_G_i}{B} \times 100$
Water coverage in a buffer (B_WC)	Percentage of water cover in a buffer	%	$\frac{\sum_{i=1}^{n} B\_W_i}{B} \times 100$
Impervious surface area coverage in a buffer (B_ISC)	Percentage of impervious surface area in a buffer	%	$\frac{\sum_{i=1}^{n} B\_IS_i}{B} \times 100$
Building area coverage in a buffer (B_BC)	Percentage of building area in a buffer	%	$\frac{\sum_{i=1}^{n} B\_B_i}{B} \times 100$

2.5. Statistical Analysis

In ArcGIS 10.6, data pertaining to park green space indicators and LST inversion were collected and analyzed. This facilitated an examination of the characteristics of park green spaces and urban heat island effects across different spatial gradients. Based on the PCI results of each park, the distribution characteristics of cool island effects in different spatial gradients were identified. After normalizing the PCI data, K-means cluster analysis was employed to classify parks in different gradients. Pearson correlation analysis was utilized to clarify the relationship between PCI and internal and external characteristic indicators of parks across different spatial gradients. Additionally, stepwise regression analysis was employed to quantify the influence of various characteristic indicators on PCI.

3. Results

3.1. Characteristics of Park Green Spaces across Different Spatial Gradients

The park area in the study area is mainly concentrated within the range of 1 ha to 10 ha, accounting for 43.99% of the total, while those less than 1 ha account for 22.20%, those between 10 ha and 25 ha account for 15.89%, and those larger than 25 ha account for 17.92%. There are also certain differences in distribution across the three gradients, with parks within the 3rd ring road mainly concentrated within 10 ha, those between the 3rd and 5th ring roads mainly concentrated within 15 ha, and those outside the 5th ring road mainly concentrated within 35 ha.

Gradient variances exist in various characteristic indicators within the parks and their buffer zones (Figure 2). The forest cover percentage peaks between the 3rd and 5th ring roads, reaches its lowest point outside the 5th ring road within the parks, and descends to its nadir within the 3rd ring road within the buffer zones. The grassland cover percentage decreases from within the 3rd ring road to outside the 5th ring road. The water cover percentage is highest within the 3rd ring road and lowest between the 3rd and 5th ring roads. The impervious surface percentage is highest outside the 5th ring road, lowest between the 3rd and 5th ring roads within the parks, and diminishes from within the 3rd ring road to outside the 5th ring road within the buffer zones. NDVI peaks between the 3rd and 5th ring roads and is lowest outside the 5th ring road. The percentage of building area within the buffer zones decreases from within the 3rd ring road to outside the 5th ring road.

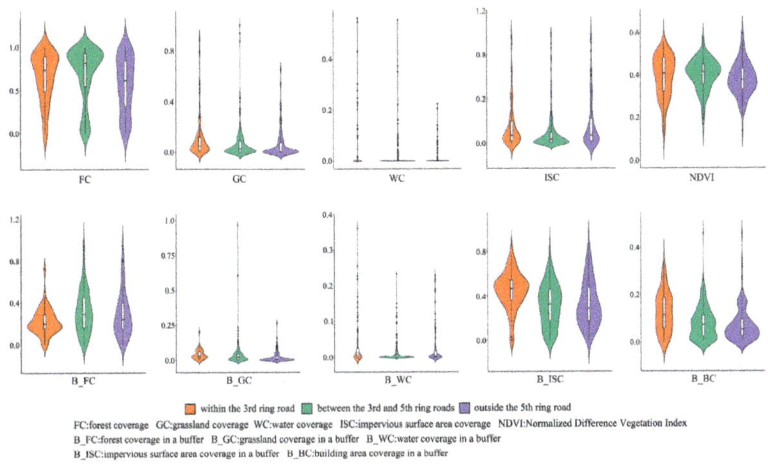

Figure 2. Various characteristic indicators within parks and their buffer zones across different spatial gradients.

3.2. Characteristics of the UHI Effect across Different Spatial Gradients

The UHI effect in the study area generally exhibits a pattern of weaker intensity in the north and stronger intensity in the south, gradually diminishing across three spatial gradients within the 3rd ring road, between the 3rd and 5th ring roads, and outside the 5th ring road (Figure 3).

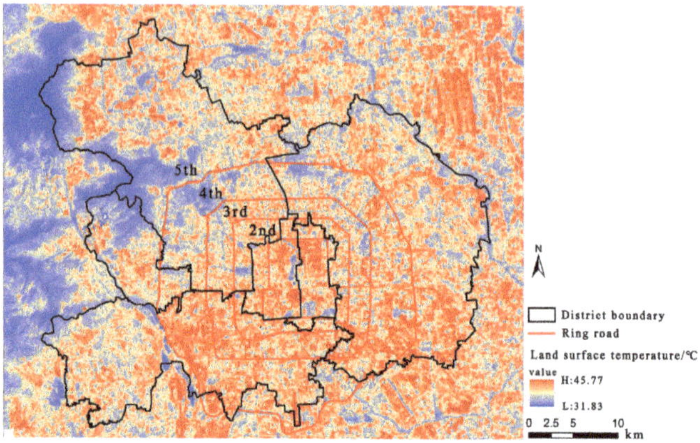

Figure 3. Distribution of LST in the study area.

Within the parks, the average and median values of LST_P are lowest in the area between the 3rd and 5th ring roads. In the buffer zones, the average and median values of LST_B are highest in the area within the 3rd ring road and lowest in the area between the 3rd and 5th ring roads (Table 3).

Table 3. Temperature variations within parks and buffer zones across different spatial gradients.

°C		3	3–5	5
LST_P	Average value	42.24	42.01	42.36
	Median value	42.35	42.09	42.33
LST_B	Mean value	43.72	43.09	43.19
	Median value	43.92	43.26	43.48

3.3. Characteristics of Park Green Space Cold Island Effect across Different Spatial Gradients

The PCI gradually decreases across three spatial gradients: within the 3rd ring road, between the 3rd and 5th ring roads, and outside the 5th ring road, with average values of 1.77, 1.46, and 1.29 °C, respectively. The corresponding median values are 1.43, 1.21, and 1.17 °C (Figure 4). Spatially, parks exhibiting stronger cold island effects are predominantly concentrated in the innermost areas, while those with weaker effects are more prevalent in the outer regions, indicating a general trend of diminishing intensity from the inner to outer areas (Figure 5).

Through cluster analysis, the parks in the study area were categorized into four groups representing different characteristics of cold island effects. The distribution of these four types of parks varies across different spatial gradients (Figure 6). Within the 3rd ring road, there are more parks classified as types 2 and 4, with type 2 being the most common. In the area between the 3rd and 5th ring roads, parks classified as types 1 and 3 are more prevalent, with type 1 being the most common. Outside the 5th ring road, parks classified as types 1 and 3 are more common, with type 3 being the most prevalent (Table 4).

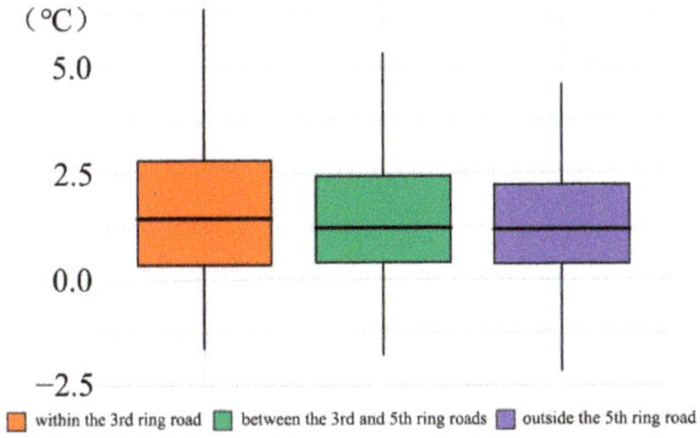

Figure 4. The PCI varies across different spatial gradients.

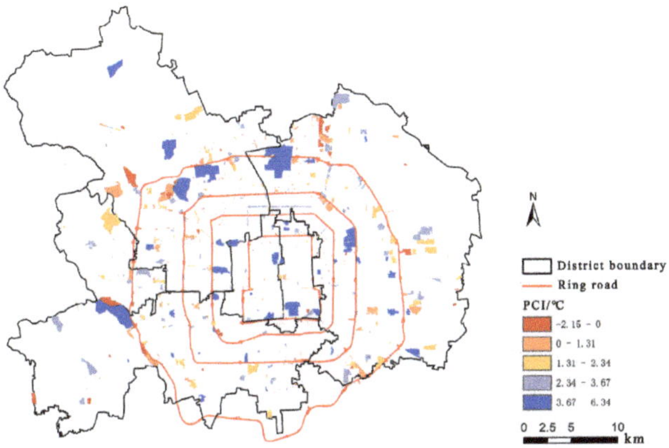

Figure 5. Distribution of cold island effect in park green spaces across different spatial gradients.

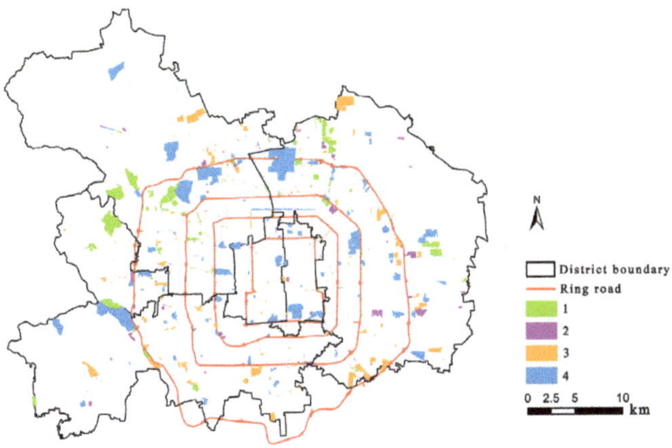

Figure 6. Distribution of the 4 types of parks with cold island effects across different spatial gradients.

Table 4. Four types of parks with cold island effect characteristics across different spatial gradients.

Type	Cluster Center			Characteristic	3	3–5	5	Total
	PCI	LST_P	LST_B					
1	0.33	0.47	0.48	The cold island effect is weak, with lower temperatures observed both within the park and in the buffer zone.	17	92	34	143
2	0.32	0.88	0.86	The cold island effect is weak, with high temperatures observed both within the park and in the buffer zone.	33	62	26	121
3	0.49	0.65	0.75	The cold island effect is moderate, with relatively high temperatures observed both within the park and in the buffer zone.	24	71	38	133
4	0.73	0.40	0.74	The cold island effect is strong, with lower temperatures observed within the park and higher temperatures in the buffer zone.	27	51	18	96

3.4. Factors Influencing the Cold Island Effect of Urban Park Green Spaces across Different Spatial Gradients

From the correlation analysis, it is evident that certain variables exhibit notable associations. Specifically, A, P, and F_MPS demonstrate significant positive correlations, albeit with varying magnitudes across different spatial gradients. PA demonstrates a significant negative correlation, with the strongest association observed within the 3rd ring road. NDVI displays a significant positive correlation between the 3rd and 5th ring roads as well as outside the 5th ring road, whereas no correlation is discernible within the 3rd ring road. In contrast, ISC exhibits a significant negative correlation within the 3rd ring road and between the 3rd and 5th ring roads, but lacks correlation outside the 5th ring road. B_ISC displays a positive correlation between the 3rd and 5th ring roads ($r = 0.21$, $p < 0.001$), while exhibiting no correlation elsewhere ($p > 0.05$). B_FC showcases a negative correlation between the 3rd and 5th ring roads ($r = -0.131$, $p < 0.05$), but does not demonstrate correlation elsewhere ($p > 0.05$). Moreover, B_GC demonstrates a negative correlation within the 3rd ring road ($r = -0.22$, $p < 0.05$), with no correlation observed elsewhere. On the other hand, B_BC presents a significant positive correlation within the 3rd ring road and between the 3rd and 5th ring roads, yet lacks correlation outside the 5th ring road. FC, GC, and B_WC exhibit no discernible correlation across the three gradients ($p > 0.05$).

The fitted curves of the factors exhibiting significant correlations across two or three spatial gradients reveal insights (Figure 7). In terms of positive correlations, variables such as A, P, WC, and F_MPS approach the PCI threshold first outside the 5th ring road, followed by the region between the 3rd and 5th ring roads, and finally, within the 3rd ring road. NDVI approaches the threshold first between the 3rd and 5th ring roads, followed by outside the 5th ring road. B_BC approaches the threshold first between the 3rd and 5th ring roads, followed by within the 3rd ring road. In terms of negative correlations, PA's impact on PCI approaches the threshold first outside the 5th ring road, followed by the region between the 3rd and 5th ring roads, and finally, within the 3rd ring road. ISC approaches the threshold first between the 3rd and 5th ring roads, followed by within the 3rd ring road.

The results of the regression analysis reveal the varying performance of the regression equations across different spatial gradients, with optimal performance observed within the 3rd ring and suboptimal performance outside the 5th ring. Specifically, within the 3rd ring, factors such as P, B_BC, and F_MPS exhibit positive effects on the dependent variable, while PA, ISC, B_FC, and GC demonstrate negative effects. In the region between the 3rd and 5th ring roads, WC, NDVI, B_ISC, and B_BC exhibit positive effects, whereas PA, ISC, and B_FC demonstrate negative effects. Outside the 5th ring, NDVI, WC, B_BC, and A show positive effects on the dependent variable (Table 5).

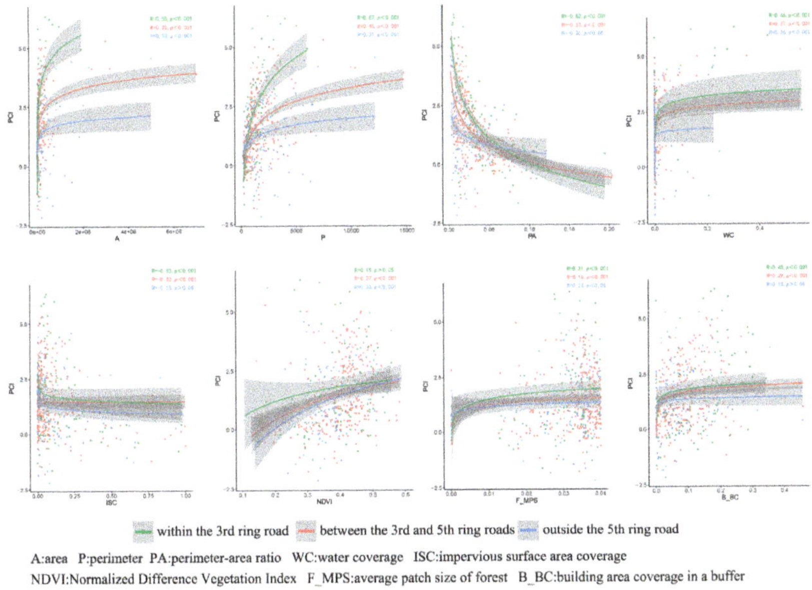

Figure 7. The fitting situation of the factors exhibiting significant correlations across two or more spatial gradients.

Table 5. Regression analysis results for different spatial gradients.

		Non-Standardized Coefficient B	p	R^2	F
3	Constant	5.262	0.000 **	0.681	F = 31.206, p = 0.000
	P	0.001	0.000 **		
	PA	−21.200	0.000 **		
	ISC	−4.758	0.000 **		
	B_FC	−3.176	0.001 **		
	B_BC	3.096	0.025 *		
	F_MPS	72.813	0.001 **		
	GC	−2.325	0.014 *		
3–5	Constant	−0.884	0.003 **	0.580	F = 48.424, p = 0.000
	PA	−13.341	0.000 **		
	WC	9.101	0.000 **		
	NDVI	5.677	0.000 **		
	B_ISC	1.648	0.000 **		
	ISC	−1.782	0.000 **		
	B_FC	−1.049	0.000 **		
	B_BC	1.883	0.043 *		
5	Constant	−1.727	0.004 **	0.264	F = 11.230, p = 0.000
	NDVI	6.404	0.000 **		
	WC	10.386	0.002 **		
	B_BC	3.840	0.012 *		
	A	4.286×10^{-7}	0.022 *		

* $p < 0.05$, ** $p < 0.01$.

4. Discussion

4.1. The Impact of Spatial Gradients on Urban Park Green Space Cold Island Effect

The study reveals that the intensity of the urban heat island effect in the study area gradually decreases from within the 3rd ring road to outside the 5th ring road. This trend is associated with the high density of buildings and roads, dense population, and relatively low vegetation coverage in the central urban area. Conversely, the peripheral areas of the city typically have more vegetation and open spaces, as well as lower-density buildings and roads, which helps mitigate the impact of the heat island effect. Similarly, the intensity of the cold island effect in urban parks also decreases gradually from within the 3rd ring road to outside the 5th ring road. This suggests that in areas where the heat island effect is more severe, urban parks have greater potential for providing cooling effects. In areas farther away from the city center, the surface temperature around urban parks is relatively lower, resulting in less noticeable temperature differences between the interior and exterior areas (Figure 8).

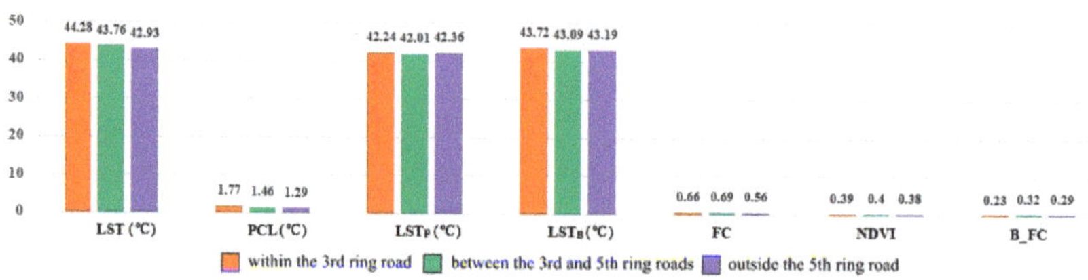

Figure 8. The difference in the mean values of each indicator on three gradients.

Moreover, the surface temperatures within the parks and buffer zones situated between the 3rd and 5th ring roads exhibit the most notable decrease compared to the other spatial gradients. Considering that the FC, NDVI, and B_FC are the highest among the three spatial gradients, while the ISC is the lowest, this observation may be related to the improvement in overall greenery due to the "First Isolated Green Belt" [47] plan implemented in this region. Under the guidance of this plan, several high-quality urban parks have been successively constructed, forming an urban park ring and significantly improving the living environment of the area [48].

4.2. Spatial Gradient Differences in the Relationship between Urban Park Green Space Characteristics and the UHI Effect

In the three spatial gradients, differences exist in the related characteristics between the urban park interiors and buffer zones, jointly determining the disparities in the park's cold island effect. Most characteristic indicators exhibit a gradual decrease in correlation with PCI from within the 3rd ring road to outside the 5th ring road. This includes six positive impact indicators: A, P, WC, F_MPS, and B_BC, as well as three negative impact indicators: PA, ISC, and B_GC. On one hand, this may be due to the higher sensitivity of cold island intensity to changes in related characteristic indicators in the 3rd ring road, where the UHI intensity is higher and the park area is smaller. As the heat island intensity decreases and the park area increases, this sensitivity diminishes. On the other hand, it may be due to the more diverse characteristics of land use in the buffer zone of urban peripheral parks, resulting in a more complex impact situation (Table 6).

The correlation between NDVI and PCI gradually strengthens from within the 3rd ring road to outside the 5th ring road, implying that the cold island effect of parks on the urban periphery is more influenced by vegetation quality. Parks with higher-quality vegetation exhibit a stronger cold island effect. B_FC and B_ISC are solely correlated with PCI between the 3rd and 5th ring roads, indicating that forest coverage and impervious surface percentage have a significant impact only in the buffer zones within this range. FC shows no correlation with PCI across the three spatial gradients, but it is significantly negatively correlated with LST_P (r = −0.27, $p < 0.001$; r = −0.31, $p < 0.001$; r = −0.39,

$p < 0.001$), indicating that the buffer zone of the park weakens the correlation between FC and PCI.

The multiple regression equations derived from the fitting analysis of pertinent characteristic indicators exhibit higher explanatory capacity within the 3rd ring road and between the 3rd and 5th ring roads. This suggests that the underlying patterns are more discernible in these two gradients, and optimization and control of indicators within these gradients yield superior outcomes compared to areas outside the 5th ring road. All three spatial gradients encompass characteristics from both park interiors and buffer zones, with park interior indicators making a more substantial contribution. Within the 3rd ring road, the most significant factor is the positive impact of the average size of forest patches within the park, signifying that the aggregation of forest patches plays a decisive and positive role in the park's cold island effect. Between the 3rd and 5th ring roads, the most notable factor is the negative influence of the perimeter–area ratio, indicating that more complex park boundaries with greater contact with the surroundings result in diminished cooling benefits.

Table 6. Correlation results of park green space-related characteristic indicators and PCI with different spatial gradients.

	PCI		
	3	3–5	5
A	0.546 **	0.319 **	0.327 **
P	0.670 **	0.459 **	0.313 **
PA	−0.623 **	−0.531 **	−0.206 *
NDVI	0.150	0.269 **	0.327 **
FC	0.119	0.026	0.151
GC	−0.147	−0.080	−0.118
WC	0.458 **	0.368 **	0.256 **
ISC	−0.332 **	−0.220 **	−0.178
F_MPS	0.312 **	0.158 **	0.228 *
F_SI	0.175	0.154 *	0.050
F_PD	−0.270	0.029	−0.142
B_FC	0.030	−0.131 *	0.061
B_GC	−0.215 *	−0.013	−0.127
B_WC	−0.117	0.014	−0.167
B_ISC	−0.132	0.205 **	0.046
B_BC	0.405 **	0.291 **	0.133

* $p < 0.05$, ** $p < 0.01$.

4.3. Different Regions Require Tailored Strategies for Optimizing and Managing Urban Park Green Spaces

Utilizing cluster analysis, tailored optimization and management strategies for urban park green spaces can be delineated according to distinct feature categories and spatial distributions. The predominant type 1 parks demonstrate a relatively subdued cold island effect, characterized by lower temperatures within the park and its surrounding buffer zones, indicative of a satisfactory cooling impact. They are predominantly situated between the 3rd and 5th ring roads and outside the 5th ring road. If the surrounding environment exacerbates the heat island effect due to urban development, it is necessary to correspondingly increase the cold island intensity of the parks' green spaces.

Type 2 parks, typified by a weak cold island effect with elevated temperatures within both the park and buffer zones, are predominantly situated within the 3rd ring road. These parks necessitate substantial optimization and management interventions, focusing on enhancing forest patch aggregation, increasing average patch size, mitigating boundary complexity, and reducing the prevalence of impervious surfaces and grassland coverage. Type 3 parks exhibit a moderate cold island effect, with relatively high temperatures within both the park and buffer zones, indicating untapped potential for improving green space cold island intensity. They are mainly distributed between the 3rd and 5th ring roads and outside the 5th ring road. Optimization strategies may entail simplifying boundary complexity, augmenting water cover proportions, and enhancing the NDVI while minimizing impervious surface percentages. Type 4 parks, representing the minority category, manifest a robust cold island effect, characterized by lower temperatures within the park and elevated temperatures in the buffer zones. Primarily located within the 3rd ring road, these parks warrant prioritized conservation endeavors. These delineated strategies aim to optimize the cooling efficacy of urban park green spaces based on their distinctive attributes and spatial distributions, thereby contributing to the overall alleviation of the UHI effect.

The cooling island effect of green spaces is significantly influenced by seasonal variations, meteorological conditions, and urban development changes. The data used in this study primarily focus on a single time period, failing to consider the comprehensive impact of seasonal changes and long-term trends on the cold island effect. In the future, using data from multiple time points can provide a more comprehensive understanding of the dynamic changes in the cold island effect. Additionally, regarding the selection of park green space characteristics, despite the consideration of multiple representative and instructive indicators based on fundamental park attributes and relevant literature, certain factors influencing the cold island effect may have been inadvertently overlooked. In terms of research and application, the next step can also combine the classification of cold island effect characteristics with the classification standard of park green spaces, and explore the differences in emphasis and tradeoff strategies of different types of parks in the optimization and improvement of cold island effect. Furthermore, this study mainly focused on Beijing, while the geographical environment, climatic characteristics, and planning and construction differ significantly in different regions. When extrapolating the research results to other cities, it is necessary to carefully consider the specificity and heterogeneity of different regions. Future research could involve selecting more urban samples to explore the more universal mechanisms of the impact of green space cold island effects.

5. Conclusions

This study focused on Beijing and conducted a comparative analysis of the cold island effect of park green spaces across different spatial gradients. The findings indicate the following: (1) Spatial gradient differences exist in both park green space characteristics and UHI, collectively influencing the cold island effect of urban park green spaces. Within the parks and their buffer zones, the lowest average and median values of LST are observed between the 3rd and 5th ring roads. The PCI gradually decreases across three spatial gradients: within the 3rd ring road, between the 3rd and 5th ring roads, and outside the 5th ring road, with median values of 1.43 °C, 1.21 °C, and 1.17 °C, respectively. (2) Through cluster analysis, four distinct types of parks with different cold island effect characteristics were identified. Over half of the parks, mainly belonging to types 2 and 3, require optimization and improvement, whereas only about one-fifth of the parks exhibit a strong cold island effect (type 4). For type 2 parks, specific optimization measures include planting more trees and vegetation, expanding existing green spaces and merging smaller patches, removing unnecessary structures that disrupt the continuity of green spaces to simplify park boundaries, replacing impermeable surfaces such as concrete or asphalt with permeable materials, and replacing some grassy areas with trees or shrubs. For type 3 parks, specific optimization measures include removing unnecessary structures that disrupt the continuity of green spaces to simplify park boundaries, expanding the area of water bodies, planting more trees and vegetation, and replacing impermeable surfaces such as concrete or asphalt with permeable materials. (3) Various park green space characteristics exhibit disparate correlations with PCI, and these correlations display discrepancies in strength and thresholds across different gradients. (4) The effectiveness of regression equations in predicting PCI declines progressively from within the 3rd ring road to outside the 5th ring road. Within the 3rd ring road, factors such as F_MPS, PA, and ISC exert substantial influences on PCI, whereas between the 3rd and 5th ring roads, PA, WC, and NDVI emerge as significant factors impacting PCI. Building upon previous studies that primarily focused on the overall perspective, this study adopted a less-explored approach by considering different spatial gradients. By investigating the influence mechanisms of the cold island effect under varying gradients of urban park green spaces and UHI patterns, it uncovered certain characteristics and patterns that were not identified in previous holistic studies. This contributes to identifying potential issues related to spatial gradient changes. The findings have expanded the understanding of the cold island effect of urban park green spaces to some extent. They aid in proposing diverse optimization management strategies for urban park green space planning and system construction in different regions, thereby enabling parks to better manifest their cold island effect and mitigate the adverse impacts of the UHI phenomenon. However, there are still limitations in the selection of park green space characteristic indicators and the scope of research application. Further research is needed to explore more urban samples, areas, regions, and time periods in order to address these limitations.

Author Contributions: Conceptualization, C.X. and H.Z.; methodology, C.X.; investigation, C.X. and W.W.; software, C.X. and W.W.; writing—original draft preparation, C.X.; writing—review and editing, C.X.; supervision, H.Z.; project administration, H.Z. All authors have read and agreed to the published version of the manuscript.

Funding: This research was funded by the Research and Development Project of the Ministry of Housing and Urban–Rural Development, a project supported by the Beijing Advanced Innovation Center for Future Urban Design, grant number 2022-K-024; Qinghai Province Science and Technology Achievement Transformation Special Project, grant number 2022-SF-159.

Data Availability Statement: Data are contained within the article.

Conflicts of Interest: Author Wenjing Wang was employed by the company China Urban Construction Design & Research Institute Co., Ltd. The remaining authors declare that the research was conducted in the absence of any commercial or financial relationships that could be construed as a potential conflict of interest.

References

1. Dong, X.; Meng, Z.; Wang, Y.; Zhang, Y.; Sun, H.; Wang, Q. Monitoring Spatiotemporal Changes of Impervious Surfaces in Beijing City Using Random Forest Algorithm and Textural Features. *Remote Sens.* **2021**, *13*, 153. [CrossRef]
2. Hendel, M.; Azos-Diaz, K.; Tremeac, B. Behavioral adaptation to heat-related health risks in cities. *Energy Build.* **2016**, *152*, 823–829. [CrossRef]
3. Levermore, G.; Parkinson, J.; Lee, K.; Laycock, P.; Lindley, S. The increasing trend of the urban heat island intensity. *Urban Clim.* **2018**, *24*, 360–368. [CrossRef]
4. Halder, B.; Bandyopadhyay, J.; Banik, P. Evaluation of the Climate Change Impact on Urban Heat Island Based on Land Surface Temperature and Geospatial Indicators. *Int. J. Environ. Res.* **2021**, *15*, 819–835. [CrossRef]
5. Kamal, A.; Mahfouz, A.; Sezer, N.; Hassan, I.G.; Wang, L.L.; Rahman, M.A. Investigation of urban heat island and climate change and their combined impact on building cooling demand in the hot and humid climate of Qatar. *Urban Clim.* **2023**, *52*, 101704. [CrossRef]
6. Yang, X.; Chen, B.; Hu, K. A Review of Impacts of Urbanization on Extreme Heat Events. *Prog. Geogr.* **2015**, *34*, 1219–1228.
7. Eugenia, K.; Ming, C. Impact of Urbanization and Land-use Change on Climate. *Nature* **2003**, *423*, 528–531.
8. Yu, Z.; Guo, Q.; Sun, R. Impacts of Urban Cooling Effect Based on Landscape Scale: A Review. *J. Appl. Ecol.* **2015**, *26*, 636–642.
9. Gao, Z.; Zaitchik, B.F.; Hou, Y.; Chen, W. Toward park design optimization to mitigate the urban heat Island: Assessment of the cooling effect in five U.S. cities. *Sustain. Cities Soc.* **2022**, *81*, 103870. [CrossRef]
10. Wang, X.; Kong, F.H.; Yin, H.; Xu, H.; Li, J.; Pu, Y. Characteristics of Vegetation Shading and Transpiration Cooling Effects During Hot Summer. *Acta Ecol. Sin.* **2018**, *38*, 4234–4244.
11. Du, H.; Song, X.; Jiang, H.; Kan, Z.; Wang, Z.; Cai, Y. Research on the Cooling Island Effects of Water Body: A Case Study of Shanghai, China. *Ecol. Indic.* **2016**, *67*, 31–38. [CrossRef]
12. Hien, W.; Liang, T.; Denia, K.D.; Hideki, T. Greenery as a Mitigation and Adaptation Strategy to Urban Heat. *Nat. Rev. Earth Environ.* **2021**, *2*, 166–181.
13. Feyisa, G.L.; Dons, K.; Meilby, H. Efficiency of Parks in Mitigating Urban Heat Island Effect: An Example From Addis Ababa. *Landsc. Urban Plan.* **2014**, *123*, 87–95. [CrossRef]
14. Yue, X.; Lin, J.; Yang, Y. Mitigation Role of Urban Green Space in Reducing Heat Island Effect—A Case Study of Central Urban Area of Baoding City. *Landsc. Arch.* **2018**, *25*, 66–70.

15. Zhao, H.; Hu, M.; Zhu, J.; Huang, T.; Zhang, Y.; Li, Y.; Yan, M. Blue-green Space Cooling Effect and Its Influencing Factors in Metropolitan Area: A Case Study on the Area Within the 5th ring Road in Beijing. *Acta Ecol. Sin.* **2023**, *43*, 4904–4919.
16. Yin, R.; Fang, J.; Zhu, D.; Hu, Y. Measurement of Cooling Efficiency in Urban Parks and Strategies for Climate Adaptation Design. *Chin. Landsc. Arch.* **2022**, *38*, 121–126.
17. Du, H.; Cai, W.; Xu, Y.; Wang, Z.; Wang, Y.; Cai, Y. Quantifying the Cool Island Effects of Urban Green Spaces Using Remote Sensing Data. *Urban For. Urban Green.* **2017**, *27*, 24–31. [CrossRef]
18. Qiu, K.; Jia, B. The Roles of Landscape Both Inside the Park and the Surroundings in Park Cooling Effect. *Sustain. Cities Soc.* **2020**, *52*, 101864. [CrossRef]
19. Su, W.; Chang, Q. Nature-based Solutions for Urban Heat Mitigation and Implementation Path Thereof: A Case Study of Beijing. *Landsc. Arch.* **2022**, *29*, 26–32.
20. García-Haro, A.; Arellano, B.; Roca, J. Quantifying the influence of design and location on the cool island effect of the urban parks of Barcelonaes. *J. Appl. Remote Sens.* **2023**, *17*, 034512. [CrossRef]
21. Zhu, W.; Sun, J.; Yang, C.; Liu, M.; Xu, X.; Ji, C. How to Measure the Urban Park Cooling Island? A Perspective of Absolute and Relative Indicators Using Remote Sensing and Buffer Analysis. *Remote Sens.* **2021**, *13*, 3154. [CrossRef]
22. Hwang, B.; Sou, H.D.; Oh, J.H.; Park, C.R. Cooling effect of urban forests on the urban heat island in Seoul, South Korea. *PLoS ONE* **2023**, *18*, e0288774. [CrossRef]
23. Xue, X.; He, T.; Xu, L.; Tong, C.; Ye, Y.; Liu, H.; Xu, D.; Zheng, X. Quantifying the Spatial Pattern of Urban Heat Islands and the Associated Cooling Effect of Blue–green Landscapes Using Multisource Remote Sensing Data. *Sci. Total Environ.* **2022**, *843*, 156829. [CrossRef]
24. Yi, Y.; Shen, G.; Zhang, C.; Sun, H.; Zhang, Z.; Yin, S. Quantitative analysis and prediction of urban heat island intensity on urban-rural gradient: A case study of Shanghai. *Sci. Total Environ.* **2022**, *829*, 154264. [CrossRef]
25. You, X.; LI, Q.; Meng, Q. Research on Spatial Patterns and Morphological Differentiation Control Strategy of Urban Heat Islands—A Case Study of Downtown Area of Guangzhou City. *Landsc. Arch.* **2021**, *28*, 74–79.
26. Kabisch, N.; Remahne, F.; Ilsemann, C.; Fricke, L. The urban heat island under extreme heat conditions: A case study of Hannover, Germany. *Sci. Rep.* **2023**, *13*, 23017. [CrossRef]
27. You, M.; Huang, J.; Guan, C. Are New Towns Prone to Urban Heat Island Effect? Implications for Planning Form and Function. *Sustain. Cities Soc.* **2023**, *99*, 104939. [CrossRef]
28. Debbage, N.; Shepherd, J.M. The urban heat island effect and city contiguity. *Comput. Environ. Urban Syst.* **2015**, *54*, 181–194. [CrossRef]
29. Wang, Q.; Peng, J.; Yu, S.; Dan, Y.; Dong, J.; Zhao, X.; Wu, J. Key Attributes of Greenspace Pattern for Heat Mitigation Vary with Urban Functional Zones. *Landsc. Arch.* **2023**, *38*, 2965–2979. [CrossRef]
30. Yang, J.; Guo, R.; Li, D.; Wang, X.; Li, F. Interval-thresholding Effect of Cooling and Recreational Services of Urban Parks in Metropolises. *Sustain. Cities Soc.* **2022**, *79*, 103684. [CrossRef]
31. Melaas, E.K.; Wang, J.A.; Miller, D.L.; Friedl, M.A. Interactions between urban vegetation and surface urban heat islands: A case study in the Boston metropolitan region. *Environ. Res. Lett.* **2016**, *11*, 054020. [CrossRef]
32. Qiu, K.; Zhang, H.; Gao, J.; Pei, W.; Zhang, B.; Wang, M.; Wang, Q. Variation of Cool Island Effect for Urban Forest Patches Across an Urban-rural Gradient in Shanghai. *Chin. J. Ecol.* **2021**, *40*, 1409–1418.
33. Algretawee, H.; Rayburg, S.; Neave, M. Estimating the effect of park proximity to the central of Melbourne city on Urban Heat Island (UHI) relative to Land Surface Temperature (LST). *Ecol. Eng.* **2019**, *138*, 374–390. [CrossRef]
34. Tian, P.; Li, J.; Pu, R.; Cao, L.; Liu, Y.; Zhang, H. Assessing the cold island effect of urban parks in metropolitan cores: A case study of Hangzhou, China. *Environ. Sci. Pollut. Res. Int.* **2023**, *30*, 80931–80944. [CrossRef]
35. Jiang, L.; Xie, M.; Chen, B.; Su, W.; Zhao, X.; Wu, R. Key areas and measures to mitigate heat exposure risk in highly urbanized city: A case study of Beijing, China. *Urban Clim.* **2024**, *53*, 101748. [CrossRef]
36. Qiao, Z.; Huang, N.; Xu, X.; Sun, Z.; Wu, C.; Yang, J. Spatio-temporal Pattern and Evolution of the Urban Thermal Landscape in Metropolitan Beijing Between 2003 and 2017. *Acta Geogr. Sin.* **2019**, *74*, 475–489.
37. Du, C.; Ren, H.; Qin, Q.; Meng, J.; Zhao, S. A Practical Split-window Algorithm for Estimating Land Surface Temperature From Landsat 8 Data. *Remote Sens.* **2015**, *7*, 647–665. [CrossRef]
38. Ren, H.; Du, C.; Liu, R.; Qin, Q.; Yan, G.; Li, Z.; Meng, J. Atmospheric Water Vapor Retrieval From Landsat 8 Thermal Infrared Images. *J. Geophys. Res.* **2015**, *120*, 1723–1738. [CrossRef]
39. Liao, W.; Guldmann, J.; Hu, L.; Cao, Q.; Gan, D.; Li, X. Linking Urban Park Cool Island Effects to the Landscape Patterns Inside and Outside the Park: A Simultaneous Equation Modeling Approach. *Landsc. Urban Plan.* **2023**, *232*, 104681. [CrossRef]
40. Peng, J.; Dan, Y.; Qiao, R.; Liu, Y.; Dong, J.; Wu, J. How to quantify the cooling effect of urban parks? Linking maximum and accumulation perspectives. *Remote Sens. Environ.* **2021**, *252*, 112135. [CrossRef]
41. Dos Santos, A.R.; de Oliveira, F.S.; da Silva, A.G.; Gleriani, J.M.; Gonçalves, W.; Moreira, G.L.; Silva, F.G.; Branco, E.R.F.; Moura, M.M.; da Silva, R.G.; et al. Spatial and temporal distribution of urban heat islands. *Sci. Total Environ.* **2017**, *605–606*, 946–956. [CrossRef] [PubMed]
42. Lin, J.; Deng, Y.; Chen, S.; Li, K.; Ji, W.; Li, W. Research Progress of Urban Park Microclimate Based on Quantitative Statistical Software. *Buildings* **2023**, *13*, 2335. [CrossRef]

43. Chang, C.R.; Li, M.H. Effects of urban parks on the local urban thermal environment. *Urban For. Urban Green.* **2014**, *13*, 672–681. [CrossRef]
44. Jiao, M.; Zhou, W.; Qian, Y.; Wang, J.; Zheng, Z.; Hu, X.; Wang, W. Influences of Patch Size on the Cooling Capacity of Urban Greenspace: Progresses, Problems and Perspectives. *Acta Ecol. Sin.* **2021**, *41*, 9154–9163.
45. Xue, S.; Yuan, L.; Wang, K.; Wang, J.; Pei, Y. Comparing the Impact of Urban Park Landscape Design Parameters on the Thermal Environment of Surrounding Low-Rise and High-Rise Neighborhoods. *Forests* **2023**, *14*, 1682. [CrossRef]
46. Liang, Z.; Li, Z.; Fan, Z. Seasonal impacts of built environment and its interactions on urban park cooling effects in Nanjing, China. *Build. Environ.* **2023**, *242*, 110580. [CrossRef]
47. Sun, Z. Thermal Environment Characteristic and Cooling Effect of Greenery in Beijing First Green Belt Area. *Chin. J. Ecol.* **2019**, *38*, 3496–3505.
48. Wang, Y.; Chang, Q.; Li, X. Promoting sustainable carbon sequestration of plants in urban greenspace by planting design: A case study in parks of Beijing. *Urban For. Urban Green.* **2021**, *64*, 127291. [CrossRef]

buildings

MDPI

Article

Resilience-Vulnerability Balance and Obstacle Factor Analysis in Urban Flooding: A Case Study in the Qinghai–Tibetan Plateau

Wenjing Wang [1], Chao Xu [2], Junchao He [1], Zhongwen Chi [1], Weilan Bai [1,*] and Rui Liu [1]

[1] China Urban Construction Design & Research Institute Co., Ltd., Beijing 100120, China; wangwenjingyuanlin@cucd.cn (W.W.); hejunchao@cucd.cn (J.H.); chizhongwen@cucd.cn (Z.C.); liurui@cucd.cn (R.L.)

[2] Beijing Advanced Innovation Center for Future Urban Design, School of Architecture and Urban Planning, Beijing University of Civil Engineering and Architecture, Beijing 100044, China; xuchao@bucea.edu.cn

* Correspondence: baiweilan@cucd.cn; Tel.: +86-18610939773

Abstract: Under the combined influence of climate change and urban development, the risk of urban flooding caused by extreme weather events has increased significantly, making assessing flood vulnerability and resilience increasingly crucial for urban flood management. With the 45 counties in Qinghai Province as the research objects, the hazard risk of flood and exposure are combined to study their vulnerability. At the same time, resilience is evaluated by the indicators selected from four dimensions (society, economy, environment, and infrastructure). Through Z-scoring, the vulnerability and resilience of each county are clustered into four groups to explore their associations from a spatial balance perspective. Obstacle factor analysis is introduced to summarize the key factors affecting the improvement of urban resilience in Qinghai Provence. The results show that the eastern areas of Qinghai experience high vulnerability to flooding because of high levels of hazard and exposure. What is more, Xining, Haidong, and Haixi experience a high level of resilience. A strong spatial mismatch between vulnerability and resilience exists in Qinghai, with 24 counties (58%) being self-adapted, 8 counties (18%) over-abundant, and 11 counties deficient in terms of nature–nurture. The length of levee and number of beds in medical institutions are the main obstacles to resilience in Qinghai. The research results can provide a theoretical and scientific basis for future urban flood management and resilience development in the Qinghai–Tibetan Plateau.

Keywords: Qinghai Province; flood disaster; resilience assessment; Z-score; obstacle factor analysis; resilience enhancement strategy

1. Introduction

The number of casualties and economic losses from natural catastrophes has increased along with urbanization, urban population growth, and economic agglomeration [1]. Globally, the frequency of natural catastrophes has increased significantly over the past three decades, with floods being the most frequent type of disaster [2]. The increased risk of climate change, coupled with ecological degradation and low coping capacities, has resulted in an increased risk of flooding globally and an increase in the number of regions affected by floods [3,4]. According to statistical information from China's flood and drought disaster bulletin, more than 100 cities in China are affected by flooding every year [5]. Furthermore, over a period of 40 years (1980–2020), China experienced economic losses equivalent to around 2% of its GDP, affecting approximately 200 million people in addition to deaths, injuries, and homelessness [6]. In addition, changes in the global climate and urban water cycle processes, leading to natural disasters such as rainstorms, flash floods, and tsunamis, have also increased the frequency of extreme rainstorms and flooding in cities and more severe disaster losses [7–9]. Therefore, improving urban flood resilience and reducing the impact of flooding has become an urgent challenge.

Resilience has become a crucial research focus to address the various threats and risks cities encounter [10]. Organizations and research institutions such as the United Nations International Strategy for Disaster Reduction (UN/ISDR), the Rockefeller Foundation, and the Intergovernmental Panel on Climate Change (IPCC) advocated for the construction of resilient cities [11]. Urban resilience refers to the ability of city systems, including their subsystems, to maintain, recover, and adapt when disturbed [12,13]. Researchers have extended the concept of flooding resilience, which refers to the ability of cities to self-organize after withstanding floods to minimize potential casualties while maintaining socio-economic characteristics [14]. Assessing and predicting urban flood resilience can facilitate proactive planning and resilient infrastructure development, enabling cities to effectively cope with and adapt to natural disasters, thereby mitigating socio-economic impacts [15]. Various institutional activities can be linked by resilience, and resilience can be promoted to help people deal with and adapt to disasters, which reduces the impact of vulnerability to them. In recent years, scholars have comprehensively assessed urban resilience from multiple perspectives. Cutter proposed a community baseline resilience evaluation index system to quantify resilience based on six dimensions: social, economic, infrastructure, institutional, community, and ecological, which is the first attempt to transition resilience from a theoretical framework to operationalized practice [16]. Several scholars have established flood resilience models based on exposure, vulnerability, and adaptability dimensions from resilience characteristics [17,18]. The main assessment methods for urban flood resilience include entropy weighting, hierarchical analysis, principal component analysis, TOPSIS, structural equation modelling, and system dynamics [19–21]. To analyze the main factors affecting urban resilience, the geographically weighted regression and obstacle model was used to quantified the degree of influence of urban resilience factors [10,22]. Several studies have explored urban flooding and climate change in high-altitude regions like the Tibetan Plateau, and revealed character of flood susceptibility, exposure, resilience and vulnerability [23–28].

Despite significant progress in understanding urban flood resilience and evolution, existing research reveals notable deficiencies in exploring the spatial balance between resilience and vulnerability. The spatial relationship of resilience–vulnerability to urban flooding emerges as a pivotal area of inquiry, aiming to dissect and understand the intricate dynamics between urban resilience—the capacity of cities to absorb, recover from, and adapt to adverse events—and vulnerability, the susceptibility of urban systems to damage and dysfunction as a result of such events. Qinghai Province is located in western China and is the main part of the Qinghai–Tibet Plateau. In the context of ongoing global climate change, Qinghai Province has experienced a significant upward trend in annual precipitation, with an increase of 15.8 mm per decade since 1961 [29]. This rate of increase substantially exceeds the national average of 5.8 mm per decade in China, and the 1.4 mm per decade observed in similar latitude regions. Qinghai's unique geography and ecosystems make it especially vulnerable to climate change, resulting in more pronounced impacts on its grasslands, glaciers, snowpack, lakes, and permafrost. This sensitivity leads to greater variability and uncertainty in predicting future climate impacts [30]. What is more, urban areas account for only 0.22% of the province's total land area and are mainly concentrated in river valleys and basins, where multi-level problems such as gully erosion hazards and geological hazards prevail [31]. Precipitation and intensity in Qinghai are unevenly distributed, gradually decreasing from east to west and from south to north [32], exacerbating the region's vulnerability to droughts and floods due to anomalous climatic conditions, resulting in concentrated disaster events. In addition, Qinghai's overall river flood control standards are inadequate, with an imperfect flood control system and various other safety hazards limiting the effectiveness of disaster reduction measures. Disparities in economic development across the province, coupled with different cultural and educational levels, conservative attitudes towards labour skills and ideological beliefs, further exacerbate the province's vulnerability to floods and significantly hamper resilience-building efforts [33,34]. Given the significant disparities in urban natural environments and unbalanced infrastructure conditions in Qinghai, mapping flood vulnerability and resilience is imperative in addressing flood issues. The increased vulnerability underscores the urgent need for detailed studies of flood vulnerability and resilience characteristics, and tailored strategies to mitigate and adapt to these environmental changes in Qinghai. The increased vulnerability underscores the critical need for detailed studies and tailored strategies to mitigate and adapt to these environmental changes in Qinghai.

This study resolved the shortage in the literature on flood vulnerability and resilience in Qinghai by proposing a framework to quantify the spatial patterns of susceptibility and resilience to urban flooding as well as investigate the spatial relationship between vulnerability and resilience. The barrier degree model was used to analyze the barrier factors influencing flood resilience. This research contributes to understanding the levels of vulnerability and resilience in Qinghai and offers targeted improvement suggestions. The results may be further applied to counties in the Qinghai–Tibet Plateau area, giving stakeholders a solid scientific foundation on which to build practical and targeted urban flood mitigation and prevention strategies. This study uses related theoretical, conceptual, and technological advances to apply the geographical balance between vulnerability and resilience to urban flooding research and practical applications. It also makes use of policy implementation, such as urban planning, to enhance urban flood resilience in the future.

2. Materials and Methods

2.1. Study Site

Qinghai Province is located in the western region of China (31°36′–39°12′ N, 89°24′–103°04′ E), on the northeastern part of the Tibetan Plateau, with an area of 722,000 km^2 and a population of 5.8 million. Regarding its administrative divisions, it governs two prefecture-level cities, Xining and Haidong, and six national autonomous prefectures: Yushu, Golog, Huangnan, Hainan, Haixi, and Haibei, and is composed of 45 counties (Figure 1). Qinghai Province plays a crucial role in conserving water resources, serving as the source of the Yellow River, Yangtze River, and Lancang River. It has 21 counties designated national key ecological function areas, making it an essential ecological function zone and ecological security barrier in China [35]. The precipitation in Qinghai gradually decreases from southeast to northwest, ranging from 15 to 750 mm. The annual precipitation is below 400 mm, with 70% of precipitation occurring between July and September. Most of Qinghai is plateaus and mountainous, and the main topographic units are the Qilian Mountains, Kunlun Mountains, Tanggula Mountains, Bayankala Mountains, Qaidam Basin, Qingnan Plateau, and Loess Plateau. Influenced by topography, extreme storms, and urban development, urban flooding disasters in Qinghai Province are increasingly intensifying.

Figure 1. Overview of counties in Qinghai Province.

2.2. The Study Framework

This study proposes a framework to explore the spatial relationship between vulnerability and resilience, and the influence mechanisms of urban resilience. Based on the results, we proposed a zoning and categorized flood resilience enhancement strategy for Qinghai Province. The whole framework is shown in Figure 2.

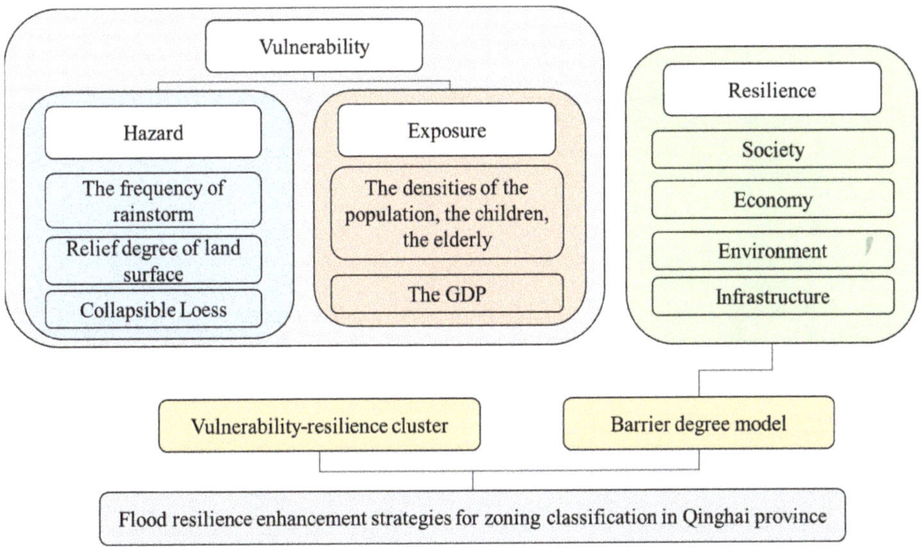

Figure 2. Overall study scheme.

2.3. Methods for the Exploration of Spatial Balance between Urban Flood Vulnerability and Resilience

2.3.1. Urban Flood Vulnerability Model

Vulnerability is commonly defined by the Third Assessment Report of IPCC as the degree of susceptibility to the damaging effects of a hazard. The vulnerability assessments focus on hazards and exposure to the socioeconomic system [36].

Rainfall is the direct factor and stimulating condition that induces heavy rainfall and flooding. The flood disaster driver index includes the amount of rainfall and frequency of rainstorms. Pearson's correlation calculations found a strong correlation exists between annual precipitation and the frequency of rainstorms. Flooding in Qinghai is mainly caused by persistent precipitation or short-term heavy precipitation, so the number of days with different intensities of precipitation is used to analyze the flooding causative factors. Therefore, the frequency of rainstorms was used in the study. Based on the meteorological disaster standards of Qinghai Province and previous studies [37], the risk grades of rainstorms and flood disasters were used to represent the factor of frequency of rainstorms. Topography is one of the key factors directly related to regional flooding. Plain areas may be flooded earlier due to the rapid flow of water from the highlands to low-lying areas [38,39]. Specifically, the Relief Degree of Land Surface (RDLS) applies the changes in altitude in a certain area to indicate the topography. The higher the RDLS, the more waterlogging and flooding are expected to occur. Collapsibility of loess is a significant factor affecting engineering construction in loess areas [40], and it can exacerbate flood disasters. Thus, this study selects frequency of rainstorms, RDLS, and collapsible loess as indicators to reflect hazards in urban planning.

Exposure can be defined as the proximity of people, property, or other elements to a hazardous area and thus suffering potential losses in any disaster [41]. In other words, exposure is the extent to which a city is plagued by catastrophic stresses. Assessment of exposure includes the number and types of people or assets exposed to a disaster. When waterlogging occurs, high population density tends to be more vulnerable to disasters. Furthermore, there is a greater chance that vulnerable populations—such as children and the elderly—will have a lower capacity to deal with urban flooding [42,43]. On the economic side, the higher the gross domestic product (GDP), the more economic damage from flooding an area will suffer. Finally, the densities of the population, the children, the elderly, and the GDP are selected as indicators to reflect the exposure to urban flooding in the area of interest.

2.3.2. Urban Flood Resilience Model

Resilience is the ability to withstand, absorb, or recover from a shock and adapt to environmental stresses such as extreme flooding [44]. According to previous studies [45,46], factors representing urban resilience can be divided into four dimensions: society, economy, environment, and infrastructure.

When selecting social resilience indicators, residents' awareness of waterlogging disasters and emergency resources are important factors to consider in urban adaptability measurements [47]. It is widely accepted that those with higher educational levels are more likely to have a high capacity for learning and innovation, as well as to be more scientific and efficient in handling disasters [48]. Health facilities are key factors in reducing the impact of disasters by reducing fatal losses through adequate and timely treatment [49]. The quantity of health facilities available, like hospital beds, can be used to assess a facility's capacity to deal with emergencies and is thus an important factor for measuring urban social resilience [50].

The term "economic resilience" describes the capacity of the local economy to withstand shocks from without. When an event strikes an urban economic system, it can bounce back fast and reach its previous level of development, if not surpass it. The tertiary industry ratio is a crucial indication of urban economic development, and the per capita GDP of urban regions reflects the comprehensive capacity of the urban economy [22]. After floods, areas with high per capita GDP and tertiary industry ratios will have more financial resources available for rehabilitation efforts [15].

The environmental resilience indicator is conceptually related to environmental quality, which enhances the absorptive capacity of coastal surges and floods [16]. Storage and infiltration capacities are significantly impacted by the percentage of urban ecological space, which could reduce peak flood flow [51]. Stronger rainwater infiltration and flood resilience are associated with higher levels of vegetation covering and a higher proportion of blue-green space [52].

Infrastructure resilience is characterized by an area's ability to respond and recover. Indicators in this dimension reflect the quality and functionality of critical infrastructure's ability to perform efficiently and in a timely manner before, during, and after an adverse event [20]. Areas with a high road density have a stronger traffic capacity and post-disaster rescue operations [53,54]. In addition, the construction of urban water-related infrastructure, such as drainage networks and levee construction, contributes to the resilience to flood risk and reduces the strength of flooding. Therefore, nine indicators, namely, education level, the number of beds in medical institutions, per capita GDP, ratio of tertiary industry, vegetation covering, proportion of blue-green space, density of roads, density of drainage networks, and the length of levee, are selected for the evaluation of resilience.

All the above-mentioned indicators were normalized by deviation maximization method:

$$r_i(x) = \frac{x_i - \min(x_i)}{\max(x_i) - \min(x_i)}, \tag{1}$$

Subjective and objective weighting are the two different approaches used for the indicators. The subjective weighting is determined by subjective factors, and the qualitative data outnumber the quantitative data. The entropy approach is an objective weighing methodology capable of capturing the utility value of the index. Compared to its weight value, the subjective weighting approach is less reliable and accurate. Thus, the entropy approach was used to determine the weight index in this study. The entropy value was calculated using the following equation:

$$w_j = \frac{1 - e_j}{\sum_{j=1}^{n}(1 - e_j)}, \quad (2)$$

$$e_j = -\frac{1}{lnm}\sum_{i=1}^{m} p_{ij} ln p_{ij}, \quad (3)$$

$$p_{ij} = \frac{x_{ij}}{\sum_{i=1}^{m} x_{ij}}, \quad (4)$$

where e_j is the the entropy of the jth criterion,p_{ij} is the normalized value of the ith alternative for the jth criterion, w_j is the weight of the index j, m is the number of indicators.

2.3.3. Identifying the Relationship of Vulnerability and Resilience

Z-scores are standard scores which represent the number of standard deviations a score is above or below the mean outcome score [55]. By standardising the distribution, comparisons can be made across different variables [55]. In the specific context of resilience and vulnerability analysis, the values of vulnerability and resilience were normalized by Z-score, and the standardized values of vulnerability and resilience were characterized on the X and Y axes, respectively, and divided into four quadrants (I, II, III, and IV). Quadrants I, II, III, and IV represent different types of zoning, such as high vulnerability–high resilience, low vulnerability–high resilience, low vulnerability–low resilience, and high vulnerability–low resilience, respectively. Such insights are instrumental in informing targeted management strategies tailored to the specific challenges and opportunities posed by different resilience-vulnerability profiles. The Z-score was calculated using the following equation:

$$x = \frac{x_i - \overline{x}}{s} \quad (5)$$

$$s = \sqrt{\frac{1}{n}\sum_{i=1}^{n}(x_i - \overline{x})^2}, \quad (6)$$

where x is the value of flood vulnerability or resiliece; $\overline{x}$ is the mean value of the sample of the indicator; s is the standard deviation of the sample of the indicator; n is the number of counties.

According to the Z-score, the flood vulnerability and resilience of different counties were combined, and these combinations are clustered into four groups, as shown in Table 1. People living in the high–high cluster have enough resources for adaptability. People in the high–low cluster lack adequate resources for adaptation and are at risk from urban flooding. While the areas in the low–high cluster are not particularly vulnerable to urban floods, the residents possess an abundance of resources for adaptation. People in low–low clusters are able to adjust to urban floods on their own since there is no risk of flooding there.

Table 1. Clusters of combinations of vulnerability and resilience.

Flood Vulnerability	Flood Resilience	Combination	Classify
≥ 0	≥ 0	High–High	Self-adaptive
≥ 0	≤ 0	High–Low	Deficient in terms of nature–nurture
≤ 0	≥ 0	Low–High	Over-abundant
≤ 0	≤ 0	Low–Low	Self-adaptive

2.4. Analysis of Obstacle Factors

Urban flood resilience is affected by society, economics, ecology, and infrastructure, and it is often difficult to consider all aspects when managing and making decisions on urban flood resilience enhancement because of limited resources. This study introduces an obstacle factor analysis to analyze and summarize the key factors impeding the improvement of urban resilience, allowing corresponding policy suggestions to be made for the improvement of various counties in Qinghai Province. In this way, prefecture-level cities can handle the urban issues brought on by waterlogging disasters, increase their regional comprehensive competitiveness, and accomplish sustainable development, improving the current state of regional urban resilience. The formula used to calculate the obstacle degree of urban resilience is expressed as follows:

$$m_j = \frac{W_j \times P_j}{\sum_{j=1}^{n}(W_j \times P_j)} \times 100\%, \quad (7)$$

$$P_j = 1 - Y_j, \quad (8)$$

where m_j is the degree of obstacle of urban resilience index; j represents the influence value of resilience evaluation index j on the regional urban resilience level; W_j is the weight of urban resilience index j; P_j is the index deviation degree and represents the difference between an urban area's resilience evaluation index factor and the total resilience of the regional urban area; Y_j is the normalized value of resilience index j.

2.5. Data Source

For the comprehensive consideration of the concept of vulnerability and resilience, seven indicators were developed from two categories of hazard and exposure, and nine indicators were developed from four categories: society, economy, environment, and infrastructure. The data sources of the evaluation index system are shown in Table 2. In this study, primary data related to hazards, exposure, and resilience were collected from 45 counties through field surveys. These data are mainly government statistics, including social, economic, and urban construction data. RDLS and Proportion of blue-green space were statistically obtained based on the collected geospatial data. In addition, some of the data were accounted for with reference to related studies in Qinghai Province, e.g., frequency of rainstorms. All raw data values were converted to comparable scales using percentages, per capita, etc. [39]. Since our variables were defined in a variety of statistical units, ranges, and scales in order to create and compute indices, these standardization techniques were crucial. Finally, ArcGIS version 10.4 was used to develop maps showing flood vulnerability and resilience in the study area.

Table 2. Indicators and data source of urban flood vulnerability and resilience.

Component	Criterion Layer	Indicator	Data Source
Urban vulnerability	Hazard	Frequency of rainstorm	Calculated based on daily precipitation from the China Meteorological Data Sharing Service System (http://data.cma.cn/site/index.html (accessed on 20 October 2023)).
		RDLS	Calculated using digital elevation model (DEM) data with 12.5 m spatial resolution from ALOS (Advanced Land Observing Satellite) (https://search.asf.alaska.edu/#/ (accessed on 21 October 2023)).
		Collapsible loess	The map of collapsible loess from Qinghai Geological Data Museum
	Exposure	The density of the population	County Statistical Yearbook in China
		The number of children	Population Census Yearbook in Qinghai
		The number of elderly	Population Census Yearbook in Qinghai
		GDP	County Statistical Yearbook in China

Table 2. *Cont.*

Component	Criterion Layer	Indicator	Data Source
Urban resilience	Society A	A1 Education level of people	Population Census Yearbook in Qinghai
		A2 Number of beds in medical institutions	Statistical Yearbook of County Construction in China
	Economy B	B1 per capita GDP	County Statistical Yearbook in China
		B2 Ratio of tertiary industry	County Statistical Yearbook in China
	Environment C	C1 Vegetation covering	Annual Report of Construction Statistics of Qinghai Province
		C2 Proportion of blue-green space	Calculated by Data of the Third National Land Survey
	Infrastructure D	D1 Density of roads	Annual Report of Construction Statistics of Qinghai Province
		D2 Density of drainage networks	Annual Report of Construction Statistics of Qinghai Province
		D3 The length of levee	Annual Report of Construction Statistics of Qinghai Province

3. Results

3.1. Urban Flood Vulnerability Assessment

3.1.1. Hazard

The frequency of rainstorms, RDLS, and collapsible loess were used to evaluate the risk levels of the hazard. The spatial distribution of hazard levels is shown in Figure 3. The hazard levels were divided into four levels of very high, high, medium, and low risk by the natural intermittent method of GIS. The amounts of very high, high, medium, and low levels were 8 (18%), 8 (18%), 14 (31%), and 15 (33%) counties, respectively.

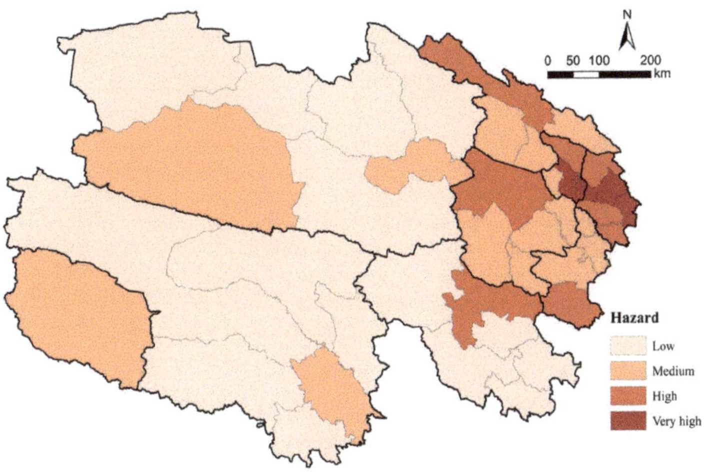

Figure 3. Spatial distribution of urban flooding hazard in Qinghai Province.

The areas with high levels of flooding hazard are found in the eastern of Qinghai. The eastern counties are confronted with rainstorms more frequently than the western regions due to the major influence of airflows and orographic lifting [56]. In addition, the eastern areas are mountainous, and urban areas tend to be situated in shallow river valleys, leading to severe secondary natural disasters like debris flows and landslides that occur after intense rains. Additionally, the counties with collapsible loess are mainly concentrated in the eastern portion of Qinghai Province, particularly Xining City and Haidong City. Our results are in line with the previous studies [57].

3.1.2. Exposure

The density of the population, numbers of children and the elderly, and the GDP were used to measure exposure to flooding. The risk levels of the exposure to flooding are shown in Figure 4. The numbers of very high-, high-, medium-, and low-exposure counties were 6 (13%), 11 (25%), 14 (31%), and 14 (31%), respectively.

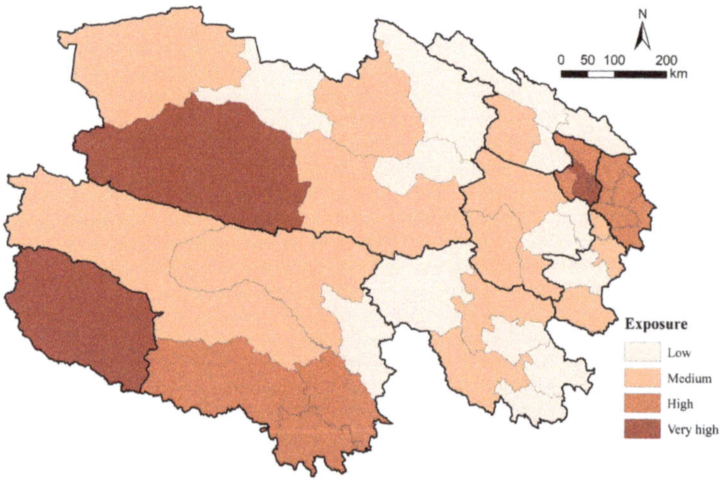

Figure 4. Spatial distribution of exposure to urban flooding.

In Qinghai Province, people exposed to flooding are more likely to reside in Xining and Haidong. Exposure is primarily determined by population density, which includes vulnerable populations like the elderly and children. This is because Xining and Haidong have experienced a rapid population expansion due to a large number of people moving there from other counties in Qinghai. The migration is primarily driven by two factors: (1) Xining and Haidong have good public infrastructure, including convenient public transportation, efficient schools, and hospitals; (2) there are more job opportunities, which attract many migrants who are young and generally bring their wives and children with them. Furthermore, Geermu, which experienced a very high level of exposure, has one of the highest GDPs in Qinghai. Notably, the high population density and high percentage of vulnerable population in Yushu City, as well as Zaduo and Nangqian counties in the south of Yushu Prefecture, are the primary causes of the high exposure levels in these areas.

3.1.3. The Spatial Distribution Characteristics of Vulnerability

The vulnerability level of Qinghai Province was assessed based on hazard and exposure. The spatial distribution of vulnerability levels is shown in Figure 5. The amounts of very high-, high-, medium-, and low-vulnerability counties were 5 (11%), 8 (18%), 17 (38%), and 15 (33%), respectively. Location with very high levels of vulnerability are scattered throughout Xining, which a has high hazard level and high exposure to flooding. Additionally, Yushu, Golmud, and Haidong City have a high level of vulnerability. There is no danger of flooding in the west of Qinghai.

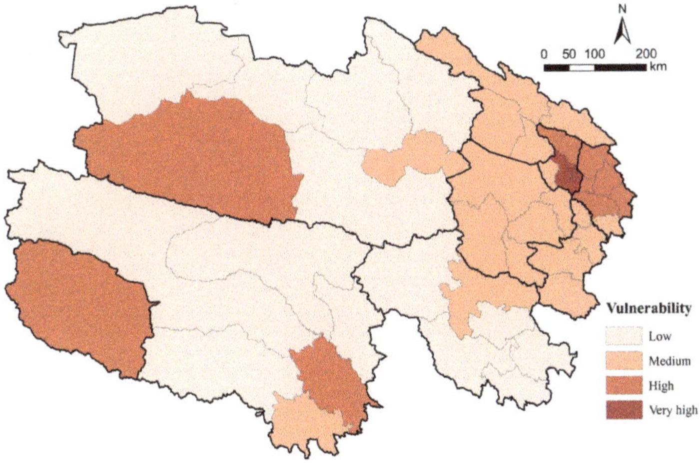

Figure 5. Spatial distribution of vulnerability to urban flooding.

3.2. Urban Flood Resilience Assessment

The nine entropy-weighted components are summarized to assess the resilience of Qinghai. The levels of resilience to flooding are spatially demonstrated in Figure 6. In Qinghai, 33 counties (73%) have low or medium levels of resilience, most of which are accumulated in the middle and south; 4 (9%) and 8 (18%) counties have very high or high resilience levels, distributed in Xining, Haidong, and Haixi.

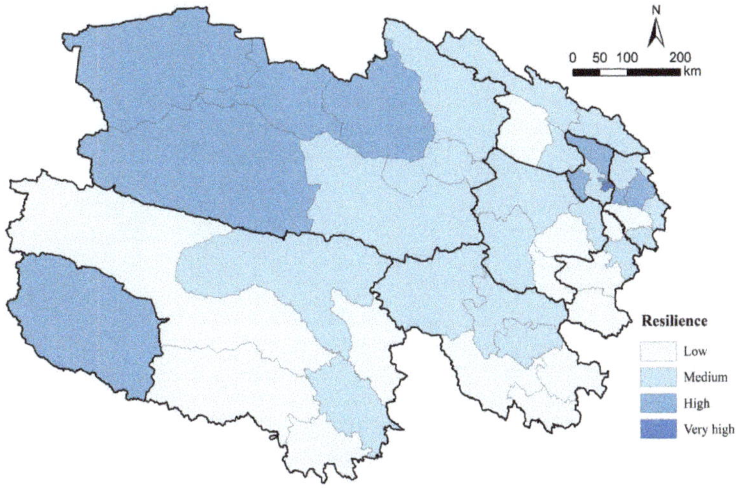

Figure 6. Spatial distribution of resilience to urban flooding.

Due to Xining's role as a political, economic, cultural, and commercial center in Qinghai, it has a very high level of resilience to disasters, mainly because of its good performance in infrastructure development, education levels, medical conditions, and industrial structure. Haixi has high social and economic resilience, but the construction level of drainage networks and levees is insufficient, and the comprehensive resilience of flooding is at a high level. The level of resilience in the central and southern regions of Qinghai is medium or low, mainly due to the following reasons: the low density of the road network and underdeveloped transport, which makes it impossible for medical teams and relief supplies to enter. Secondly, weak economic development makes the level of recovery from disasters less favorable.

3.3. Spatial Balance between Vulnerability and Resilience

This study identifies clustering patterns combined with the vulnerability and resilience to flooding. Based on four clustering patterns, three scenarios are categorized as follows: (1) self-adaptive, where vulnerability and resilience are matched; (2) over-abundant, where resources for resilience are excessive; and (3) deficient nature–nurture, where resilience is insufficient to overcome high vulnerability to flooding (Figure 7).

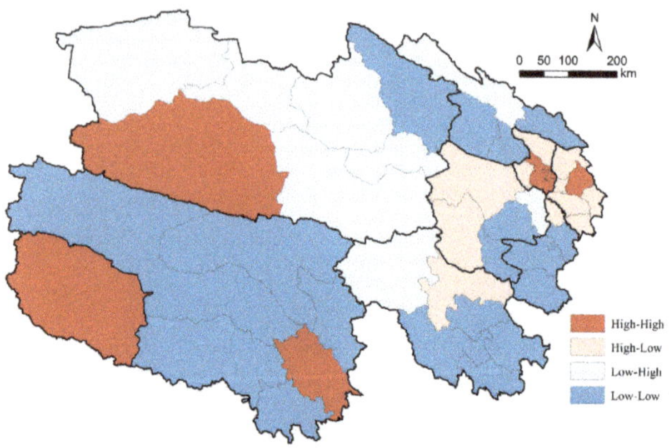

Figure 7. Spatial distribution of combination between vulnerability and resilience.

The spatial distribution of flood vulnerability and resilience in Qinghai is spatially unbalanced (Figure 7). The results show that 8 counties (18%) belong to the high–high cluster, and 18 counties (40%) belong to the low–low cluster. Thus, 24 counties (58%) can self-adapt to urban flooding by using resources that match their requirements. These areas can adapt to flooding hazard by using resilience resources that meet their needs, which is an ideal state for urban sustainable development. In addition, 8 counties (18%) in Qinghai belong to the low–high cluster, which has surplus flood adaptation resources, mainly concentrated in the north. Meanwhile, 11 counties (24%) in the high–low cluster have insufficient resilience resources and are hence vulnerable to flooding. Those areas are mainly distributed in the middle and east of Qinghai.

Therefore, vulnerability and adaptation resource allocation in Qinghai is spatially uneven. The majority of Qinghai Province's counties are in a state of self-adaptiveness, meaning that the resilience resources possessed by local residents match the level of flood vulnerability, and residents are able to receive timely assistance in the event of a disaster and do not suffer from a lack of resources. Several counties in Haidong and Hainan lack adaptation resources, while almost all excess adaptation resources are located in Haixi. This indicates that despite that the Chinese government playing a dominant role in adaptation resource distribution for the reduction of equity gaps [58], resources for mitigating the effects of urban flooding remain superfluous in economically well-developed areas. Thus, the methods for balancing such distributing resources must be investigated and deliberated.

3.4. Obstacle Factor Analysis

The primary obstacle to urban resilience in Qinghai Province is the insufficient length of flood control levees, impacting 25 counties (Figure 8). Due to the inadequate flood control infrastructure in Qinghai Province, 18 counties have yet to establish flood control levees, consequently diminishing urban resilience against rainfall and floods. Moreover, the availability of hospital beds, which rank first and second in 18 (40%) and 19 (42%) counties, respectively, emerges as the second most important factor impacting flood resilience. The reason for this is due to the fact that Qinghai Province, with the exception of Xining, has relatively low rates of urbanization and regional economic development, and the urban social security system is directly impacted by the underfunding of medical resources. The allocation of medical resources in the province is also imbalanced. The infrastructure resilience rankings for road density (D1) and drainage network density (D2) are fourth and fifth, respectively. Infrastructure has a substantial effect on resilience. Regarding economic resilience, per capita GDP (B1) is ranked third, while the percentage of tertiary industry (B2) is ranked ninth. This is mostly because agriculture dominates the urban economic structure and there is little urban development, which results in the region's overall slowly economic growth. The quantity of blue-green areas (C2) and vegetation coverage (C1) of environmental resilience have a lesser degree of resilience hurdles, typically placing seventh and eighth, respectively. In recent years, Xining has actively engaged in afforestation, ecological restoration, and tree planting, thereby enhancing the quality of the ecological environment. In other counties, there is a high rate of ecological resource preservation and a high degree of ecological resilience because of the low levels of urbanization and development [59,60].

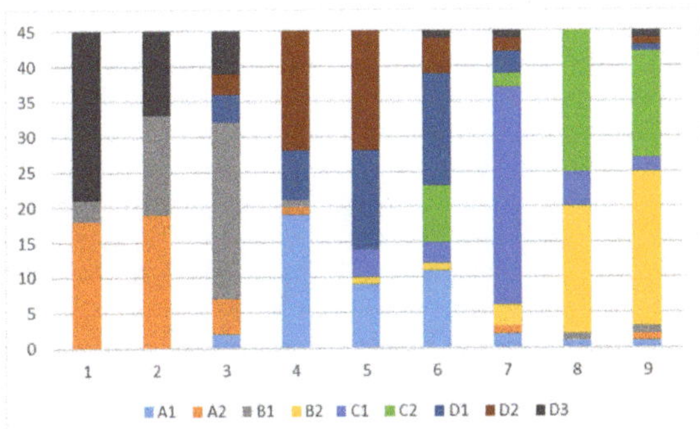

Figure 8. Ranking frequency of resilience indicators.

4. Discussion

4.1. Assessment of Urban Flood Vulnerability and Resilience

Nowadays, cities are challenged by multiple external shocks (e.g., global warming) and internal pressure (e.g., urbanization and human activities) that contribute to increased flood risks [61]. Investigating cities' susceptibility to flooding and their ability to adapt and explore the spatial structure of urban flood vulnerability and resilience can lead to effective targeted urban planning to mitigate the flood risk and increase resilience. Prior researches have evaluated vulnerability and resilience to natural disasters, incorporating social, economic, and environmental factors to assess hazard risks in particular locales [62–64], with existing literature often missing concrete strategies for enhancing resilience. This study conceptualizes vulnerability to hazards and exposure and resilience to urban flooding and explores the spatial pattern between vulnerability and resilience from the perspective of spatial equilibrium. Employing a barrier degree model identifies vital factors influencing urban flood resilience. The findings offer valuable insights for refining resilience metrics and fostering interregional collaboration to strengthen resilience. Further, it suggests expanding the research scope to cover broader areas in future studies.

Scholars have explored and discussed the concepts of vulnerability and resilience in recent works [54,65]. Some scholars see resilience and vulnerability as constituent elements of each other [66], and others consider one or both of these concepts as constituent elements of broader notions such as risk [54]. Resilience, broadly defined as the ability to withstand and recover from loss, needs to address two components: the losses and the recovery process [67,68]. Losses, which include physical, economic and social dimensions, provide critical insights into the impact of hazards, such as damaged buildings, loss of assets, and affected populations. The recovery process is an integral part of efforts to build resilience. Key aspects of the recovery process include adequate financial resources, infrastructural facilities and logistical support for timely and coordinated responses. Compared to existing studies, this study proposed an assessment framework with the dimensions of hazard, exposure and resilience: (1) hazard, which is urban flooding caused by storms, including frequency and severity; (2) exposure, which is the situation of population casualties and asset losses caused by rainfall; and (3) resilience, which refers to the allocation of resources and socio-economic systems for post-disaster recovery and adaptation to urban flooding. The framework covers hazards, disaster losses and the recovery process of flooding and can inform risk assessment, promote sustainable development and enhance adaptive capacity to reduce vulnerability.

It is difficult to find a single indicator or set of indicators for measuring urban vulnerability and resilience due to the significant influence of geography and climate. In contrast to previous studies [19,69,70], the framework and indicator system of this study takes into account the unique soil and climatic conditions of the Tibetan Plateau and ensures that the indicators can be applied robustly under conditions of high altitude, extreme climate, and terrain diversity. Through collaboration with local stakeholders and professionals, this study has accessed and integrated local knowledge and experience to enrich and refine the study framework to ensure the feasibility and applicability of our approach in practice. It acknowledges the unique issue of loess collapsibility in Qinghai, which leads to structural damage like house tilting and road collapses—urgent problems in Chinese central and western areas. Under climatic humidification, the collapsibility of loess will exacerbate the flood disaster. Thus, the collapsibility of loess was considered the vulnerability factor. Furthermore, flooding disasters in Qinghai Province are caused by persistent or short-term excessive precipitation. Despite official meteorological criteria in Qinghai province (DB63/T 372-2011) [71] deeming rainstorms as events with over 50 mm of rainfall in 24 h or 30 mm in 12 h, such definitions do not align with Qinghai's flood occurrences. Therefore, this research utilizes risk grading from related studies to represent rainstorm frequency in Qinghai more accurately [37]. Our research methodology not only accurately assesses urban vulnerability and resilience in the Tibetan Plateau region, but can also provide practical recommendations and guidance for resolving regional development and sustainable management under conditions of increased flooding risk of climate change.

4.2. Strategies for Qinghai Province to Improve Urban Flood Resilience

The spatial matching perspective of the vulnerability-resilience correlation exploration not only enriches our knowledge of urban flood vulnerability and resilience but also simulates practical ways to reduce the impacts of hazardous sites through urban planning, primarily through optimising the allocation of adaptation resources. This study has found spatial imbalances in the vulnerability and resilience of counties in Qinghai Province (Figure 7). Specifically, 24 counties (58%) are self-adaptive to urban flooding, 8 counties (18%) have surplus resources for flood adaptation, and 11 counties (24%) have insufficient resources for resilience and are therefore vulnerable to flooding. The disparities between counties are significant and require targeted solutions [72]. Our findings have several policy relevance for enhancing resilience in different regions to reduce the impact of urban flood vulnerability.

In self-adaptive areas, urban sustainability can be maintained through urban blue and green line control, limiting urban expansion, and protecting existing natural resources. Areas with insufficient urban resilience resources prioritize impact factors related to urban resilience, such as levee construction and allocation of healthcare resources. In summary, urban resilience can be increased through engineering and non-engineering measures.

(1) Improving the local economy is crucial to increasing counties' resilience to floods. The driving force behind the development of counties and the well-being of their inhabitants depends mainly on the local economy. Simultaneously, urban flood mitigation facilities and post-flood reconstruction efforts cannot be separated from the city's financial support.

Qinghai Province should increase the radiation-driven role of Xining's capital city to the surrounding area and promote urbanization construction and economic development.

(2) The integrated green–grey–blue (IGGB) system has gained interest worldwide in mitigating flood issues [73,74]. The IGGB urban infrastructure system for flood management should be strengthened to construct different "watershed, city, drainage zone levels". A sustainable and resilient urban environment can be created with the aid of IGGB system measures like enhancing the implementation of natural storage space and rainwater drainage channels, coordinating ecological adjustments for water conservation, optimizing the functional layout of blue and green spaces, and strategically placing grey facilities for drainage. The key to implementing an IGGB system for managing flooding is combining various technologies, including planning, water conservation, drainage, ecology, and landscape, as well as above-ground and subsurface facilities, natural space, and artificial facilities at different scales. The specialized technique of reducing urban waterlogging and establishing a habitable environment through the IGGB system is used in the construction of sponge city in China. In other words, the complete control of the total amount, peak value, frequency, and ecological and environmental pollution of urban stormwater runoff can be achieved through the comprehensive application of "green" + "blue" +" grey" engineering measures and intelligent regulation. This creates cities with the "resilience" to withstand flooding disasters.

(3) An intelligent information management platform is being built. Emerging urban planning technologies such as big data, cloud computing, artificial intelligence, mobile Internet, and so on are constantly updated while putting forward new requirements for urban flood risk management and resilience construction. In the future, counties should take urban safety technology as the core and build an integrated flood risk factor intelligent information management platform driven by interdisciplinary technology to achieve efficient and intellectual development of urban resilience management.

(4) Ecological cities are the main direction of sustainable and green development. As urban areas expand and urban populations increase, the contradiction between environmental protection and development is becoming increasingly evident. Ecological restoration of lakes, rivers, and forests in Qinghai Province should be carried out to enhance water absorption and fixation capacity during heavy rains and floods.

4.3. Contribution and Implication

This study constructed the index system of spatial balance between urban flood vulnerability and resilience in Qinghai Province. The method in this study improves the shortcomings of the previous urban flood resilience assessment in index data collection, index weighting coupling, and barrier degree assessment and can effectively assess the flood resilience of various cities in Qinghai Province without losing its accuracy [75,76]. Compared with the urban flood resilience assessment framework commonly used in existing studies, it simplifies the previous complex assessment framework from pre-disaster exposure, disaster sensitivity, adaptability, and post-disaster resilience [18]. The evaluation framework with exposure, disaster, and resilience in the present study can more intuitively correspond to the resilience characteristics in the three periods before, during, and after the flood. In addition, the research scope of the existing literature mostly focuses on the comprehensive assessment results of urban-scale flood resilience [19,57]. In this study, the spatial balance between urban flood vulnerability and resilience and the degree of barrier assessment were conducted for each administrative region on the provincial scale, which was more conducive to identifying regional differences in flood resilience levels and analyzing the causes from the level of important barrier factors. This study can further expand the resilience assessment and presentation of dynamic data and can provide better guidance for the formulation of regional flood management and resilient urban planning.

Our results indicate that the number of hospital beds, the length of flood levees, and per capita GDP have significant impacts on urban flood resilience. Strengthening these aspects would assist cities both within and beyond the Qinghai–Tibetan Plateau more effectively cope with natural disasters such as floods. Other cities can also adopt corresponding methods and measures to identify the main obstacles and improve urban flood management. We believe that our research is valuable for global urban planning and disaster reduction efforts. Many countries' cities face similar natural disaster challenges, and our spatial modeling framework proposed can help identify critical vulnerable areas and formulate targeted response strategies. This framework provides a basis for the effective allocation and prioritization of adaptive resources, thereby comprehensively enhancing disaster resilience, reducing disaster losses, and promoting sustainable urban development.

5. Conclusions

This study investigates the spatial patterns between vulnerability and resilience and analyzes the obstacles to urban resilience using the case of Qinghai Province. The results show that the spatial inequality of resilience resources exacerbates the level of flood vulnerability in some counties of Qinghai. Therefore, the Qinghai–Tibet Plateau's distinctive urban flood mitigation solutions are given. The conclusions indicate the following: (1) The counties with high level of vulnerability are scattered throughout Xining, which has high hazard levels and exposure to flooding. The main factor contributing to the high level of hazard of the eastern counties is that these areas are mountainous, with collapsible loess and more frequent rainfall. Meanwhile, the factor contributing to exposure in the east is the combination of high GDPs and dense populations, including vulnerable groups. (2) A total of 33 counties (73%) have low or medium levels of resilience, and 4 (9%) and 8 (18%) counties have very high or high resilience levels, respectively, in Qinghai. (3) The spatial distribution of flood vulnerability and resilience in Qinghai is spatially unbalanced. The main factors affecting urban resilience in Qinghai include the number of hospital beds, the length of flood levees, and per capita GDP. Due to the active ecological restoration work in Qinghai, the environmental resilience obstacle factor is minimal. (4) Local governments follow these conclusions while constructing and preventing urban flooding. It is crucial to manage risk through engineering and non-engineering means, such as economy, ecology, IGGB system, and intelligent information management platform, in order to fully strengthen urban resilience to flooding.

Certain limitations must be addressed to enhance the quality and impact of our future work. Firstly, while this study establishes a robust framework for assessing vulnerability and resilience, it encounters constraints due to data availability, computational complexity, and the complexity of the concepts of vulnerability and resilience. The indicators may not fully reflect the level of vulnerability and resilience in the study area. Critical indicators, such as the spatial distribution of buildings and the number of emergency pumping stations within the city, which are challenging to quantify or access, were not sufficiently represented. Future research should strive to incorporate and analyze these indicators more comprehensively. Secondly, the entropy weighting method was chosen to determine the weights of the indicators, and the stakeholders were not considered. In future research, the accuracy of the model should be improved by combining expert consultation and questionnaire surveys with subjective and objective methods. Thirdly, the data processing process may cause biases in the results. For example, the distribution map of collapsible loess was obtained by manual vectorization in GIS, which may lead to inaccuracy. Additionally, the strategy proposed in this study predominantly reflects a governmental viewpoint. Subsequent research should explore the dynamics between various social actors throughout the disaster management process. It is advisable to devise enhancement strategies from the perspective of key stakeholders in urban flood management, including governments, residents, businesses, and volunteers, focusing on a collaborative approach to bolster urban resilience.

Author Contributions: Conceptualization, W.W., C.X. and W.B.; methodology, W.W. and J.H.; investigation, W.W., Z.C., C.X. and R.L.; software, W.W., J.H. and Z.C.; writing—original draft preparation, W.W.; writing—review and editing, W.W., Z.C., J.H. and R.L.; supervision, W.B.; project administration, W.B. All authors have read and agreed to the published version of the manuscript.

Funding: This research was funded by Qinghai Province Science and Technology Achievement Transformation Special Project, grant number 2022-SF-159. Research and Development Project of the Ministry of Housing and Urban–Rural Development, Project supported by Beijing Advanced Innovation Center for Future Urban Design, grant number 2022-K-024.

Data Availability Statement: The original contributions presented in the study are included in the article, further inquiries can be directed to the corresponding authors.

Conflicts of Interest: Authors Wenjing Wang, Junchao He, Zhongwen Chi, Weilan Bai and Rui Liu were employed by the company China Urban Construction Design & Research Institute Co., Ltd. The remaining authors declare that the research was conducted in the absence of any commercial or financial relationships that could be construed as a potential conflict of interest.

References

1. Kotzee, I.; Reyers, B. Piloting a social-ecological index for measuring flood resilience: A composite index approach. *Ecol. Indic.* **2016**, *60*, 45–53. [CrossRef]
2. Abbas, A.; Bhatti, A.S.; Ullah, S.; Ullah, W.; Waseem, M.; Zhao, C.; Dou, X.; Ali, G. Projection of precipitation extremes over South Asia from CMIP6 GCMs. *J. Arid. Land* **2023**, *15*, 274–296. [CrossRef]
3. Ullah, W.; Karim, A.; Ullah, S.; Rehman, A.U.; Bibi, T.; Wang, G.; Ullah, S.; Bhatti, A.S.; Ali, G.; Abbas, A. An increasing trend in daily monsoon precipitation extreme indices over Pakistan and its relationship with atmospheric circulations. *Front. Environ. Sci.* **2023**, *11*, 1228817. [CrossRef]
4. Ullah, W.; Wang, G.; Lou, D.; Ullah, S.; Bhatti, A.S.; Ullah, S.; Karim, A.; Hagan, D.F.T.; Ali, G. Large-scale atmospheric circulation patterns associated with extreme monsoon precipitation in Pakistan during 1981–2018. *Atmos. Res.* **2021**, *253*, 105489. [CrossRef]
5. Bin, L.; Xu, K.; Pan, H.; Zhuang, Y.; Shen, R. Urban flood risk assessment characterizing the relationship among hazard, exposure, and vulnerability. *Environ. Sci. Pollut. Res.* **2023**, *30*, 86463–86477. [CrossRef]
6. Mu, Y.; Li, Y.; Yan, R.; Luo, P.; Liu, Z.; Sun, Y.; Wang, S.; Zhu, W.; Zha, X. Analysis of the Ongoing Effects of Disasters in Urbanization Process and Climate Change: China's Floods and Droughts. *Sustainability* **2024**, *16*, 14. [CrossRef]
7. Eccles, R.; Zhang, H.; Hamilton, D.; Trancoso, R.; Syktus, J. Impacts of climate change on streamflow and floodplain inundation in a coastal subtropical catchment. *Adv. Water Resour.* **2021**, *147*, 103825. [CrossRef]
8. Ghaderpour, E.; Vujadinovic, T.; Hassan, Q.K. Application of the Least-Squares Wavelet software in hydrology: Athabasca River Basin. *J. Hydrol.-Reg. Stud.* **2021**, *36*, 100847. [CrossRef]
9. Zhang, J.X.; Wang, H.M.; Huang, J.; Sun, D.C.; Liu, G.F. Evaluation of Urban Flood Resilience Enhancement Strategies-A Case Study in Jingdezhen City under 20-Year Return Period Precipitation Scenario. *ISPRS Int. J. Geo-Inf.* **2022**, *11*, 285. [CrossRef]
10. Lin, Y.Z.; Peng, C.; Shu, J.F.; Zhai, W.; Cheng, J.Q. Spatiotemporal characteristics and influencing factors of urban resilience efficiency in the Yangtze River Economic Belt, China. *Environ. Sci. Pollut. Res.* **2022**, *29*, 39807–39826. [CrossRef]
11. Coaffee, J.; Therrien, M.C.; Chelleri, L.; Henstra, D.; Aldrich, D.P.; Mitchell, C.L.; Tsenkova, S.; Rigaud, É. Urban resilience implementation: A policy challenge and research agenda for the 21st century. *J. Contingencies Crisis Manag.* **2018**, *26*, 403–410. [CrossRef]
12. Bozza, A.; Asprone, D.; Manfredi, G. Developing an integrated framework to quantify resilience of urban systems against disasters. *Nat. Hazards* **2015**, *78*, 1729–1748. [CrossRef]
13. Tate, E.; Strong, A.; Kraus, T.; Xiong, H.Y. Flood recovery and property acquisition in Cedar Rapids, Iowa. *Nat. Hazards* **2016**, *80*, 2055–2079. [CrossRef]
14. Duy, P.N.; Chapman, L.; Tight, M.; Thuong, L.V.; Linh, P.N. Urban resilience to floods in coastal cities: Challenges and opportunities for Ho Chi Minh city and other emerging cities in southeast Asia. *J. Urban Plan. Dev.* **2018**, *144*, 5017018. [CrossRef]
15. Song, J.; Chang, Z.; Li, W.F.; Feng, Z.; Wu, J.S.; Cao, Q.W.; Liu, J.Z. Resilience-vulnerability balance to urban flooding: A case study in a densely populated coastal city in China. *Cities* **2019**, *95*, 102381. [CrossRef]
16. Cutter, S.L.; Ash, K.D.; Emrich, C.T. The geographies of community disaster resilience. *Glob. Env. Chang.* **2014**, *29*, 65–77. [CrossRef]
17. Shi, C.; Zhu, X.; Wu, H.; Li, Z. Assessment of Urban Ecological Resilience and Its Influencing Factors: A Case Study of the Beijing-Tianjin-Hebei Urban Agglomeration of China. *Land* **2022**, *11*, 921. [CrossRef]
18. Sun, R.; Shi, S.; Reheman, Y.; Li, S. Measurement of urban flood resilience using a quantitative model based on the correlation of vulnerability and resilience. *Int. J. Disaster Risk Reduct.* **2022**, *82*, 103344. [CrossRef]

19. Chen, X.; Guo, Z.; Zhou, H.; Qian, X.; Zhang, X. Urban Flood Resilience Assessment Based on VIKOR-GRA: A Case Study in Chongqing, China. *KSCE J. Civ. Eng.* **2022**, *26*, 4178–4194. [CrossRef]
20. Moghadas, M.; Asadzadeh, A.; Vafeidis, A.; Fekete, A.; Kötter, T. A multi-criteria approach for assessing urban flood resilience in Tehran, Iran. *Int. J. Disaster Risk Reduct.* **2019**, *35*, 101069. [CrossRef]
21. Pathak, S.D.; Kulshrestha, M. Assessment of Flood Resilience Using RAAAR Framework: The Case of Narmada River Basin, India. *Environ. Eng. Manag. J. (EEMJ)* **2021**, *20*, 1263–1276.
22. Chen, X.; Jiang, S.; Xu, L.; Xu, H.; Guan, N. Resilience assessment and obstacle factor analysis of urban areas facing waterlogging disasters: A case study of Shanghai, China. *Env. Sci. Pollut. Res.* **2023**, *30*, 65455–65469. [CrossRef] [PubMed]
23. Tempa, K. District flood vulnerability assessment using analytic hierarchy process (AHP) with historical flood events in Bhutan. *PLoS ONE* **2022**, *17*, e270467. [CrossRef] [PubMed]
24. Yazdani, M.; Loosemore, M.; Mojtahedi, M.; Sanderson, D.; Haghani, M. An integration of operations research and design science research methodology: With an application in hospital disaster management. *Prog. Disaster Sci.* **2023**, *20*, 100300. [CrossRef]
25. Yazdani, M.; Mojtahedi, M.; Loosemore, M.; Sanderson, D. A modelling framework to design an evacuation support system for healthcare infrastructures in response to major flood events. *Prog. Disaster Sci.* **2022**, *13*, 100218. [CrossRef]
26. Yazdani, M.; Mojtahedi, M.; Loosemore, M.; Sanderson, D.; Dixit, V. An integrated decision model for managing hospital evacuation in response to an extreme flood event: A case study of the Hawkesbury-Nepean River, NSW, Australia. *Saf. Sci.* **2022**, *155*, 105867. [CrossRef]
27. Yazdani, M.; Haghani, M. Elderly people evacuation planning in response to extreme flood events using optimisation-based decision-making systems: A case study in western Sydney, Australia. *Knowl.-Based Syst.* **2023**, *274*, 110629. [CrossRef]
28. Kreibich, H.; Di Baldassarre, G.; Vorogushyn, S.; Aerts, J.C.; Apel, H.; Aronica, G.T.; Arnbjerg Nielsen, K.; Bouwer, L.M.; Bubeck, P.; Caloiero, T. Adaptation to flood risk: Results of international paired flood event studies. *Earth's Future* **2017**, *5*, 953–965. [CrossRef]
29. Bai, D.; Cao, J.; Zhang, Y. Temporal and spatial distribution of extreme precipitation in Qinghai Province in recent 60 years. *Yangtze River* **2022**, *53*, 59–64. (In Chinese)
30. Han, Z.; Song, W.; Deng, X.Z.; Xu, X.L. Trade-Offs and Synergies in Ecosystem Service within the Three-Rivers Headwater Region, China. *Water* **2017**, *9*, 588. [CrossRef]
31. Guo, J.; Wei, Z.; Ren, J.; Luo, Z.; Zhou, H. Early-warning measures for ecological security in the Qinghai alpine agricultural area. *Int. J. Environ. Res. Public Health* **2020**, *17*, 9292. [CrossRef]
32. Yang, H.; Shen, H.; Ma, Y.; Yang, Y.; Wang, X.; Feng, G. Research on the Characteristics and Influence Factors of Autumn Continuous Rain in Qinghai Province. *Front. Earth Sc.* **2022**, *10*, 801075. [CrossRef]
33. Ye, S.; Ge, Y.; Xu, S.; Ma, X. Measurement and prediction of coupling coordination level of economic development, social stability and ecological environment in Qinghai—Thoughts on sustainable societal safety. *Sustainability* **2022**, *14*, 10515. [CrossRef]
34. Bruwier, M.; Maravat, C.; Mustafa, A.; Teller, J.; Pirotton, M.; Erpicum, S.; Archambeau, P.; Dewals, B. Influence of urban forms on surface flow in urban pluvial flooding. *J. Hydrol.* **2020**, *582*, 124493. [CrossRef]
35. Ouyang, Z.Y.; Song, C.S.; Zheng, H.; Polasky, S.; Xiao, Y.; Bateman, I.J.; Liu, J.G.; Ruckelshaus, M.; Shi, F.Q.; Xiao, Y.; et al. Using gross ecosystem product (GEP) to value nature in decision making. *Proc. Natl. Acad. Sci. USA* **2020**, *117*, 14593–14601. [CrossRef]
36. Yang, W.; Xu, K.; Lian, J.; Bin, L.; Ma, C. Multiple flood vulnerability assessment approach based on fuzzy comprehensive evaluation method and coordinated development degree model. *J. Environ. Manag.* **2018**, *213*, 440–450. [CrossRef]
37. Li, W.; Yu, D.; Feng, X.; Zhang, T. Risk assessment of rainstorm and flood disasters based on the hazard grades/indices in Qinghai Province. *J. Glaciol. Geocryol.* **2019**, *41*, 680–688. (In Chinese)
38. Hemmati, M.; Ellingwood, B.R.; Mahmoud, H.N. The role of urban growth in resilience of communities under flood risk. *Earth's Future* **2020**, *8*, e2019EF001382. [CrossRef]
39. O'Donnell, E.C.; Thorne, C.R. Drivers of future urban flood risk. *Philos. Trans. R. Soc. A* **2020**, *378*, 20190216. [CrossRef]
40. Zhao, Q.; Li, X.; Cao, Y.; Li, Z.; Fan, J. Prediction of Collapsibility of Loess of Construction Sites in Xining Based on Machine Learning Methods. *Res. Sq.* **2021**. [CrossRef]
41. Bosher, L.; Dainty, A.; Carrillo, P.; Glass, J.; Price, A. Attaining improved resilience to floods: A proactive multi-stakeholder approach. *Disaster Prev. Manag. Int. J.* **2009**, *18*, 9–22. [CrossRef]
42. McCarthy, J.F. Using community led development approaches to address vulnerability after disaster: Caught in a sad romance. *Glob. Environ. Chang.* **2014**, *27*, 144–155. [CrossRef]
43. Namdeo, A.; Tiwary, A.; Farrow, E. Estimation of age-related vulnerability to air pollution: Assessment of respiratory health at local scale. *Environ. Int.* **2011**, *37*, 829–837. [CrossRef] [PubMed]
44. Miguez, M.G.; Veról, A.P. A catchment scale Integrated Flood Resilience Index to support decision making in urban flood control design. *Environ. Plan. B-Urban* **2017**, *44*, 925–946. [CrossRef]
45. Cutter, S.L. The landscape of disaster resilience indicators in the USA. *Nat. Hazards* **2016**, *80*, 741–758. [CrossRef]
46. Song, J.L.; Huang, B.; Li, R.R. Measuring Recovery to Build up Metrics of Flood Resilience Based on Pollutant Discharge Data: A Case Study in East China. *Water* **2017**, *9*, 619. [CrossRef]
47. Abubakar, I.R.; Aina, Y.A. The prospects and challenges of developing more inclusive, safe, resilient and sustainable cities in Nigeria. *Land Use Policy* **2019**, *87*, 104105. [CrossRef]

48. Zhang, X.W.; Song, J.; Peng, J.; Wu, J.S. Landslides-oriented urban disaster resilience assessment-A case study in ShenZhen, China. *Sci. Total Environ.* **2019**, *661*, 95–106. [CrossRef] [PubMed]
49. Hoque, M.A.; Tasfia, S.; Ahmed, N.; Pradhan, B. Assessing Spatial Flood Vulnerability at Kalapara Upazila in Bangladesh Using an Analytic Hierarchy Process. *Sensors* **2019**, *19*, 1302. [CrossRef]
50. Liao, K.H. A Theory on Urban Resilience to Floods-A Basis for Alternative Planning Practices. *Ecol. Soc.* **2012**, *17*, 48. [CrossRef]
51. Farley, K.A.; Jobbágy, E.G.; Jackson, R.B. Effects of afforestation on water yield: A global synthesis with implications for policy. *Glob. Chang. Biol.* **2005**, *11*, 1565–1576. [CrossRef]
52. Ruan, J.E.; Chen, Y.B.; Yang, Z.W. Assessment of temporal and spatial progress of urban resilience in Guangzhou under rainstorm scenarios. *Int. J. Disaster Risk Reduct.* **2021**, *66*, 102578. [CrossRef]
53. Wood, N.J.; Burton, C.G.; Cutter, S.L. Community variations in social vulnerability to Cascadia-related tsunamis in the US Pacific Northwest. *Nat. Hazards* **2010**, *52*, 369–389. [CrossRef]
54. Cutter, S.L.; Barnes, L.; Berry, M.; Burton, C.; Evans, E.; Tate, E.; Webb, J. A place-based model for understanding community resilience to natural disasters. *Glob. Environ. Chang.* **2008**, *18*, 598–606. [CrossRef]
55. Andrade, C. Z scores, standard scores, and composite test scores explained. *Indian J. Psychol. Med.* **2021**, *43*, 555–557. [CrossRef] [PubMed]
56. Liu, Y.; Gao, G.; Li, H.; Liu, L.; Fan, Z.; Wen, T. Spatiotemporal Variations and Causes of Wind/Rainfall Erosion Climatic Erosivity in Qinghai Province, China. *Atmosphere* **2022**, *13*, 1649. [CrossRef]
57. Tayyab, M.; Zhang, J.; Hussain, M.; Ullah, S.; Liu, X.; Khan, S.N.; Baig, M.A.; Hassan, W.; Al-Shaibah, B. Gis-based urban flood resilience assessment using urban flood resilience model: A case study of peshawar city, khyber pakhtunkhwa, pakistan. *Remote Sens.* **2021**, *13*, 1864. [CrossRef]
58. Witten, K.; Exeter, D.; Field, A. The Quality of Urban Environments: Mapping Variation in Access to Community Resources. *Urban Stud.* **2003**, *40*, 161–177. [CrossRef]
59. Han, Z.; Song, W.; Deng, X. Responses of Ecosystem Service to Land Use Change in Qinghai Province. *Energies* **2016**, *9*, 303. [CrossRef]
60. Fan, Y.; Fang, C. Evolution process and obstacle factors of ecological security in western China, a case study of Qinghai province. *Ecol. Indic.* **2020**, *117*, 106659. [CrossRef]
61. Li, W.; Jiang, R.; Wu, H.; Xie, J.; Zhao, Y.; Song, Y.; Li, F. A System Dynamics Model of Urban Rainstorm and Flood Resilience to Achieve the Sustainable Development Goals. *Sustain. Cities Soc.* **2023**, *96*, 104631. [CrossRef]
62. Frigerio, I.; Ventura, S.; Strigaro, D.; Mattavelli, M.; De Amicis, M.; Mugnano, S.; Boffi, M. A GIS-based approach to identify the spatial variability of social vulnerability to seismic hazard in Italy. *Appl. Geogr.* **2016**, *74*, 12–22. [CrossRef]
63. Shepard, D.P.; Humphreys, T.E.; Fansler, A.A. Evaluation of the vulnerability of phasor measurement units to GPS spoofing attacks. *Int. J. Crit. Infrastruct. Prot.* **2012**, *5*, 146–153. [CrossRef]
64. Song, G.; Chen, Y.; Tian, M.; Lv, S.; Zhang, S.; Liu, S. The Ecological Vulnerability Evaluation in Southwestern Mountain Region of China Based on GIS and AHP Method. *Procedia Environ. Sci.* **2010**, *2*, 465–475. [CrossRef]
65. Miller, F.; Osbahr, H.; Boyd, E.; Thomalla, F.; Bharwani, S.; Ziervogel, G.; Walker, B.; Birkmann, J.; Van der Leeuw, S.; Rockström, J. Resilience and vulnerability: Complementary or conflicting concepts? *Ecol. Soc.* **2010**, *15*, 11. [CrossRef]
66. Turner, B.L.; Kasperson, R.E.; Matson, P.A.; McCarthy, J.J.; Corell, R.W.; Christensen, L.; Eckley, N.; Kasperson, J.X.; Luers, A.; Martello, M.L. A framework for vulnerability analysis in sustainability science. *Proc. Natl. Acad. Sci. USA* **2003**, *100*, 8074–8079. [CrossRef]
67. Kourehpaz, P.; Molina Hutt, C. Machine learning for enhanced regional seismic risk assessments. *J. Struct. Eng.* **2022**, *148*, 4022126. [CrossRef]
68. Forcellini, D. An expeditious framework for assessing the seismic resilience (SR) of structural configurations. *Structures* **2023**, *56*, 105015. [CrossRef]
69. Zhang, H.; Liu, X.; Xie, Y.; Gou, Q.; Li, R.; Qiu, Y.; Hu, Y.; Huang, B. Assessment and Improvement of Urban Resilience to Flooding at a Subdistrict Level Using Multi-Source Geospatial Data: Jakarta as a Case Study. *Remote Sens.* **2022**, *14*, 2010. [CrossRef]
70. Li, Z.; Zhang, X.; Ma, Y.; Feng, C.; Hajiyev, A. A multi-criteria decision making method for urban flood resilience evaluation with hybrid uncertainties. *Int. J. Disaster Risk Reduct.* **2019**, *36*, 101140. [CrossRef]
71. *DB63/T 372-2011*; Qinghai Province Bureau of Quality and Technology Supervision. Local Standards of Qinghai Province: Meteorological Disasters Standards. Qinghai Province Bureau of Quality and Technology Supervision: Xining, China, 2011.
72. Zhu, S.; Feng, H.; Arashpour, M.; Zhang, F. Enhancing urban flood resilience: A coupling coordinated evaluation and geographical factor analysis under SES-PSR framework. *Int. J. Disaster Risk Reduct.* **2024**, *101*, 104243. [CrossRef]
73. Yin, D.; Zhang, X.; Cheng, Y.; Jia, H.; Jia, Q.; Yang, Y. Can flood resilience of green-grey-blue system cope with future uncertainty? *Water Res.* **2023**, *242*, 120315. [CrossRef]
74. Zhang, J.X.; Wang, H.M.; Huang, J.; Wang, Y.X.; Liu, G.F. A study on dynamic simulation and improvement strategies of flood resilience for urban road system. *J. Environ. Manag.* **2023**, *344*, 118770. [CrossRef]

75. Qasim, S.; Qasim, M.; Shrestha, R.P.; Khan, A.N.; Tun, K.; Ashraf, M. Community resilience to flood hazards in Khyber Pukhthunkhwa province of Pakistan. *Int. J. Disaster Risk Reduct.* **2016**, *18*, 100–106. [CrossRef]
76. Zhang, H.; Yang, J.; Li, L.; Shen, D.; Wei, G.; Dong, S. Measuring the resilience to floods: A comparative analysis of key flood control cities in China. *Int. J. Disaster Risk Reduct.* **2021**, *59*, 102248. [CrossRef]

buildings

MDPI

Article

Adaptability and Interconnectedness in Latin American Innerburbs: An Assessment of the Public/Private Interface in Informal Settlements

Carlos Augusto Martínez Muñoz * and Francisco Javier Maroto Ramos

Departamento de Proyectos Arquitectónicos, Escuela Técnica Superior de Arquitectura de Madrid, Universidad Politécnica de Madrid, 28040 Madrid, Spain; franciscojavier.maroto@upm.es
* Correspondence: carlosaugusto.martinez.munoz@alumnos.upm.es

Abstract: An "Innerburb" is an urban structure that emerged between the 1950s and 1980s, settled in rural areas, and is considered the first periphery of the Latin American city. This structure results from socio-spatial and territorial evolutionary processes, constituting the pinnacle of informal evolution. However, despite offering a comprehensive perspective on the informal problem, innerburbs have been scarcely reviewed in the literature. This article explores the Latin American innerburb by adopting as a method a comparative study of the public/private interface in the cases of San Cosme in Lima, Villa Rodolfo Ricciardelli in Buenos Aires and Vila Heliopolis in São Paulo, evaluating their adaptation and interconnection with the city through morphological indicators, using morphological mapping through satellite images as an analytical tool, and using Space Syntax as a topological approach in the analysis of connectivity and visibility indicators. The objective of the research is the detection of morphological patterns that alter the functioning of the public/private interface in innerburbs. The results show that the existence of impermeable facades, the lack of public spaces and the illegal appropriation in the development of informal practices are transgressive adaptability patterns at the micro-scale that affect the interface, drastically limiting the interconnection between the innerburb and the formal fabric, restricting its morphological openness and affecting the development of socioeconomic activities. As a discussion and conclusion, corrective measures for progressive improvement in innerburbs are established, focusing on the adaptability of housing and road space as a means of interconnection between the innerburb and the city.

Keywords: public/private interface; informal interstices; innerburbs; inner peripheries; Latin America

1. Introduction

Innerburbs were the first suburbs in cities of the global north and south, initially developed between 1950 and 1980 [1–3]. The term, coined by Peter Ward, originally described 1950s North American suburban development for the middle class and its progressive decline due to factors linked to immigration, high poverty levels, land use changes and a weak housing market [3]. Thus, according to Ernest Burguess' concentric theory, these structures would be in the second ring of the city, between the transition zone and the residential zone [4]. In Latin America, innerburbs were formed by immigration from rural areas, causing large-scale suburbanisation, initially turning rural and agricultural territories into active interstitial areas characterised by a precarious informal economy [5]. The slow development processes of these structures produced their morphological and social isolation, becoming enclosed in the middle of the city by the proliferation of new encroachments [6] and by impassable geographical features, progressively becoming "inner peripheries". In that sense, self-built housing became the main evolving component of innerburbs, according to the population's needs [7,8]. The morphology of innerburbs is mostly irregular in a constant state of intensification (intensification: a process of infilling in which the settlement gains density as spaces or plots are filled, streets/roads are encroached upon and floors are added [9]), with an average height of three levels and lots between 70 and 200 square metres, oriented to a real estate market linked to generational housing tenure, land trafficking mafias and informal renting [3,10]. However, while informal consolidation processes are observed, there is also evidence of physical deterioration of the urban fabric due to overcrowding, economic inequality, lack of access to essential services and socio-spatial segregation, leading to slumification (slumification: process that occurs when a consolidated neighbourhood starts to degrade due to a lack of government or private upgrading actions on the settlement [11]). Although these aspects have been widely studied in the literature, few works investigate the innerburb as a segregated transitional structure in continuous transformation, with autonomous processes of adaptability and interconnection. Issues such as interior and exterior spatial configuration and their socio-spatial relationships are analysed from a morphological perspective by studying the public/private interface. This instrument identifies morphological components in patterns that produce cohesion, morphological fragmentation and socio-spatial dynamics. However, they are scarcely disseminated in the literature as components capable of directing the interactions of the settlement towards its success or failure in the processes of public/private interface with the formal city, filling a gap in the body of knowledge on the Latin American innerburb. Thus, although the body of scientific literature presents essential works on the morphology and forms of operation in informal settlements, its conclusions focus on exploring urban form and the typological categorisation of interfaces without answering the problem of morphological interaction between the formal and informal fabric. In this sense, the research aims to reveal the main morphological patterns that alter the public/private interface, restricting the inner and outer interconnection of the innerburb, taking as case studies the innerburbs of San Cosme in Lima, Villa Rodolfo Ricciardelli in Buenos Aires and Vila Heliopolis in São Paulo, which have morphogenic characteristics and socio-spatial and productive dynamics close to the central areas. As a hypothesis, it is proposed that detecting these patterns in the inner and outer spheres of the innerburbs can enable the application of differentiated morphological corrections in domestic settlements and road infrastructure as systems of interconnection with formal fabrics. The researched patterns are essential to advancing knowledge on interface dynamics in the innerburbs, producing new lines of research that can be extrapolated to other Latin American informal contexts.

1.1. *Innerburbs and Urban Informality*

The processes of urban informality-causing structures such as innerburbs have resulted from factors such as poverty, encouraging migration to cities as a response to a search for economic opportunities close to urban centres. The difficulty of access to urban land, the scarcity of housing alternatives due to their high cost, and the lack of adequate housing policies forced massive settlement in remote or inaccessible areas [12–14]. These developments are the product of squatting, experiencing rapid processes of land intensification and urban consolidation [9,15]. Thus, urban informality is encouraged as a mode of production by the need to develop economic activities as a form of survival for the population, involuntarily establishing a morphogenic assemblage with formal structures with the capacity to create new informal fabrics according to their needs [16,17]. Thus, urban planning actions in the face of these processes are reduced to morphological "healing" actions with little impact on the regeneration of informal fabrics [18,19].

1.2. *Public/Private Interface*

The public/private interface is a transitional space where the private domain connects with the public sphere, revealing the existence of socio-spatial dynamics and economic exchanges [20,21]. Its study allows us to measure the interaction between the urban morphology configured by the façade and the transitional space, composed of the street and the activities carried out in its ambit. The study of interfaces has been a central theme in urban design discourse to foster urban diversity and increased social interactions in streets through a classification of typologies of street-building interfaces in planned cities [21–25]. However, the exploration of interfaces in informal settlements is recent, with studies developing morphological analyses that have led to typological classifications. However, given the complex dynamics of transformation in informal settlements [20,21], establishing a general typology requires a comprehensive global study. Nevertheless, several studies have made significant findings in identifying recurrent morphological interaction components in informal space, such as permeability, interconnectedness, mobility and adaptability. Thus, interconnectedness and adaptability in informal settlements have been reviewed in the exploration of informal configurations based on initial studies of incremental organisation and codes [26–29] and as an instrument for measuring street vitality and spatial permeability [30,31]. Kamalipour has identified interface typologies according to their degrees of connectivity (impermeable, accessible, porous) and their proximity, with levels of enclosure according to interface transition (adjacent and distant) [20]. Jones developed a combination of typologies based on proximity and remoteness (aligned, setback and set-forward) identified in the informal context of Lebak Siliwangi in Indonesia [32]. Van Oostrum has explored the limited relationships between streets and buildings in Beijing, Delhi, Guangzhou and Bangalore, identifying their narrowness as a morphological characteristic of informal space (accessible, cantilever and sawn) responding to an increasing intensity of urban living [30]. Peimani and Kamalipour incorporate the mobility component into the study of interfaces through the behaviour of ambulatory street shops, identifying three temporal conditions of interaction (fixed, semifixed and unfixed) providing micro-scale dynamics capable of activating pedestrian streets through social, economic and spatial dimensions [33].

2. Materials and Methods

2.1. Case Selection

The proposed methodology (Figure 1) is a comparison of case studies by mapping the informal innerburbs of San Cosme in Lima (Figure 2), Villa Rodolfo Ricciardelli in Buenos Aires (Figure 3), and Vila Heliopolis in São Paulo (Figure 4) from a synchronic approach. The selection of the cases responds to characteristics that place these structures as representative Latin American models:

- Because of their age, they are recognisable as urban structures produced in the 1950s and 1960s.
- By its informal morphogenesis, following a characteristic informal evolutionary process (invasion-occupation-intensification-consolidation).
- Because of their condition as inner peripheries close to the central nuclei of Lima, Buenos Aires and São Paulo, respectively.
- For the existence of formalised fabrics within their morphological limits, configuring socio-spatial and economic dynamics is identified as an informal production system.
- The difference in the scale of the cases allows us to understand informal behaviour from a multi-scale perspective. Thus, each case configures three recurrent urban phenomena in Latin America: San Cosme is a reference to urban practices of informal trade; La Villa Rodolfo Ricciardelli presents critical processes of informalisation in state housing; and Vila Heliopolis is a reference to the intention to establish a connection between informal housing and public intervention in social housing.

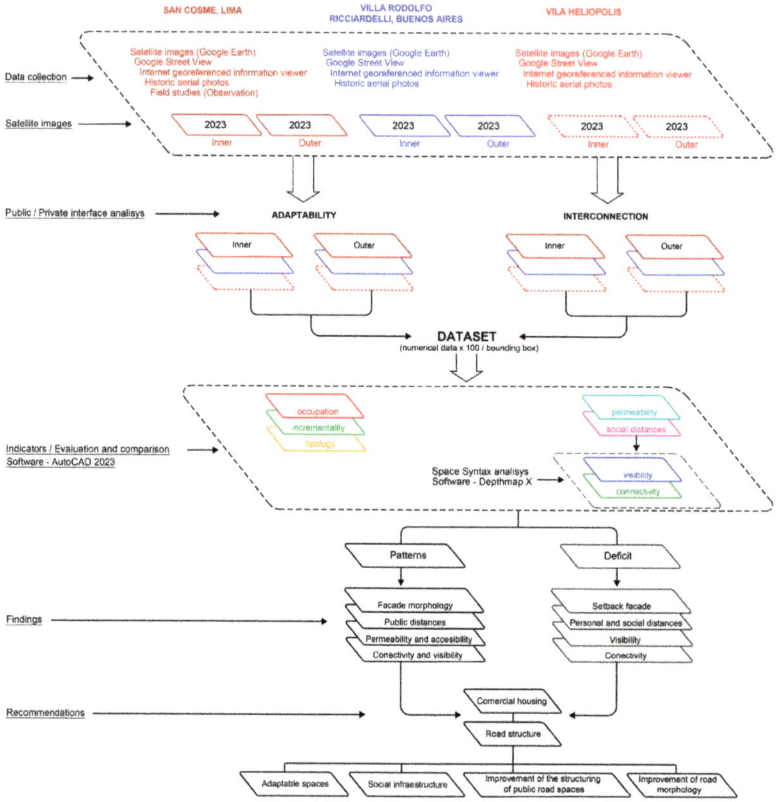

Figure 1. Research design and methodological structure.

Figure 2. San Cosme and the inner and outer sample areas.

Figure 3. Villa Rodolfo Ricciardelli and the inner and outer display areas.

Figure 4. Vila Heliopolis and the inner and outer display areas.

2.1.1. San Cosme, Lima

San Cosme (SC) (Figure 2) constitutes an urban island within an inner periphery [6] located in the district of La Victoria in Lima, starting in 1946 with the establishment of fruit markets and "La Parada" to its new location, near the slope of San Cosme hill, attracting settlers and traders and initiating the invasions on the slope of the mountain [34]. SC, with a population of more than six thousand inhabitants, represents the formation of the slum by commercial interests, produced mainly by the markets. This innerburb presents two important commercial dynamics around housing; its existence as a "dormitory town" encourages the adaptation of domestic spaces for the development of commerce and the use of housing for the rental of rooms near the wholesale market, the most critical need in Lima [35].

2.1.2. Villa Padre Rodolfo Ricciardelli, Buenos Aires

The Villa Rodolfo Ricciardelli (VRR) (Figure 3), located in the Autonomous City of Buenos Aires (CABA), began in the mid-1950s, constituting Villa 31, the first innerburbs in the city. VVR originated from the foreign immigration of Peruvians, Paraguayans and Bolivians. Its rapid consolidation initiated the informalisation of the social housing previously located within its boundaries through tenancy as a commercial practice [36,37]. VRR is a high crime rate neighbourhood, with a periphery consolidated from the incremental development of housing as the most common infrastructural element thanks to the commercial activities on its boundaries (mechanic workshops, drug dealing and remiserias (Remisería is a business in charge of providing taxi services).

2.1.3. Vila Heliopolis, São Paulo

Vila Heliopolis (VH) (Figure 4) was formed as a villa hacienda until 1971, when the occupation of the villa began with the relocation of 153 families as part of the Vilas de Habitação Provisoria programme. VH was subject to the occupation of some sectors by Petrobras and SABESP and new constructions from other favelas [38]. Its location close to the centre of São Paulo and industrial axes led to its rapid urbanisation, making it one of the largest favelas in the city [39]. The settlement was subject to the illegal commercialisation of land in the face of state neglect, consolidating itself as an inner periphery. Since the 1980s, VH has been subject to urban reintegration with the Companhia Metropolitana de Habitação de São Paulo (COHAB) and the Secretaria Municipal de Habitação (SEHAB). VH has since become an intervention laboratory for eradicating favelas and developing new social housing.

2.2. Mapping

The mapping design (Figure 1) drives through the conceptual differentiation of two spheres of action in the innerburbs: the inner zone, characterised by its social segregation and morphological compactness, with limited social practices, and the outer zone, the innerburb boundary area where economic practices and interaction with the formal fabric of the city take place. As the primary mapping technique, 270.00 m $\times$ 70.00 m boundary squares were established for the inner and outer areas. The squares have been referenced from studies with similar comparative intentions [15,30,40,41]. For the comparative development of the cases, satellite images obtained from Google Earth and Google Street View were used as the main tools to determine typologies in existing informal buildings. San Cosme was the only case where observing the interface relationships in the interior environment with satellite images was impossible. Therefore, it was necessary to carry out a direct observation in this informal settlement in March 2023 through photographs and field studies "in situ" through the measurement of the plots included in the analysed areas, developing facade surveys to observe the height of the dwellings (incrementality), the illegal occupation of spaces (appropriation) and their morphology (interface typologies), given the morphological compactness of the settlement. Three people participated in the field study, accompanied by two residents, for our safety. In the cases of Buenos Aires and São Paulo, the settlements have high-resolution images and the possibility of observing heights from Google Earth in three dimensions, making direct observation in situ unnecessary. However, the complex morphology of the cases made it necessary to use geo-referenced viewers and cadastres of the settlements in Buenos Aires and São Paulo to obtain the exact delimitation of the lots. Although most of the indicators used synchronic information, the appropriation indicator required a compilation of historical photos for the diachronic observation of squatting using chronological images from Google Earth, the National Aero Photographic Service in Peru, the Ministry of Urban Development and Transport of the city of Buenos Aires and the Prefeitura de São Paulo. Once the information had been constructed in AutoCAD 2023 software, the methodological analysis was complemented with Space Syntax for topological analysis of connectivity and visibility indicators. For this purpose, the "DepthmapX 8.0" program was used, using the parameters "Axial Map—Connectivity" and "Visibility Graphic Analysis". The pre-processing methodology consisted of a calculation of areas and perimeters, obtaining percentages (numerical data x 100/bounding box) for each indicator and converting the information into linear diagrams to enable the normalisation of results and allow a comparison between cases (Figure 1). The results were filtered using a proposed protocol for detecting predominant interface components, where values above 40.00% in one indicator were considered as interface patterns in the analysed areas. Additionally, values below 5.00% are presented to identify morphological components of low presence at the public/private interface.

2.3. Space Syntax

The development of new analysis platforms such as GIS and the Space Syntax approach has added valuable information to the body of literature on transformation processes in informal settlements, focusing on their connections, integration and morphology. These researchers develop studies based on the use of axial axes and angular segments obtained through GIS [42]; other researchers analyse urban connectivity and integration using visibility maps [43,44]; incremental informal morphogenesis in Nigeria from Space Syntax [45]; the identification of new centralities from the assessment of integration and connectivity in informal settlements from a macro-urban perspective [46]. These studies present Space Syntax as a valuable tool for analysing morphological and social spatial structures. However, its main disadvantage is the impossibility of measuring degrees of permeability, producing problematic results in axial measurements on a micro-scale [47]. This disadvantage is a significant issue in the morphological analysis of informal settlements, being necessary for the construction of detailed maps and the support of other tools for a correct interpretation. In that sense, several researchers have complemented the use of Space Syntax with different approaches in the study of urban forms, such as the Space Matrix and mixed-use index, to diagnose the functioning of the settlement [48,49]. However, most of these works explore cases in European and Asian contexts, with few studies addressing the Latin American context [50,51]. The application of DepthmapX 8.0 software in the analysis of connectivity and visibility has been used in several investigations to improve resilience in informal settlements with restricted essential services in vulnerable areas or to reveal informal regions of high economic production. [52,53]. The use of the parameters "Visibility Graphic Analysis" (VGA) and "Axial Map—Connectivity" of the DepthmapX software applied on the roads and streets of the cases allows identifying and evidencing the existence of isolated sub-zones in areas of informal appropriation, as well as the proliferation of morphological intersections of low social and economic interaction arising from the spontaneous organisation of the initial population in the settlements; the established informal practices; and a segregative governmental housing design.

2.4. Description of the Indicators Used in the Analysis

The study uses the concepts of adaptability and interconnectedness to identify visible socio-spatial dynamics that foster resilience to external factors as an inherent characteristic of informal fabrics. The results of these dynamics are reflected in the public/private interface. Adaptability (Adaptability is an inherent process in the evolution of informal settlements. It is a response to external phenomena (environmental, physical, political, geographical changes, etc.) that produces resilience [54]) and is obtained through incrementality indicators, studying levels of growth, successive appropriation processes and typological identification of facades based on previous studies, which aim to establish variations in connectivity and proximity [21,27,30,55,56]. The second dimension is interconnectedness (interconnectedness: the socio-spatial dynamics and practices that assemble the informal settlement with the formal fabric), identified at the outer boundary of the innerburb with systems linking to adjacent areas. Permeability and socio-spatial distances are reviewed from an approach based on the studies of Dovey and Wood, Kamalipour, Hall [21,27,57]. Visibility and connectivity indicators were obtained using Space Syntax to measure and establish a comparison between cases. A more extensive evaluation of interfaces is proposed in the study, resulting in two linear comparison diagrams for inner and outer areas. The study describes the indicators analysed in the research:

a. Incrementality. Incremental processes in housing are an informal morphological characteristic that modifies the spatial relationship between the street and the building through illicit appropriations using horizontal and vertical additions that progressively improve living conditions inside the dwelling but detract from the spatial quality on the outside. The study focuses on mapping the constructive conditions, height, permanence and temporality of the dwellings that establish a relationship with the street, identifying their degrees of consolidation and degradation (Figure 5).
b. Appropriation. In informal urbanism, appropriation is the progressive exclusion of users of public space, which begins temporarily and ends with the permanent occupation of a given territory [58]. The study presents the appropriation and spatial privatisation of public spaces in settlements through urban voids and current infrastructure, as well as the processes of road transgression at the innerburb limits. The aim is to identify a pattern of behaviour in informal occupation that undermines the production of a transitional avenue/street—building interface (Figure 6).
c. Façade typologies. This mapping reveals a pattern of façade design, configured according to their predominance in the interior and exterior realms in the cases studied and their compatibility or incompatibility as means of interface with commercial practices and social interaction in innerburbs (Figure 7).
d. Permeability. Permeability measures road "porosity" in pedestrian streets and avenues. In that sense, the objective is to identify the openings and enclosures produced by the processes of appropriation and incrementality in the different types of façades, as well as social and commercial activities in innerburb environments, extending or reducing the transitional interface spaces (Figure 8).
e. Socio-spatial distances. The analysis aims to measure the morphology of streets as an element of interaction in proxemic relations of personal proximity and distance. This study indicates areas where spaces for social interaction are predominant and scarce (Figure 9).
f. Visibility. This analyses the visual openness of the settlement's internal and external roads. The analysis used the "Visibility Graphical Analysis" (VGA) parameter of the DepthmapX software, identifying the road morphological openness and the quality of the visual fields configured by the innerburbs. The study makes it possible to determine the degree of inner compactness and the level of morphological spatial openness with the outer formalised fabric (Figure 10).
g. Connectivity. It allows for establishing the degree of road segregation based on the settlement's topological connections. The DepthmapX 8.0 software was used in the analysis, using axial road lines identified under the parameter "Axial Map—Connectivity". This analysis considers all the direct connections that a street has with other streets in its closest range. In that sense, a street with many connections has a high degree of connectivity, while a street with few connections has a low degree [59,60]. The measurement is set from grade 1 (low) to grades 6 and 13 (high and very high) (Figure 11).

2.5. Methodological Procedure

The research procedure consisted of processing the collected data in AutoCAD 2023 software, allowing the construction of indoor and outdoor sample tables for the comparison of cases (Figure 1). This numerical processing divided the study into two types of data obtained: surfaces in square metres (m^2) and linear metres (mL). The incrementality, appropriation, socio-spatial distances and visibility were measured by the percentage between the square metres obtained in each indicator and the total square metres (m^2) of the bounding box in each case study. On the other hand, the permeability, façade typologies and connectivity indicators were obtained by calculating the percentage of linear metres drawn for each indicator and the number of linear metres (mL) existing in the streets of the sample in each case. The results were represented in linear diagrams (Figures 5–11) to standardise the results and offer a better understanding of the case comparison process. Finally, the results were filtered using a proposed interface component detection protocol (Figure 12). Values above 40.00% in an indicator were considered public/private interface patterns in the analysed zones, while values below 5.00% were considered morphological components in deficit in the interface processes. This protocol type was used in similar research [27,61].

2.6. Limitations

The lack of previous evolutionary information on the settlements studied constitutes the main limitation in the study, making it challenging to obtain historical information on the transformation processes and the progressive increase of informal practices in the three cases, with the collection and subsequent cataloguing of evolutionary data in informal settlements being a pending work agenda. On the other hand, the scarce literature on socio-spatial functioning in Latin American innerburbs constitutes a critical gap, as does the lack of detailed information through geo-referenced viewers in Lima, making in situ documentation necessary. These limitations and gaps in information constitute a line of research pending development and dissemination.

3. Results

3.1. Adaptability

3.1.1. Incrementality

Inner

Inner pedestrian walkways in SC have irregular facades with a standard height of three to five levels and overhangs at their upper levels that reduce street lighting and ventilation due to a lack of municipal regulation. In comparison, the incremental façade in VRR has a greater height facing the main avenue and, to a lesser extent, the street. The social housing system maintains a profile that regulates building size. In contrast, in VH, the height of social housing is predominant, with illegal growth being observed on the first level (Figure 5a).

Outer

The proximity of the metropolitan markets to SC promotes using the first level as warehouses, adaptable shops and a higher building height. In VVR, the facades are modified for the development of commercial activities (workshops, taxi shops, room rentals and shops). These activities increase the informality of adjacent state-owned housing, adding new flats for rent. In VH, incremental housing is in the upper levels, as the first ones are occupied with commerce, establishing contact with the formalised limit of Estrada das Lagrimas Avenue, with an entirely commercial profile (Figure 5b).

3.1.2. Appropriation

Inner

Two processes divide the informal appropriation in SC: the occupations of 1950 that precariously structured the inner roads and their later delimitation. In VRR, the occupation of social housing developed in urban voids and commercial squatting. In VH, the occupation is similar, with the appropriation of openings for private use and the eradication of uninhabitable informal dwellings to construct social housing (Figure 6a).

Outer

The process of squatting in SC is an extension of the innerburb towards markets as a form of street trade development. In this sense, the transition of interfaces is diffuse, hindering the fluidity of pedestrian and vehicular movement. In VRR, Bonorino Avenue is the nexus of informalisation between the innerburb and the adjacent social housing. In VH, the Sacomã neighbourhood is a morphological border with commercial activities and metropolitan equipment, limiting its extension. However, the informal zonal commerce establishes the socio-spatial connection with Sacomã (Figure 6b).

3.1.3. Façade Typologies

Inner

In SC, the "cantilevered" façade predominates (54.57%). This typology drastically reduces the visual perception of the street, reducing illumination and increasing the feeling of narrowness. In the case of VRR, "set-forward/direct" predominates in squatter buildings. This façade, in combination with the "aligned/direct" typology (34.89%), produces interfaces with little road interaction. In VH, the aligned/direct facades have a higher percentage (51.77%). This typology comes from the beginnings of social housing, characterised by low visual interaction with the surrounding public spaces (Figure 7a).

Outer

In SC, the "aligned/direct" typology predominates (71.52%). Set-forward/direct and cantilevered typologies prevail in the interior. In VRR, the "set-forward/direct" typology (43.68%) is recurrent, especially in squatting. The "aligned/direct" typology (34.89%) predominates in buildings in front of main roads. In VH, the "setback" typology (26.19%) is found in formalised areas, while the "aligned/direct" type corresponds to the innerburb boundary. However, the intercalation between these typologies encourages more significant social interaction (Figure 7b).

3.2. *Interconnection*

3.2.1. Permeability

Inner

In SC, the openings (doors and windows) allow minimal accessibility (53.16%) without connecting elements or are impermeable (46.84%). The scarcity of permeable facades is a characteristic associated with the functioning of the innerburb and its territorial segregation. In VRR, impermeable facades (77.31%) are a product of monolithic social housing design, encouraging private squatting and reducing social interaction. In VH, there is a contrast between the porosity of shops in front of arterial roads (11.46%) and the impenetrable facades of social housing, appropriating common spaces to privatise them (garages), eliminating visibility and accessibility on the facades (Figure 8a).

Outer

The facades in SC are adapted as product stores and shops of relative accessibility (45.84%) due to their proximity to the wholesale market in front of San Pablo Avenue. However, SC's occupation form produces impermeable access (48.43%). In VRR, the innerburb accesses towards Bonorino Avenue are composed of impermeable façades (50.04%), while in the avenue, accessible and permeable façades predominate with shops that invade the road profile. In VH, the exterior façade is accessible and porous, favouring socio-spatial contact with the Estrada das Lagrimas Avenue (Figure 8b).

3.2.2. Socio-Spatial Distances

Inner

In SC, the interconnection is pedestrian, with wide pedestrian-scale access streets that function as social spaces. SC lacks interior public spaces. Thus, street trading is located at road intersections to increase their visibility. In VRR, the general distances near and far on the main roads are maintained despite the reduction of the road profile on Charrúa Street due to squatting. In VH, public lengths on Da Silva and Almirante Delamare avenues vary near and far. However, the design of public spaces adjoining social housing is restricted to social and personal distances, isolating the street-building interaction (Figure 9a).

Outer

Street trading in SC and VRR modifies the social distances produced by urban morphology. The itinerant trade zones in SC create sub-zones with social and personal spaces for moving goods by privatising them. The case of VRR is similar, with the creation of privatised sub-zones (car parks) of a permanent character on the innerburb frontages. In VRR, the innerburb accesses are narrow, with distances between 0.70 and 1.20 mts. In comparison, the innerburb boundary of VH is marked by the Estrada das Lágrimas Avenue with public socio-spatial spaces (13.50 mts.) without ambulatory shops, homogenising the road profile (Figure 9b).

3.2.3. Visibility

Inner

In SC, the low road visibility of the corridors constitutes the highest percentage (31.77%), along with the lack of inner road intersections. The area with the most increased visibility (26.65%) is a public space with no interior access. In VRR, the junction of Riestra and Charrúa avenues has intermediate visibility (3.76%) but is limited by social housing buildings. In VH, Almirante Delamare Avenue and its public space present higher visibility at their intersections. However, illegal parking squats drastically reduce road visibility (Figure 10a).

Outer

In SC, San Pablo Avenue allows an intermediate visibility (19.79%) towards the Lima wholesale market. However, street commerce reduces its visibility (14.21%). In VRR, the intersection of the pedestrian streets of the Rivadavia I neighbourhood visually ends in public spaces towards Bonorino Avenue. Similar to SC, squatting on the outer façade reduces visibility. Compared to SC and VRR, Estrada das Lágrimas Avenue in VH presents the highest visibility rate without visual obstacles (73.30%) (Figure 10b).

3.2.4. Connectivity

Inner

The existence of irregular roads causes the low connectivity of SC without intersections and the identification of dead-end streets. Thus, grades 1 (24.92%), 2 (51.12%) and 3 (16.31%) are the central values detected. In VRR, 10.87% of the roads lack connectivity (grade 0), while Charrúa Street has grade 3 connectivity due to its scarce road intersections. The roads with the highest connectivity in VH are pedestrian inside the public space, surrounded by social housing buildings reaching grade 5 (3.49%), while in the road intersections, grade 3 predominates (14.05%) (Figure 11a).

Outer

In this sense, it is proposed to fill this gap by studying interfaces as a tool capable of identifying with urban morphology the socio-spatial behaviour as a system of interaction between the formal and informal urban fabric, improving physical interventions in these settlements. This section discusses patterns and deficits in the form of interfaces, which establish a greater or lesser adaptability and spatial interconnection between the innerburb and the formal city (Figure 11b).

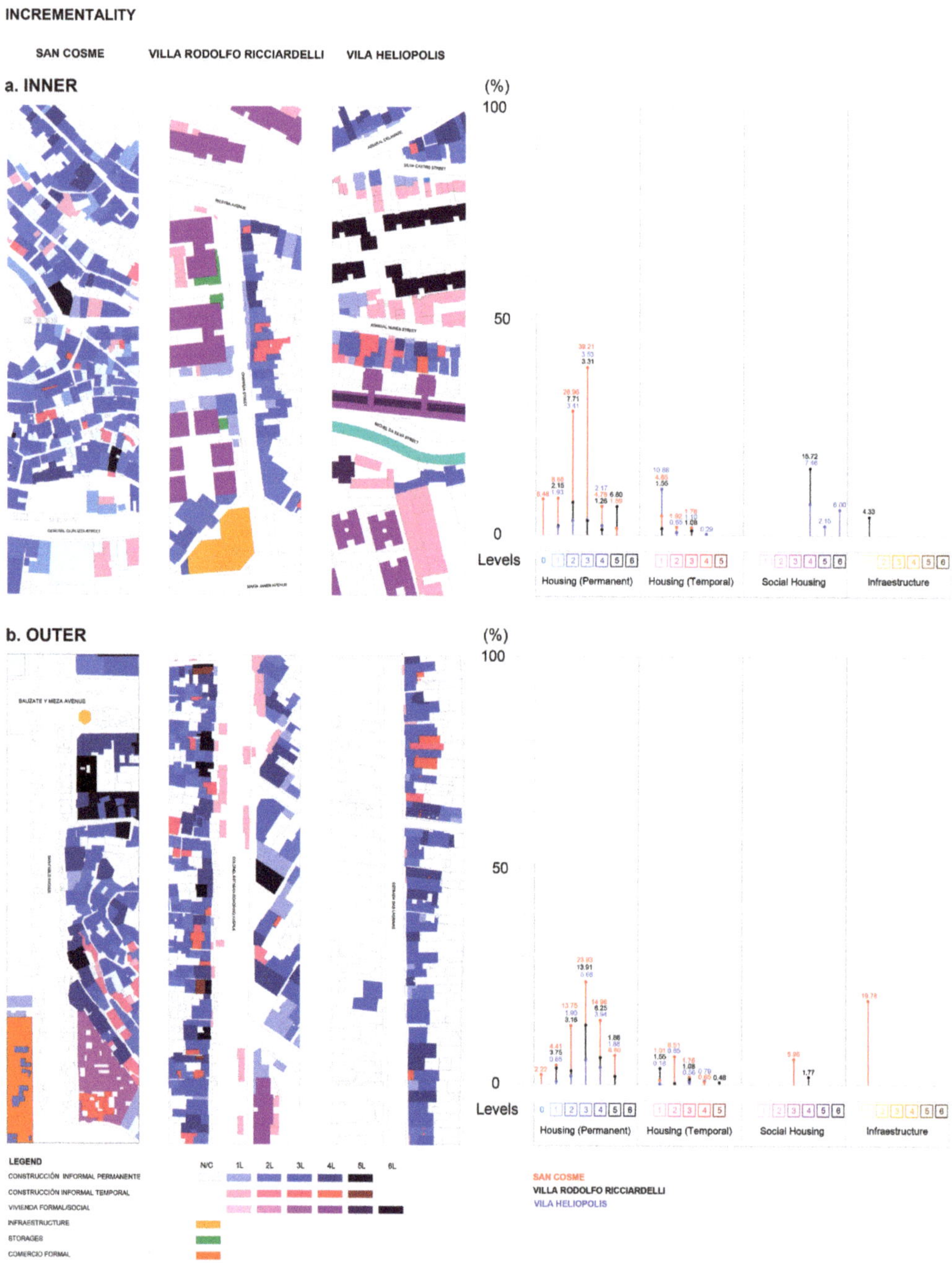

Figure 5. Comparison of incrementality processes of the study areas. The linear diagrams represent a comparison between the indicators.

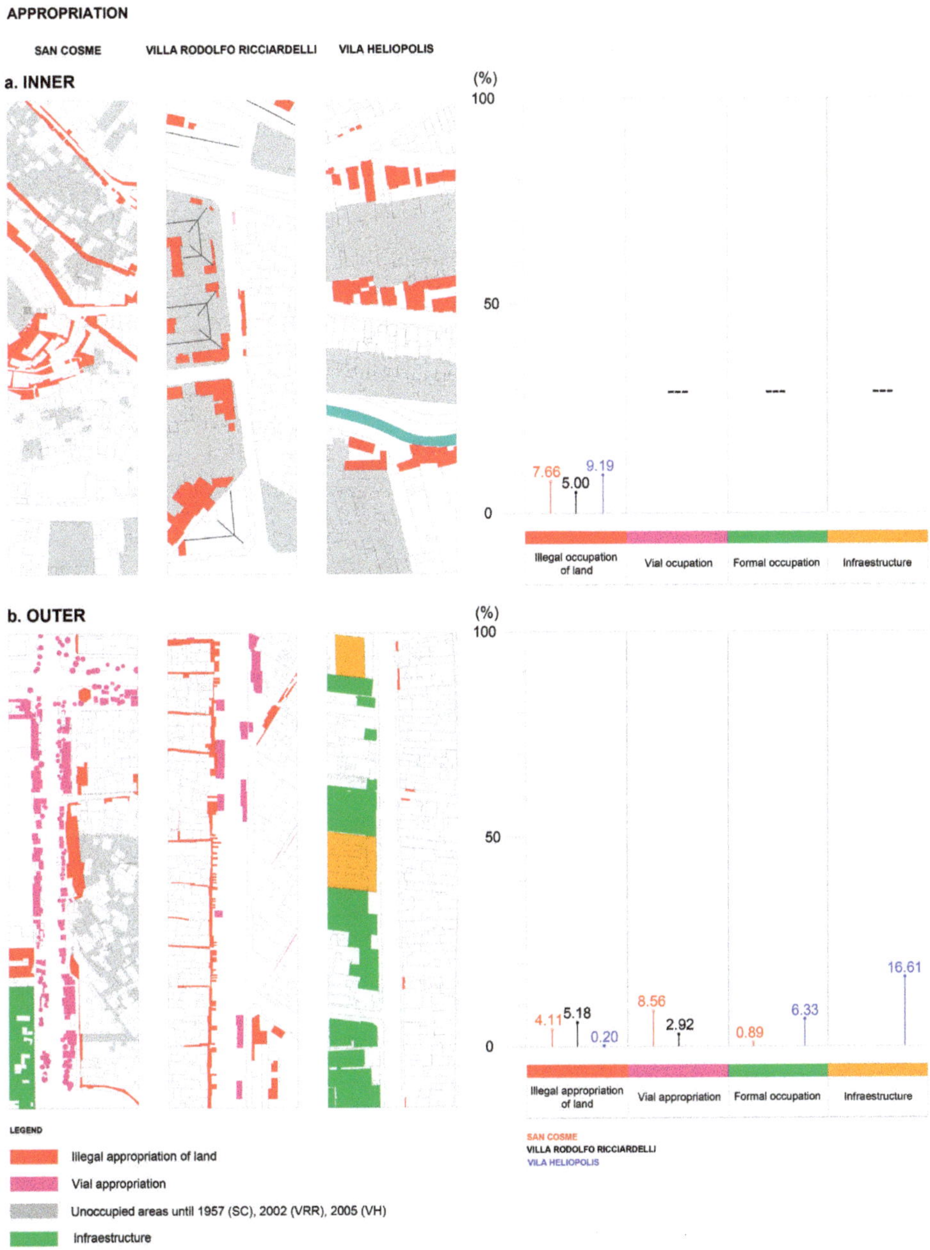

Figure 6. Comparison of appropriation processes of the study areas. The linear diagrams represent a comparison between the indicators.

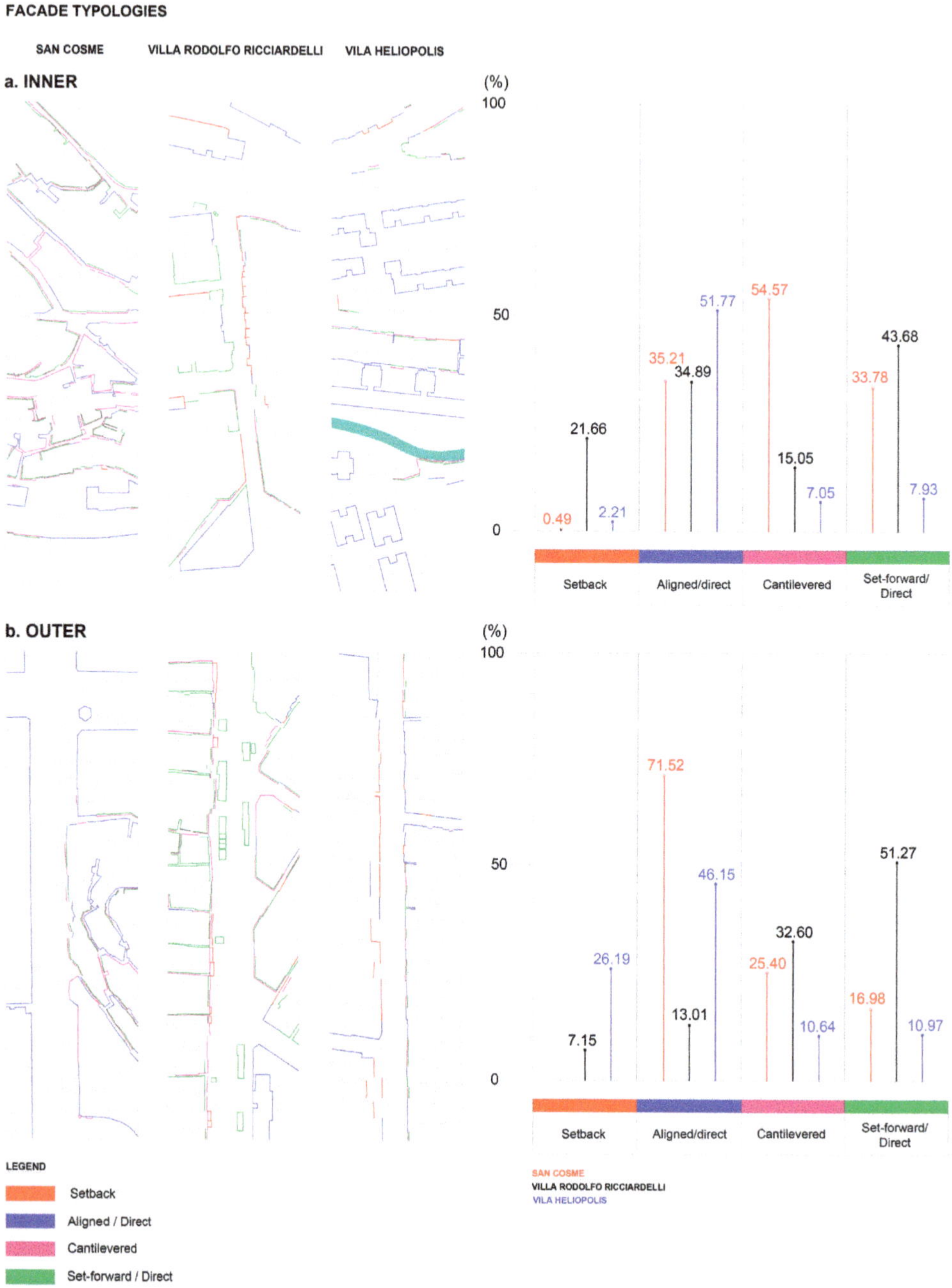

Figure 7. Comparison of façade typologies of the study areas. The linear diagrams represent a comparison between the indicators.

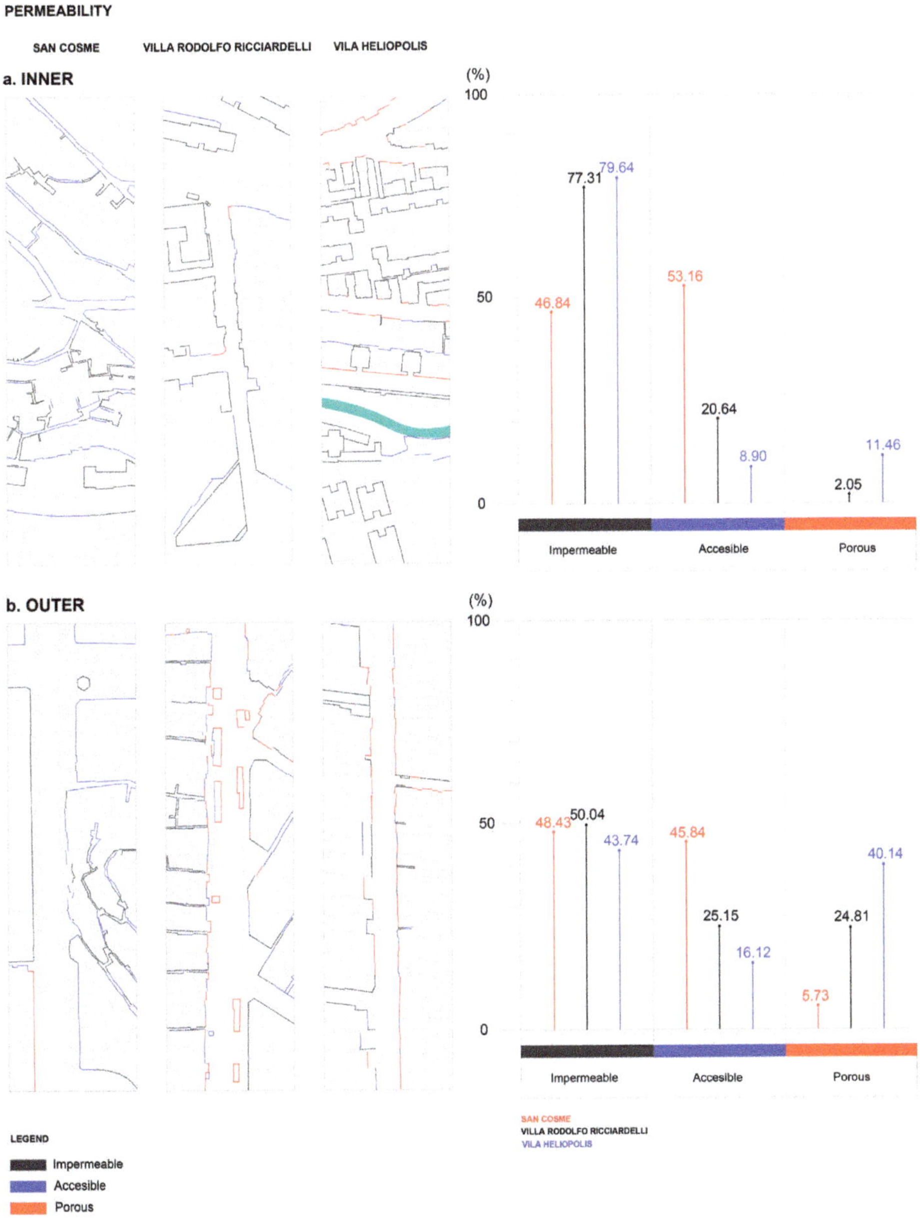

Figure 8. Comparison of permeability of the study areas. The linear diagrams represent a comparison between the indicators.

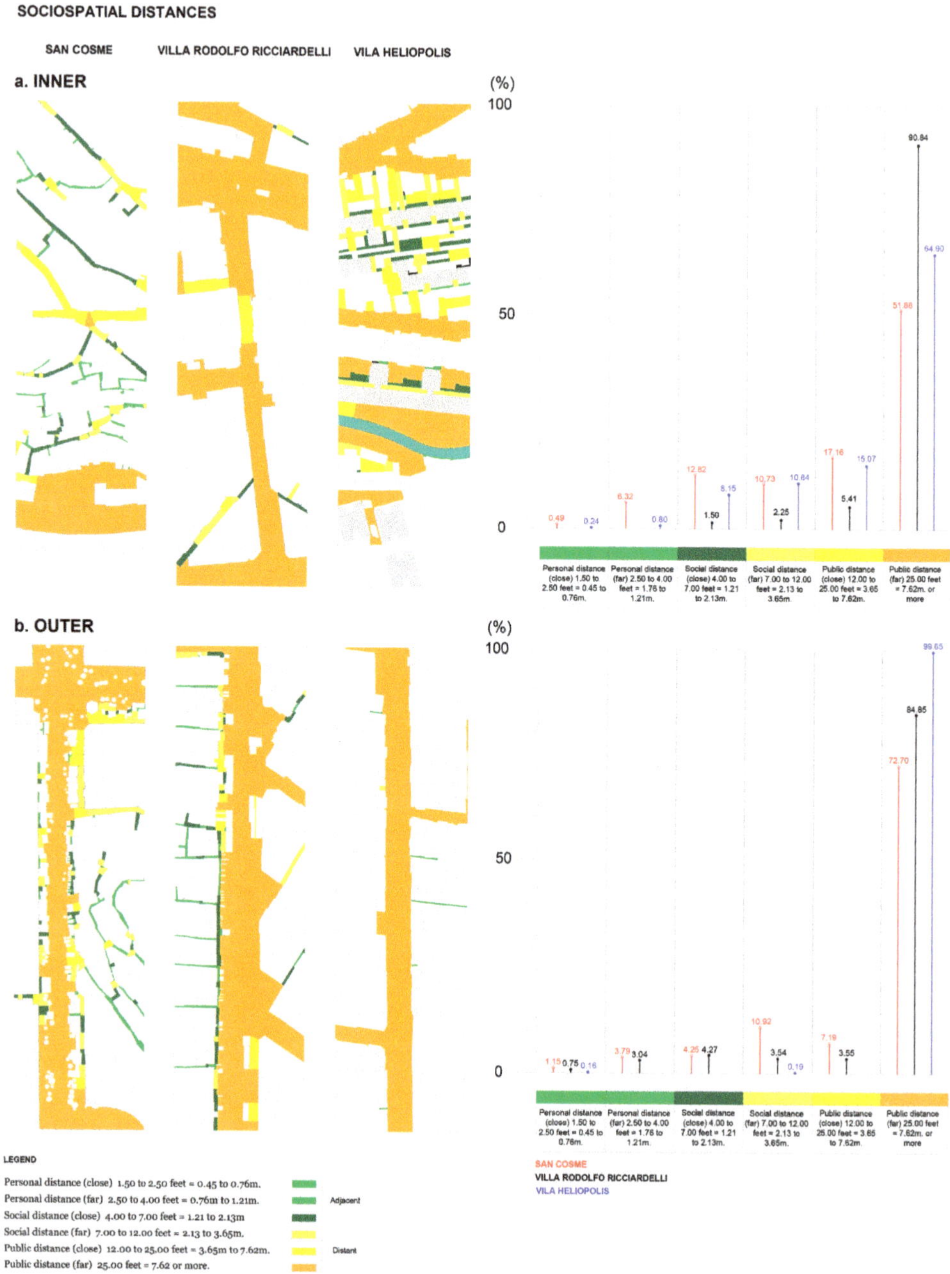

Figure 9. Comparison of the socio-spatial distances of the study areas. The linear diagrams represent a comparison between the indicators.

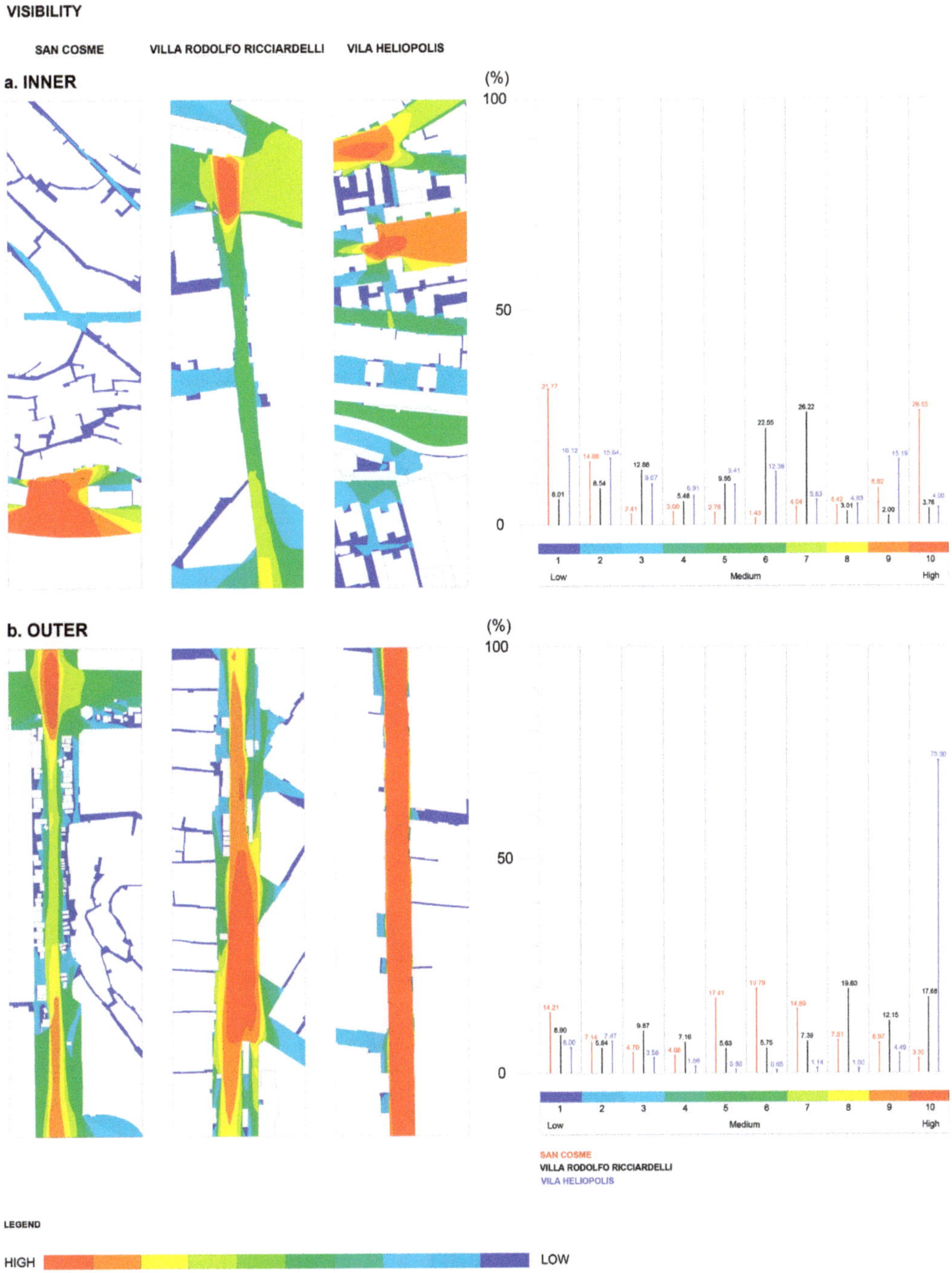

Figure 10. Comparison of visibility of the study areas obtained with DepthmapX 8.0. The linear diagrams represent a comparison between the areas with the highest and lowest visibility.

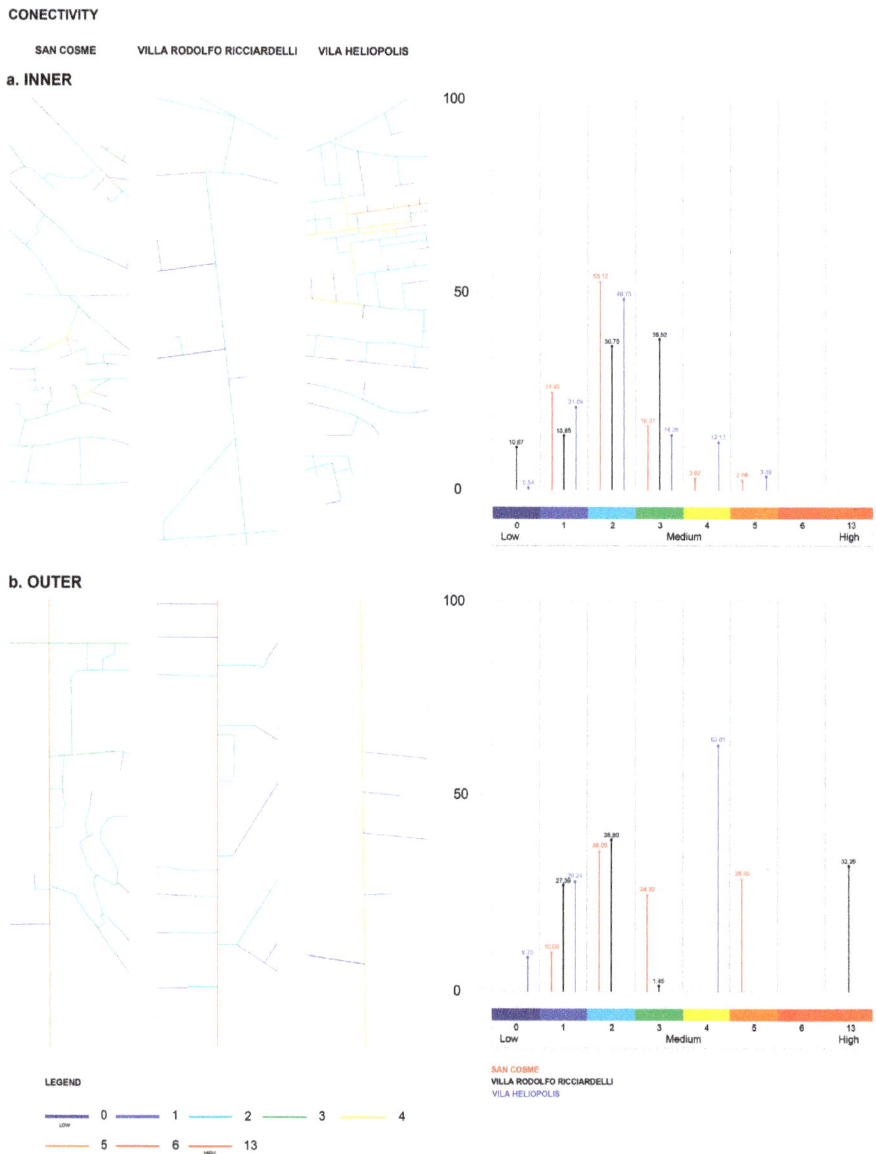

Figure 11. Comparison of the connectivity of the study areas obtained with DepthmapX 8.0. The linear diagrams represent a comparison between the paths with the highest and lowest topological connection.

3.3. *Results of the Analysis*

The results reveal a series of factors that provoke and intensify the socio-spatial fragmentation of innerburbs, limiting their public/private interface. Rapid informal urbanisation driven by governmental and private entities enclosed these suburban structures that managed to exist due to their specific socio-spatial dynamics, such as informal trade, the informal housing market due to migration and the need for housing. However, this enclosure led to slow slumification. Thus, aspects such as the temporary renting of slum spaces and illegal trade related to migration processes from the interior in developing countries caused residential segregation and the "ghettoisation" of the innerburb, encouraging the emergence of practices such as illegal spatial appropriation, producing a progressive modification of housing in these structures and further residential segregation. These modifications are observable in the study of the façades on interior and exterior roads, with constant morphological transformations in the fronts that limit the innerburb with the adjacent formal fabric, evidencing the appropriation of interstitial spaces that limit the formal fabric with the innerburb, drastically diminishing the possibilities of social interaction between both structures. Spatial permeability is non-existent in the interior, established by the proliferation of blind façades and the lack of social and commercial activities due to the excessive road narrowness and consequent proximity to other neighbouring housing fronts. Connectivity and visibility in innerburbs are conditioned by perimeter arterial roads as the only possibility of interaction and urban interconnection with formalised fabrics. In this sense, a similar problem is observed among the case studies, identifying patterns and deficits in the morphological development of the innerburbs analysed (Figure 12).

3.4. Identified Patterns

a. Typology of façades: they vary according to their location concerning the innerburb, with cantilevered interior façades predominating. On the outside, aligned façades are predominant. This phenomenon is caused by continuously modifying frontages and transforming the facades to obtain greater commercial visibility or more saleable areas. However, the most apparent transformations occur at the boundaries. This pattern raises the economic value of the building due to its proximity to the street, taking advantage of its location to carry out informal commercial activities and implementing the first level as flexible spaces adaptable to the economic needs of their owners.
    - Aligned/direct: They lack courtyards or entrance halls. They are located on narrow pedestrian thoroughfares. This façade is predominant on the innerburb boundary.
    - Set-forward/direct: They are configured inside and outside the innerburb in narrow corridors. Their configuration is similar to an aligned façade. However, the first level is used for retail.
    - Cantilevered: This is the most widespread typology at an advanced consolidation stage. Its configuration includes several levels. The first floor is generally a cellar; the following floors are used for housing and room rentals. These facades predominate in the interior.

b. Public distance: The interior and exterior boundaries structure their spatial organisation with arterial perimeter roads as the only form of socio-spatial interconnection, linking informal occupations to economic production. These occupations diminish public distance on the avenues and eliminate interface transitions, becoming unused sub-spaces and privatised dead-end thoroughfares. These appropriations occur spatially at the boundaries of the innerburb.

c. Permeability and accessibility: The access roads at the innerburb limits contain a high percentage of impermeable facades from informal buildings and social housing, eliminating the interface and acting as a negative pattern limiting its integration with the city by generating barriers.

d. Connectivity and visibility: Grade 2 roads have a predominance of 50.00% (SC/VH), showing a low connectivity and development pattern in the innerburb typologies studied. On the outside, only VH presents grade 4 connectivity on the Estrada das Lágrimas road and grade 10 visibility. However, the areas with the highest visibility occur near social housing buildings and public spaces, which do not promote interconnection with other informal locations.

3.5. Deficits in Adaptability and Interconnection Interface Components

Values below 5.00% (Figure 12) reveal a lack of dynamism in components that promote the public/private interface, such as setback façades that favour street-to-building spatial transitions and social distances that encourage social interaction. These results also expose morphological problems, such as the functioning of connectivity (grades 4 and 5 below 5.00%), causing spatial segregation. The lack of visibility in all three cases is notorious, a fact related to the connectivity deficit. However, it is necessary to interpret the information presented. The values of incrementality and occupation present reduced percentages because the innerburbs are consolidated structures, making it impossible to observe the processes of occupation and incrementality in previous periods with synchronic analysis. The following components favour the interface, but they are scarce:

a. Setback: Its configuration promotes a transitional space between the street and the building, being used as an extension of the dwelling to develop shops and used for private use. It has the lowest percentage of predominance in the study.
b. Social and personal distances: The compactness of the innerburbs causes a lack of neighbourhood-scale public spaces and road profiles with social interaction (social distance). Although pedestrian routes in the inner city are numerous, their narrowness and irregularity increase the feeling of insecurity (personal space).
c. Visibility: Access in innerburbs lacks visibility due to the scarcity of public spaces and elements of urban articulation, promoting interior self-segregation. There is a greater visual field in areas of state intervention due to the implementation of public spaces that connect streets and avenues, which are not used as formal-informal assembly zones. On the other hand, the illegal appropriation of public areas in the arteries and unregulated incrementality considerably reduce the horizontal and vertical field of vision, eliminating contact with the innerburb fronts as a form of transition of interfaces with the city.
d. Connectivity: The study reveals low levels of grade 3, 4 and 5 connectivity, which is sufficient to reinforce the interfaces inside and outside the innerburbs. The main problem identified is the lack of intersections and irregular dead-end roads. On the outside, the primary connections consist of longitudinal roads with a few perpendicular to the main avenues.

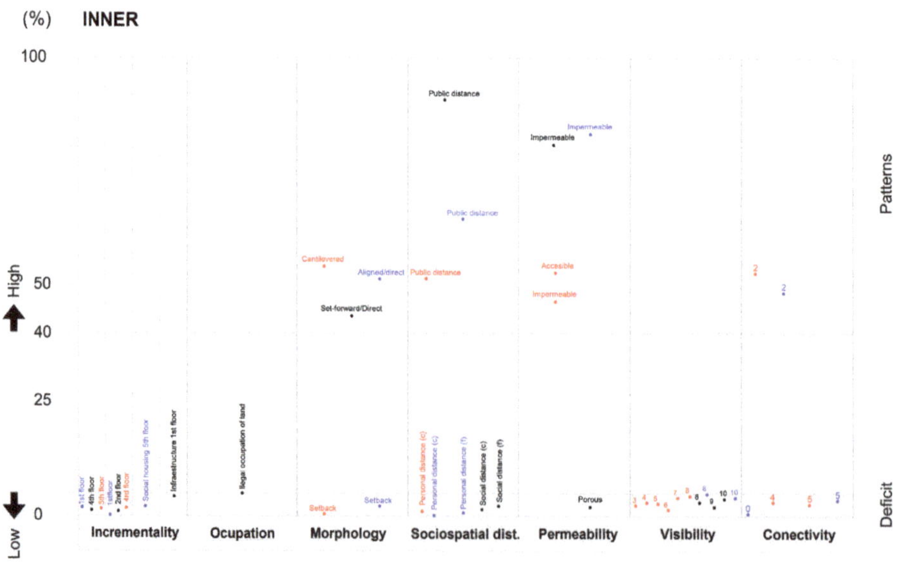

Figure 12. *Cont.*

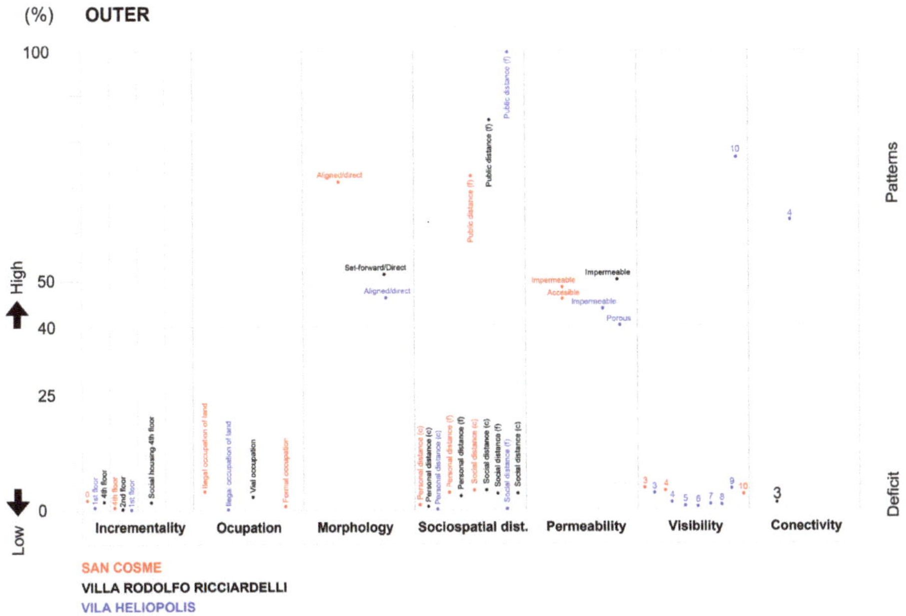

Figure 12. Detection of the predominant and deficit interface components in the cases studied.

4. Discussion

The research addresses the study of the dimensions of adaptability and interconnectedness by analysing indicators that provide new information to the body of literature on the morphological, socio-spatial and transformational functioning of the public/private interface in Latin American informal innerburbs, enhancing the understanding of these urban structures in contexts of the global south. The comparison among the cases reveals contradictory dynamics between adaptability and interconnectedness. Thus, the processes of adaptability identified inside and outside the innerburbs function in isolation, manifesting themselves visibly in the façade typologies through processes of incrementality lacking transitional spaces capable of enhancing the interface. Although the scalar morphological difference in the cases analysed is evident, one of the main findings in the research identifies micro-scale transformations as the primary triggers in the low interconnectivity between the formal and informal fabric, producing morphological alterations through forms of transgression such as the appropriation and privatisation of public domain spaces that promote informalised socio-economic activities, prompting degraded areas of domestic character. Thus, findings such as the illegal appropriation of avenues and streets with a comprehensive road profile for informal economic activities as the primary form of interconnection and social interaction between the innerburb and the city and the spatial subdivision of public space caused by informal appropriation are underlying mechanisms of appropriation processes as one of the most common forms of informal development, observable from the initial processes of invasion and squatting. Another finding is the emergence of hermetic facades during the incremental transformation of dwellings through processes of addition and renovation inside and outside the innerburb, fostered by weak public policies of regularisation and code. The incorporation of Space Syntax in the study as a system for measuring visibility and connectivity has revealed the existence of a precarious network of interior connections whose morphology fosters segregation and the loss of opportunities for socio-economic development. Thus, on the roads that connect with the exterior, informal, productive activities proliferate, appropriating the road profile and considerably reducing pedestrian visibility, being notorious for the spatial ghettoisation between the interior and exterior zones caused by a chaotic and disorganised configuration produced during the process of illegal occupation of the innerburb. As argued in the Introduction Section, section, several works have examined the interface in informal settlements. However, they have focused on typological identification as a starting point for more complex future studies. The findings presented in the study on interconnectivity and visibility in innerburbs are comparable to research that has analysed the public/private interface in informal settlements from a morphological perspective [30], identifying typologies characteristic of informal development in slums in China similar to the typologies specified in the study; other work has explored the study of interfaces from the lens of adaptability [58,62] differentiating from research on the use of an incremental approach to adaptability in urban interaction and extension networks in Asian urban slums and research on the creeping occupation of public spaces, the illegal appropriation of spaces adjacent to government-sponsored housing in Nairobi and the temporary occupation of street trading in easily visible areas in different contexts in the global south [17,63]. The incorporation of Space Syntax in the study as a system for measuring visibility and connectivity has revealed a precarious network of connections that foster social isolation and the loss of opportunities for social and economic development caused by a chaotic and disorganised configuration produced by the illegal inception of these settlements. In this sense, this work is a valuable tool in the diagnosis and identification of socio-spatial phenomena of interaction between the formal and informal urban fabric from a morphological perspective, establishing recommendations for physical action to promote a progressive improvement in consolidated settlements with similar characteristics to the innerburbs analysed.

Recommendations

The recommendations are aimed at regenerating informal forms of economic production, spatial quality and road reorganisation, improving the levels of adaptability and interconnection to integrate the innerburb with the city to avoid its segregation. These solutions can be extended to the problem of isolation in other informal settlements in the global south.

Adaptability: The study exposes a more significant commercial proliferation in the outer limits as a basic system of formal-informal interconnection. In that sense, the incorporation of flexible spaces for the development of high-economic-value commercial activities in informal self-construction (air projection for future processes of vertical incrementality) and social interest infrastructure (housing and commerce) may be able to promote spatial porosity by increasing the transitional street-housing space for the improvement of the public/private interface and thereby improve the fluidity of economic exchange at the edges. Applying these two typologies at the edges aims to control the processes of progressive informal appropriation and the development of socio-economic activities for population integration with the economic dynamics of the formal city. This type of solution has been recently implemented in informal contexts [64,65]. However, it has been scarcely reviewed in governmental interventions, so its adoption can be beneficial as a system of urban assemblage between formal and informal fabrics at a local and intermediate scale, promoting its adaptability.

Interconnection: The study shows that given the scarcity of public spaces such as parks and squares, local and arterial roads become the main elements of interconnection between the innerburb and the city, having a decisive impact on the development of the settlement. Establishing a perimeter road renewal between the innerburb and the city is fundamental for promoting urban interfaces. The identified roads function simultaneously as roads and public spaces. Therefore, delimiting commercial zones (permanent and itinerant commerce) and pedestrian traffic on these roads can improve their spatial organisation, eliminating the possibility of fragmentation and privatisation of road space through illegal appropriations. On the other hand, a road transformation is also necessary to improve visibility and the provision of visual auctions that allow for better orientation and public convergence towards arterial roads. Applying similar solutions in programmes like Favela Bairro in

Brazil has been beneficial [66,67]. Finally, including economic dynamics at innerburb intersections and incorporating regulations for street trading activities on public roads are necessary to avoid affecting interface conditions at their boundaries.

5. Conclusions

This paper extends the study of the public/private interface in informal settlements, adding indicators linked to socio-spatial dynamics that allow us to measure the dimensions of adaptability and interconnectedness of the Latin American innerburbs vis à vis formal fabrics. The research provides important lessons that can be considered in future public policies for territorial planning and physical intervention in consolidated and newly formed settlements. Innerburbs are dichotomous interstitial structures characterised by the generation of commercial practices and socio-spatial dynamics at their boundaries as a system of informal economic production. This factor triggers transformations such as the continuous modification of facades, the privatisation of public space, the ambulatory appropriation of streets and vertical incrementality. Meanwhile, spatial segregation and ghettoisation within the innerburb are the result of the initial lack of organisation, a product of the cumulative process of informal housing during squatting as a form of protection against eviction, producing impermeable structures. Similarities such as low connectivity, scarce permeability and the lack of community facilities contribute to their degradation and isolation, relating the case studies to a problem representative of the Latin American context. In this sense, the research is a tool for detecting morphological and socio-spatial phenomena at interfaces between recently formed and consolidated settlements, which can be replicated in similar contexts for diagnosis and improvement. However, adding other indicators can complement the detection of new informal phenomena as possible lines of future research pending development. The cases of Lima, Buenos Aires and Sao Paulo reveal three typologies of spatial dynamics according to the degree of governmental interference, with the construction of social housing buildings being the most extensive form of intervention in the innerburbs analysed. However, the impact of these interventions has been discreet, serving only as containers for population absorption and as impermeable urban barriers without contributing to the improvement of interfaces. The recommendations presented are aimed at enhancing economic practices from a socio-spatial perspective and at improving metropolitan assembly with formal fabrics, optimising the adaptability and interconnection characteristics of the innerburb to reduce its segregation, offering initial responses to a recurrent problem in informal settlements in the global south, which can be applied independently through comprehensive or local planning.

Author Contributions: Conceptualization, C.A.M.M. and F.J.M.R.; methodology, C.A.M.M. and F.J.M.R.; software, C.A.M.M.; validation, C.A.M.M. and F.J.M.R.; formal analysis, C.A.M.M.; investigation, C.A.M.M.; resources, C.A.M.M.; data curation, C.A.M.M. and F.J.M.R.; writing—original draft preparation, C.A.M.M. and F.J.M.R.; writing—review and editing, C.A.M.M. and F.J.M.R.; visualization, C.A.M.M.; supervision, C.A.M.M. and F.J.M.R.; project administration, C.A.M.M.; funding acquisition, C.A.M.M. and F.J.M.R. All authors have read and agreed to the published version of the manuscript.

Funding: Funding for open access charge: Universidad Politécnica de Madrid.

Data Availability Statement: The authors declare that the data obtained, sampling and construction of the information presented in the research are the intellectual property of the authors. However, they can be consulted at any time through your contact and consent.

Acknowledgments: We thank all the people involved in the research during the process of collecting field information and for providing us with access to historical and statistical data in Peru, Brazil and Argentina.

Conflicts of Interest: The authors declare no conflicts of interest.

References

1. Ward, P. Urban Regeneration in Latin America. *Int. Encycl. Hous. Home* **2012**, 219–227. [CrossRef]
2. Ward, P.; Jimenez, H.; Di Virgilio, M. *Housing Policy in Latin American Cities. A New Generation of Strategies and Approaches for 2016 UN HABITAT III*; Routledge: New York, NY, USA, 2015; ISBN 9781138776869.
3. Ward, P. Housing rehab for consolidated informal settlements: A new policy agenda for 2016 UN-Habitat III. *Habitat Int.* **2015**, *50*, 373–384. [CrossRef]
4. Park, R.; Burguess, E.; McKenzie, R. *The City*; The University of Chicago Press Ltd.: London, UK, 1925.
5. Davis, M. *Planet of Slum*; Verso: New York, NY, USA, 2006; ISBN 1-84467-022-8.
6. Kapstein, P.; Aranda, E. Lima's inner peripheries: Locating and identifying neighbourhoods that are focal points of vulnerability. The case of San Cosme. *INVI J.* **2014**, *29*, 19–62. [CrossRef]
7. Kamalipour, H.; Dovey, K. Incremental production of urban space: A typology of informal design. *Habitat Int.* **2020**, *98*, 102133. [CrossRef]
8. Silver, J. Incremental infrastructures: Material improvisation and social collaboration across post-colonial Accra. *Urban Geogr.* **2014**, *35*, 788–804. [CrossRef]
9. Dovey, K.; Van Oostrum, M.; Chatterjee, I.; Shafique, T. Towards a morphogenesis of informal settlements. *Habitat Int.* **2020**, *104*, 102240. [CrossRef]
10. Gilbert, A. Rental housing: The international experience. *Habitat Int.* **2015**, *54*, 173–181. [CrossRef]
11. Azhar, A.; Buttrey, H.; Ward, P. Slumification of Consolidated Informal Settlements: A Largely Unseen Challenge. *Curr. Urban Stud.* **2021**, *93020*, 315–342. [CrossRef]
12. Clichevsky, N. *Pobreza y Acceso al Suelo Urbano. Algunas Interrogantes Sobre las Políticas de Regularización en América Latina*; CEPAL, Divisíón de Desarrollo Sostenible y Asentamientos Humanos, Naciones Unidas: Santiago, Chile, 2003.
13. Smolka, M.; Mullahy, L. *Urban Perspectives: Critical Issues in Land Policy in Latin America*; Library of Congress Cataloging-in-Publication Data; Lincoln Institute of Land Policy: Cambridge, MA, USA, 2017; ISBN 13-978-1-55844-163-7.
14. Fernandes, E. *Consideraciones Generales Sobre las Políticas Públicas de Regularización de Asentamientos Informales en América Latina*; Revista Eure: Santiago de Chile, Chile, 2008; Volume XXXIV, pp. 25–38.
15. Thinh, N.; Kamalipour, H. The morphogenesis of villages-in-the-city: Mapping incremental urbanism in Hanoi city. *Habitat Int.* **2022**, *130*, 102706. [CrossRef]
16. Dovey, K. Informal Settlement as a Mode of Production. In *The New Companion to Urban Design*; Bannerjee, T., Loukatou-Sideris, Eds.; Routledge: New York, NY, USA, 2019.
17. Kamalipour, H.; Peimani, N. Negotiating Space and Visibility: Forms of Informality in Public Space. *Sustainability* **2019**, *11*, 4807. [CrossRef]
18. Fernandes, E. *Regularization of Informal Settlements in Latin America*; Policy Focus Report/Code PF023; Lincoln of Land Policy Institute: Cambridge, MA, USA, 2011; ISBN 978-1-55844-202-3.
19. Magalhães, F. *Slum Upgrading and Housing in Latin America*; Inter-American Development Bank: Washington, DC, USA, 2016; ISBN 978-1-59782-269-5.
20. Kamalipour, H. Mapping Urban Interfaces: A Typology of Public/Private Interfaces in Informal Settlements. *Spaces Flows Int. J. Urban Extra Urban Stud.* **2016**, *8*, 1–12. [CrossRef]
21. Dovey, K.; Wood, S. Public/private urban interfaces: Type, adaptation, assemblage. *J. Urban.* **2015**, *8*, 1–16. [CrossRef]
22. Jacobs, J. *The Death and Life of Great American Cities*; Random House: New York, NY, USA, 1961; ISBN 0-679-74195-X.
23. Gehl, J. *Life Between Buildings: Using Public Space*; Van Nostrand Reinhold: New York, NY, USA, 1987.
24. Gehl, J.; Kaefer, L.J.; Reigstad, S. Close encounters with buildings. *Urban Des. Int.* **2006**, *11*, 29–47. [CrossRef]
25. Bobic, M. *Between the Edges: Street-Building Transition as Urbanity Interface*; Thoth Publishers: Bussum, The Netherlands, 2004.
26. Suhartini, N.; Jones, P. Better Understanding Self-Organising Cities: A Typology of Order and Rules in Informal Settlements. *J. Reg. City Plan.* **2020**, *31*, 237–263. [CrossRef]
27. Kamalipour, H. Forms of informality and adaptations in informal settlements. *Archnet-IJAR* **2016**, *10*, 60–75. [CrossRef]
28. Kamalipour, H.; Iranmanesh, A. Morphogenesis of Emerging Settlements: Mapping Incremental Urbanism. *Land* **2021**, *10*, 89. [CrossRef]
29. Sawira, S.; Rahman, T. An observation on the quality of interfaces to understand the complexity and coherence of informal settlement: A study on Tamansari Kampung in Bandung. *IOP Conf. Ser. Earth Environ. Sci.* **2018**, *158*, 012002. [CrossRef]
30. Van Oostrum, M. Informal laneway encroachment: Reassessing public/private interface transformation in urban villages. *Habitat Int.* **2020**, *104*, 102259. [CrossRef]
31. Pafka, E.; Dovey, K. Permeability and interface catchment: Measuring and mapping walkable Access. *J. Urban.* **2016**, *10*, 150–162. [CrossRef]
32. Jones, P. The shaping of Form and Structure in Informal Settlements: A Case Study of Order and Rules in Lebak Siliwangi, Bandung, Indonesia. *J. Reg. City Plan.* **2019**, *30*, 43–61. [CrossRef]
33. Peimani, N.; Kamalipour, H. Access and Forms of Urbanity in Public Space: Transit Urban Design Beyond the Global North. *Sustainability* **2019**, *12*, 3495. [CrossRef]
34. Matos Mar, J. *Estado Desbordado y Sociedad Nacional Emergente. Historia Corta del Proceso Peruano: 1940–2010*; Second reprint; Editorial Universitaria, Ricardo Palma University: Lima, Peru, 2016.

35. Muñoz, C.A.M.; Ramos, F.J.M. Urban Informality and State intervention: The case of Lima. *Cities* **2022**, *127*, 103728. [CrossRef]
36. Cravino, M. *Las Villas de la Ciudad. Mercado e Informalidad Urbana*; Universidad Nacional de General Sarmiento: Buenos Aires, Argentina, 2006; ISBN 9789879300862.
37. Di Virgilio, M. Urbanisations of informal origin in Buenos Aires. Logics of urban land production and access to housing. *Demogr. Urban Stud.* **2015**, *30*, 651–690.
38. Sampaio, M. Heliópolis: O percurso de uma invasão. In *Tese de Livre Docência em Arquitetura e Urbanismo*; FAU/USP: São Paulo, Brazil, 1991.
39. Saraiva, C.; Brajato, D.; Moreira, F.; Denaldi, R. Recovering the past to think about the future: Urbanisation of favelas in São Paulo and post-pandemic challenges. In Proceedings of the II Encuentro de la Red de Asentamientos Populares 2021, Resistencia, Argentina, 20–22 May 2021.
40. Dovey, K.; Alegría, V. Morphogenesis of contemporary informal settlement in Chile. *Urban Des. Int.* **2022**, 1–17. [CrossRef]
41. Dovey, K.; Recio, R. Harnessing settlement re-informalization: Morphogenesis of on-site upgrading at Sitio Pajo. *Habitat Int.* **2023**, *131*, 102724. [CrossRef]
42. Cutini, V.; Di Pinto, V.; Rinaldi, A.; Rosini, F. Informal Settlements Spatial Analysis Using Space Syntax and Geographic Information Systems. In *Computational Science and its Applications ICCSA 2019*; Springer: Berlin/Heidelberg, Germany, 2019.
43. Greene, M.; Link, F.; Mora, R.; Figueroa, C. De la casa al barrio. *ARQ* **2014**, 78–87. [CrossRef]
44. Mohamad, W.; Said, I. A review of variables of urban street connectivity for spatial connection. *IOP Conf. Ser. Earth Environ. Sci.* **2014**, *18*, 012173. [CrossRef]
45. Iranmanesh, A.; Kamalipour, H. A configurational morphogenesis of incremental urbanism: A comparative study of the access network transformations in informal settlements. *Cities* **2023**, *140*, 104444. [CrossRef]
46. Tufek-Memisevic, T. Evaluating Integration of Informal Settlements in Sarajero Through Space Syntax Analysis. *Lect. Notes Netw. Syst.* **2023**, *529*, 253–263. [CrossRef] [PubMed]
47. Pafka, E.; Dovey, K.; Aschwanden, G. Limits of space syntax for urban design: Axiality, scale and sinuosity. *Environ. Plan. B Urban Anal. City Sci.* **2022**, *47*, 508–522.
48. Mohamed, A.; Ubareviciene, R.; van Haam, M. Morphological evaluation and regeneration of informal settlements: An experience-based urban design approach. *Cities* **2022**, *128*, 103798. [CrossRef]
49. Ye, Y.; van Nes, A. Quantitative tools in urban morphology: Combining space syntax, space matrix and mixed-use index in a GIS framework. *Urban Morphol.* **2014**, *18*, 97–118. [CrossRef]
50. Greene, M. Housing and community consolidation in informal settlements: A case of movement economy. In Proceedings of the 4th International Space Syntax Symposium, London, UK, 17–19 June 2003.
51. Barbosa, V.; Suárez, M.M. Identifying the Social Urban Spatial Structure of Vulnerability: Towards Climate Change Equity in Bogotá. *Urban Plan.* **2021**, *6*, 365–379. [CrossRef]
52. Tambunan, L.; Donny, M.; Arslana, N.; Izmi, F. The Study of Connectivity and Network Degree toward Mitigation Strategy for Resilient Kampung in Indonesia (Case Study: Kampung Taman Sari, Bandung). *IOP Conf. Ser. Earth Environ. Sci.* **2021**, *738*, 012080. [CrossRef]
53. Mahdzar, S.; Sulkernain, M.; Jaberolansar, S. Enhancing the spatial visibility of pedestrians to the hidden eating place through vertical spaces at jalan ampang, Kuala Lumpur. In Proceedings of the 11th International Space Syntax Symposium, Lisbon, Portugal, 3–7 July 2017.
54. Testi, A. Exploring the relationship between 'sites and services' projects and urban adaptability. A case study of Charkop (Mumbai). *Habitat Int.* **2023**, *139*, 102091. [CrossRef]
55. Jones, P. Distance and proximity matters: Understanding housing transformation through micromorphology in informal settlements. *Int. J. Hous. Policy* **2020**, *21*, 169–195. [CrossRef]
56. Alian, S.; Wood, S. Stranger adaptations: Public/private interfaces, adaptations, and ethnic diversity in Bankstown, Sydney. *J. Urban. Int. Res. Placemaking Urban Sustain.* **2018**, *12*, 83–102. [CrossRef]
57. Hall, E. *The Hidden Dimension*; Anchor Books Editions: New York, NY, USA, 1990; ISBN 0-385-08476-5.
58. Van Oostrum, M. Appropriating public space: Transformations of public life and loose parts in urban villages. *J. Urban. Int. Res. Placemaking Urban Sustain.* **2021**, *15*, 84–105. [CrossRef]
59. Van Nes, A.; Yamu, C. *Introduction to Space Syntax in Urban Studies*; Springer: Berlin/Heidelberg, Germany, 2021; ISBN 978-3-030-59140-3. [CrossRef]
60. Hillier, B. *Space Is the Machine. A Configurational Theory of Architecture*; Press Syndicate of the University of Cambridge: London, UK, 1996; ISBN 978-0-9556224-0-3.
61. Gao, Y.; Shabab, S.; Ahmadpoor, N. Morphology of Urban Villages in China: A Case Study of Dayuan Village in Guangzhou. *Urban Sci.* **2020**, *4*, 23. [CrossRef]
62. Thinh, N.; Gao, Y. Understanding the Informal Morphology of Villages-in-the-City: A Case Study in Hanoi City, Vietnam. *Sustainability* **2021**, *13*, 13136. [CrossRef]
63. Van Oostrum, M. Informal extension of public housing estates in Nairobi—An appraisal of historical typologies and emergent spatial patterns. *J. Urban Des.* **2023**, *28*, 663–681. [CrossRef]
64. Cruz, T.; Forman, F. *Socializing Architecture. Top-Down/Bottom-Up*; MIT Press: Cambridge, MA, USA, 2022; ISBN 10-0262545187.

65. Urban Think Tank. UTT-EMPOWER NPC. An Integrated Development for the Upgrade of BT Section C, Khayelitsha Cape Town, Sudafrica 2013–2023. Available online: https://drive.google.com/file/d/1YGrF__nl-Uj-GfqZ5XjdMEmvDltust1R/view (accessed on 23 April 2024).
66. Segre, R. Formal-Informal Connections in the Favelas of Rio de Janeiro: The Favela-Bairro Programme. In *Rethinking the Informal City. Critical Perspectives from Latin America*; Berghahn Books: New York, NY, USA, 2012; pp. 181–204; ISBN 978-0-85745-607-6.
67. Fiori, J.; Brandao, Z. Spatial Strategies and Urban Social Policy: Urbanism and Poverty Reduction in the Favelas of Rio de Janeiro. In *Rethinking the Informal City. Critical Perspectives from Latin America*; Berghahn Books: New York, NY, USA, 2012; pp. 181–204; ISBN 978-0-85745-607-6.

buildings

MDPI

Article

A Synergetic Perspective on the Planning of the "City in a Park": A Case Study of the Sichuan Tianfu New Area, China

Siyu Chen [1,2], Jian Qiu [1,*], Mei Yang [2] and Jing Li [1,3]

1 School of Architecture, Southwest Jiaotong University, Chengdu 611756, China; chen_siyu@my.swjtu.edu.cn (S.C.); lijing2011@mail.xhu.edu.cn (J.L.)
2 College of Art, Sichuan Tourism University, Chengdu 610100, China; yangmei@sctu.edu.cn
3 School of Architecture and Civil Engineering, Xihua University, Chengdu 610039, China
* Correspondence: qiujian@home.swjtu.edu.cn

Abstract: The "City in a Park" (CIP) is a new concept of urban transformation and development proposed in China in recent years, guiding the construction of healthy and sustainable living environments. This paper analyzes urban planning based on the CIP concept from a synergetic perspective, aiming to explore how the integrated planning of ecological spaces and built environments can promote systematic sustainable development in ecology, economy, and society. This research employs methods including document collection, unstructured interviews, field observations, and participatory observation, focusing on a case study of the Sichuan Tianfu New Area (STNA), a demonstration zone for the CIP. The study finds that the planning of the STNA extends the planning scope of urban ecological spaces beyond the traditional urban construction boundaries, not only preserving the natural resources but also enhancing the city's overall sustainability through regional ecological services. By designing ecological spaces as green infrastructure that connects urban and rural areas, the primary sector is more readily integrated with the secondary and tertiary sectors, facilitating the integration of the urban and rural infrastructure and industries. The STNA integrates urban and rural administrative divisions, builds a cross-departmental collaborative management platform, and guides public participation in the planning process, ensuring the efficiency and effectiveness of planning implementation and enhancing the equitable sharing of social services. This research provides new insights into comprehensive, cross-disciplinary, and ecology-oriented urban planning. It offers evidence for an understanding of the application pathways and effects of the CIP concept in urban planning practice and provides valuable experience for other cities to promote harmonious coexistence between the city and nature.

Keywords: City in a Park; synergetics; urban planning; sustainability; Sichuan Tianfu New Area

1. Introduction

Cities are currently grappling with the monumental challenge of transitioning from a traditional industrial civilization to a more sustainable ecological civilization. While the conventional models of urban development have contributed to economic prosperity, they have concurrently given rise to pressing issues such as environmental degradation, resource depletion, and social inequality. The rapid urbanization and industrialization in China over the past few decades, while driving robust economic growth, have also brought forth a range of undeniable adverse effects. The recent attention to extreme high-temperature weather has raised widespread concern, not only posing threats to residents' health but also negatively impacting the urban image [1]. Furthermore, high housing prices and the concentration of educational and medical resources have intensified the life pressures experienced by some low-income groups, leading to an expanding social gap within the city [2]. With the growing emphasis on health and sustainable living environments, cities are facing an urgent need for transformation.

Urban ecological spaces play an indispensable role in the transformative development of cities. Traditional urban development models have emphasized economic growth and infrastructure construction, often neglecting environmental sustainability and the quality of residents' lives. Urban transition seeks a more comprehensive, sustainable, and human-centered development path. As a core component of urban systems, ecological spaces play a pivotal role in achieving goals such as improving the ecological environment, enhancing citizens' health, and promoting overall well-being [3–5].

Since the twentieth century, the paradigm of urban development in China has gradually changed. Since the 18th National Congress of the Communist Party of China in 2012, ecological civilization has been integrated into the national economic, social, cultural, and political systems, termed the "Five-in-One" overall layout. This aims to establish a long-term environmental protection mechanism, elevating environmental protection to a new level in national policy areas [6,7].

In 2018, the new urban development concept of a "City in a Park" (CIP) was proposed by General Secretary Xi Jinping during his inspection of the Sichuan Tianfu New Area (STNA). The CIP emphasizes the diverse value of the city, including ecological, aesthetic, cultural, and economic value, while exhibiting a high degree of inclusivity [8,9]. It serves as a new concept for the development of green cities in China and an urban expression of ecological civilization [10,11]. The concept plays a crucial role in leading the comprehensive functions of ecological spaces within the urban context.

The essence of the CIP is fundamentally an open and complex social–ecological system, influenced by various factors, such as the natural environment, social structure, and economic development [12]. Synergetics is a theoretical framework that studies the interactions and synergy among various elements within a system [13]. In recent years, synergetics has gradually been applied to urban studies, including regional planning [14], transportation planning [15], and planning decision-making [16]. However, there is still a lack of research on the new urban development concept of the CIP from the perspective of synergetics. Guided by this theoretical framework, we endeavor to comprehend how the sustainable development of the CIP is propelled through the synergistic interaction between the urban ecological space and the built environment.

The STNA is a comprehensive functional zone approved by the State Council of China, tasked with significant national development and reform missions. Its planning and construction prominently reflect China's strong emphasis on urban sustainable development and ecological civilization construction in the New Era. As the pioneer and demonstration area for the CIP, the STNA prioritizes ecological considerations in its planning. It actively practices this development concept, achieving preliminary success over the past five years. The practices of the CIP, green infrastructure, and the low-carbon city in the STNA not only have a significant impact on urban development in Sichuan Province and the western region of China but also provide valuable insights for global sustainable urban development. However, the STNA still faces several challenges in promoting the CIP. These challenges include how to protect and utilize natural resources effectively in the context of rapid urbanization, balancing the relationship between economic development and ecological conservation, and enhancing the urban governance capabilities to address complex urban issues. Therefore, this study selects the STNA as a case study, aiming to explore how the integrated planning of ecological spaces and built environments can promote systematic sustainable development in ecology, economy, and society.

2. Theoretical Foundation

2.1. The Concept of the CIP

2.1.1. Previous Related Theories

Research on the integrated planning of ecological spaces and built environments within global urban planning theories has continually developed over the past one to two centuries. In 1989, the Garden City theory, proposed in the UK by Howard, emphasized considering the countryside surrounding the city as an integral part of the urban area, while introducing green spaces and gardens into the cityscape. This theory influenced a series of representative planning theories, including organic dispersal theory, broadacre city, and landscape urbanism [17]. Patrick Geddes critiqued British industrial cities from a biological and ecological perspective, proposing that natural regions be the fundamental framework for planning, as discussed in the literature on evolving cities [18]. To reduce urban sprawl and leave more space for nature, Western urban scholars have proposed theories on the optimization of urban functions. Notable among these are Le Corbusier's "Urban Planning and Design" [19], Perry's neighborhood unit concept [20], and the "Athens Charter" [21]. In 1969, Ian McHarg, in California, USA, published "Design with Nature", emphasizing the use of ecological principles in urban planning and design by carefully considering the ecological and landscape characteristics of regions [22]. Contemporary scholar Richard T.T. Forman has revealed the patterns and ecological processes of urban landscapes from the perspective of landscape ecology in "Urban Ecology: Science of Cities" [23]. Concepts such as the eco-city, sustainable city, green cities, and city in a garden propose development models that enhance urban ecological service functions, promoting the coordinated development of "environment–economy–society" [24–26]. As urban social issues continue to emerge, the equity issues of urban public spaces have gained attention. Among these, Lefebvre and David Harvey have extensively speculated on the inequalities in power and space among urban and rural residents [27].

In China, the concept of the "Shanshui City", proposed by scientist Xuesen Qian in 1990, carries forward the traditional Chinese views of natural landscapes and the philosophical concept of unity between heaven and humans. This concept has laid a theoretical foundation for the planning of ecological cities [28]. The natural environment of cities in China has increasingly gained attention in urban development. Based on different core value orientations towards the environment in urban planning, urban theories related to ecology in China can be divided into three stages.

The first stage is the landscape garden city stage [29]. During this stage, planning is focused on the construction of parks, with the goals of increasing greenery, beautifying the environment, controlling pollution, and providing recreational and entertainment facilities. The second stage is the ecological garden city stage. Planning in this stage emphasizes the construction of ecological network systems and the enhancement of ecological functions. Specific regulations have been established for environmental monitoring indicators, such as air, water, noise, and heat island effects, and certain requirements have been set for urban biodiversity. The third stage is the low-carbon city stage [30]. In this stage, low-carbon and ecological concepts are integrated into more comprehensive urban planning, effectively promoting green urban development. In urban development assessments, standards related to energy conservation and emission reduction play a significant role, such as mixed land use, a low-carbon economy, green transportation, green buildings, and the utilization of renewable energy [31]. In response to the improvement in the urban water environment, China has proposed the concept of a sponge city, committed to constructing ecological spaces that mimic natural water circulation patterns [32].

2.1.2. The CIP

The concept of the CIP has evolved from these urban planning theories guided by ecological principles, providing a new perspective on urban ecological environment protection, green industry development, and the enhancement of people's well-being. The CIP goes beyond being a city with numerous parks; instead, it embodies the idea of an entire city as a "big park", forming a large, park-like system that covers the entire urban landscape [33]. The ecological spaces constitute the environmental foundation of this "big park", with the city emerging as a series of buildings within this expansive concept. Consequently, the ecological spaces of the CIP extend beyond traditional urban development areas, encompassing regional park systems, urban park systems, and ecological corridor systems—a multi-scale, cross-regional park network [34]. Simultaneously, the economic development of the CIP is fueled by the innovative drive of green industries [35]. On the one hand, the ecological and livable environment attracts innovative businesses and talent, propelling the transformation of traditional industries. On the other hand, it activates the ecological value of traditional industries, such as leisure agriculture and high-end manufacturing, guiding the consumption of green industries. Finally, the CIP emphasizes the public attributes inherent in parks, stressing the enhancement of fairness in social services. For instance, spatial planning is centered around the preservation and restoration of the regional ecology, alleviating the primary contradiction of China's current "unbalanced and inadequate" development, thereby promoting the improvement of people's well-being [36].

2.2. *Synergetics of Urban Systems*

Synergetics, a theory proposed by the renowned physicist Haken in the 1970s, is an important branch of systemics. Synergetics studies the mechanism by which systems composed of various quantum subsystems, under certain conditions, exhibit self-organizing structures with specific functions on a macroscopic scale, surpassing the sum of the individual subsystem functions, through the synergistic interaction of matter, energy, and signals among the subsystems [13,37]. Synergetics has been widely applied in the natural sciences, such as physics, biology, and chemistry, as well as in sociology and economics, and it has gradually been introduced into urban planning [38]. In the field of urban planning, the theory of synergetics provides a framework for the interpretation of cities as complex systems. Cities are not just collections of physical spaces; they are open, complex systems where social activities, economic development, and the natural environment interact. The synergies within urban systems exhibit the following characteristics.

2.2.1. Resilience

The urban system possesses the capability to dynamically adapt to environmental changes. The concept of adaptability was initially introduced in ecology to describe the ability of an ecosystem to recover its original structure or function after a disturbance [39]. Holling emphasized the concept of dynamic stability in ecosystems, introducing the "adaptive cycle" to explain how ecosystems maintain their complexity and adaptability through different phases of growth, accumulation, reorganization, and release [40]. The concept of adaptability was later extended to social–ecological systems (SES), which is an integrated framework capturing the interactions between human society and ecological systems. Berkes and Folke explored how social–ecological systems enhance adaptability through learning, self-organization, and multi-level governance [41]. Folke et al. further introduced concepts such as "resilience", "adaptability", and "transformability", emphasizing the system's responsiveness to change [42].

In recent years, resilience theory has found application in urban systems, particularly in the fields of urban planning and sustainable development [43,44]. Urban resilience refers to a city's ability to continuously adjust its structure and functions through feedback mechanisms to adapt to external changes, such as climate change and economic crises [45,46]. Concepts such as urban green infrastructure and nature-based solutions, as strategies to enhance urban resilience, work by reinforcing ecosystem services to support the sustainability of cities [47–49].

2.2.2. Self-Organization

The self-organizing capacity of urban systems reflects their inherent learning, complexity, and evolutionary characteristics, enabling cities to spontaneously form ordered new structures through local interactions among their internal subsystems without central control. This self-organization is the foundation for urban adaptation to environmental changes and internal disturbances, manifesting as the formation of functional zones, development of transportation networks, and evolution of community structures [12]. Initially derived from the study of natural systems, such as ecosystems and economic systems, self-organization was later introduced into urban research. The self-organizing processes of urban systems demonstrate their learning ability, wherein cities continuously learn from their experiences and adjust their behaviors and structures through ongoing interaction and feedback loops. The decisions and actions of urban actors, such as residents, businesses, and governments, interact to form complex networks. Through learning and adaptation, these networks can generate new spatial patterns and social structures [43,50,51]. The self-organizing processes of urban systems lead to the complexity of their structure and functions. As a dynamic complex system, the complexity of the city arises from numerous interdependent and interacting components. Moreover, these interactions are often nonlinear, meaning that small, local changes can trigger large-scale impacts across the system, rendering it unpredictable [12]. This complexity enables cities to spontaneously generate new organizational structures and stable operational patterns across multiple scales, allowing them to more effectively respond to environmental changes and societal demands [52].

2.2.3. Multi-Level Structure

As cities continuously generate new structures and gradually increase in complexity, self-organizing systems produce multi-level structures, including individuals, communities, cities, and urban clusters. Each level has its specific organizational methods and operational rules, and the interactions between levels influence the behavior and development of the entire urban system [52]. However, the internal correlations within subsystems are often more numerous and stronger than those between subsystems. Various subsystems, such as transportation networks, residential areas, and commercial centers, form tight interactions and feedback loops internally. While interactions between these subsystems exist, they tend to be relatively weak. Due to differences in economic activities, social networks, and cultural preferences, urban spaces exhibit varying degrees of differentiation and specialization, leading to a high concentration and interdependence among the functions and activities within each region [53]. Research indicates that the connections within the city's transportation network and functional zones are usually denser and more frequent than those between regions. These internal connections support the specialization and efficiency improvement of regional functions [54]. At the community level, studies on neighborhood effects reveal that social interactions and network relationships among residents primarily form within geographically proximate communities. These internally tight social connections have a significant impact on community cohesion and identity [55]. Carlos Moreno, in his concept of the "15 min city", suggests creating an environment where residents can easily reach most places necessary for daily life within a 15 min walk or bike ride, such as workplaces, schools, shopping centers, medical facilities, entertainment venues, and parks. This approach aims to enhance urban convenience and improve the quality of life of residents [56].

2.3. *Multi-Dimensional Benefits of Urban Ecological Spatial Planning*

The benefits of urban ecological spatial planning consists of three parts, including an ecological built environment, integrated industrial functions, and fair governance mechanisms. The material spatial foundation of the CIP system is composed of ecological elements such as forests, wetlands, and parks, along with architectural elements like buildings, roads, and squares. The spatial functions of the CIP system are multifaceted, integrating various socio-economic activities from primary, secondary, and tertiary industries, serving as a driving force for the achievement of sustainable development. To ensure the operation of the CIP system, governance mechanisms such as fair public services and public participation processes are crucially needed.

2.3.1. Ecological Built Environment

The ecological space system and built environment spontaneously interact with each other. The ecological benefits of green space extend beyond the system itself, enhancing the surrounding built environment. The effective integration of ecological space and the built environment can significantly enhance the ecological performance of urban environments, including improving the air quality, mitigating climate change, reducing stormwater runoff, and enhancing biodiversity [57–59]. For instance, urban trees and green spaces can absorb air pollutants generated by the built environment and provide shade to lower the temperature of moving air, thereby alleviating the urban heat island effect [60–62]. Even small-scale ecological spaces, such as green roofs, can reduce the average flood depth by up to 13% in critical areas of the built environment [57]. Designers can create diverse ecological spaces around buildings, such as vertical gardens and urban gardens, to provide habitats for various species and increase the opportunities for residents to connect with nature [63]. Urban landscape ecological planning establishes ecological networks and connects ecological corridors to protect and restore ecological functions [64]. Large-scale park systems, conservation areas, and green infrastructure in urban areas contribute to maintaining biodiversity and providing essential ecological services [65].

2.3.2. Integrated Industrial Functions

Industrial functional diversity refers to the diversification and complementarity of the urban industrial layout, encompassing the spatial agglomeration of economic activities and the enhancement of ecological and social value. The ecological space system serves not only as an ecological infrastructure but also as a crucial resource for the promotion of economic innovation, creating employment opportunities, and enhancing urban competitiveness. The integration of ecological space and the built environment provides opportunities for leisure and tourism, enhances the urban attractiveness, and improves the quality of living environments, directly and indirectly fostering urban economic development [66,67]. Research indicates that a well-designed urban ecological environment can attract talent and investment and stimulate economic diversification, thereby strengthening the urban economic resilience and sustainable development capabilities [68,69]. Ecological spaces are considered part of the urban innovation system, offering the ecological services and creative inspiration needed for innovative activities. For example, suburban farmland exhibits multifunctionality. The development of green industries such as urban agriculture and ecotourism not only provides new points of economic growth for the city but also drives technological innovation and the emergence of new business models [70,71]. Ecological spaces play a bridging role in urban spatial restructuring, promoting the optimization of industrial layouts and the rebalancing of regional economies through ecological restoration and spatial reshaping. In industrial parks, engaging in ecological industries reduces the ecological footprint of the park [72]. Planning green belts and water bodies around industrial areas can provide ecological services, improve working environments, and promote sustainable industrial development [73]. In business districts, sustainable landscapes offer leisure and relaxation spaces for white-collar workers and attract commercial investments [74,75].

2.3.3. Equitable Social Services

Ecological spaces, considered as public spaces within the city, serve as a lever in achieving the fair governance of urban spaces, playing a crucial role in enhancing social well-being, promoting public health, and strengthening public participation [76–78]. Parks, gardens, green belts, and similar areas provide valuable leisure and relaxation spots for urban residents, contributing to improved quality of life and happiness [79,80]. These spaces encourage outdoor activities, offering opportunities for interaction with nature, thereby reducing stress and enhancing individual mental health [81]. Ecological spaces have a significantly positive impact on public health. Research indicates that exposure to natural environments can reduce psychological stress and anxiety, lowering the risk of chronic diseases such as heart disease, obesity, and diabetes [82,83], and can even extend life expectancy [84]. Spaces like parks and green belts encourage residents to engage in physical activities like walking, running, and cycling, contributing to maintaining a healthy lifestyle [85]. The planning, design, and management processes of ecological spaces provide crucial opportunities to promote public participation [86,87]. Through engagement in these processes, residents can voice their opinions on their living environments and participate in decision-making, thereby strengthening their sense of belonging and responsibility to the community. Community gardens and urban farming projects not only enhance the ecological quality of the city but also foster the active participation of residents in environmental protection and sustainable development [88].

3. Materials and Methods

3.1. Conceptual Framework

Based on an analysis of the related theories, this study adopts the perspective of synergetics and focuses on urban planning that practices the concept of the CIP. We built a conceptual framework (Figure 1). It seeks to explain how to plan an integrated spatial pattern comprising urban ecological spaces and built environments to establish ecological and equitable adaptation mechanisms. This approach aims to activate the multifunctionality of urban spaces and promote sustainable development.

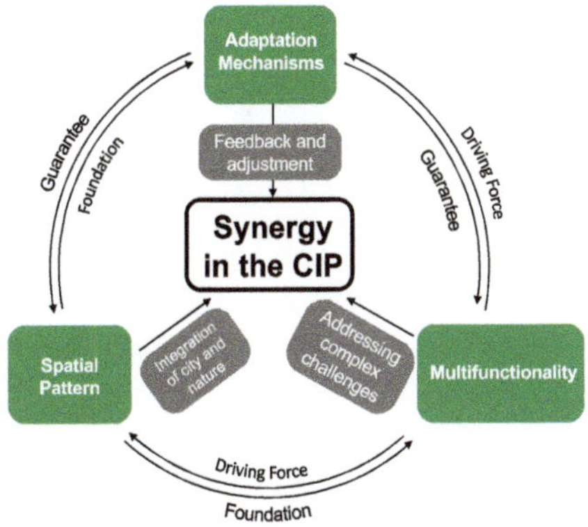

Figure 1. Conceptual framework of synergy in the CIP.

Multifunctionality refers to the multiple dimensions of urban spaces, such as ecological, economic, and social functions. Spatial planning not only focuses on enhancing a single function but also aims to enhance the overall urban sustainability by integrating multiple functions, thereby addressing the complex social challenges faced in urban development. These challenges include climate change, the development of green industries, and public health, etc.

The spatial pattern is the integrated system of urban ecological spaces and built environments. This system can be understood and constructed through macro, meso, and micro scales and conceptualized through points, lines, and planes. Each spatial element plays a unique role within the system and works in synergy with others. For example, at the neighborhood scale, a point could be a small spatial node like a street corner park, a green building, or a garden-style community.

Adaptation mechanisms focus on establishing dynamic feedback systems to monitor, assess, and adjust various practices in urban planning and management, ensuring that cities can continually adapt to changing environmental, social, and economic conditions. For instance, in terms of the environment, environmental monitoring equipment is utilized to collect data on ecological conditions, analyze the performance of spatial functions, and then feed the evaluation results back to the planning and management departments to adjust urban planning and environmental policies.

3.2. Case Study: Sichuan Tianfu New Area

Officially established in 2014, the STNA is situated to the southeast of the central urban area of Chengdu, Sichuan Province. The region spans seven counties (cities and districts) under the jurisdictions of Chengdu, Meishan, and Ziyang, covering a planned area of 1578 km^2, with an anticipated population of 3.5 million (Figure 2). The urban planning and design of the STNA adhere to the principles of rational ecological planning, emphasizing priorities such as environmental protection, the integration of industry and city, and cultural heritage [89,90]. Inspired by the rational ecological planning theory, the area represents a prototype of an advanced urban form of the CIP.

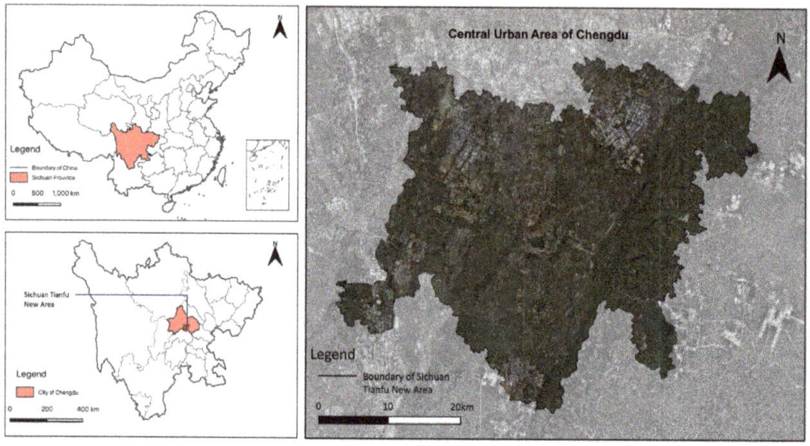

Figure 2. Location of Sichuan Tianfu New Area. Source: administrative boundary data derived from the National Earth System Science Data Center. The aerial photo is derived from Google Earth in 2024.

Since the concept of the CIP was proposed in February 2018, the STNA has actively explored new planning theories and practices. It has followed the technical guidance pathway of the "Park City Index (Framework System)" [8], employing green building, nature-based solutions (NbS), and smart city planning and management technologies. Having been implemented for over five years, it has achieved demonstrative results. Therefore, this study selects the STNA as a case study for the analysis of the planning of a CIP.

3.3. *Data Sources and Collection Methods*

3.3.1. Document Collection

This subsection describes the methods employed to collect textual and imagery data relevant to the study. The collected texts serve as a foundation for an understanding of the theoretical and practical aspects of the CIP planning in the Tianfu New Area and Chengdu.

- Urban Planning and Policy Documents: This includes comprehensive urban master plans, urban design guidelines, and special planning documents for the STNA. These materials are sourced from publicly released documents and planning materials provided by government urban planning agencies, as well as internal technical documents provided by urban planning and design firms. They are utilized to understand the objectives, strategies, and schemes of CIP planning from a policy perspective.
- Academic Papers and Monographs: This category encompasses scholarly evaluations, case analyses, and theoretical discussions regarding the CIP planning of the STNA and Chengdu. These materials are obtained from databases such as the China National Knowledge Infrastructure (CNKI), thematic reports from academic conferences, and publications from book publishers. They are used to acquire third-party assessments and theoretical support for CIP planning in Chengdu.
- Media Reports: This includes news articles, public comments, and featured discussions reporting on CIP planning in Chengdu. These materials are sourced from news media, blogs, and social media platforms and are employed to comprehend the

implementation of CIP planning, as well as public opinions and reactions towards CIP planning in Chengdu.

3.3.2. Unstructured Interviews

In February 2021, the research team organized a collective discussion lasting approximately two hours. The participants included professionals from multiple relevant fields, such as officers from urban planning and management departments, planners from planning and design institutes, and researchers from research institutes and universities. The participants shared the specific practices that their organizations implemented in promoting the CIP concept, the challenges encountered, and the strategies to address these challenges. This information could be used to understand and analyze the experiences and perspectives of various stakeholders involved in the management, planning, and research processes of a CIP.

In addition to organizing a collective discussion, we extensively used unstructured interviews as a method for data collection, to delve deeper into the implementation of the planning in the STNA. These interviews included not only conversations with management and design personnel from the planning departments but also daily interactions with residents living in the STNA. Interviewees were allowed to freely express their views and feelings about their living environments, community facilities, and the implementation of the CIP concept. Through these diverse perspectives, we were able to collect comprehensive information ranging from policy-making to everyday life experiences.

3.3.3. Field Observations

The research team resided in the city where the STNA is located. Since the introduction of the CIP concept in 2018, over approximately five years, the researchers have used informal observation methods in their work and daily lives. They employed direct observation, photography, and note-taking to collect data on the utilization of spaces, environmental quality, and residents' activities. These data could be used to reveal the impact of urban planning on residents' lives and understand how residents adapt to and utilize urban spaces.

3.3.4. Participatory Observation

The corresponding author of this paper, who has served as the Head of Planning Technology for the STNA since its inception in 2010, participated in the preparation process and assumed responsibilities for planning design, technical organization, and implementation management until 2019. Additionally, some authors, as design technicians, were involved in the process from design to the cooperative construction of community gardens in the STNA in June 2023. These experiences and insights are utilized to deeply understand the dynamic changes in the CIP planning of the STNA and the reasons behind these changes.

4. Results

4.1. Spatial Planning of the "Great Park"

4.1.1. Site Selection Avoiding Prime Farmland

Food security is a fundamental guarantee for the sustainable development of a CIP. In 2013, when selecting a site for the STNA, there was a decision to be made about whether to develop towards the north, south, east, or west. From a construction perspective, the area west of Chengdu's main urban zone is the flattest, and the northern area has a solid foundation in advanced manufacturing and urban infrastructure, making it a prime candidate for development. However, an analysis of the natural resources in the overall planning documents revealed that these two regions had rich water systems and abundant fertile farmland, making them valuable for agricultural preservation [91].

The southern region, home to Shuangliu International Airport and bordered by several major railways and highways, connects with the high-tech and modern service industry-focused High-Tech Zone. Its terrain comprises mountains, low hills, and plateaus, which are less suitable for agriculture than the western and northern areas. Balancing ecological value and construction costs and adhering to the site selection principles of "protecting fertile farmland, ensuring urban safety, maintaining ecological livability, leveraging regional advantages, and integrating new and old urban areas", the southern area was ultimately chosen for development [92]. The selection of this site not only preserved the high-quality agricultural land but also laid the foundation for the STNA to be approved in 2021 to start a pilot project for standardized CIP development.

4.1.2. Spatial Layout with Ecological Priority

Creating an ecologically livable environment is one of the primary construction goals of a CIP. The planning for the STNA prioritized the layout of ecological functional areas and ecological corridors, dividing urban construction land into several spatial units with multifunctional capabilities. This approach not only avoided the sprawl of "pancake-style" urban expansion but also protected ecological spaces. Ultimately, the planning areas for ecological spaces and urban construction zones are controlled at a ratio of 7:3 [92].

Urban units are designed to follow the natural mechanisms of the terrain and water systems, with a focus on protecting the ecologically sensitive Longquan Mountain in the east and the Pengzu Mountain low hills in the southwest. Along major waterways, railways, expressways, and fast roads, "four horizontal and seven vertical" ecological corridors are formed [91]. These interwoven ecological corridors and urban transport systems lay the foundation for the district's ecologically livable environment.

The ecological space planning reserves ecological buffer zones and corridors for the city. These spaces become the city's ventilation corridors and "green lungs", also serving as strategically significant "reserved" areas. At the same time, ecological functional areas divide the city into a pattern of "one belt, two wings, one city, six areas" (see Figure 3). In addition to providing urban ecological conservation functions, these ecological functional areas also serve the residents of adjacent urban functional areas, offering multiple functions such as recreation, entertainment, and science education. For example, the Luxi River Ecological Zone, one of the main ecological corridors of the STNA, is located between the science city cluster and the central business district cluster. The ecological zone, while restoring forest and wetland spaces, also provides riverside greenways and water paths. On sunny holidays, a large number of citizens visit for camping, hiking, cycling, and other activities, enhancing the social value of the ecological spaces (see Figure 4).

4.1.3. Construction of Integrated Urban–Rural Green Space System

Equity is one of the objectives of a CIP. Ecological spaces are public spaces that can provide social welfare. The planning for the STNA adopted the strategy of "green wedges, green belts, and green hearts", creating a balanced distribution of green spaces [91]. "Green wedges" located in suburban areas separate urban functional zones, creating ecological corridors and shortening the distance between urban and rural areas. For example, the remnants of Pengzu Mountain extend deep into the southern side of the STNA. By preserving the large-scale woodlands that have already formed and protecting the adjacent Jinjiang and Luxi Rivers, ecological resources are safeguarded. Fresh air can easily be channeled into urban functional areas, and city residents have more convenient access to nearby ecological parks for leisure and recreation. In an interview with a woman who had lived in Chengdu for over ten years, she mentioned, "Since Chengdu started building the CIP, I've noticeably found more places to take my child. The variety of activities available in these large parks has also increased".

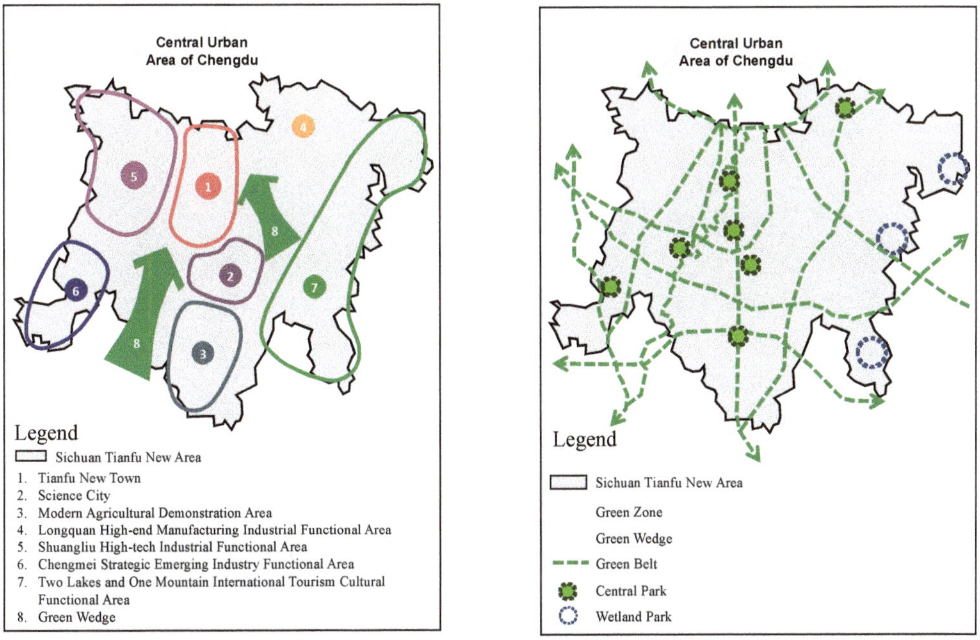

(a) (b)

Figure 3. Spatial structure layout of STNA: (**a**) urban functional areas divided by ecological functional areas; (**b**) ecological network. Source: authors' redrawing based on the Master Plan of the STNA (2010–2030) [92].

Inside urban functional zones, ecological "green belts" divide the area into multiple urban clusters and connect the interior parks and green spaces with the "green wedges" around the city (see Figure 5). The landscape design of the "green belts" is meticulous, featuring seasonally changing plants, convenient walking paths, bicycle lanes, and fitness equipment, providing residents with a variety of services, including rest, entertainment, and fitness. Urban public service facilities are arranged around parks of various levels within urban clusters and communities, forming the "green hearts" of the clusters and communities. The urban cluster-level "green heart" is a comprehensive "central park" that houses cluster-level public service facilities, including cultural activity centers, sports fields, and community health service centers. The community park-level "green heart" incorporates community-level public service facilities built in a complex style, including comprehensive community service centers, community rooms, farmers' markets, kindergartens, and fitness facilities. This arrangement creates an ecologically livable public space system for community residents, with "green within 500 m and water within 1000 m".

Figure 4. Photos of Luxi Ecological Functional Area: (**Top Left**) residents camping; (**Top Right**) trails along wetland; (**Bottom**) overall aerial photo.

Figure 5. Photos of green space between urban clusters in the STNA.

4.2. Development of Green Industry

4.2.1. Integration of Diverse Industries

The integration of primary, secondary, and tertiary industries is one of the pathways to achieving economic green development in a CIP. The planning for the STNA included spatial units for coordinated urban–rural development, promoting the integration of diverse industrial functions within the region. Towns and villages have enhanced their infrastructure for roads, transportation, and communication, connecting the "last mile" to the city's infrastructure and thereby improving the accessibility and convenience of urban–rural transport. Improved infrastructure creates a favorable material foundation for the introduction of secondary and tertiary industries to rural areas.

For example, Guantang Village in Zhengxing Street, located adjacent to the Science City and West Expo City, serves as a rural pilot in the CIP demonstration area. Guantang Village has renovated roads connecting to the city and adopted a "small-scale, clustered, micro-pastoral, ecological" model. It has integrated idle and fragmented agricultural land resources, enhancing the land utilization and overall value. The village has constructed a group of Sichuan-style rural dwellings with traditional cultural characteristics and designed landscapes according to the requirements of tourist attractions, integrating the residences into a cohesive whole. On specific holidays, Guantang Village organizes activities featuring rural and traditional folk customs.

While preserving agricultural production, Guantang Village uses its farmland landscape and traditional folk experiences as attractions, incorporating homestays, intangible cultural heritage studios, hotels, and restaurants into its tourism offerings, thus promoting the integration of urban and rural industries. This has expanded the income sources for villagers and provided city dwellers with richer and more convenient rural tourism experiences (see Figure 6).

Figure 6. Photos of Guantang New Village.

4.2.2. Attraction of Innovative Enterprises

A CIP leverages industrial innovation as the driving force for economic development, which hinges on innovative talent and enterprise. In the master plan of the STNA, a central business district focused on information services and finance has been strategically positioned at the core of the center. This area is designed with an ecologically favorable living environment and is complemented by high-quality public services including commerce, education, and healthcare.

For instance, in the core area of Luxi Intelligent Valley within the Science City, located next to Xinglong Lake, the region's largest man-made lake, the area boasts rich landscapes, convenient transportation, and a full suite of public service facilities. It has attracted a host of state-owned enterprises and national research institutions. Among these, the National Chengdu Agricultural Science and Technology Center has planned and designed a 300-acre cluster for research, office, and residential support facilities, as well as a 3000-acre modern agricultural technology integration demonstration zone. This zone provides multifunctional services including experimental demonstrations, results transformation, and leisure experiences, offering a high-quality environment for the work and living of research personnel.

The influx of innovative enterprises and talent into the STNA has enhanced the region's innovative capacity. During random interviews with innovative talent who had relocated to the STNA from other cities, the respondents frequently highlighted the ecological environment as a major attraction. One researcher who moved from Chongqing commented, "The ecological environment in Tianfu New Area is exceptionally good, it's like a big park" and added, "I was amazed when I first arrived!"

4.2.3. Shaping of Cultural Characteristics

The CIP respects diverse cultures. In the planning of the STNA, distinctive cultural characteristics have been sculpted through urban design guidelines based on the different industrial functions of the seven urban functional zones, combined with each zone's historical, cultural, and natural environmental endowments.

For example, the central business district (CBD) is designed to reflect an urban landscape with natural scenery. The central area predominantly features modern-style office buildings. Between these skyscrapers, Tianfu Central Park has been planned and designed, integrating modern skyscrapers with natural landscapes to create a cutting-edge business atmosphere for financial, computer, and other innovative service enterprises (see Figure 7). In the modern agricultural demonstration area, the plan involves a clustered layout of modern multi-story office buildings with farmland planned around the periphery. Large, well-organized fields facilitate the use of modern agricultural machinery, creating a modern pastoral urban landscape. The "Two Lakes and One Mountain" tourism area preserves and restores historical buildings and plans designs in the traditional Western Sichuan residential style, creating a historical and cultural ambiance that attracts tourists.

The unique cultural characteristics of each urban functional zone foster differentiated competition within industries, enhancing the comprehensiveness and balance of the industries in the STNA.

Figure 7. Aerial photo of central park in the Tianfu New Town functional area.

4.3. Equitable Social Services

4.3.1. Integration of Urban and Rural Administrative Divisions

The CIP emphasizes the integration of urban and rural areas. The STNA has adjusted its administrative divisions to unify the planning, construction, and governance of urban and rural areas. The administrative division of the STNA, directly managed by the city of Chengdu, breaks traditional urban–rural administrative boundaries. It has been reorganized based on the resources, functions, environment, and location of each area, resulting in nine integrated urban–rural street governance units. This new administrative division has promoted the balanced development of infrastructure such as transportation, education, and healthcare, allowing rural residents to enjoy public services similar to those available to urban residents. This administrative and governance innovation effectively facilitates the optimal allocation of resources and the balanced development of service facilities within the region, thereby strengthening the functional complementarity and socio-economic integration between urban and rural areas.

4.3.2. Establishment of Interdepartmental Collaboration Platform

The CIP faces complex social challenges. To better address these challenges, effective cross-departmental collaboration is often required. The STNA has established a flat management collaboration mechanism. Under this mechanism, the Management Committee plays a leading role, the CIP Construction Management Bureau is responsible for work coordination, and the CIP Research Institute provides technical support and leadership. This has resulted in a collaborative working model among the departments, ensuring smooth planning and construction processes.

In terms of projects' collaborative approval processes, the STNA has established a digitalized smart planning management network platform. This platform makes full use of advanced technological tools such as big data and artificial intelligence to realize a complete digital planning management process from project planning to completion and acceptance. This not only simplifies the approval process but also significantly enhances the efficiency of project approvals. Through such a platform, different departments and management functions achieve efficient collaborative work, further advancing the integration of urban and rural governance.

4.3.3. Guidance on Public Participation Processes

The starting point of a CIP is to fulfill people's aspirations for a better life. The planning and implementation of the STNA adhere to the principles of scientific and democratic governance, establishing mechanisms for public participation. In the early stages of planning, draft plans are publicly released through various channels, including government websites, community bulletin boards, and local media, to ensure that all stakeholders have timely access to planning information. The STNA has set up an online feedback platform and social media accounts. Citizens can express their opinions through phone calls, web message boards, and social media.

Community committees organize workshops that invite community members, urban planners, environmental experts, and policymakers to discuss the planning proposals. However, the government still holds the ultimate authority in the planning process, and public engagement is not very high. For example, a member of our research team, as a design professional, participated in the design and construction of a community garden in the Jianghua Community of Longquanyi District. Although community residents participated in discussing the plans, designing public art, and planting in the garden, the final decision-making power remained with the government. Residents' participation was focused on enriching their experiences rather than influencing outcomes.

5. Discussion

Through a case study of the Sichuan Tianfu New Area (STNA), this paper explores the practices of City in a Park (CIP) planning, finding that the CIP concept facilitates the synergy between ecological spaces and the built environment, promoting the transformation of cities towards ecological civilization and sustainable development.

Firstly, in the process of promoting the synergy between ecological spaces and the built environment, the planning of the STNA goes beyond the traditional scope of urban construction, including surrounding rural areas. This approach ensures that the planned area of the STNA includes a higher proportion of ecological spaces, accounting for about 70%. The ratio is similar to the Xiong'an New Area near Beijing, which is an important strategic project for China's future development [93]. In the suburbs, the STNA has planned "green wedges" and "green corridors", creating a broader ecological space system together with the city's green infrastructure, presenting a city within a park. Ecological spaces serve multiple functions, such as providing ecological resources, regulating the ecological environment, and offering recreational activities. Therefore, urban ecological areas act as "purifiers" for urban functional zones, improving the air quality and mitigating urban heat island effects. Although both traditional "garden cities" and the CIP incorporate urban–rural integration [17], the STNA places more emphasis on the ecological value of rural areas, such as biodiversity and urban climate regulation, providing new planning ideas and tools to address environmental challenges on a larger scale.

Secondly, the ecological functional areas in the countryside have introduced more industries, enhancing the integration of the primary, secondary, and tertiary sectors. The agricultural fields in the STNA are small and fragmented, a layout linked to historical farming methods and land distribution systems. Historically, agriculture in the Chengdu area has focused on small-scale, high-efficiency farming, which is not conducive to the use of large modern agricultural equipment. However, as the infrastructure connection between rural and urban areas strengthens and commercial facilities increase, these rural ecological spaces create new investment opportunities, such as rural tourism, including country hotels, natural education bases, and intangible cultural heritage workshops. Singapore, following the "City in a Garden" concept, also retains large forested areas within its urban planning zone [26], but these ecological spaces mainly serve ecological functions like water and air pollution control, as well as global climate adaptation, with limited recreational functions [94]. In contrast, the rural ecological spaces in the STNA not only preserve local residential and agricultural production functions but also create leisure opportunities, providing more income sources for farmers. This offers practical pathways and experiences to improve the rural industrial levels and integrate the three economic sectors.

Lastly, the planning of the STNA emphasizes the public service attributes of ecological spaces. The balanced layout of ecological spaces and other basic service infrastructures ensures the more equitable sharing of public resources. Although the residents of the STNA participate in planning, compared to public participation in developed countries' cities, such as Stockholm, Sweden [86], the voices and enthusiasm of the STNA's community residents are relatively weak. Fortunately, the STNA has established an integrated multi-level green space system across urban and rural areas, placing public service facilities in functional areas, clusters, and communities based on public green spaces, achieving "green within 500 m, water within 1000 m" and enhancing the equity and public nature of services. The expansion of urban public service facilities into surrounding rural areas, such as medical, educational, and transportation services, effectively narrows the accessibility gap in public services between urban and rural areas, providing planning tools and methods to improve the fairness of social services.

6. Conclusions

Urban development is always accompanied by spatial expansion. However, the question of how to expand without harming the ecological environment poses a challenge for urban planning. The planning of the Sichuan Tianfu New Area (STNA) extends the scope of urban ecological spaces beyond traditional urban construction boundaries. Based on ecology, it creates a pattern of "green wedges" and "green corridors", which are connected to the city's internal green space system. The city is divided into clustered urban functional areas by ecological spaces, resembling buildings in a large "city-like park". This ecology-centric planning strategy not only preserves the natural resources but also enhances the overall sustainability of the city through regional ecological services. By designing ecological spaces as green infrastructure connecting urban and rural areas, the STNA's planning promotes urban–rural integration. This design not only addresses the issue of uneven development between urban and rural areas but also strengthens the physical and social connections between them through shared infrastructure and services.

Traditional economic growth faces challenges due to natural resource consumption. However, thanks to the spatial pattern integrating ecological spaces and built environments, the primary industry in the STNA's ecological spaces can more easily interact with the secondary and tertiary industries, forming new green industries and distinctive urban landscapes. Moreover, an ecologically livable urban environment attracts innovative enterprises and talent, playing a key role in enhancing the city's economic competitiveness and innovative capacity.

The nature of land in rural and urban areas is complex and it is managed by multiple departments, which causes the planning and implementation of urban–rural integration to face coordination issues among different interest groups. The STNA addresses this by integrating the administrative divisions of urban and rural areas, allowing public services to be shared across urban and rural scopes. The construction of a cross-departmental collaborative management platform significantly improves the efficiency of project approvals. Guiding public participation in the planning process also enables residents to better understand planning policies, express their needs, and provide feedback on issues during planning. This not only enhances the transparency of planning and residents' satisfaction but also allows residents to directly influence and shape their living environment, increasing the acceptance and effectiveness of the planning.

The STNA's planning practices offer valuable experiences for global cities, showcasing the great potential and benefits of sustainable urban development. These strategies not only promote the protection of urban ecological environments and the enhancement of ecological services but also strengthen the city's social and economic structure through promoting economic diversification and community participation. This article presents preliminary thoughts on the STNA's practice of the CIP concept from a synergistic perspective. The analysis in this study is primarily qualitative. For similar comprehensive, cross-disciplinary, and ecology-oriented planning research, further detailed studies incorporating quantitative data and addressing specific complex social challenges are needed to deepen the understanding of different types and scales of urban space planning.

Author Contributions: Conceptualization, S.C. and J.Q.; methodology, S.C. and J.Q.; software, S.C. and M.Y.; investigation, S.C. and J.L.; writing—original draft preparation, S.C.; writing—review and editing, S.C., M.Y. and J.L.; visualization, S.C.; supervision, J.Q. All authors have read and agreed to the published version of the manuscript.

Funding: This research was funded by the National Natural Science Foundation of China (grant number: 52078423) and the Key Research and Development Project of Science and Technology Plan of Sichuan Province (grant number: 2020YFS0054).

Data Availability Statement: The original contributions presented in the study are included in the article, further inquiries can be directed to the corresponding authors.

Conflicts of Interest: The authors declare no conflicts of interest.

References

1. Zhang, Q.; Jiang, X.J.; Tong, D.; Davis, S.J.; Zhao, H.Y.; Geng, G.N.; Feng, T.; Zheng, B.; Lu, Z.F.; Streets, D.G.; et al. Transboundary health impacts of transported global air pollution and international trade. *Nature* **2017**, *543*, 705–709. [CrossRef] [PubMed]
2. Wang, H.M.; Cheng, Z.; Zhu, D.J. Striving for global cities with governance approach in transitional China: Case study of Shanghai. *Land Use Policy* **2020**, *90*, 104288. [CrossRef]
3. Hansen, R.; Buizer, M.; Buijs, A.; Pauleit, S.; Mattijssen, T.; Fors, H.; van der Jagt, A.; Kabisch, N.; Cook, M.; Delshammar, T.; et al. Transformative or piecemeal? Changes in green space planning and governance in eleven European cities. *Eur. Plan. Stud.* **2023**, *31*, 2401–2424. [CrossRef]
4. Bekele, N.K.; Hailu, B.T.; Suryabhagavan, K.V. Spatial patterns of urban blue-green landscapes on land surface temperature: A case of Addis Ababa, Ethiopia. *Curr. Res. Environ. Sustain.* **2022**, *4*, 100146. [CrossRef]
5. Kolokotsa, D.; Lilli, A.A.; Lilli, M.A.; Nikolaidis, N.P. On the impact of nature-based solutions on citizens' health & well being. *Energy Build.* **2020**, *229*, 110527. [CrossRef]
6. Hansen, M.H.; Li, H.; Svarverud, R. Ecological civilization: Interpreting the Chinese past, projecting the global future. *Glob. Environ. Chang.-Hum. Policy Dimens.* **2018**, *53*, 195–203. [CrossRef]
7. Gu, Y.; Wu, Y.; Liu, J.; Xu, M.; Zuo, T. Ecological civilization and government administrative system reform in China. *Resour. Conserv. Recycl.* **2020**, *155*, 104654. [CrossRef]
8. Shi, N.; Wang, B.; Qu, C.; Hu, B. A General Framework for the Park City Index. *City Plan. Rev.* **2022**, *46*, 7–11, 45.
9. Peng, N.; Wang, K.; Zhang, Y.; Li, X. A New Proposition of Urban Development in the Context of Ecological Civilization: Exploring the Path of Park City Construction in New Age. *Urban Stud.* **2022**, *29*, 21–25.
10. Han, R.; Wang, K.; Zhang, Y.; Li, X. The Road of Green and High-quality Urban Development Since the Reform and Opening up: Historical Logic and Development Path of the Concept of Park City in the New Era. *Urban Stud.* **2021**, *28*, 28–34.
11. Sun, Z.; Sun, S.; Li, C. Exploration of Park City: Connotation, Concept and Development. *Chin. Landsc. Archit.* **2021**, *37*, 14–17.
12. Portugali, J. Self-organizing cities. *Futures* **1997**, *29*, 353–380. [CrossRef]
13. Haken, H. From Cybernetics to Synergetics. *Cybernetica* **1994**, *37*, 273–290.
14. Chen, Q.; Song, Y. Coordination mechanism and coordination modelof urban agglomerations in China. *J. Grad. Sch. Acad. Sci.* **2010**, *27*, 356–363.
15. Gao, P.; Yang, Z. A Dynamic Traffic Coordination Control Model of Congestion Based on Synergetics. In Proceedings of the International Conference on Materials Engineering for Advanced Technologies (ICMEAT2011), Singapore, 5–6 May 2011; pp. 1039–1042.
16. Shu, C.; Bai, Q. Synergetics-based Government Department Integration Decision. *J. Tongji Univ. Nat. Sci.* **2009**, *37*, 700–703.
17. Huang, M.; Hui, Q. A Pastoral City? A Garden City? Re-Understanding of Howard's Garden City. *City Plan. Rev.* **2018**, *42*, 9–17.
18. Geddes, P. *Cities in Evolution: An Introduction to the Town Planning Movement and to the Study of Civic*; Williams & Norgate: London, UK, 1915; p. 391.
19. Moulis, A. An exemplar for modernism: Le Corbusier's Adelaide drawing, urbanism and the Chandigarh plan. *J. Archit.* **2012**, *17*, 871–887. [CrossRef]
20. Byun, N.; Choi, Y.; Choi, J. The Neighborhood Unit: Effective or Obsolete? *J. Asian Archit. Build. Eng.* **2014**, *13*, 617–624. [CrossRef]
21. Lawton, P. Tracing the Provenance of Urbanist Ideals: A Critical Analysis of The Quito Papers. *Int. J. Urban Reg. Res.* **2020**, *44*, 731–742. [CrossRef]
22. McHarg, I.L. *Design with Nature*; American Museum of Natural History: Los Angeles, CA, USA, 1969; p. 197.
23. Forman, R.T.T. *Urban Ecology: Science of Cities*; Cambridge University Press: Cambridge, UK, 2014.
24. Register, R. *Ecocities: Building Cities in Balance with Nature*; Berkeley Hills Books: Berkeley, CA, USA, 2001; p. 320.

25. Gordon, D. *Green Cities*; Black Rose Books: Montreal, QC, Canada, 1990; p. 308.
26. Friess, D.A. Singapore as a long-term case study for tropical urban ecosystem services. *Urban Ecosyst.* **2017**, *20*, 277–291. [CrossRef]
27. Wong, T.C.; Liu, R. Developmental Urbanism, City Image Branding and the "Right to the City" in Transitional China. *Urban Policy Res.* **2017**, *35*, 210–223. [CrossRef]
28. Liu, W.; Li, X. Research on the construction strategy of human settlement environment in shanshui city. *Ind. Constr.* **2018**, *48*, 7–11.
29. Chen, C. From Building Parks in Cities to How to Making Cities into Parks. *Urban Probl.* **1993**, *52*, 5–6.
30. Cheng, J.; Yi, J.; Dai, S.; Xiong, Y. Can low-carbon city construction facilitate green growth? Evidence from China's pilot low-carbon city initiative. *J. Clean. Prod.* **2019**, *231*, 1158–1170. [CrossRef]
31. Zhang, Q.; Ye, X.; Chen, G. Low-carbon Urban Planning: A New Vision. *City Plan. Rev.* **2010**, *34*, 13–18, 41.
32. Qi, Y.; Chan, F.K.S.; Thorne, C.; O'Donnell, E.; Quagliolo, C.; Comino, E.; Pezzoli, A.; Li, L.; Griffiths, J.; Sang, Y.; et al. Addressing Challenges of Urban Water Management in Chinese Sponge Cities via Nature-Based Solutions. *Water* **2020**, *12*, 2788. [CrossRef]
33. Wang, Z.; Wu, Y.; Jing, Z. Park Integrating City and Scenes Constructing City: New Thoughts on the Construction Model of "Park City". *Chin. Landsc. Archit.* **2021**, *37*, 7–11.
34. Ye, J.; Zhang, Y.; Wang, H. The New Mode of the Construction and Development of Park City from the Perspective of Humanism in the New Era. *Chin. Landsc. Archit.* **2021**, *37*, 24–28.
35. Li, X.; Zhang, Y. A New Approach in Urban Green Development for the New Era—Strategies for Building Park Cities. *Chin. Landsc. Archit.* **2018**, *34*, 38–43.
36. Yuan, L. City Park Systems and Human Wellbeing: On Park City for People. *Chin. Landsc. Archit.* **2018**, *34*, 39–44.
37. Haken, H. *Synergetics: An Introduction*; Springer: Berlin/Heidelberg, Germany, 1983.
38. Meijers, E. Polycentric urban regions and the quest for synergy: Is a network of cities more than the sum of the parts? *Urban Stud.* **2005**, *42*, 765–781. [CrossRef]
39. Holling, C.S. Resilience and Stability of Ecological Systems. *Annu. Rev. Ecol. Syst.* **1973**, *4*, 1–23. [CrossRef]
40. Holling, C.S. Understanding the Complexity of Economic, Ecological, and Social Systems. *Ecosystems* **2001**, *4*, 390–405. [CrossRef]
41. Berkes, F.; Folke, C. *Linking Social and Ecological Systems: Management Practices and Social Mechanisms for Building Resilience*; Cambridge University Press: Cambridge, UK, 2000.
42. Folke, C.; Carpenter, S.R.; Walker, B.; Scheffer, M.; Chapin, T.; Rockstrom, J. Resilience Thinking: Integrating Resilience, Adaptability and Transformability. *Ecol. Soc.* **2010**, *15*, 20. [CrossRef]
43. Pickett, S.T.A.; Cadenasso, M.L.; Grove, J.M. Resilient cities: Meaning, models, and metaphor for integrating the ecological, socio-economic, and planning realms. *Landsc. Urban Plan.* **2004**, *69*, 369–384. [CrossRef]
44. Sun, H.; Zhen, F.; Lobsang, T.; Li, Z. Spatial characteristics of urban life resilience from the perspective of supply and demand: A case study of Nanjing, China. *Habitat Int.* **2019**, *88*, 101983. [CrossRef]
45. Meerow, S.; Newell, J.P.; Stults, M. Defining urban resilience: A review. *Landsc. Urban Plan.* **2016**, *147*, 38–49. [CrossRef]
46. Chen, X.; He, B.-J. Planning for heat-resilient 15 min-cities: Opportunities, measurement, mechanism, and pathways. *Environ. Impact Assess. Rev.* **2024**, *105*, 107406. [CrossRef]
47. Meerow, S. A green infrastructure spatial planning model for evaluating ecosystem service tradeoffs and synergies across three coastal megacities. *Environ. Res. Lett.* **2019**, *14*, 125011. [CrossRef]
48. Ahern, J. From fail-safe to safe-to-fail: Sustainability and resilience in the new urban world. *Landsc. Urban Plan.* **2011**, *100*, 341–343. [CrossRef]
49. Beceiro, P.; Brito, R.S.; Galvao, A. Assessment of the contribution of Nature-Based Solutions (NBS) to urban resilience: Application to the case study of Porto. *Ecol. Eng.* **2022**, *175*, 106489. [CrossRef]
50. Li, X.; Tang, J.; He, B.-J. Highly Heterogeneous Heat-Related Health Impacts in China and Implications for Sustainable Urban Heat Governance. *Adv. Sustain. Syst.* **2024**, *8*, 2300538. [CrossRef]
51. Mitincu, C.-G.; Nita, M.-R.; Hossu, C.-A.; Ioja, I.-C.; Nita, A. Stakeholders's involvement in the planning of nature-based solutions: A network analysis approach. *Environ. Sci. Policy* **2023**, *141*, 69–79. [CrossRef]
52. Batty, M. *Cities and Complexity*; The MIT Press: Cambridge, MA, USA, 2007.
53. Fainstein, S.S. Cities and Diversity: Should We Want It? Can We Plan For It? *Urban Aff. Rev.* **2005**, *41*, 3–19. [CrossRef]
54. Porta, S.; Crucitti, P.; Latora, V. The network analysis of urban streets: A dual approach. *Phys. A Stat. Mech. Appl.* **2006**, *369*, 853–866. [CrossRef]
55. Sampson, R.J.; Morenoff, J.D.; Gannon-Rowley, T. Assessing "Neighborhood Effects": Social Processes and New Directions in Research. *Annu. Rev. Sociol.* **2002**, *28*, 443–478. [CrossRef]
56. Moreno, C.; Allam, Z.; Chabaud, D.; Gall, C.; Pratlong, F. Introducing the "15-Minute City": Sustainability, Resilience and Place Identity in Future Post-Pandemic Cities. *Smart Cities* **2021**, *4*, 93–111. [CrossRef]
57. Twohig, C.; Casali, Y.; Aydin, N.Y. Can green roofs help with stormwater floods? A geospatial planning approach. *Urban For. Urban Green.* **2022**, *76*, 127724. [CrossRef]
58. Alberti, M.; Marzluff, J.; Shulenberger, E.; Bradley, G.; Ryan, C.; ZumBrunnen, C. Integrating Humans Into Ecology: Opportunities and Challenges for Studying Urban Ecosystems. *BioScience* **2009**, *53*, 1169–1179. [CrossRef]
59. Grimm, N.; Faeth, S.; Golubiewski, N.; Redman, C.; Wu, J.; Bai, X.; Briggs, J. Global Change and the Ecology of Cities. *Science* **2008**, *319*, 756–760. [CrossRef] [PubMed]

60. Semeraro, T.; Gatto, E.; De Bellis, L.; Luvisi, A.; Emmanuel, R.; Buccolieri, R. A decision-making framework for promoting the optimum design and planning of Nature-based Solutions at local scale. *Urban For. Urban Green.* **2023**, *84*, 127945. [CrossRef]
61. Zhou, T.; Vermaat, J.E.; Ke, X. Variability of agroecosystems and landscape service provision on the urban-rural fringe of Wuhan, Central China. *Urban Ecosyst.* **2019**, *22*, 1207–1214. [CrossRef]
62. Dutta, K.; Basu, D.; Agrawal, S. Synergetic interaction between spatial land cover dynamics and expanding urban heat islands. *Environ. Monit. Assess.* **2021**, *193*, 184. [CrossRef] [PubMed]
63. Barron, S.; Rugel, E.J. Tolerant greenspaces: Designing urban nature-based solutions that foster social ties and support mental health among young adults. *Environ. Sci. Policy* **2023**, *139*, 1–10. [CrossRef]
64. Egerer, M.; Anderson, E. Social-Ecological Connectivity to Understand Ecosystem Service Provision across Networks in Urban Landscapes. *Land* **2020**, *9*, 530. [CrossRef]
65. Pauleit, S.; Ambrose-Oji, B.; Andersson, E.; Anton, B.; Buijs, A.; Haase, D.; Elands, B.; Hansen, R.; Kowarik, I.; Kronenberg, J.; et al. Advancing urban green infrastructure in Europe: Outcomes and reflections from the GREEN SURGE project. *Urban For. Urban Green.* **2019**, *40*, 4–16. [CrossRef]
66. Rice, L. Nature-based solutions for urban development and tourism. *Int. J. Tour. Cities* **2020**, *6*, 431–448. [CrossRef]
67. Heslinga, J.; Groote, P.; Vanclay, F. Towards Resilient Regions: Policy Recommendations for Stimulating Synergy between Tourism and Landscape. *Land* **2020**, *9*, 44. [CrossRef]
68. Fan, P.; Ouyang, Z.; Basnou, C.; Pino, J.; Park, H.; Chen, J. Nature-based solutions for urban landscapes under post-industrialization and globalization: Barcelona versus Shanghai. *Environ. Res.* **2017**, *156*, 272–283. [CrossRef] [PubMed]
69. Chen, T.; Zhao, Y.; Yang, H.; Wang, G.; Mi, F. Recreational Services from Green Space in Beijing: Where Supply and Demand Meet? *Forests* **2021**, *12*, 1625. [CrossRef]
70. Wang, X.; Wang, D.; Wu, S.; Yan, Z.; Han, J. Cultivated land multifunctionality in undeveloped peri-urban agriculture areas in China: Implications for sustainable land management. *J. Environ. Manag.* **2023**, *325*, 116500. [CrossRef] [PubMed]
71. Li, W.; Zhou, Y.; Zhang, Z. Culture-Led Plan for Peri-Urban Agricultural Sustainability: A Case of Pu'an Village in China. *Land* **2021**, *10*, 242. [CrossRef]
72. Fan, Y.; Fang, C. Assessing environmental performance of eco-industrial development in industrial parks. *Waste Manag.* **2020**, *107*, 219–226. [CrossRef] [PubMed]
73. Zheng, S.; Cui, K.; Sun, S.; Wu, J.; Qiu, Q.; Tian, Y. Planning and design based on landsenses ecology: The case study of Chongming Island Landsenses Ecol-industrial Park. *Int. J. Sustain. Dev. World Ecol.* **2020**, *27*, 435–442. [CrossRef]
74. Nappi-Choulet, I.; Decamps, A. Can Sustainability Enhance Business District Attractiveness? A Survey of Corporate Property Decisions in France. *Urban Stud.* **2013**, *50*, 3283–3304. [CrossRef]
75. Xie, X.; Zhou, H.; Gou, Z.; Yi, M. Spatiotemporal Patterns of the Use of Green Space by White-Collar Workers in Chinese Cities: A Study in Shenzhen. *Land* **2021**, *10*, 1006. [CrossRef]
76. Tzoulas, K.; Korpela, K.; Venn, S.; Yli-Pelkonen, V.; Kazmierczak, A.; Niemela, J.; James, P. Promoting ecosystem and human health in urban areas using Green Infrastructure: A literature review. *Landsc. Urban Plan.* **2007**, *81*, 167–178. [CrossRef]
77. Song, J.; Lu, Y.; Zhao, Q.; Zhang, Y.; Yang, X.; Chen, Q.; Guo, Y.; Hu, K. Effect modifications of green space and blue space on heat–mortality association in Hong Kong, 2008–2017. *Sci. Total Environ.* **2022**, *838*, 156127. [CrossRef] [PubMed]
78. Xiong, G.; He, R.; Wang, G.; Hong, J.; Jin, Y. Environmental Inequalities in Ecosystem Services Benefits of Green Infrastructure: A Case Study from China. *Forests* **2024**, *15*, 73. [CrossRef]
79. Hunter, R.F.; Cleland, C.; Cleary, A.; Droomers, M.; Wheeler, B.W.; Sinnett, D.; Nieuwenhuijsen, M.J.; Braubach, M. Environmental, health, wellbeing, social and equity effects of urban green space interventions: A meta-narrative evidence synthesis. *Environ. Int.* **2019**, *130*, 104923. [CrossRef]
80. Wu, W.; Yao, Y.; Song, Y.; He, D.; Wang, R. Perceived influence of street-level visible greenness exposure in the work and residential environment on life satisfaction: Evidence from Beijing, China. *Urban For. Urban Green.* **2021**, *62*, 127161. [CrossRef]
81. Cox, D.T.C.; Shanahan, D.F.; Hudson, H.L.; Fuller, R.A.; Anderson, K.; Hancock, S.; Gaston, K.J. Doses of Nearby Nature Simultaneously Associated with Multiple Health Benefits. *Int. J. Environ. Res. Public Health* **2017**, *14*, 172. [CrossRef] [PubMed]
82. Mitchell, R.; Popham, F. Effect of exposure to natural environment on health inequalities: An observational population study. *Lancet* **2008**, *372*, 1655–1660. [CrossRef] [PubMed]
83. Hu, K.; Zhang, Z.; Li, Y.; Wang, S.; Ye, T.; Song, J.; Zhang, Y.; Wei, J.; Cheng, J.; Shen, Y.; et al. Urban overall and visible greenness and diabetes among older adults in China. *Landsc. Urban Plan.* **2023**, *240*, 104881. [CrossRef]
84. Wei, D.; Lu, Y.; Wu, X.; Ho, H.C.; Wu, W.; Song, J.; Wang, Y. Greenspace exposure may increase life expectancy of elderly adults, especially for those with low socioeconomic status. *Health Place* **2023**, *84*, 103142. [CrossRef] [PubMed]
85. Mytton, O.T.; Townsend, N.; Rutter, H.; Foster, C. Green space and physical activity: An observational study using Health Survey for England data. *Health Place* **2012**, *18*, 1034–1041. [CrossRef] [PubMed]
86. Page, J.; Mörtberg, U.; Destouni, G.; Ferreira, C.; Nässtrom, H.; Kalantari, Z. Open-source planning support system for sustainable regional planning: A case study of Stockholm County, Sweden. *Environ. Plan. B-Urban Anal. City Sci.* **2020**, *47*, 1508–1523. [CrossRef]
87. Zingraff-Hamed, A.; Huesker, F.; Lupp, G.; Begg, C.; Huang, J.; Oen, A.; Vojinovic, Z.; Kuhlicke, C.; Pauleit, S. Stakeholder Mapping to Co-Create Nature-Based Solutions: Who Is on Board? *Sustainability* **2020**, *12*, 8625. [CrossRef]

88. Saldivar-Tanaka, L.; Krasny, M.E. Culturing community development, neighborhood open space, and civic agriculture: The case of Latino community gardens in New York City. *Agric. Hum. Values* **2004**, *21*, 399–412. [CrossRef]
89. Qiu, J.; Zeng, J. *Planning of Tianfu New Area-Theoretical and Practical Exploration of Ecological and Rational Planning*; China Building Industry Press: Beijing, China, 2021.
90. Qiu, J. Main Concepts of Tianfu New District Planning in Sichuan Province. *City Plan. Rev.* **2014**, *38*, 84–89.
91. China Academy of Urban Planning and Design; Chengdu Institute of Planning and Design; Sichuan Research Center of Urban and Rural Planning. *The Master Plan of Sichuan Province Chengdu Tianfu New District (2010–2030)*; Sichuan Provincial Department of Housing and Urban-Rural Development: Chengdu, China, 2015.
92. Qiu, J. The Plan of Sichuan Tianfu New Area in Retrospect—A Tribute to 40th Anniversary of Reform and Opening up. *J. Hum. Settl. West China* **2018**, *33*, 12–18. [CrossRef]
93. Peng, J.; Li, H.; Liu, Y.; Hu, Y.; Yang, Y. Identification and optimization of ecological security pattern in Xiong'an New Area. *Acta Geogr. Sin.* **2018**, *73*, 701–710.
94. Alemu, I.J.B.; Richards, D.R.; Gaw, L.Y.F.; Masoudi, M.; Nathan, Y.; Friess, D.A. Identifying spatial patterns and interactions among multiple ecosystem services in an urban mangrove landscape. *Ecol. Indic.* **2021**, *121*, 107042. [CrossRef]

buildings

MDPI

Article

Research on Work Performance of Monolithic Precast Concrete Shear Walls with Post-Cast Epoxy Resin Concrete

Peiqi Chen [1,2], Zihao Liu [1,2], Xiaojie Zhou [1,2,*], Shuo Xu [1,2] and Junyi Wang [1,2]

1 Tianjin Key Laboratory of Civil Buildings Protection and Reinforcement, Tianjin 300384, China; cpq@tcu.edu.cn (P.C.); 18534059276@163.com (Z.L.); 17302290313@163.com (S.X.); jinjin19961210@126.com (J.W.)
2 School of Civil Engineering, Tianjin Chengjian University, Tianjin 300384, China
* Correspondence: zhouxj@tcu.edu.cn

Abstract: Precast concrete structures are popular in the building industry because of their high efficiency and environmental friendliness. In this paper, the U-type reinforcement ferrule connection technique was applied to study the seismic performance of precast concrete shear walls. Five shear wall finite element models and four shear wall specimens were prepared. Both experiments and finite element analysis were conducted to explore the impact of parameters on the work performance of precast reinforced concrete shear walls, such as the variety of post-cast concrete, the form of horizontal joints, and the buckle length of U-type reinforcements. On this basis, the mechanism of failure as well as the characteristics of hysteresis, ductility, and energy dissipation capacity were analyzed. According to the analytical results, the cast in situ reinforced concrete shear wall is inferior to the precast shear wall with post-cast epoxy resin concrete in terms of seismic performance. In addition, the specimen with a keyway on the horizontal joint interface outperforms the specimen without a keyway. With an increase in the buckle length of the U-type reinforcement, there is a rise in the sectional height and stiffness of the hidden beam at the bottom of the wall, while the horizontal load-bearing capacity of the wall is improved. However, its ductility and energy dissipation capacity are decreased. As revealed by a thorough analysis, the construction scheme most suitable for precast shear wall horizontal joints adopts epoxy resin concrete as the post-cast material, the buckle length of U-type reinforcements is approximately one-third the height of the horizontal joint, and there is a keyway at the interface of the joint.

Keywords: monolithic precast concrete shear wall; U-type reinforcement ferrule; epoxy resin concrete; work performance; test; finite element analysis (FEA)

1. Introduction

Precast construction represents a significant trend in the development of the building industry, showing various advantages, such as integration, industrialization, high efficiency, environmental friendliness, high energy efficiency, and reduced emissions. Shear wall structures are the most common types of precast concrete construction; their seismic performance is largely determined by the stability of connection between the precast walls. Currently, there are some popular methods of connection for shear walls, such as adopting dependable steel bar connections [1], creating a keyway at the joint [2,3], and using superior materials at the joint, such as fiber concrete, polymer concrete, ultra-high-performance concrete material, etc. [4–6].

The outcome of connection can be guaranteed by traditional steel bar connection methods like the grout sleeve splicing of bars, bolt anchoring, and rebar lapping in grout-filled holes. However, it is difficult to install nodes in the proper position. Also, the construction is high in cost but low in efficiency. Because of this, Yu et al. [7] proposed to simplify the method of steel bar connection for precast shear walls through the U-type reinforcement ferrule connection technique. To evaluate the reliability of the U-type reinforcement ferrule connection technique, Zhou et al. [8] assessed the anchorage performance of the U-type reinforcement ferrule. Lu et al. [9] and Jiao et al. [10] conducted a plane pseudo-static experiment using the U-type reinforcement ferrule connection technique for shear walls, and Gao et al. [11] investigated the bending performance of the out-of-plane shear wall. According to these studies, the precast shear wall connected through a U-type reinforcement ferrule meets the relevant seismic standards. It is easy to achieve U-type reinforcement through ferrule connection, as only the U-type reinforcement of the upper precast wall and the lower precast wall are lap-spliced to each other to form closed loops. Then, the horizontal bars are inserted at the corners of the closed loops, with the horizontal bars tied with a U-type reinforcement, as shown in Figure 1 [12]. Finally, the concrete is poured to form the post-cast horizontal joint for connecting the two precast walls, which is referred to as a precast concrete shear wall with a horizontal joint.

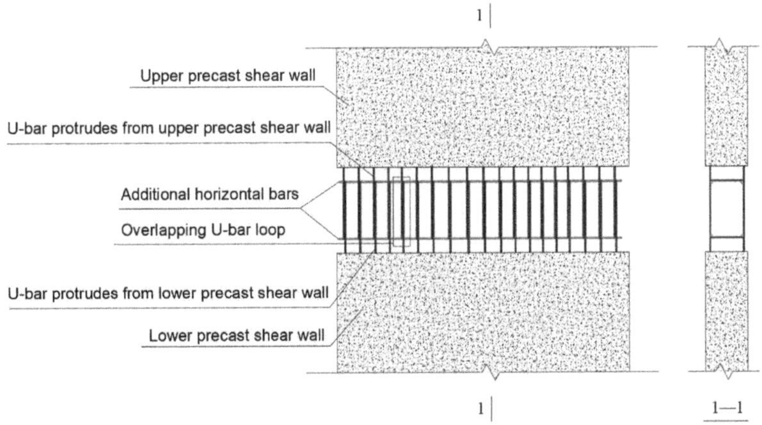

Figure 1. Schematic diagram of U-typed reinforcement connection technique.

Epoxy resin concrete is a variety of composite material that contains epoxy resin, a curing agent, sand, gravel aggregate, and other ingredients [13]. It is advantageous over conventional silicate concrete in terms of tensile strength, crack resistance, material uniformity, and integrity. Natarajan et al. [14] replaced a portion of cement by weight with epoxy resin, revealing that the compressive strength of epoxy resin concrete is improved. Qian et al. [15] conducted axial tension testing to investigate the tensile properties of non-reinforced and reinforced epoxy resin concretes. According to the test results, the tensile strength of epoxy resin concrete is higher than that of ordinary concrete. Epoxy resin concrete performs well in crack resistance, exhibiting strong bond properties between the epoxy resin concrete and the reinforcement. Also, the specimen showed high integrity. Sun et al. [16] explored the bond performance between ordinary strength concrete and epoxy resin concrete, showing that the bond strength increases as the epoxy resin concrete ages. Liang et al. [17] studied the impact of different epoxy resin dosages on the mechanical properties of modified cement concrete. The results show that epoxy resin accelerates the hydration reaction. Compared with the concrete containing no epoxy resin, compressive strength and flexural strength was increased. Due to its high performance in interface bonding strength with concrete, thermal stability, chemical resistance, and mechanical properties [18], epoxy resin matrix is widely used in the context of reinforcement engineering for such purposes as concrete crack and structural repair and road construction [19,20].

For the treatment of the bonding interface between old and new concrete, Rizkalla et al. [21] studied various horizontal joints of precast shear walls, reporting that the mechanical properties of shear walls with keyways are superior to those of ordinary horizontal joint shear walls. Julio et al. [22] explored different technologies applicable to treat the bonding interface between old and new concretes. According to the results, the bonding strength of the interface treated by sandblasting is the highest, but the construction efficiency is lower, which makes the reserved keyway more suitable for precast structures. Jang et al. [23] conducted tests on the flexural performance of precast concrete members with joint grouting. The test results show that the precast members performed better in bonding and bending resistance when polymer cement mortar was taken as the grouting material. Wu et al. [24] analyzed the impact of keyways on the interfacial bond strength between UHPC-NC through shear testing, the results of which showed that the interfacial bond strength can be improved by setting a single keyway at the interface. Zhang et al. [25] analyzed the impact of keyway parameters on the interfacial shear performance of UHPC-NC. According to the analytical results, trapezoidal multi-keyway treatment maximizes the interfacial shear strength, and planting reinforcement prevents the keyway from brittle failure. Jiang et al. [26] investigated the mode of shear failure occurring at the UHPC-NC interface given different keyway numbers. It was demonstrated that the shear stiffness of the specimen reaches its maximum when the keyway number is three. Moreover, as this number increases, the overall shear capacity of the specimen improves.

The finite element analysis method is also applicable to the performance analysis of precast shear wall structures and epoxy resin concrete. Zhou et al. [27] employed ABAQUS (2020) software to investigate the mechanical mechanism underlying the anchorage performance of U-type reinforcement ferrule connections, subsequently proposing a suitable anchorage length. A bearing capacity model was proposed considering the bond and dowel actions simultaneously, and it could predict the bearing capacity well. Jin et al. [28] analyzed the mechanical performance of joint connections in epoxy resin concrete truss structures using ABAQUS finite element simulation. Their results indicate that the numerical model can accurately simulate and predict the strength and failure behavior of the specimens. The interface between the new and old concrete in precast concrete structures can be simulated using surface-to-surface contact, and the tangential action is based on the Coulomb friction model. Ref. [29] and EU Norms [30] provide recommendations for the value of the friction coefficient.

This paper aims to improve the shear resistance of the precast shear wall with horizontal joints through the U-type reinforcement ferrule connection technique. Four shear wall specimens and five finite element models are prepared with the buckle length of the U-type reinforcement ferrule, joint materials, and joint form (with or without keyway) as the design parameters. Both quasi-static testing and finite element simulation are performed to analyze the influence of each parameter on the bending and shear performance of the precast shear wall with a horizontal joint. On this basis, the optimal construction scheme is developed for the horizontal joint of the precast shear wall. This study provides guidance on the application of epoxy resin concrete in prefabricated concrete structures, which is conducive to improving the work performance of precast concrete structures. The research methodology and steps of this paper are outlined in Figure 2.

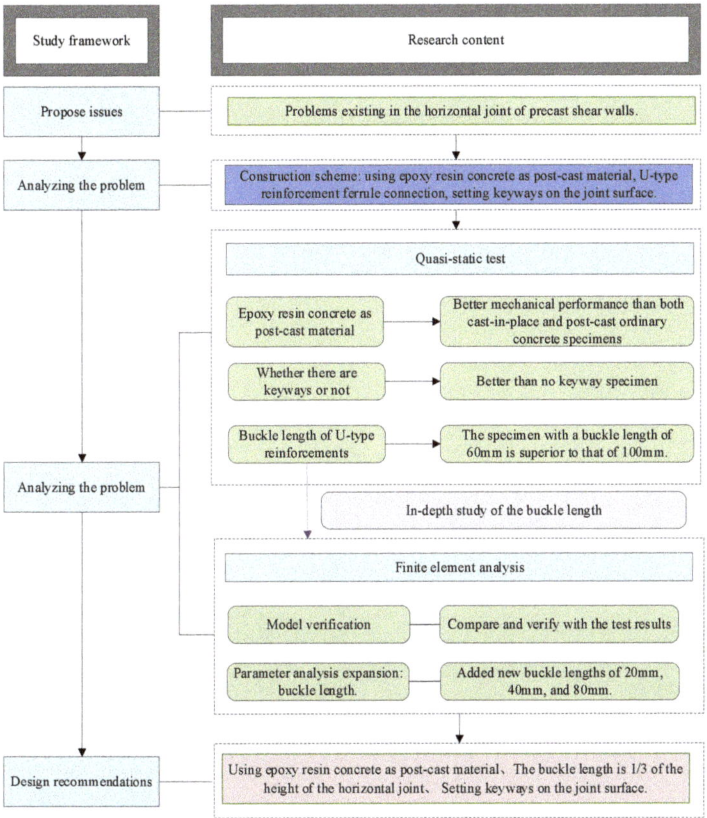

Figure 2. Flowchart of research work.

2. Test Summary

2.1. Test Piece Design

Four shear wall specimens are prepared to investigate the effect of three parameters on the seismic performance of the shear wall: the specimen fabrication method, the buckle length of the U-type reinforcement, and the type of joint. The details of the specimens are presented in Table 1.

Table 1. Details of the specimens.

Specimen Number	Type of Post-Cast Concrete	Keyway Dimension/mm	Buckle Length of U-Type Reinforcements/mm	Remark
EHY−1−60	Epoxy resin concrete	width 100 mm, depth 35 mm	60 mm (33.3% of the height of the horizontal joint)	---
EHY−1−100	Epoxy resin concrete	width 100 mm, depth 35 mm	100 mm (55.6% of the height of the horizontal joint)	---
EHY−2−60	Epoxy resin concrete	----	60 mm (33.3% of the height of the horizontal joint)	---
EZJ	---	---	---	Cast in situ specimen

The specimen consists of a shear wall, a loading beam, and a ground beam. The wall is 680 mm in net height, 750 mm in width, 150 mm in thickness, 500 mm in precast wall height, 180 mm in horizontal joint height, 100 mm in the width of the reserved keyway, and 35 mm in depth. The shear span ratio λ is approximately 1.05. The horizontal reinforcement in the precast wall is 8 mm in diameter. The vertical reinforcement (U-type reinforcement) is 8 mm and 12 mm in diameter (configured at the end of the wall). The U-type reinforcement extended from the ground beam has a diameter of 8 mm and 12 mm. Figure 3 shows the dimensions of the specimens.

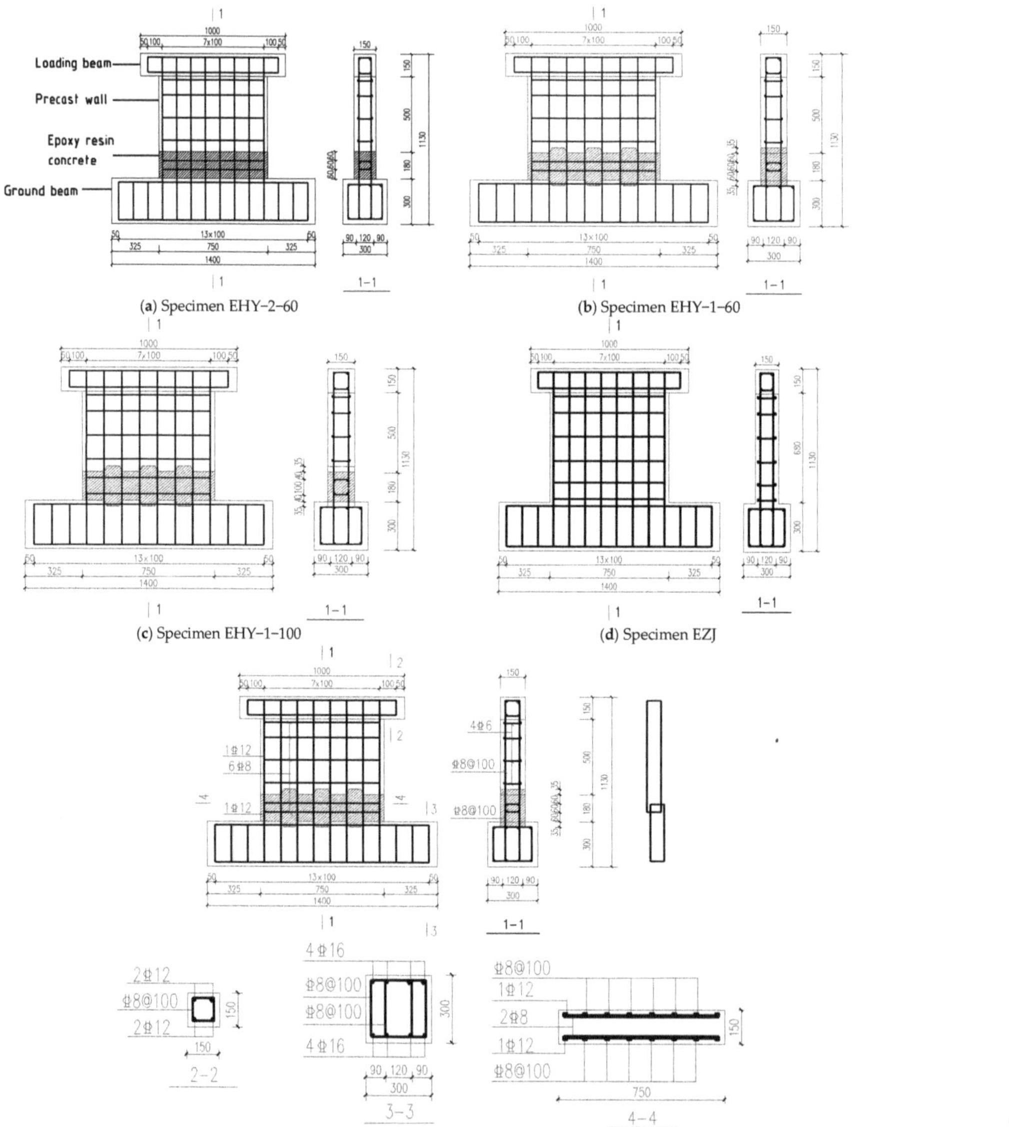

(e) Reinforcement details
(Taking the specimen EHY-1-60 as an example.)

Figure 3. Specimen dimensions and reinforcement (in mm).

2.2. Basic Mechanical Properties of Materials

According to the method proposed in Reference [31], the strengths of concretes and steel bars were measured separately. On average, the cube compressive strength of the precast concrete is measured to be 42.1 MPa. The mix ratio of epoxy resin concrete is shown in Table 2. The average cube compressive strength is measured to be 55.9 MPa. The average yield strength of 8 mm steel bar is measured to be 361.1 MPa. The average yield strength of 12 mm steel bar is measured to be 360.0 MPa. In addition, the average elastic modulus of the steel bar is determined as 2.0×10^5 MPa.

Table 2. Epoxy resin concrete mix ratio [31].

Epoxy Resin (kg/m^3)	Curing Agent (kg/m^3)	Cement (kg/m^3)	Cobble (kg/m^3)	Sand (kg/m^3)	Diluent (kg/m^3)
800	320	1200	2800	3200	80

2.3. Loading Device and Loading Scheme

Figure 4 illustrates the facility used for load test. The horizontal reciprocating load is applied by a 20,000 kN horizontal actuator, and a vertical load of 200 kN is applied by the loading beam to the wall, at an axial compression ratio of 0.09. This test consists of two stages: load control and displacement control. Before the specimen yields, load control is exercised, and the load is applied progressively. The amount of loading control is a multiple of the predicted cracking load, such as 0.2 *P*cr and 0.4 *P*cr. After the specimen yields, displacement control load is applied. At this time, the amount of loading control is a multiple of the yield displacement Δy, such as 2Δy, 3Δy, etc. Each stage of loading is repeated three times. When the load-bearing capacity of the specimen drops below 85% of the peak load, the test is terminated [32].

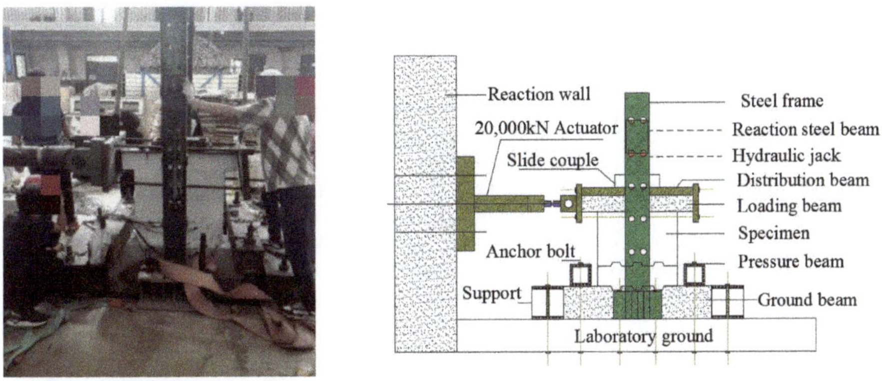

(**a**) Loading device (**b**) Elements of loading device

Figure 4. Loading test device.

3. Failure Characteristics of Specimen

(1) Specimen EHY−1−60

When the load reaches 102.8 kN, horizontal cracks first emerge at the point of connection between the post-cast horizontal joint and the precast wall. Then, the crack develops rapidly. With the increase in reciprocating load, horizontal cracks appear at the bottom of the precast wall and then develop diagonally. When the load rises to 285.4 kN, the specimen yields. The yield displacement denoted as Δy is 4.36 mm. Subsequently, the load of displacement control is applied. With the increase of displacement, diagonal cracks further develop on the wall. The horizontal cracks emerging at the upper joint surface gradually develop toward the perimeter of the keyway, which leads to the connection with some diagonal cracks of the wall. When Δ = 3Δy, the diagonal cracks of the wall are extended to the top of the wall, with the maximum width of cracks reaching 4.5 mm. When the specimen is subject to the peak load of 449.0 kN, the concrete of the wall shows signs of spalling, and the post-cast epoxy resin concrete joint is slightly damaged. When Δ = 4Δy, the horizontal load drops to 85% of the peak load. At this point, the test is terminated. At this time, the angle at which the main diagonal crack (i.e., the red crack marker line) develops is about 35°~41°, and the angle at which the vice diagonal crack (i.e., the blue crack marker line) develops is about 34°~39°. The horizontal crack developing at the upper joint surface is visible, indicating the bending failure of the specimen. In this case, the failure mode is bending-shear failure. The ultimate state of the failed specimen is shown in Figure 5.

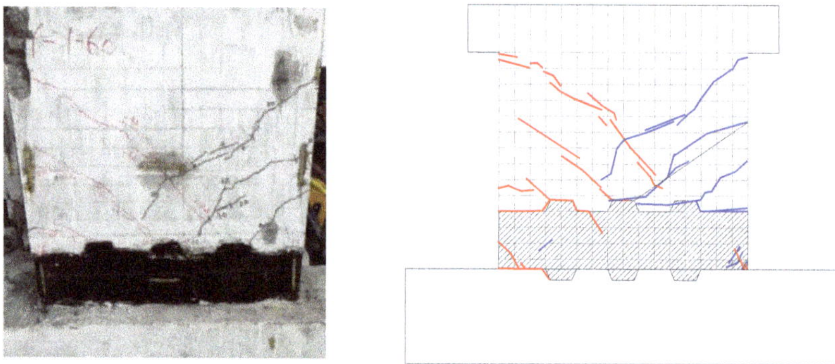

(**a**) Specimen EHY-1-60 (**b**) Crack pattern of EHY-1-60

Figure 5. Failure condition of specimen EHY−1−60.

(2) Specimen EHY−1−100

The failure of specimen EHY−1−100 follows a similar process to specimen EHY−1−60. When the load reaches 123.9 kN, horizontal cracks first emerge at the point of connection between the post-cast horizontal joint and the precast wall. As the reciprocating load increases, the cracks develop gradually into cross oblique cracks. When the load rises to 336.6 kN, the specimen yields, and the yield displacement denoted as Δy is 4.79 mm. When Δ = 2Δy, X-shaped shear oblique cracks develop along the main and vice diagonals of the wall, with a width of 0.2 mm. When Δ = 3Δy, the specimen is subject to the peak load of 504.7 kN. Afterwards, the cracks emerging at the keyway between the upper and lower joint surfaces of the horizontal joint develop rapidly, while the cracks emerging at the keyway on the upper joint surface penetrate the horizontal joint and meet some diagonal cracks in the wall. In this case, the post-cast epoxy resin concrete is damaged to a greater extent. At this time, the angle at which the crack develops in the main diagonal direction is about 28°~54°, and the angle at which the crack develops in the vice diagonal direction is about 26°~48°. The angle range of wall oblique crack distribution is wider, the spacing between cracks is smaller, and oblique cracks penetrate the horizontal joint. This indicates that the shear failure of the specimen is more significant. The failure mode is shear-bending failure. Figure 6 shows the ultimate state of the failed specimen.

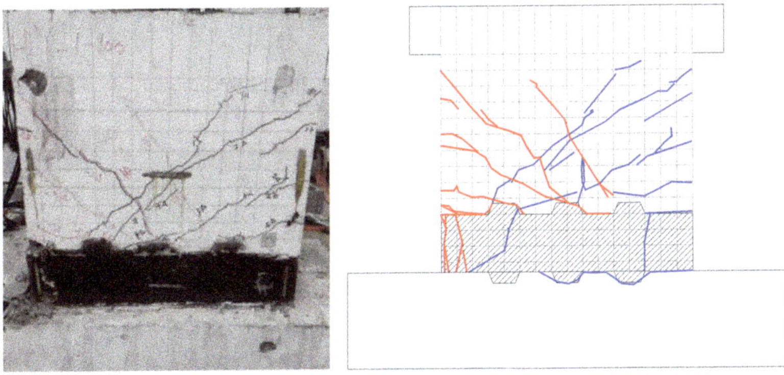

(a) Specimen EHY-1-100 (b) The crack pattern of EHY-1-100

Figure 6. Failure of the specimen EHY−1−100.

(3) Specimen EHY−2−60

When the load reaches 123.3 kN, cracks first emerge at the point of connection between the post-cast horizontal joint and the precast wall. With the increase of the reciprocating load, horizontal cracks emerge at the bottom of the precast wall. Then, the cracks develop diagonally. When the load rises to 308.5 kN, the horizontal crack emerging at the interface of the upper joint continues to expand. When the specimen yields, the yield displacement denoted as Δy is 6.88 mm. Then, the load of displacement control is applied. With the increase of displacement, the diagonal cracks on the precast wall further develop. At this time, the interface between the ground beam and the horizontal joint cracks. When $\Delta = 2\Delta y$, the specimen is subject to the peak load of 430.4 kN. At this time, the cracks on the surface of the horizontal joint develop with a width of 0.1 mm, meet some diagonal cracks of the wall, and expand to the post-cast joint area. When $\Delta = 3\Delta y$, the load drops to 85% of the peak load, and the test is terminated. At this time, the angle at which the crack develops along the main diagonal direction is about 20°~40°, and the angle at which the crack develops along the vice diagonal direction is about 20°~38°. The cracks are mostly concentrated on the upper surface of the horizontal joint. The specimen shows visible bending failure, and the failure mode is bending-shear failure. Figure 7 shows the ultimate state of the failed specimen.

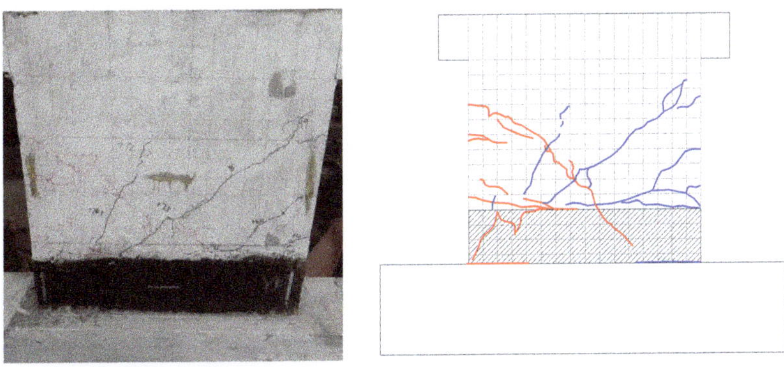

(a) Specimen EHY-2-60 (b) The crack pattern of EHY-2-60

Figure 7. Failure of the specimen EHY−2−60.

(4) Specimen EZJ

When the load reaches 30.0 kN, cracks first appear at the bottom of the cast in situ wall. With the increase of the reciprocating load, horizontal cracks emerge at the point of connection between the bottom of the wall and the ground beam. Then, diagonal cracks develop due to diagonal extension. When the load is raised to 347.1 kN, the specimen yields, and the yield displacement denoted as Δy is 5.43 mm. Then, the load of displacement control is applied. With the increase of displacement, diagonal cracks increasingly emerge and expand. When the displacement is 7.61 mm, the specimen is subject to a peak load of 377.4 kN. When $\Delta = 2\Delta y$, the width of the main and vice diagonal cracks is 0.1 mm, and the concrete on the left side of the bottom of the wall crumbles. At this point, the load declines to 85% of the peak load, and the test is terminated. At this time, the angle at which the crack develops along the main diagonal direction is about 35°~46°, and the angle at which the crack develops along the vice diagonal direction is about 30°~46°. The cracks are evenly distributed mainly near the bottom of the wall. The specimen shows signs of bending failure, and the failure mode is bending-shear failure. The ultimate state of the failed specimen is shown in Figure 8.

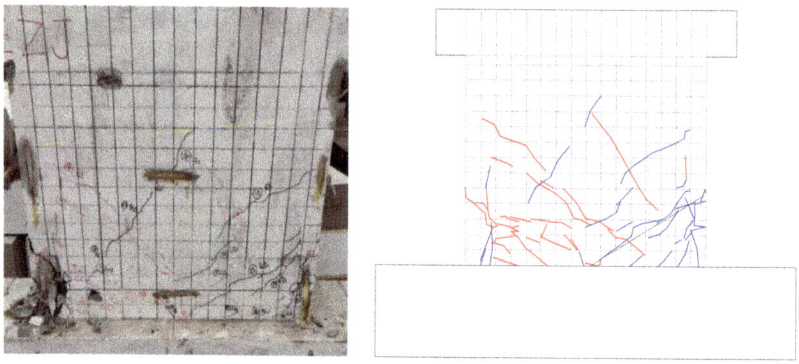

(**a**) Specimen EZJ (**b**) Crack pattern of EZJ

Figure 8. Failure of the specimen EZJ.

By comparing the process of how each specimen fails, the following conclusions can be drawn.

(1) Compared to the cast in situ specimen, there is an improvement in the cracking load, yield load, peak load, and ultimate load of the precast specimens with the epoxy resin concrete cast. In addition, the number of total loading cycles is greater relative to the cast in situ specimen, indicating the advantage of the precast shear wall with epoxy resin concrete cast in mechanical properties over the cast in situ specimen.

(2) Compared to the precast shear wall without keyways at the joint surface, the precast shear wall with keyways is outperformed in load-bearing capacity and the capacity to withstand deformation. Also, the failure of the keyway-free specimen occurs mainly at the horizontal joint and its surface.

(3) Among these specimens, the cracking load, yield load, peak load, and ultimate load of the precast shear wall EHY−1−100 are invariably the highest, indicating that the buckle length of the U-type reinforcements is one of the factors affecting the load-bearing capacity of the wall. A greater buckle length of the U-type reinforcements means a higher load-bearing capacity of the wall.

4. Experimental Results and Analysis

4.1. Hysteresis Curve

Figure 9 shows the hysteresis curve of each specimen.

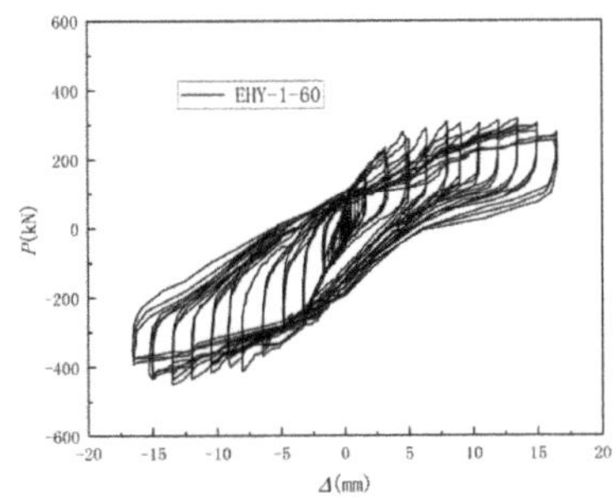

(**a**) Specimen EHY-1-60

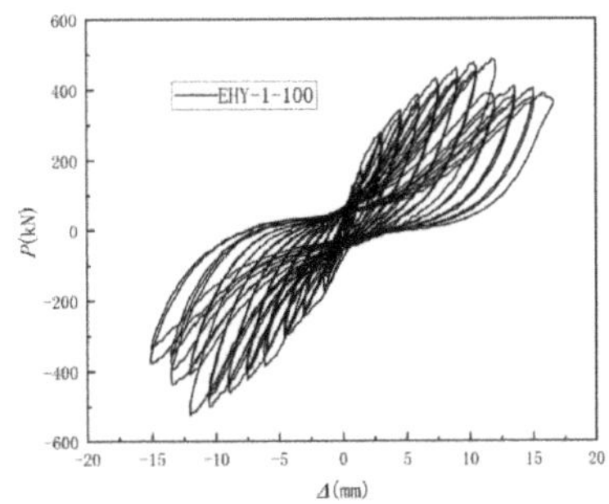

(**b**) Specimen EHY-1-100

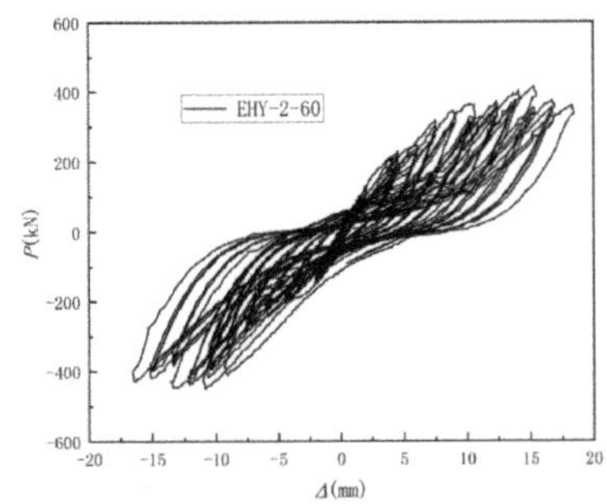

(**c**) Specimen EHY-2-60

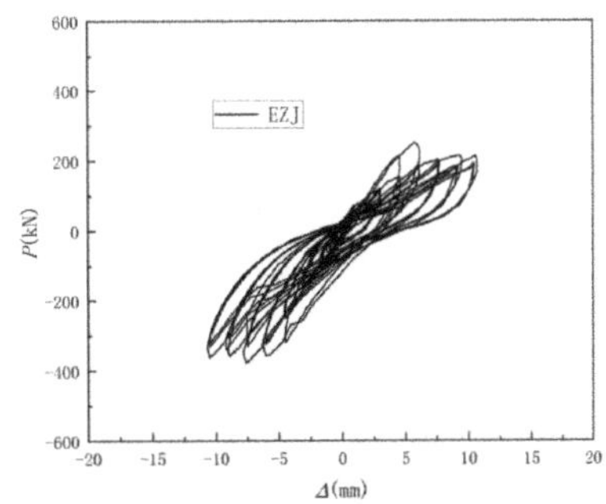

(**d**) Specimen EZJ

Figure 9. Hysteresis curve of the specimens.

During the test, the loading beam end of the specimen EHY−1−60 is unevenly loaded in the later stage of loading, which causes localized crushing of concrete at the loading beam end and the slight slippage of the ground beam. Consequently, there are errors in the positive hysteresis curve of the test. When the specimen EZJ is loaded, the ground beam undergoes slight slippage, thus leading to errors in the positive hysteresis curve of the test. For this reason, the negative hysteresis curves of the specimens EHY−1−60 and EZJ are analyzed.

According to Figure 8, the following results are found:

(1) The hysteresis curve of the specimen EHY−1−60 is relatively enriched, indicating an excellent performance of the specimen in terms of plastic deformation and energy dissipation. The hysteresis curves of the other three specimens show an obvious sign of pinching, implying the poor performance of the specimen in ductility and energy dissipation.

(2) Compared with the cast in situ specimen EZJ, the precast shear wall specimens EHY−1−60, EHY−2−60, and EHY−1−100 have a larger hysteresis loop and perform better in deformation and energy dissipation. It is indicated that the precast shear wall of the horizontal joint concrete post-cast with epoxy resin outperforms the cast in situ specimen in terms of mechanical properties.

(3) Compared with the precast shear wall specimen EHY−2−60, the hysteresis curve of the specimen EHY−1−60 is more enriched, and it outperforms the specimen EHY−2−60 in deformation and energy dissipation, indicating that the keyway at the interface of the horizontal joint plays a role in improving the performance of the shear wall in ductility and energy dissipation.

(4) Compared with the precast shear wall specimens EHY−1−60 and EHY−2−60, the peak load of specimen EHY−1−100 is greater, indicating that a greater buckle length of the U-type reinforcements leads to a higher horizontal load-bearing capacity of the wall.

4.2. Skeleton Curve

Figure 9 presents the skeleton curves of each specimen. Table 3 lists the data of each characteristic point of the skeleton curve. Below are the findings from Figure 10 and Table 3:

Table 3. The measurement data on feature points of skeleton curve for each specimen.

Specimen Number	Cracking Load P_{cr}/kN	Cracking Displacement Δ_{cr}/mm	Yield Load P_y/kN	Yield Displacement Δ_y/mm	Peak Load P_{max}/kN	Peak Displacement Δ_{max}/mm	Ultimate Load P_u/kN	Ultimate Displacement Δ_u/mm	Ductility Coefficient $\mu = \Delta_u/\Delta_y$
EHY−1−60	102.8	0.52	285.4	4.36	449.0	13.45	389.3	16.46	3.78
EHY−1−100	123.9	1.37	336.6	4.79	504.7	11.88	390.2	15.01	3.13
EHY−2−60	123.3	1.53	308.5	6.88	430.4	13.06	365.8	17.55	2.55
EZJ	30.0	0.45	347.1	5.43	377.4	7.61	363.2	10.54	1.94

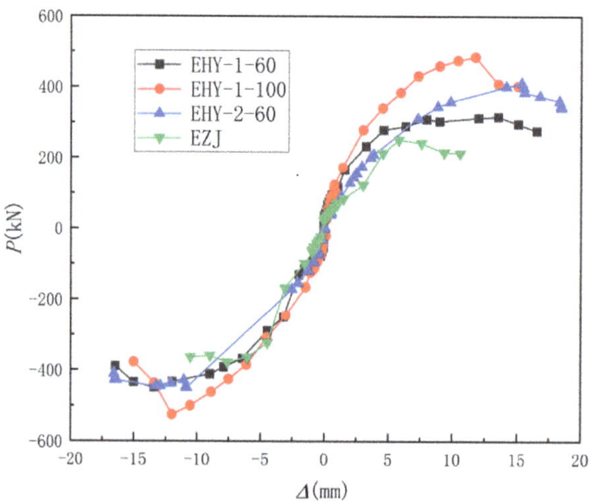

Figure 10. Skeleton curve.

(1) Compared with the cast in situ specimen EZJ, there is a significant increase in the cracking load, peak load, and ultimate load of the shear wall specimens with a horizontal joint post-cast with epoxy resin concrete. In reference [33], the same method of steel reinforcement connection as proposed in this paper is used, but the post-cast concrete is ordinary micro-expansion concrete. Also, the strength of the post-cast concrete is 22.8% higher relative to the prefabricated wall. According to the experimental results, the maximum load-bearing capacity and ductility coefficient of prefabricated shear walls with a U-type reinforcement ferrule are lower compared to cast-in-place specimens. This illustrates the advantage of post-cast epoxy resin concrete in prefabricated structures.

(2) Compared to the specimens EHY−1−60 and EHY−2−60, the load values at each characteristic point of the specimen EHY−1−100 are the largest, with its peak load reaching 504.7 kN, which is 12.4% higher compared to the specimen EHY−1−60 (with keyway, buckle lengths of 60 mm) and 17.3% higher relative to the specimen EHY−2−60 (without keyway, buckle lengths of 60 mm).

(3) The ductility coefficient of the specimen EHY-1-60 is the largest. Comparatively, the ductility coefficient of the specimens EHY−1−100, EHY−2−60 and EZJ is reduced by 17.2%, 32.5%, and 48.7%, respectively.

Obviously, the best solution for improving the horizontal load-bearing capacity of the wall is to "use epoxy resin concrete for horizontal joints, set keyways at the joint surface of horizontal joints, and use the U-type reinforcement ferrule with a buckle length of 100 mm", despite the limited ductility of the wall. Also, the best solution for improving the ductility of wall displacement is to "use epoxy resin concrete for horizontal joints, set keyways at the joint surface of horizontal joints, and use U-type reinforcements buckle lengths of 60 mm", with a higher load-bearing capacity reached. In summary, setting keyways at the joint surface and post-casting epoxy resin concrete at the horizontal joint is beneficial to improving the load-bearing capacity and ductility of the wall, while increasing the buckle length of U-type reinforcement enhances the load-bearing capacity of the wall, although the ductility of the wall declines.

There are four longitudinal steel bars passing through the buckle range of U-type reinforcement, with the longitudinal steel bars and U-type reinforcement connected to form a reinforcement skeleton. The post-casting of epoxy resin concrete plays a similar role to the construction of a hidden beam. When the buckle length of U-type reinforcement increases, there is a rise in the section height and stiffness of the hidden beam. Due to the high strength and high coherence of epoxy resin concrete, as well as the favorable factors such as keyways at the joint surface, the hidden beam is approximately an embedded end of the upper precast shear wall. If the hidden beam (horizontal joint) is taken as a fixed end, the net height of the wall decreases, the shear span ratio declines, and the horizontal load-bearing capacity and the initial stiffness are improved, but the performance in energy dissipation deteriorates.

4.3. Stiffness Degradation Curve

Figure 11 shows a comparison of the stiffness degradation curve of the specimens. In this figure, K refers to the secant stiffness of the specimen [31]. From Figure 11, the following conclusions can be drawn.

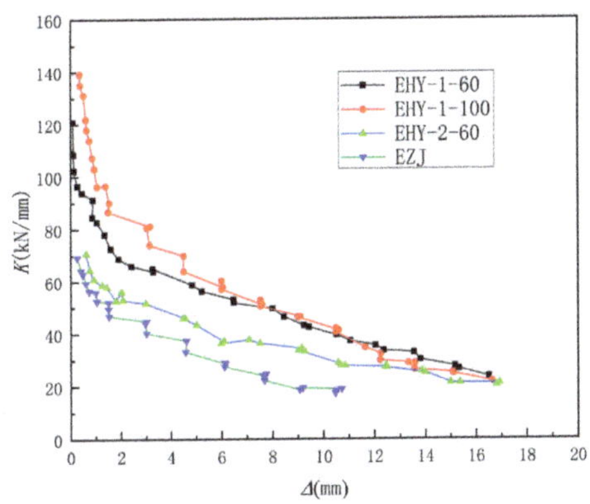

Figure 11. Stiffness degradation curve.

(1) The initial stiffness of the cast-in-place specimen EZJ is 69.4 kN/mm. Compared with the specimen EZJ, the initial stiffness of the specimens EHY−1−100, EHY−1−60, and EHY−2−60 increases by approximately 102%, 72.9%, and 2.2%, respectively. The initial stiffness of the shear wall with post-pouring epoxy resin concrete is greater compared to the cast-in-place specimen, while that of the keyway-free specimen shows no significant improvement, which is almost the same as that of the cast-in-place specimen.

(2) Compared with the specimen EHY−1−100, the initial stiffness of the specimen EHY−1−60 is slightly lower, but it declines at a lower rate.

(3) The stiffness of the specimen EHY−1−60 with the keyway is consistently higher than that of the keyway-free specimen EHY−2−60. Also, the stiffness declines at a relatively low rate in the later stage.

4.4. Energy Dissipation Performance

Figure 12 shows the energy dissipation properties of each specimen. The data on cumulative energy dissipation are listed in Table 4. Below are the main findings.

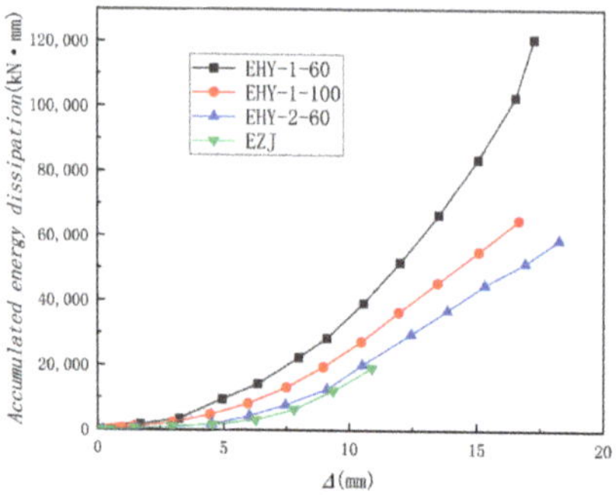

Figure 12. Cumulative energy dissipation by the specimens.

Table 4. Cumulative energy dissipation by each specimen.

Specimen Number	EZJ	EHY−1−60	EHY−1−100	EHY−2−60
E/kN·mm	19,386.0	119,787.5	65,031.1	58,970.0
Relative value of Cumulative energy dissipation	1	6.18	3.36	3.04

(1) Compared with the cast in situ specimen EZJ, the shear wall specimens with a horizontal joint post-cast with epoxy resin concrete perform better in energy dissipation. Compared with the specimen EZJ, the cumulative energy dissipation of EHY−1−60, EHY−1−100, and EHY−2−60 increases by 518%, 236%, and 204%, respectively. In reference [7], the steel bars are connected in the same way as described in this paper. However, the test results show that the precast shear wall specimens with post-cast ordinary concrete but without keyways at the joint perform worse in energy dissipation capacity than the cast-in-place shear wall specimens. In this paper, the energy dissipation capacity of the prefabricated shear wall specimens is higher than that of the cast-in-place specimens. This is because the post-cast ordinary concrete is difficult to pour, and defects exist [7]. Epoxy resin concrete can effectively prevent this because of its high bonding performance with ordinary concrete. In addition, the keyways at the joint can improve the overall performance as well.

(2) Compared with the specimen EHY−2−60 without keyways at the joint surface, the specimens EHY−1−60 and EHY−1−100 with keyways at the joint surface perform better in cumulative energy dissipation. Therefore, setting keyways at the joint surface can improve the capacity of cumulative energy dissipation by the specimens.

(3) Relative to the specimen EHY−1−100 with U-type reinforcement that has a buckle length of 100 mm, the specimen EHY−1−60 performs better in cumulative energy dissipation. Therefore, the optimal buckle length of U-type reinforcement is set to one-third the height of the horizontal joint.

5. Finite Element Analysis

5.1. Establishment of Model

In this paper, the finite element software ABAQUS is used to model and analyze the wall. Concrete is modeled using the CPS4-type two-dimensional shell unit, while rebars are modeled using the T2D2-type two-dimensional truss unit. A structured mesh technique is used to divide the mesh, with the loading beam and ground beam are treated as secondary parts of the specimen. The mesh size is 100 mm. The mesh size at the precast wall and horizontal joint is set to 50 mm, and the mesh size of the reinforcement is set to 60 mm, as shown in Figure 13. The rebar model is integrated into the concrete model. The interface contact between precast concrete and post-cast epoxy resin concrete is modeled using tie constraints and surface-to-surface contact models. The normal action is governed by "hard contact", and the tangential action is reliant on the Coulomb friction model. According to the calculation results, the friction coefficient μ between the epoxy resin concrete and the ordinary concrete interface is set to 1.4.

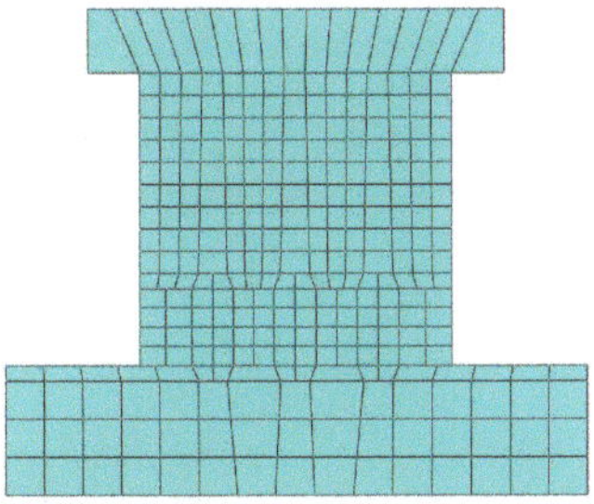

(a) Mesh division of concrete

(b) Mesh division of reinforcement

Figure 13. Mesh division of the model (with EHY−1−60 as an example).

The boundary conditions of the model are consistent with those set for the test, and the bottom of the ground beam is taken as the fixed constraint. At the same time, the center of the top surface of the loading beam is taken as the reference point to prevent the loading beam from stress concentration during the loading process. The analysis is conducted in two steps. Firstly, a 200 kN axial force is applied at the reference point. Then, a reciprocating load is applied. The loading system is the same as used in the test. The boundary conditions and loading methods under the finite element model are shown in Figure 14.

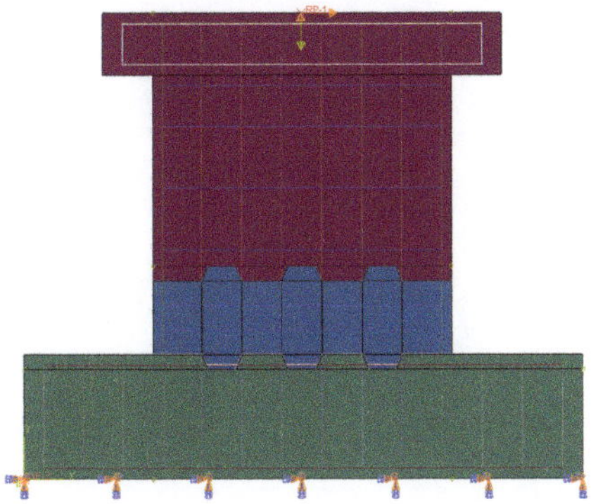

Figure 14. Boundary conditions and loading methods.

5.2. Constitutive Models of Materials

The constitutive model of normal concrete and rebar is simulated using the subroutine developed by Fang Zihu through ABAQUS [34], and normal concrete is simulated using the MCFT model under cyclic loading as proposed by Fang Zihu [35]. Figure 15a shows not only the compressive and tensile stress–strain relationships of the concrete in the MCFT model under cyclic loading but also the loading–unloading path of cyclic loading. The parameters that need to be inputted by the subroutine include the compressive strength of the concrete (f_c), the ratio of minimum residual strength to compressive strength of the concrete (α), and the strain corresponding to the starting point of the minimum residual strength of the concrete (ε_m). In this paper, the number of state variables dependent on the solution in the non-independent variables of the concrete is 10. $f_c = 32$ MPa, $\alpha = 0.3$, and $\varepsilon_m = 0.04$. The hysteretic model of steel bar takes into account the interaction between the steel bar and the concrete. The steel bar model used in the subroutine is shown in Figure 15b. The physical parameters that the subroutine needs to input include the initial stiffness of steel bar $E_s\left(\mathrm{N/mm^2}\right)$, the yield strength of steel bar f_y(MPa), and the ratio of the hardening stiffness of the steel bar to the initial stiffness E_{sh}/E_s. In this paper, the number of state variables dependent on the solution in the non-independent variables of the steel bar is 10. $E_s = 200000$ MPa, $f_y = 400$ MPa, and $E_{sh}/E_s = 0.01$. Epoxy resin concrete is simulated using the Concrete Damage Plasticity model, and the stress–strain relationship is determined by the fitting function in the early stage of research (Equation (1)). The undetermined parameters a and b are determined using the mass ratio of epoxy resin to curing agent. In this paper, a = 0.7 and b = 5. Table 5 lists the main parameters of the epoxy resin concrete model.

$$y = \begin{cases} ax + (4.9 - 4.23a)x^2 + (-4.67 + 6.67a)x^3 + (-0.27 - 4.74)x^4 + (1.07 + 1.27a)x^5,\ 0 \le x \le 1 \\ \frac{x}{b(x-1)^2 + x},\ x > 1 \end{cases} \tag{1}$$

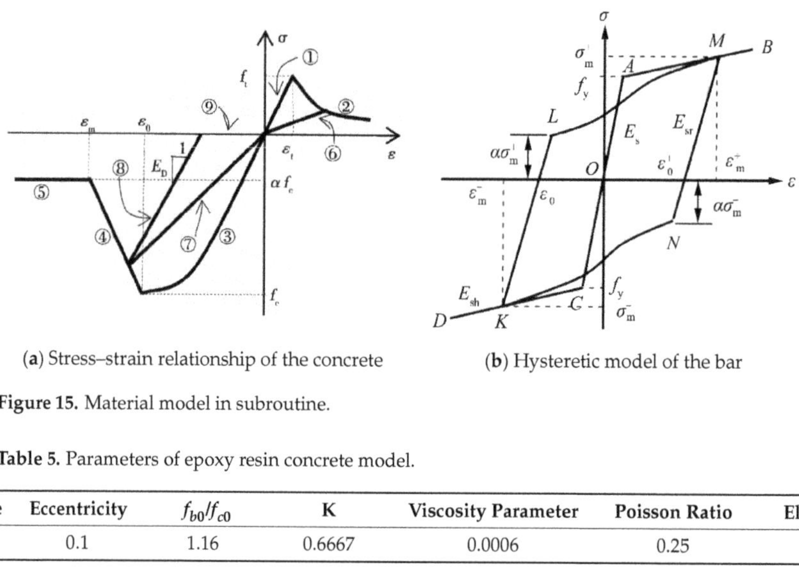

(a) Stress–strain relationship of the concrete (b) Hysteretic model of the bar

Figure 15. Material model in subroutine.

Table 5. Parameters of epoxy resin concrete model.

Dilation Angle	Eccentricity	f_{b0}/f_{c0}	K	Viscosity Parameter	Poisson Ratio	Elastic Modulus
40°	0.1	1.16	0.6667	0.0006	0.25	18,000 Mpa

5.3. Model Verification

The stress distribution of the wall and rebar of the specimens EHY−1−60 and EHY−1−100 is captured through finite element analysis, as shown in Figures 16 and 17. In the specimen EHY−1−60, the higher stress of the wall is distributed at the bottom of the shear wall and the keyway of the horizontal joint, while the highest stress of the rebar arises at the left side of the lower joint surface of the horizontal joint. In the specimen EHY−1−100, a higher stress of the wall concentrates at the bottom of the wall and the post-cast horizontal joint, while the highest stress of the rebar arises at the vertical rebar of the hidden column at the joint surface of the upper part of the horizontal joint. The stress distribution of the above specimens shows the same pattern of distribution as the test results.

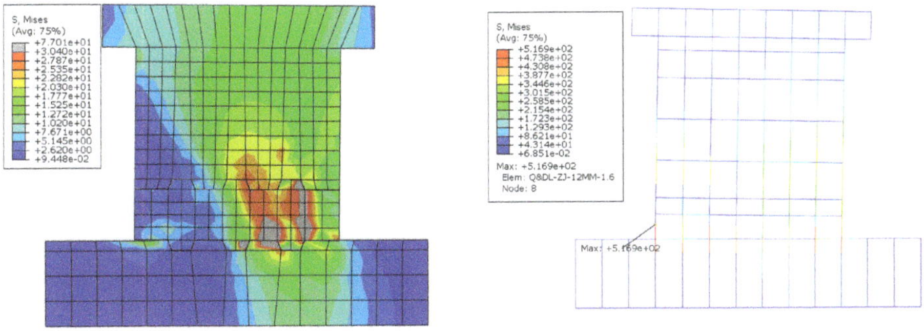

(**a**) Stress nephogram of the wall (**b**) Stress nephogram of the rebars

Figure 16. Stress distribution of the specimen EHY−1−60.

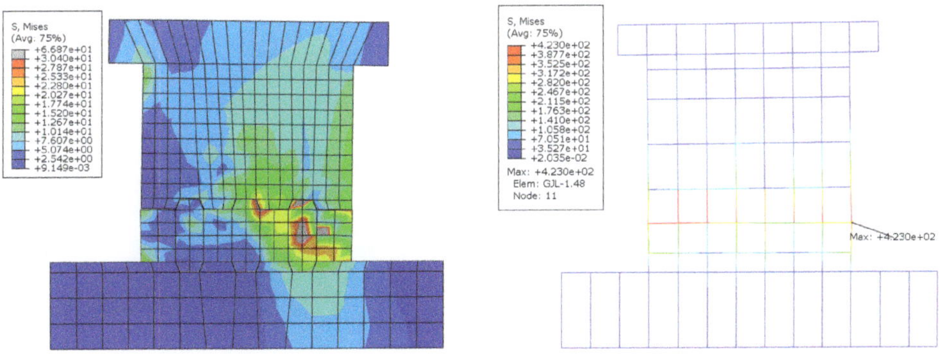

(**a**) Stress nephogram of the wall (**b**) Stress nephogram of the rebars

Figure 17. Stress distribution of the specimen EHY−1−100.

As shown in Figures 18 and 19, a comparison is performed between the test results and the results of simulating the hysteresis curves and the skeleton curves for specimens EHY−1−60 and EHY−1−100. The statistical comparison is detailed in Table 6. The simulation result differs from the test result of the peak load by approximately 5%. On average, the difference in maximum displacement is roughly 10%. The positive fitting of the hysteresis curve of the specimen EHY−1−60 is slightly worse, for the reasons mentioned above. The skeleton curve of the specimen EHY−1−60 is basically consistent with the test results in the negative direction. Overall, the results of finite element analysis are basically consistent with the test results, and the finite element model established in this paper is applicable to approximately simulate the mechanical properties of precast shear walls with horizontal joints.

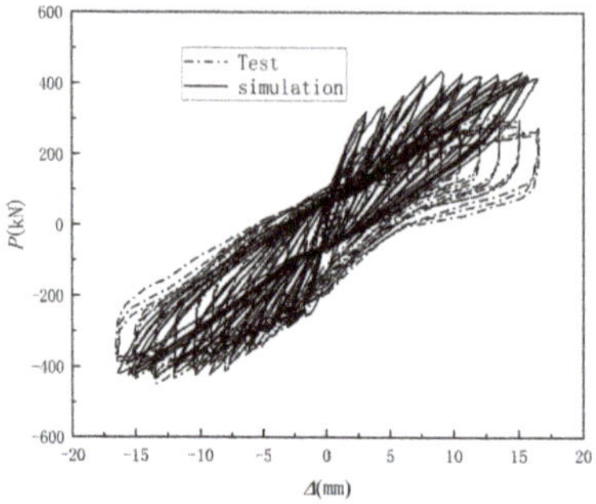

(a) EHY−1−60

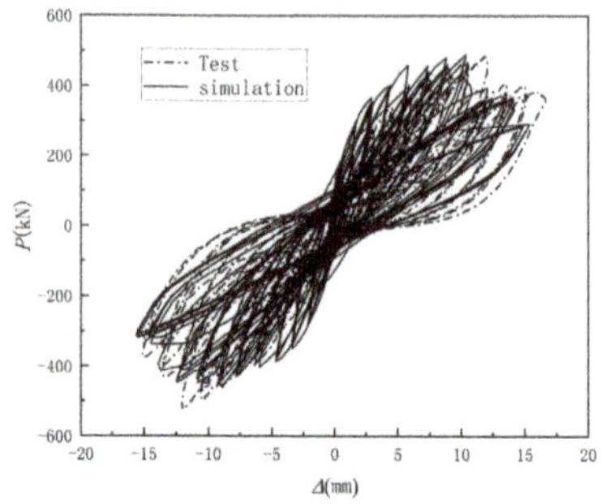

(b) EHY−1−100

Figure 18. Comparison between simulated hysteresis curve and tested hysteresis curve.

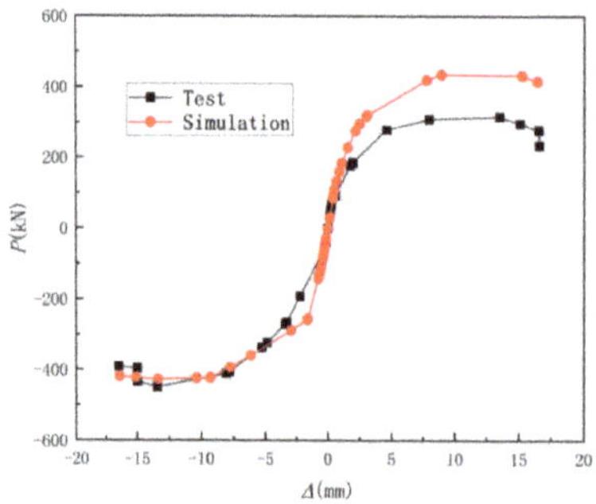

(a) EHY−1−60

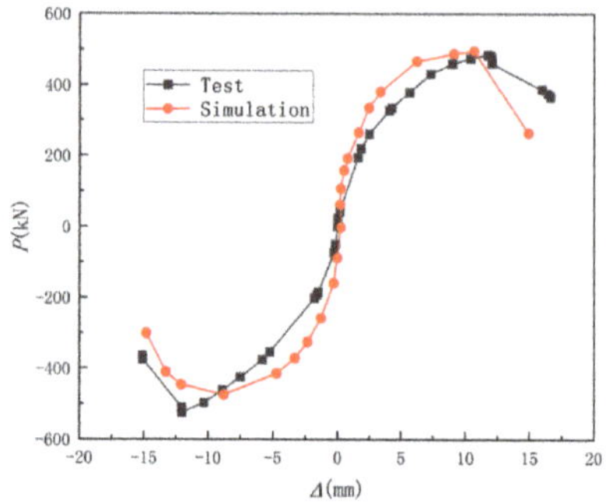

(b) EHY−1−100

Figure 19. Comparison between simulated skeleton curve and tested skeleton curve.

Table 6. Comparison between experimental strength and simulated strength.

Specimen Number	P_t/kN	Δ_t/mm	P_{nu}/kN	Δ_{nu}/mm	P_{nu}/P_t	Δ_{nu}/Δ_t
EHY−1−60	449.0	13.45	427.7	13.38	0.95	0.99
EHY−1−100	504.7	11.88	484.7	9.71	0.96	0.82

Note: P_t and Δ_t are the peak load and peak displacement measured in the test; P_{nu} and Δ_{nu} are the peak load and peak displacement measured in numerical simulation. For the specimen EHY−1−60, negative values are taken; for the specimen EHY−1−100, both positive and negative average values are taken.

5.4. Effect of Buckle Length of the U-Type Reinforcements on Seismic Performance of the Shear Wall

To fully reveal the impact of the buckle length of the U-type reinforcement on the seismic performance of precast shear walls with post-cast epoxy resin concrete, the previous experimental study is referenced to establish the finite element model of shear walls with a buckle length of 20 mm, 40 mm, and 80 mm. They are denoted as EHY−1−20, EHY−1−40, and EHY−1−80, accordingly. The loading system is identical to that used for the test.

5.4.1. Comparison and Analysis of Stress and Plastic Damage Distribution of Each Specimen

Figure 20 shows the distribution of stress and epoxy resin concrete damage in each specimen. The distribution of stress in each specimen is comparable, with higher stresses concentrated at the horizontal joint and keyways. With the increase of the fastening length, the higher stress in the horizontal joint gradually shifts from the keyway of the lower joint surface to the keyway of the upper joint surface. Meanwhile, the maximum stress shows a trend of oblique development. From the distribution of damage in each specimen, it can be seen that the damage caused to the epoxy resin concrete of the specimens EHY−1−20, EHY−1−40, and EHY−1−60 is mainly concentrated at the keyway of the lower surface of the joint, while the damage caused to the specimens EHY−1−80 and EHY−1−100 is mainly concentrated at the keyway of the upper surface of the joint, showing a trend of gradual oblique development. With the increase of buckle length, the horizontal joint acts as a constraint at the fixed end of the wall, which causes a shift of the damage to the upper joint surface. Also, the shear failure becomes increasingly obvious.

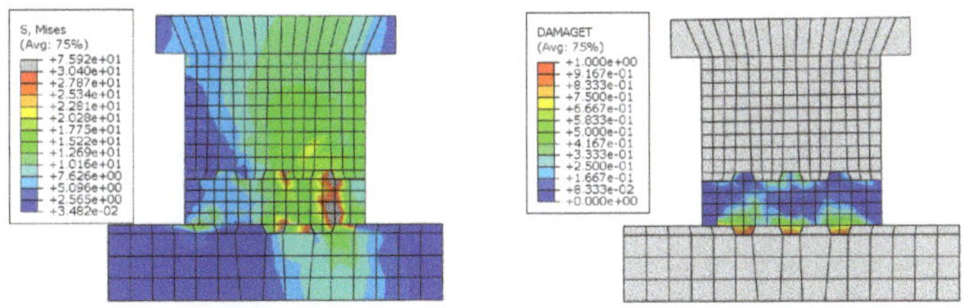

(**a**) EHY−1−20

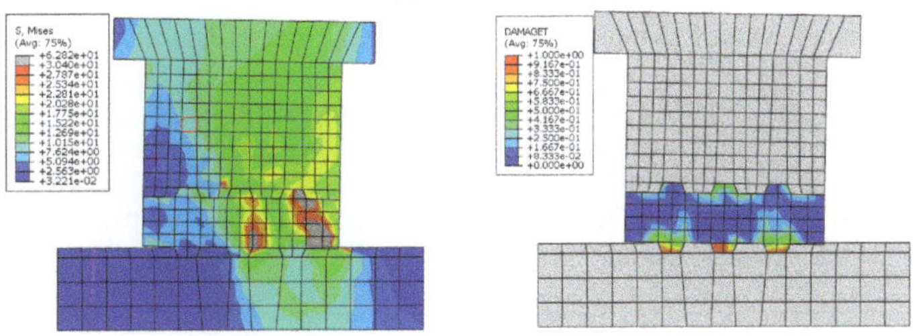

(**b**) EHY−1−40

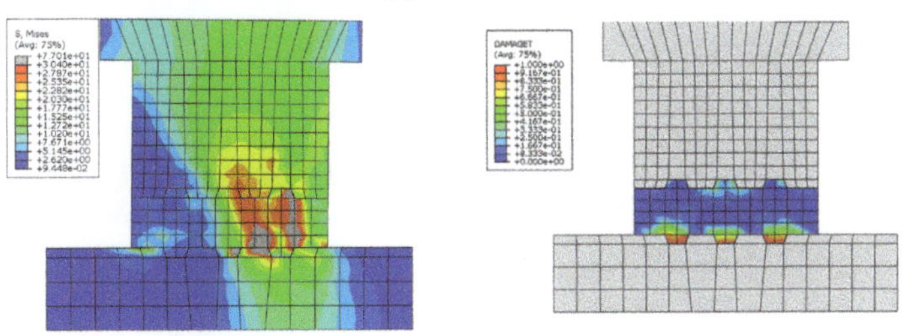

(**c**) EHY−1−60

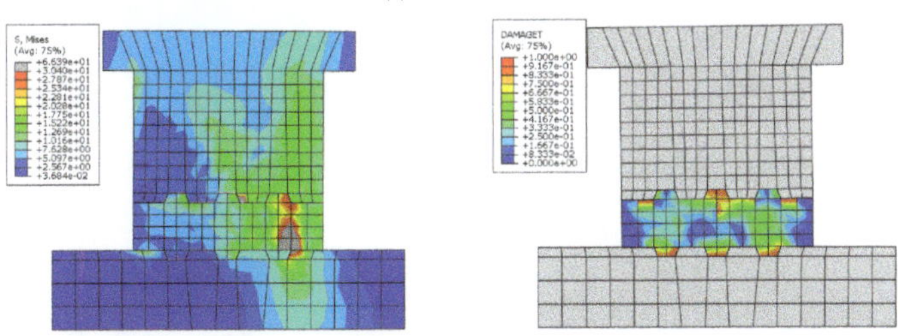

(**d**) EHY−1−80

Figure 20. *Cont.*

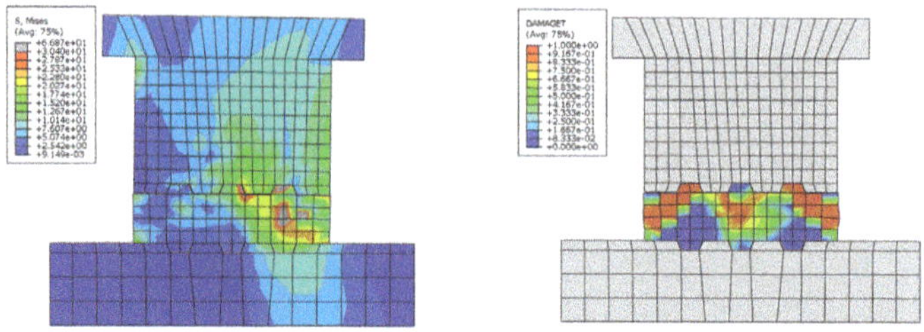

(**e**) EHY−1−100

Figure 20. Stress distribution and epoxy resin concrete plastic damage distribution.

5.4.2. Comparison and Analysis of Skeleton Curves of Each Specimen

Figure 21 shows the skeleton curve of each specimen. The data of each feature point are shown in Table 6. Apparently, with the increase in buckle length of the U-type reinforcement, the initial stiffness and load-bearing capacity of the specimen is improved gradually. Compared with the specimen EHY−1−20, the peak load of the specimens EHY−1−40, EHY−1−60, EHY−1−80, and EHY−1−100 increases by 9.3%, 12.7%, 21.1%, and 26.5%, respectively. Relative to the specimen with a 60 mm buckle length of the U-type reinforcement, the load-bearing capacity is lower among the specimens with a buckle length of 20 mm or 40 mm, although the capability to withstand deformation is comparable. Also, the load-bearing capacity increases when the buckle length of the specimens increases to 80 mm and 100 mm. However, the descending section of the curve becomes steeper, the capability to withstand deformation is weakened, and brittle failure occurs in the specimens abruptly after the peak load is reached. Therefore, the specimen with a 60 mm buckle length of the U-type reinforcement achieves the best seismic performance. Roughly, the optimal buckle length of the U-type reinforcement is one-third the height of the horizontal joint.

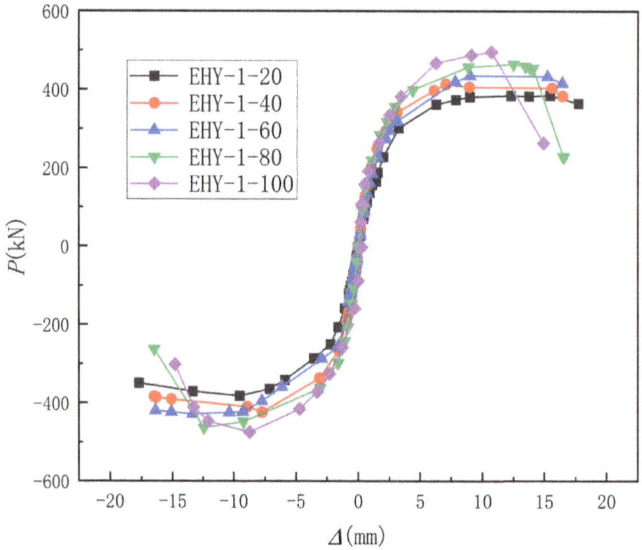

Figure 21. Comparison of the skeleton curves of each specimen.

5.4.3. Comparison and Analysis of Ductility of Each Specimen

Table 7 lists the ductility coefficients of each specimen. Obviously, the ductility coefficient of the specimen decreases progressively with the increase in buckle length of the

U-type reinforcement. Meanwhile, the performance of the shear wall in withstanding deformation deteriorates continuously. Compared with the specimen EHY−1−20, the ductility coefficient of the specimens EHY−1−40, EHY−1−60, EHY−1−80, and EHY−1−100 is reduced by 8.2%, 33.1%, 33.3%, and 44.7%, respectively. The main reason for this is observed in the test; that is, the greater the buckle length of the U-type reinforcement, the greater the section height and stiffness of the hidden beam at the bottom of the wall, and the more similar the role of the horizontal joint to the new fixed end of the wall. If the horizontal joint is treated as a fixed end, the net height of the wall decreases, and the horizontal load-bearing capacity increases. By contrast, the ability to withstand deformation is weakened.

Table 7. Ductility coefficient of each specimen.

Specimen Number	Load Direction	Yield Load P_y/kN	Yield Displacement Δ_y/mm	Peak Load P_{max}/kN	Peak Displacement Δ_{max}/mm	Ultimate Load P_u/kN	Ultimate Displacement Δ_u/mm	Ductility Coefficient $\mu = \Delta_u/\Delta_y$
EHY−1−20	Positive	305.1	3.42	384.6	15.45	353.8	18.9	5.53
	Negative	279.0	3.29	381.7	9.54	351.1	17.4	5.28
	AVG	292.1	3.35	383.1	12.50	352.5	18.1	5.41
EHY−1−40	Positive	348.7	3.44	414.0	7.04	351.9	17.8	5.15
	Negative	349.5	3.82	423.1	7.75	359.7	18.3	4.78
	AVG	349.1	3.63	418.6	7.40	355.8	18.0	4.97
EHY−1−60	Positive	354.6	4.71	435.8	8.98	370.4	18.1	3.84
	Negative	339.0	5.26	427.7	13.38	418.7	17.9	3.40
	AVG	346.8	4.99	431.7	11.18	349.6	18.0	3.62
EHY−1−80	Positive	415.0	3.59	464.9	12.47	395.2	14.6	4.08
	Negative	380.0	4.40	463.0	12.45	393.5	13.8	3.13
	AVG	397.5	4.00	463.9	12.46	394.4	14.2	3.61
EHY−1−100	Positive	410.4	2.59	495.7	10.67	446.1	11.6	2.67
	Negative	388.2	3.86	473.7	8.75	426.3	12.7	3.30
	AVG	399.3	3.23	484.7	9.71	436.2	12.2	2.99

6. Conclusions and Prospects

6.1. Conclusions

In this paper, seismic performance testing, finite element simulation, and parameter analysis are conducted to study the precast shear walls with horizontal joints post-cast with epoxy resin concrete. The main conclusions of this paper are as follows:

(1) The bonding performance between epoxy resin concrete and ordinary concrete is excellent, and the seismic performance of the shear wall with post-cast epoxy resin concrete is better than the cast in situ concrete shear wall. Compared with the cast-in-place concrete shear wall, the horizontal load-bearing capacity of the post-cast epoxy resin concrete shear wall increases by 14.0–33.7%, the ductility coefficient increases by 31.4–94.8%, and the cumulative amount of energy consumption increases by 204–518%. Therefore, epoxy resin concrete is applicable as the post-cast material for the horizontal joint of the precast shear wall.

(2) The load-bearing capacity, ductility, and energy dissipation capacity of the shear wall with keyways on the horizontal joint interface are higher than those without keys on the joint interface, by about 4.3%, 48.2%, and 314%, respectively. Therefore, the keyways on the joint interface are effective in improving the seismic performance of the shear wall.

(3) Regarding the U-type reinforcement ferrule connection technique applied to precast shear walls, the buckle length of the U-type reinforcement is one of the critical factors affecting the seismic performance of the shear wall. With a rise in the buckle length of the U-type reinforcement, there is an increase in the section height and stiffness of the hidden beam at the bottom of the wall, and the horizontal load-bearing capacity of the wall increases. However, the ability to withstand deformation is weakened. For example, relative to the specimen with the U-type reinforcement that is 60 mm in buckle length, the peak load of the specimen with the buckle length of 100 mm increases by 12.4%. By

contrast, the ductility coefficient and cumulative energy dissipation are reduced by 20.8% and 84.2%, respectively.

(4) An excellent performance in horizontal load-bearing capacity and withstanding deformation is achieved by the precast shear wall with post-cast epoxy resin concrete at the horizontal joint, keyways at the joint interface, and a 60 mm buckle length of the U-type reinforcement. It produces the best seismic performance. Therefore, the optimal buckle length of the U-type reinforcement is approximately one-third the height of the horizontal joint.

6.2. Prospects

(1) Considering the poor fluidity of epoxy resin, it is necessary to strictly control the temperature during the production of epoxy resin concrete. Meanwhile, it should be ensured that the epoxy resin is fully mixed with the aggregate. Further research will be focused on how to simplify the preparation process, improve production efficiency, and reduce production costs.

(2) The analysis conducted in this paper is limited to the impact of buckle length of the U-type reinforcement on the work performance of the precast shear wall. Therefore, it is worth further exploring other influencing factors, such as mechanical strength, the position of the post-pouring area, the method of treatment, the axial compression ratio, and so on. Notably, the difference in strength between precast concrete and post-cast epoxy resin concrete has a significant impact on the mechanical properties of precast components. According to a previous study [36], when the strength of post-cast epoxy resin concrete approaches or slightly exceeds that of precast concrete, precast components perform best in mechanical properties. As revealed by this test, the difference in strength between the two is approximately 33%. However, due to space limitations, a detailed analysis was not carried out; this will require further research.

Author Contributions: Conceptualization, X.Z.; Methodology, P.C.; Software, Z.L.; Writing—original draft, Z.L. and J.W.; Writing—review & editing, P.C. and S.X.; Project administration, X.Z. All authors have read and agreed to the published version of the manuscript.

Funding: This research was funded by the National Natural Science Foundation of China, grant number 51678389.

Data Availability Statement: Data are contained within the article.

Conflicts of Interest: The authors declare no conflict of interest.

References

1. Fan, H.; Zhang, Y.; Zhang, Y.; Jiang, H.; Du, L.; Liu, Q. Mechanical Properties and Energy Conservation Analysis for The New Connection of The Fabricated Shear Wall. *IOP Conf. Ser. Earth Environ. Sci.* **2021**, *804*, 022–094. [CrossRef]
2. Hu, R.; Fang, Z.; Shi, C.; Benmokrane, B.; Su, J. A review on seismic behavior of ultra-high performance concrete members. *Adv. Struct. Eng.* **2021**, *24*, 1054–1069. [CrossRef]
3. Basereh, S.; Okumus, P.; Aaleti, S. Seismic retrofit of reinforced concrete shear walls to ensure reparability. In *Structures Congress*; American Society of Civil Engineers: Reston, VA, USA, 2020; pp. 498–509.
4. Choi, W.C.; Yun, H.D.; Cho, C.G.; Feo, L. Attempts to apply high-performance fiber-reinforced cement composite (HPFRCC) to infrastructures in South Korea. *Compos. Struct.* **2014**, *109*, 211–223. [CrossRef]
5. Sharma, S.; Aaleti, S.; Okumus, P. Experimental testing of RC shear wall seismic retrofit using selective weakening, self-centering, and Ultra High-performance concrete. *Resilient Cities Struct.* **2023**, *2*, 76–90. [CrossRef]
6. Gopal, B.A.; Hejazi, F.; Hafezolghorani, M.; Voo, Y.L. Shear Strength of Dry and Epoxy Joints for Ultra-High-Performance Fiber-Reinforced Concrete. *ACI Struct. J.* **2020**, *117*, 279–288. [CrossRef]
7. Yu, Z.W.; Peng, X.D.; Guo, W.; Peng, M.P. Seismic performance of precast concrete shear wall with U-type reinforcements ferrule connection. *J. Zhejiang Univ. (Eng. Sci.)* **2015**, *49*, 975–984. (In Chinese)
8. Zhou, J.; Zhi, X.; Fan, F.; Jiao, A.; Qian, H. Experimental and numerical investigation on failure behavior of ring joints in precast concrete shear walls. *Adv. Struct. Eng.* **2020**, *23*, 118–131. [CrossRef]
9. Lu, X.; Wang, L.; Wang, D.; Jiang, H. An innovative joint connecting beam for precast concrete shear wall structures. *Struct. Concr.* **2016**, *17*, 972–986. [CrossRef]
10. Jiao, A.L.; Zhang, Z.S.; Gao, Y.F. Quasi-static test of full-scale structure on precast shear walls with annular closed reinforcements. *World Earthq. Eng.* **2017**, *33*, 211–221. (In Chinese)

11. Gao, Y.F.; Jiao, A.L.; Zhang, Z.S.; Liu, J.; Zhi, X.D. Tests on out-of-plane bending of precast shear walls with annular closed reinforcements. *Build. Struct.* **2016**, *2*, 439–442. (In Chinese)
12. Chen, F.; Yu, Z.; Yu, Y.; Zhai, Z.; Liu, Q.; Li, X. Experimental Investigation of Seismic Performance of Precast Concrete Wall–Beam–Slab Joints with Overlapping U-Bar Loop Connections. *Materials* **2023**, *16*, 3318. [CrossRef] [PubMed]
13. Beeldens, A.N.N.E.; Monteny, J.; Vincke, E.; De Belie, N.; Van Gemert, D.; Taerwe, L.; Verstraete, W. Resistance to biogenic sulphuric acid corrosion of polymer-modified mortars. *Cem. Concr. Compos.* **2001**, *23*, 47–56. [CrossRef]
14. Natarajan, S.; Neelakanda Pillai, N.; Murugan, S. Experimental investigations on the properties of epoxy-resin-bonded cement concrete containing sea sand for use in unreinforced concrete applications. *Materials* **2019**, *12*, 645. [CrossRef] [PubMed]
15. Qian, Y.; Li, Z.; Jin, Y.; Wang, R. Experimental Study on Axial Tension Members of a New Epoxy Resin Concrete. *Sci. Adv. Mater.* **2021**, *13*, 2005–2015. [CrossRef]
16. Sun, N.; Song, Y.; Hou, W.; He, S.; Lin, W. Experimental study on factors influencing ERC-NSC interface bonding performance. *Eng. Fail. Anal.* **2022**, *142*, 106819. [CrossRef]
17. Liang, X.J. Preparation and durability experimental analysis of epoxy resin modified cement concrete. *J. Funct. Mater.* **2023**, *54*, 3217–3223. (In Chinese)
18. Xiang, Q.; Xiao, F. Applications of epoxy materials in pavement engineering. *Constr. Build. Mater.* **2020**, *235*, 117529. [CrossRef]
19. Rahman, M.M.; Akhtarul Islam, M. Application of epoxy resins in building materials: Progress and prospects. *Polym. Bull.* **2022**, *79*, 1949–1975. [CrossRef]
20. Sneed, L.H.; Verre, S.; Ombres, L.; Carloni, C. Flexural behavior RC beams strengthened and repaired with SRP composite. *Eng. Struct.* **2022**, *258*, 114084. [CrossRef]
21. Rizkalla, S.H.; Serrette, R.L.; Heuvel, J.S.; Attiogbe, E.K. Multiple shear key connections for precast shear wall panels. *PCI J.* **1989**, *34*, 104–120. [CrossRef]
22. Julio, E.N.; Branco, F.A.; Silva, V.D. Concrete-to-concrete bond strength. Influence of the roughness of the substrate surface. *Constr. Build. Mater.* **2004**, *18*, 675–681. [CrossRef]
23. Jang, J.; Lee, H.; Ha, K.; Kim, H.K.; Yang, B. Flexural Strength of Precast Concrete Segments with Joint Grouts. In Proceedings of the 22nd Conference on Structural Mechanics in Reactor Technology, San Francisco, CA, USA, 18–23 August 2013.
24. Wu, X.G.; Zhang, X.C. Investigation of short-term interfacial bond behavior between existing concrete and precast ultra-high performance concrete layer. *J. Build. Struct.* **2018**, *39*, 156–163. (In Chinese) [CrossRef]
25. Zhang, Y.; Wu, J.; Shao, X.; Hou, C. Experiment on interfacial shear properties between ultra-high performance concrete and normal strength concrete. *China Civ. Eng. J.* **2021**, *54*, 81–89. (In Chinese)
26. Jiang, X.; Song, H.; Li, K.; Qiang, J.; Tian, J. Experimental Investigation on Shear Strength between Ultra-High-Performance Concrete and Normal Concrete Substrates. *Adv. Mater. Sci. Eng.* **2023**, *2023*, 3169912. [CrossRef]
27. Zhou, J.; Zhi, X.; Fan, F.; Jiao, A.; Qian, H. Anchor design of a ring joint based on reliability in a precast shear wall structure. *Appl. Sci.* **2019**, *9*, 3361. [CrossRef]
28. Jin, Y.J.; Zhao, Y.; Xie, X.Y.; Zhang, X.; Cai, J.w. Simulation Analysis of Joint Connection of New Epoxy Resin Concrete Truss Structure. *Int. J. Perform. Eng.* **2021**, *1*, 143–154.
29. LV, X.; Yu, Z.; Shan, Z. Seismic behaviour of frame structures with assembly of prefabricated concrete beam. *J. Build. Eng.* **2021**, *40*, 102765. [CrossRef]
30. *EN1992-1-2*; European Standard. Euro-Code2: Design of Concrete Structures-Part1: General Rules and Rules for Buildings. European Committee for Standardization: Bruxelles, Belgium, 2021.
31. Wang, J.Y. *Research and Application on Horizontal Joint Structural of Precast Reinforced Concrete Shear Wall*; Tianjin Chengjian University: Tianjin, China, 2022. (In Chinese)
32. *JGJ/T 101-2015*; Specification for Seismic Test of Buildings. China Architecture & Build Press: Beijing, China, 2015.
33. Jiao, A.L.; Zhang, P.S.; Li, Y.H. Tests on seismic behavior of pre-cast shear walls with annular closed reinforcements. *J. Build. Struct.* **2015**, *05*, 103–109. [CrossRef]
34. Fang, Z.H.; Zhen, Y.; Li, X.P. Steel hysteretic model of reinforced concrete structures. *Eng. J. Wuhan Univ.* **2018**, *7*, 613–619. (In Chinese) [CrossRef]
35. Fang, Z.H.; Zhou, Y.; Jian, X.Y. MCFT model of concrete structures subjected to cyclic loading. *Earthq. Eng. Eng. Vib.* **2015**, *35*, 73–78. (In Chinese) [CrossRef]
36. Chen, P.Q.; Xu, S.; Zhou, X.J.; Xu, D.Z. An Experimental Study on Flexural-Shear Behavior of Composite Beams in Precast Frame Structures with Post-Cast Epoxy Resin Concrete. *Buildings* **2023**, *13*, 3137. [CrossRef]

buildings

MDPI

Article

Adaptive Flexibility of the City's Complex Systems

Sara Abdul Aalie Rasheed [1], Wahda Shuker Al-Hinkawi [1] and Mohammed Fareed Sherzad [2,*]

1 Department of Architectural Engineering, University of Technology, Baghdad 10063, Iraq; ae.20.03@grad.uotechnology.edu.iq (S.A.A.R.); wahda.s.mahmoud@uotechnology.edu.iq (W.S.A.-H.)
2 Department of Architecture, College of Architecture, Art and Design, Ajman University, Ajman P.O. Box 346, United Arab Emirates
* Correspondence: msherzad@ajman.ac.ae

Abstract: The city represents a complex adaptive system consisting of a set of interconnected subsystems within self-organization and flexible structures that give it the ability to adapt to continue operating over time. This research discusses the adaptive flexibility of the complex urban system by measuring the degree of complexity and adaptation of secondary systems within the general system depending on its kinetic and visual system, making it self-organized and within a coherent and harmonious unit. Thus, the research problem arises from "the lack of knowledge on how to determine the appropriate degree of complexity and adaptation for urban systems, providing them with adaptive flexibility that allows them to maintain their energy and ensure their survival and continuity over time". This research aims to determine the adaptive flexibility of the complex urban system, which keeps it self-organized and able to survive depending on its kinetic and visual system that links the secondary systems of the general system based on the degree of complexity and adaptation ideal. The research adopted quantitative analysis, which includes calculating the adaptive dimensions of the complex system based on the kinetic and visual system, as well as the nodes connecting the parts of the secondary system, in addition to determining both the area and the perimeter of the study area. The historic Karkh area in Baghdad was selected for implementation because it includes a model of the complex system that originated and developed within different time stages. The system's adaptive flexibility was calculated to determine its ability to survive and continue. This research found that it is possible to determine the strength of the urban format and its ability to survive or not, as well as the possibility of developing it, by understanding the ideal degree of adaptive flexibility for the format. This ideal degree is 1, achieved through visual and kinetic coherence among the nodes that make up the parts of the format. The farther the format is from this degree, the less adaptive flexibility it has, which affects its ability to survive and continue. Therefore, it is necessary to maintain the strength of the interconnections between the parts of the kinetic and visual system and between the parts of the complex format, which directly affects its ability to self-regulate and survive.

Keywords: adaptive system; complex system; entropy; systematic correlation; visual and kinetic axis; complex system flexibility

1. Introduction

Marshall revealed that complex urbanization arises from the bottom up (from the emergence and development of secondary parts, represented by buildings, urban spatial tissues, and the kinetic system linking the parts). The process of understanding it depends on understanding the parts that formed this system, analyzing the emerging systems formed over time, and knowing the causes and motives of their emergence, which in turn directly affects the analysis of the complex system and determines the appropriate ways to address its parts. When studying and analyzing complex systems, their origins must be taken into account, which can be either as a result of the natural evolution of the urban system (spontaneously) or as a result of partial introduction and development processes or upper development processes that arise from top to bottom. In both cases, hierarchically differentiated system structures emerge, which may suggest central planning. However, it is not due to the presence of individual elements and parts within them that compete with each other. This includes what adapts, changes, decays, and ends. Therefore, it is important to study the relationship between the secondary parts formed within the general system and to understand the strength of the relationship between them [1].

The measurement of the strength of the sequence interconnection depends on the creation of elements and rules of the system interaction and the analysis of the emerging impact. Thus, a city consists of multiple sequences of hierarchical systems that evolve and change according to respective contextual rules and conditions.

Alexander described this concept as a semi-lattice concept and the evolution of consistency as a result of the cumulative collection and removal of parts and the representation of the changing needs of individuals [2]. It is necessary to determine the complex system's ability to adapt and self-organize permanently and continuously because it grows and develops constantly as a result of changing its morphology, the essential elements that make it up, or the relations between its parts. This format change comes to meet cultural or economic needs or support different urban functions and may appear as a result of interventions for urban development projects of the city and the addition of parts or elements within the different systems that increase the degree of complexity of the system. Therefore, it was necessary to identify the relationship between secondary systems and their interconnections with each other, which directly affect the strength of the general urban system and facilitate the development processes of how to deal with the structural parts and strengthen the weak ones and strengthen them to ensure their continuity and preservation from extinction [3]. Since the secondary systems are interconnected with each other through kinetic and visual systems, this makes them of great importance in achieving general coordinate coherence; therefore, measuring the ability of complex systems to adapt requires knowledge of the strength of interdependence and finding elements and rules that allow the best interaction of the parts of the system. It also requires the analysis of the emerging influence and, thus, the understanding of the city, which consists of more than one complex system carrying sequential hierarchical systems that evolve and change according to the specific contextual rules and conditions, and this is what Christopher Alexander described within the concept of semilattice, where the systems develop as a result of the cumulative addition and deletion of the parts, the deletion or addition of appropriate kinetic associations between the parts, and the representation of the changing needs of individuals [4].

Thus, there is an intrasystem part correlation, which not only impacts the overall system but also acts as a basic or secondary factor. The factors share knots and joints, thus making the system coherent [5].

A complex system depends on the way that interpart relationships are interconnected and shaped. Within the system, the strength of the links between the nodes gives the entire system sufficient strength, flexibility, specificity, and spatial strength [6].

A complex system is influenced by the street structure, the road network, and knot-to-knot movement, while the strength of the system depends on the coherence of the kinetic axes, which increases the system's strategic value [7].

Stereotypical behavior depends on unexpected interactions between autonomous parts. These parts additionally adapt to each other and, thus, regulate themselves, which enables the system to persist and evolve [6].

System flexibility is defined as adaptability without becoming a new coordinated identity and, thus, represents a dynamic ability to shift systems between adaptive situations without exceeding the threshold of irreversibility. Flexibility, in this sense, seeks to stabilize the system through diversity and part-wise interconnectedness [8].

The complex system is influenced by both the street structure and the traffic system (road network and traffic flow from one node to another). Such interdependence creates interconnected urban systems, which, in turn, contain intrasystem visual and kinetic relationships and associations [9]. Flexible consistency is adaptable only if it contains a diverse mix of internal and peripheral connections between the interconnection contract of its parts consisting of complex structures that are geometrically and topologically adapted and coherent because of their potential to maintain their entropy and bioenergy.

Within this consistency, flexibility heavily depends on its ability to relate and communicate with every possible number of contacts between components, and, therefore, the more points of contact there are, the more vivid and adapted this consistency [10]. Furthermore, each system needs appropriate, fragmented, and self-regulated complexities so that it can maintain its internal structure and balance. Moreover, these complexities can help the system work and adapt to external conditions [11].

The kinetic and visual system directly affects the coordinate adaptation, as it affects the relationship of secondary systems with each other, and flexible systems are adaptable only if they contain a diverse mix of internal and terminal connections between the connecting nodes of their parts. The flexibility of these systems depends largely on their ability to interconnect and connect with every possible number of connections between components; therefore, the more and larger the number of points of contact, the more and more these systems will be alive and adapted [12].

This system also possesses different degrees of complexity, indicators of change-oriented flexibility. Therefore, its balance and flexibility depend on the level of self-regulation generated by the part interconnectedness and the visual and kinetic hierarchies [13].

It is clear from the above that the complex adaptive system consists of a group of parts whose impact on the city varies due to the different types of kinetic and visual relations between the centers and nodes, which requires knowledge of the strength of the coordinate bonding that affects the adaptation and flexibility of these systems. Therefore, this research seeks to study the visual and kinetic system and its role in achieving adaptive flexibility and maintaining the ideal complexity of the urban system by knowing the appropriate degree of complexity and adaptation based on the kinetic and visual relationships and the nodes that make up the system and are responsible for linking its parts. The complex system is affected by the street structure, the road network, and the flow of traffic from one node to another, which gives the strength of the internal system that ensures the best self-organization and gives it the adaptive flexibility that makes it able to continue. The research aims to "determine the adaptive flexibility of the complex urban system, which keeps the system self-organized and viable depending on the kinetic and visual system that links the secondary systems of the general format based on the degree of complexity and optimal adaptation". To achieve the objective of the research, the critical realism approach will be adopted, which relies on the use of reliable measurement to calculate the adaptive dimensions of the complex system based on the associative system of the coordinate parts (the relationship between the visual and kinetic axes between the nodes). We determine both the area and perimeter of the study area and then apply them to a selected sample that meets the requirements of the study and calculate the adaptive flexibility of the system by adopting the following steps:

1. Build a theoretical framework and extract indicators to measure the adaptation of the complex system
2. Determine the method of measuring indicators and apply them to an area characterized by a complex system to achieve the research objective.

2. Previous Studies

Different studies have been conducted on adaptable systems in urban planning, link-based measurements of complexity and adaptation, and kinetic and visual internode communications that form the intrasystem joining parts:

Boeing (2017) [14] showed that node-linking pathways in urban systems are indicators of the sequentially integrated parts, as the flexibility of adaptable consistency depends on the emerging and ongoing networking of the harmonious entirety to keep the system balanced.

System adaptability is measured using the shortest physical bonds with the least change in direction versus the optical linkages between the nodes as representations of adaptation and when the urban sequence is studied over different periods. Changeable interconnectedness in intrasystem parts can be observed as new completely different linkages may emerge. In this case, kinetic and visual links are very important in determining system adaptability [14].

According to Jiang (2018), an accurate analysis of urban systems depends on the mathematical description and quantitative analysis, which depends on the analysis of the nodes and the internode linkages. All these parameters make up the whole system, determine its length, and recognize relevant points. Each part of the system demonstrates, as generated, the number of nodes, the internode interconnection, and the interaxis distances. All these demonstrate how much interdependence and harmonious coherence have been achieved [15].

Al-Kasibi (2020) [12] proposed a way of calculating urban system complexity and adaptability by creating new adaptabilities, measuring them against two-dimensional structures, detecting their behaviors, and enabling them to control parts.

This method differs from fractional dimensions in that it is a calculation process shaped according to adaptive dimensions by focusing on the living processes that operate at the same scale level and therefore depend on the size rather than the scale, as in fractional dimensions, and it, therefore, examines the nodes and the kinetic and visual links that arise between these nodes [12].

In the same vein, in another study, Al-Kasibi (2021) posited that changing consistency over time needs to regulate its complexity for survival by maintaining levels of interaction between kinetic and visual axes and focusing on the adaptive complexities of its flexible structures. He further emphasized that consistency required optimal compositions of physical and visual associations that would enhance adaptability by relying on the construction of the space sentence, and he noted that each existing urban fabric had a structure with local, global, and comparable adaptive dimensions [16]. Ali & others found that maintaining or developing any system depends mainly on the study of kinetic relationships and access points to the parts of the system, and therefore it achieves the coherence of the system with the rest of the city and makes it capable of spatial attraction [17].

An adaptable urban system depends on the interconnection of the visual and kinetic axes, as well as the connecting nodes. Additionally, these interconnections demonstrate the degree of perfection that the system is meant to achieve to achieve adaptive flexibility and maintain ideal complexity, as shown in Figure 1.

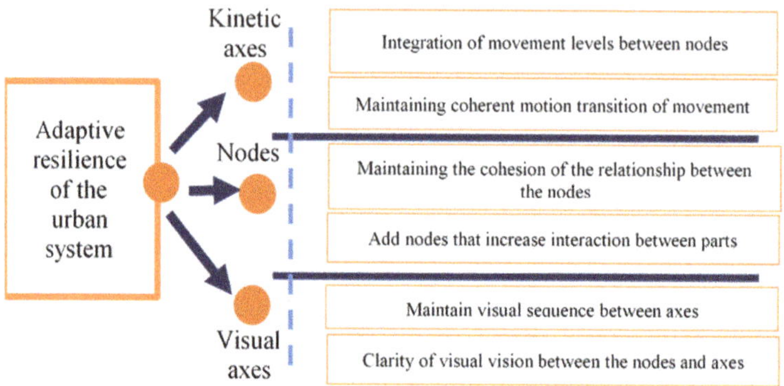

Figure 1. The effect of kinetic interconnectedness on the adaptive flexibility of the urban system along with the most essential effective indicators.

3. Methodology and Tools

Critical realism is followed in this paper to approach the problem of the research and to achieve the aims of the study. Critical realism depends on reality-based theorization, previous studies, the determination of a factual problem, and the adoption of a previous tool to solve a new problem. Such an approach is applied and tested to obtain findings, analyze data, and discuss results. This methodology is chosen because it uncovers the invisible structures that generate city-affecting events, as shown in Figure 2.

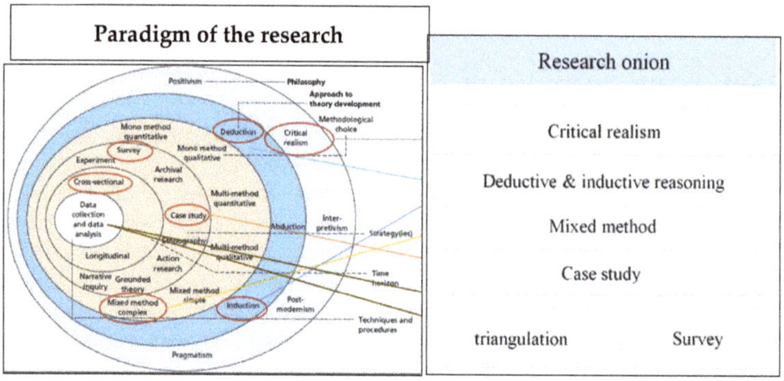

Figure 2. Onion-based research methodology.

Regarding the tools adopted for analyses, the observation method was relied upon to collect data. This research relied on indirect observations based on documents, records, and development reports of the area examined, prepared by the Baghdad Municipality, in addition to direct observations made by visiting the site and identifying factors that affect the nature of the system and development. Then, the maps obtained were drawn in AutoCAD software 2007, and the kinematic and visual axes of the place were analyzed while calculating the existing nodes, area, and perimeter and the number of axes and their lengths within the software.

In this methodology, a set of steps is followed according to critical realism and adopted as a basis in the research, which starts from the first step, which is to identify the concrete problem from reality. The research steps are as follows: as shown in Figure 3.

Figure 3. *Cont.*

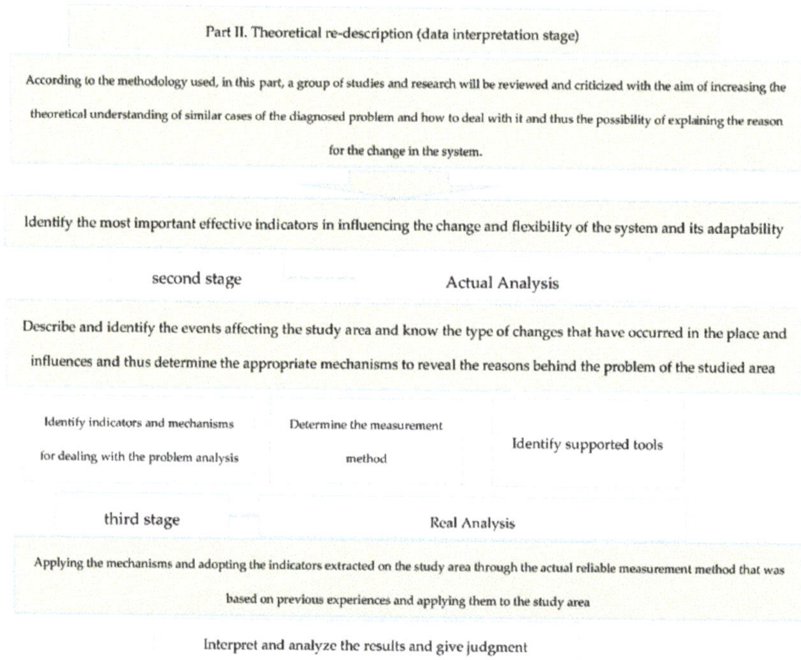

Figure 3. The research steps.

4. Study Population

Al-Karkh area was selected to measure the adaptive flexibility of the complex system found in this area located in the city of Baghdad (Figure 4).

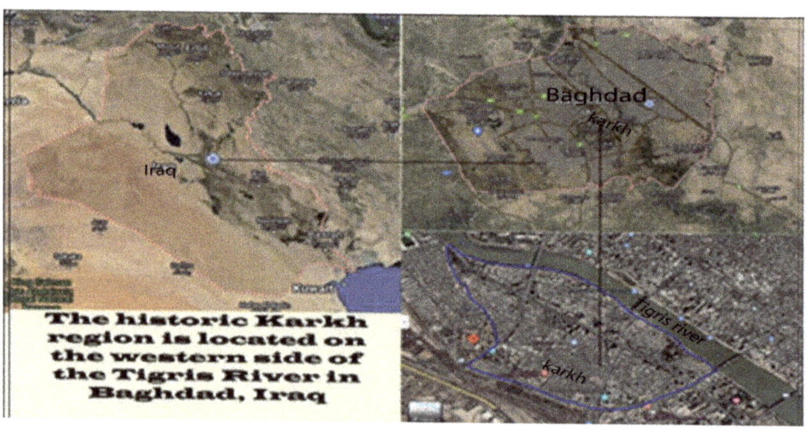

Figure 4. An aerial view of the Karkh area, Google Map.

5. Data and Source

The sample comprises nonrandomized, intentional data representing Baghdad's Karkh, specifically an area overlooking the Tigris between the Ahrar Bridge and the 14 July Bridge.

Given the many changes made to the system, this study attempted to reveal whether the structures are flexible enough to adapt. With that part of Baghdad being under study, pre- and post-urban changes were taken into account.

This part has been home to multiple urban, kinetic, visual, and other activities; thus, different texture-impacting changes and trans-systems have taken place. These changes have also resulted in a complex system that requires an analysis to determine how flexibly adaptable the contained parts are, as shown in Figures 5 and 6.

Figure 5. Map of Baghdad Karkh region in the 19th century, source [6].

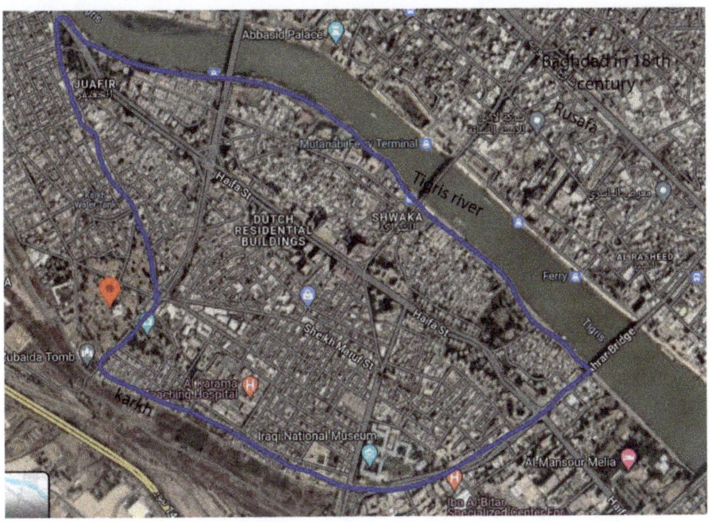

Figure 6. Map of Baghdad Karkh area at the present Google Earth.

6. Measurement Method

To achieve interaxis integration, some formulas, approved by [12] and previously applied to multiple cases, were used. This method calculates system adaptability based on the interaxis linkages existing between the nodes, in addition to determining both the area and perimeter. This determination designed the ideal adaptability of the system, represented by a value of 1. This method, therefore, was used and applied to Baghdad's Karkh. The formulae are stated in [16].

$$DA = (\log(N \times L \div S + V)^2 - \log A) - (\log P - (\log N \times L \div S + V))$$

A: Total area within the whole visual field (zone or field of influence).

N: Total number of vertices within the whole visual field.

L: Total connections' lengths (excluding those which surround solids or penetrate internal open spaces).

V: Number of all probable, extended, and visual connections that link vertices altogether (excluding those that surround solids or penetrate internal open spaces holes).

S: Number of all probable physical segments of direct connections that link each pair of adjacent vertices within the whole visual field (excluding those which surround solids or penetrate internal open spaces holes).

P: Perimeter length (excluding those that surround solids or penetrate internal open spaces).

DA: Adaptive dimensions (DA).

7. Practical Application

After an AutoCAD-assisted identification and demarcation of the sampled area, the requirements of the formula set to measure the flexible, adaptable system (the number and the lengths of the kinetic and visual axes, the number of urban-forming nodes, and the area and perimeter of the studied area) were determined.

Then, the formula was applied to realize the system's adaptability for two time spans, during which potential changes were physically introduced into the system. The values obtained from the two situations were contrasted, and the system (in)adaptability was verified based on the ideal value of the adaptable system, which is 1, as stated in Al-Kasibi [12].

7.1. Nineteenth Century-Era Karkh

7.1.1. Calculating the Complex Adaptive System of the Karkh Region

This is the period during which Karkh was populated and urbanized. The area was analyzed after the kinetic/visual axes, as well as linkage nodes, were defined. In addition, interaxis correlations were calculated as per the formula and the value was reached to determine the then-system adaptability.

The axes were calculated as shown in Figure 7, which is an AutoCAD-drawn map.

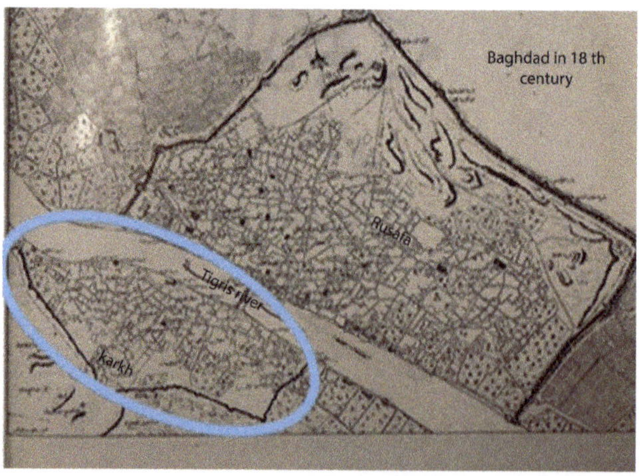

Figure 7. Nineteenth century-era Karkh [6].

7.1.2. Calculation of the System Adaptability of the Emerging Karkh

The system's adaptability was analyzed and calculated based on the identification of the intrasystem kinetic axes. The numbers and lengths of the axes and the interaxis linkages were measured using AutoCAD.

Additionally, intra-axis visual nodes were determined, measured, and gathered for processing using the formula of system adaptability and flexibility, as shown in Figure 8.

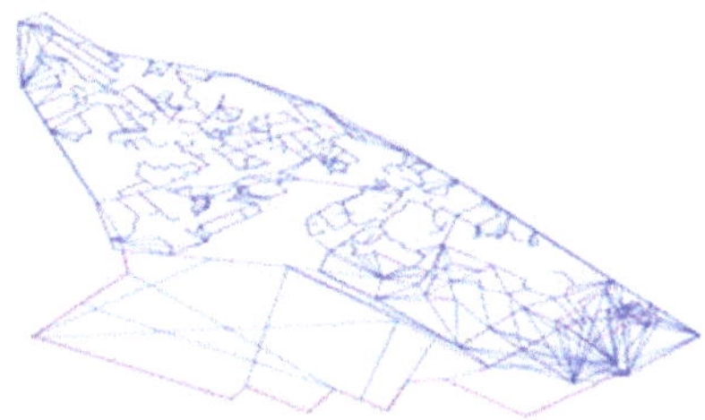

(**A**) shows both the kinetic and visual axes and the nodes within the study area, where the blue color represents the kinetic axes, while the red color is for the optical axes, and the meeting points represent the connecting nodes between them.

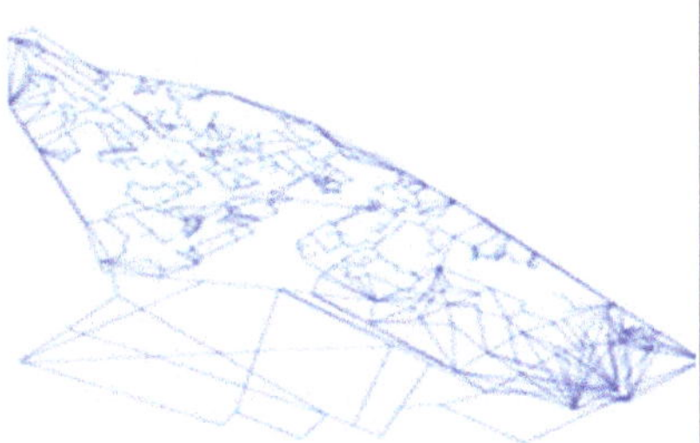

(**B**) The kinetic axes of the region and the interconnection points between them, which represent the nodes and the integration of movement between the parts of the system.

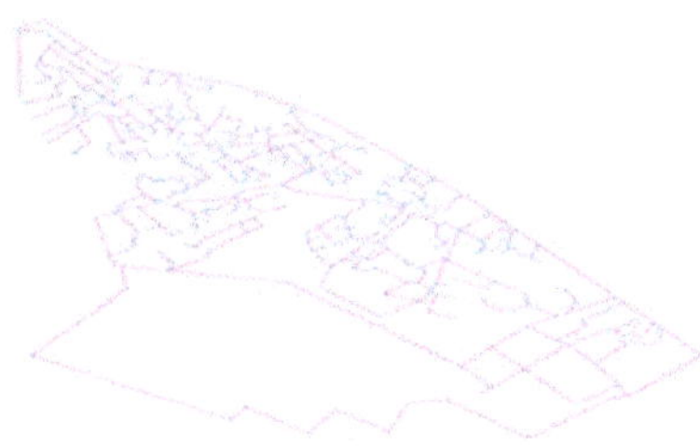

(**C**) The visual axes of the area.

550 nodes, TOTAL L = 32,588.1, $S + V$ = 2246

LOG A = 6.3, LOG P = 3.3, LOG $N \times L \div S + V$ = 3.9

LOG $(N \times L \div S + V)^2$ = 7.9

In the system for the region, where each of the total lengths of the kinetic and visual axes was calculated by adopting the AutoCAD program with the calculation of the number of nodes, perimeter, and area of the system, and these indicators represent the basis for the application of the equation to calculate the flexibility of the system.

$$DA = (\log(N \times L \div S + V)^2 - \log A) - (\log P - (\log N \times L \div S + V))$$

$$DA = (7.9 - 6.3) + (3.3 - 3.9) \rightarrow DA = 1$$

Figure 8. The stages of calculating the elasticity of the adaptive system of the Karkh region in the period of evolution.

7.2. Present-Day Karkh

7.2.1. Calculating the Complex Adaptive System of the Karkh Region Currently

In this stage, the same system was analyzed after having undergone many changes, such as the construction of new roads, changes in kinetic pathways, and the addition of buildings and elements. These developments accordingly led to a change in some kinetic pathways and the interruption of some visual axes. Therefore, there is a need to analyze the area and determine how impactful these factors are, as shown in Figure 9.

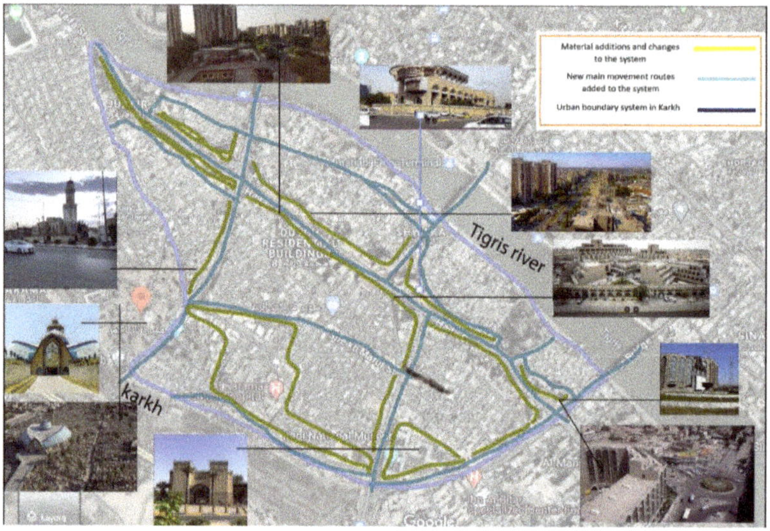

Figure 9. The analysis at this stage: The same system was analyzed after experiencing many changes, which were represented by the construction of new roads, changing movement paths, and adding buildings and elements that led to changing some kinetic paths and interrupting some visual axes due to the added elements, which led to the need to analyze the area and determine the impact of these changes.

7.2.2. Calculation of the System Adaptability of the Emerging Karkh

The system's adaptability was analyzed and calculated based on the identification of the intrasystem kinetic axes. The numbers and lengths of the axes and the interaxis linkages were measured using AutoCAD.

Additionally, intra-axis visual nodes were determined, measured, and gathered for processing using the formula of system adaptability and flexibility, as shown in Figure 10.

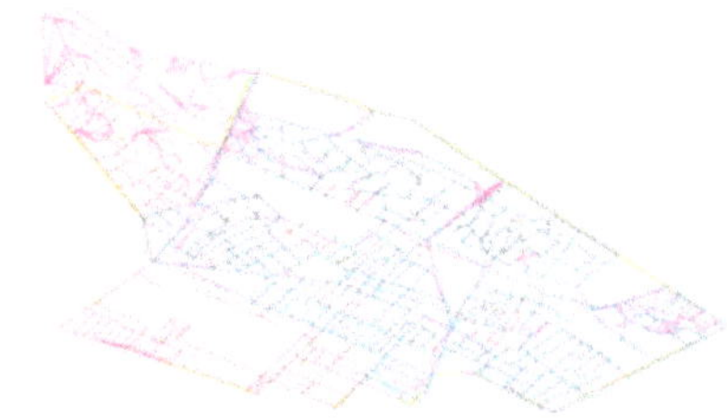

(**A**) shows both the kinetic and visual axes and the nodes within the study area, where the blue color represents the kinetic axes, while the red color is for the optical axes, and the meeting points represent the connecting nodes between them.

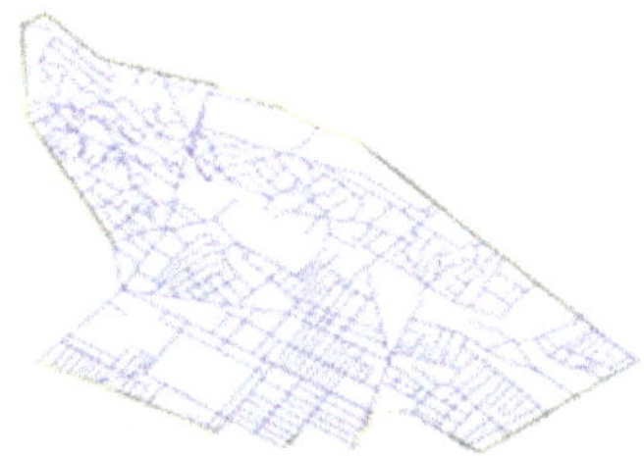

(**B**) The kinetic axes of the region.

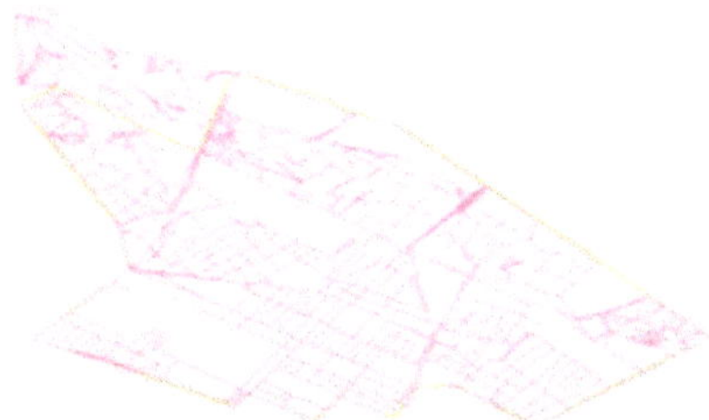

(**C**) The visual axes of the area.

$N = 608$, TOTAL $L = 148{,}267.2$, $S + V = 2465$

LOG $A = 6.5$, LOG $P = 3.9$, LOG $N \times L \div S + V = 4.5$

$$\text{LOG } (N \times L \div S + V)^2 = 9$$

In the system for the region, where each of the total lengths of the kinetic and visual axes was calculated by adopting the AutoCAD program with the calculation of the number of nodes, perimeter, and area of the system, and these indicators represent the basis for the application of the equation to calculate the flexibility of the system.

$$DA = (\log(N \times L \div S + V)^2 - \log A) - (\log P - (\log N \times L \div S + V))$$

$$DA = (9 - 6.5) + (3.9 - 4.5) \rightarrow DA = 1.9$$

Figure 10. The stages of calculating the elasticity of the adaptive system of the Karkh region in present-day Karkh.

8. Statistical Computation

Following an AutoCAD-assisted creation of drawings, a mathematical formula, stated earlier in this paper, was used, and then the formula-realizing values were processed in a Microsoft Excel spreadsheet.

9. Results

This study found that system constituent parts featured different adaptations over time, owing to morphological changes in the interconnectedness of the nodes and the hierarchies of their parts.

Self-regulation varied over time as a result of new buildings, which directly affected the interconnected and complete axes.

The results indicate that the pre-modern early system of Baghdad's historical Karkh was adaptably interconnected, meaning that the system was relevant and interdependent, having carried adaptive dimensions ($DA = 1$).

This study found that the system fell within the ideal coordinated adaptation. The adaptability suggests dynamic part-oriented integration and internode interconnectedness. The axis-level interventions, and the enforced innovations and structures, such as many physical elements and buildings, dominated the system and the intrasystem visual axes, overlooked the inherited parts, and severed interpart physical and visual relationships. This associative system led to the strengthening of the system so that it became integrated parts, regular and adaptive to each other.

These changes cut short the flexibility of urban complexity, which reached $DA = 1.9$. In other words, the system, due to interventions, is no longer adaptable. Therefore, the associative system of the complex urban system directly affects the determination of the adaptive state of the system and, thus, determines its ability to survive and continue, as it greatly affects the statement of the strength of this system and its ability to develop and contributes to the possibility of strengthening the relations between its parts and identifying the best flows that contribute to the interdependence of the parts of the system and their self-organization.

10. Conclusions

Structural flexibility is greatly affected by the nature of the associative system between the secondary parts that make up the general urban system, as it expresses the adaptive state through the degree of relational integration between the nodes and the axes linking them, which expresses the extent of the ability of the system to survive, develop, and continue in order to maintain the complexity characteristics. This is achieved by maintaining optimal coherence between the nodes that make up the parts of the system, resulting in adaptive structures capable of self-regulation for survival and continuity. As a result, it is necessary to consider the sudden changes and unstudied interventions that occur in the format. This includes closing or opening the new axes and adding structures that obscure or hinder the continuity of the relationship between the parts within the same format, which can cause the format to move away from its adaptive state by measuring the complexity of the system and knowing the impact of new additions on the strength or weakness of the interconnection between the parts. Therefore, it is essential to consider the evolutionary age of the format and the relationships between its parts, as well as the relationships between visual and kinetic axes, as a key point in maintaining systemic flexibility. This process also involves reading the format from bottom to top and through the parts down to the overall format across multiple time periods to identify the points of imbalance. It is important to identify pathways and unadapted parts to partially address them rather than interventions that may affect the urban system as a whole. All of this depends on determining the degree of complexity and appropriate adaptation, which is determined through the interconnection system between the subparts that make up the complex system. Where the kinetic and visual system affects the cohesion of urban parts and their interdependence with each other, enhancing their ability to survive and develop and providing the possibility of dealing with development projects in terms of strengthening their link with other formats or integrating them within the general system is important.

Author Contributions: Conceptualization, S.A.A.R. and W.S.A.-H.; methodology, S.A.A.R. and W.S.A.-H., software, S.A.A.R.; validation, S.A.A.R., W.S.A.-H. and M.F.S.; formal analysis, S.A.A.R.; investigation, S.A.A.R. and W.S.A.-H.; resources, S.A.A.R., W.S.A.-H. and M.F.S.; data curation, S.A.A.R.; writing—original draft preparation, S.A.A.R.; writing—review and editing, S.A.A.R., W.S.A.-H. and M.F.S.; visualization, S.A.A.R., W.S.A.-H. and M.F.S.; supervision, W.S.A.-H. and M.F.S.; project administration, S.A.A.R. and W.S.A.-H. All authors have read and agreed to the published version of the manuscript.

Funding: The APC was funded by Ajman University.

Data Availability Statement: The original data are available from the authors upon request.

Acknowledgments: The authors would like to express their gratitude to Ajman University for APC support.

Conflicts of Interest: The authors declare no conflict of interest.

References

1. Marshall, S. *Cities Design and Evolution*, 1st ed.; Routledge: London, UK; New York, NY, USA, 2008; p. 130. [CrossRef]
2. Hillier, B. The Now and Future of Space Syntax: From Structures and Models to Theory. In Proceedings of the The Ninth International Space Syntax Symposium, Seoul, Republic of Korea, 31 October–3 November 2013; Volume 33, pp. 65–70.
3. Miguel, M.A.G. Explorations on an Urban Interventions Management System. 2019. Available online: https://rgu-repository.worktribe.com/output/841507 (accessed on 11 July 2019).
4. Nes, V.; Nes, A.V. What Is the Explanatory Power of Space Syntax Theory? The Application of Modal Logics from the Theory of Science What is the Explanatory Power of Space Syntax Theory? The Application of Modal Logics from Theory of Science. In Proceedings of the 11th Space Syntax Symposium, Lisbon, Portugal, 3–7 July 2017.
5. Al-Hinkawi, W.S.; Youssef, S.S.; Abd, H.A. Effects of Urban Growth on Street Networks and Land Use in Mosul, Iraq: A Case Study. *Civ. Eng. Archit.* **2021**, *9*, 1667–1676. [CrossRef]
6. Al-Saaidy, H.J. Perspective Chapter: A Conceptualization of Measuring People's Activity in Sustaining Urban Life. In *Sustainable Regional Planning*; IntechOpen: London, UK, 2022. [CrossRef]
7. Nilsson, F. A Complex Adaptive System Approach on Logistics. 2003. Available online: https://lucris.lub.lu.se/ws/portalfiles/portal/5572329/1030967.pdf (accessed on 11 March 2024).
8. Dovey, K. Informal urbanism and complex adaptive assemblage. *Int. Dev. Plan. Rev.* **2012**, *34*, 349–367. [CrossRef]
9. Hammoodi, S.A.; Al-Hinkawi, W.S. The role of spatial value in the reconstruction of religious buildings Mosul city—A case study. *Ain Shams Eng. J.* **2023**, *14*, 102164. [CrossRef]
10. Walker, B.; Salt, D. *Resilience Thinking, Sustaining Ecosystems and People in a Changing World*; Island Press: Washington, DC, USA, 2006. Available online: https://faculty.washington.edu/stevehar/Resilience%20thinking.pdf (accessed on 11 March 2024).
11. Neal, Z.; Rozenblat, C. *Handbook of Cities and Networks*; Edward Elgar Publishing: Cheltenham, UK, 2021; ISBN 978. [CrossRef]
12. Al-Kasibi, H.A. A new concept of adaptive complexity. *Period. Eng. Nat. Sci.* **2020**, *8*, 1687–1700. [CrossRef]
13. Salingaros, N. Complexity in Architecture and Design. *Oz* **2014**, *36*, 4. [CrossRef]
14. Boeing, G. OSMnx: New methods for acquiring, constructing, analyzing, and visualizing complex street networks. *Comput. Environ. Urban Syst.* **2017**, *65*, 126–139. [CrossRef]
15. Jiang, B. New Paradigm in Mapping: A Critique on Cartography and GIS. *Abstr. Int. Cartogr. Assoc.* **2019**, *1*, 150. [CrossRef]
16. Al-Guesbi, H.A. Well-Organized Complexities: A New Theory of Life. *IOP Conf. Ser. Mater. Sci. Eng.* **2021**, *1090*, 012134. [CrossRef]
17. Ali, S.H.; Sherzad, M.F.; Kadhim, A.H. Managing Strategies to Revitalize Urban Cultural Heritage after Wars: The Center of the Old City of Mosul as a Case Study. *Buildings* **2022**, *12*, 1298. [CrossRef]

buildings

MDPI

Article

Perceptual Evaluation of Street Quality in Underdeveloped Ethnic Areas: A Random Forest Method Combined with Human–Machine Confrontation Framework Provides Insights for Improved Urban Planning—A Case Study of Lhasa City

Chong Liu , Yang Yu * and Xian Yang

School of Architecture, Southwest Jiaotong University, Chengdu 611756, China; lncj371@my.swjtu.edu.cn (C.L.); 2020300362@my.swjtu.edu.cn (X.Y.)
* Correspondence: yuyang@swjtu.edu.cn

Abstract: The utilization of street view big data is increasingly being used to uncover visual characteristics and spatial perceptions of urban streets. However, there is a lack of studies that combine street view big data for perceptual evaluation in underdeveloped ethnic areas and better street quality. This study integrates deep learning methods to create a human–computer confrontational model for perception score, with a focus on the central city of Lhasa in Tibet. Pearson correlation analysis was conducted on six dimensions of perception data (beautiful, wealthy, safe, lively, boring and depressing) and visual elements. The streets in the top 20% for both visual elements and perceptual scores were identified to reveal areas with high visual element proportions and high perceptual scores. The spatial distribution characteristics and correlation between visual elements and street perceptions were thoroughly analyzed. The findings of this study reveal that the central city of Lhasa exhibited high percentages of visual elements in buildings (88.23%), vegetation (89.52%), and poles (3.14%). Out of the six perceptions examined, the highest scores were for boring (69.70) and depressing (67.76) perceptions, followed by beautiful (60.66) and wealthy (59.91) perceptions, with lively (56.68) and safe (50.64) perceptions receiving the lowest scores. Visual components like roads (−0.094), sidewalks (−0.031), fences (−0.036), terrain (−0.020), sky (−0.098), cars (−0.016), and poles (−0.075) were observed to have a significant deterring effect on the boring perception, while other visual elements showed a positive influence. This investigation seeks to provide valuable insights for the design and advancement of urban streets in marginalized ethnic localities, addressing a void in perception research of urban streets in such areas.

Keywords: deep learning; human–computer confrontational model; underdeveloped ethnic areas; street quality; perceptual evaluation

1. Introduction

The problem of unequal urbanization in developing countries has become increasingly apparent during the process of global urbanization [1]. This imbalance is primarily characterized by hindered progress in underdeveloped regions due to various development challenges [2]. From the perspective of development levels in developing nations, it is clear that underdeveloped regions generally exist in the inner periphery of these countries. These regions face numerous difficulties, including unfavorable migration balance, lower living standards, aging populations, limited job opportunities, lower education levels, and reduced regional investments [3]. The conventional model of urbanization encounters hindrances and drawbacks such as 'risky urbanization', 'pseudo-urbanization', 'urban diseases', and 'urban–rural imbalance' [3,4]. In the contemporary context, streets are recognized as crucial urban elements that play a significant role in driving economic growth and fostering local development. This is particularly evident in marginalized ethnic regions, such as Tibetan areas. Therefore, utilizing city streets as a driving force for urbanization presents a favorable option for the progress of underdeveloped ethnic regions [5–7]. In underdeveloped ethnic regions, such as Lhasa City streets, there is a significant presence of Tibetan culture and ambiance. Assessing the quality of streets in Lhasa is essential for the progression of urban areas in underdeveloped ethnic regions and improving the perception of streets in these localities.

Experts have recently conducted extensive investigations in underdeveloped ethnic regions that lack adequate resources. For instance, Zhou [8] employed the TOPSIS entropy weighting method and an integrated coordination model to evaluate the resilience of six cities and states in Sichuan and Tibet between 2015 and 2020. Their research focused on a range of factors, including the economy, society, environment, and infrastructure, with the aim of identifying strategies to enhance urban resilience. They accomplished this by utilizing correlation analysis and gray correlation analysis to identify the primary factors that influence resilience in underdeveloped regions. In another study, Zhang [9] directed their investigation towards the mountainous region located on the western border of Yunnan, which serves as a representative underprivileged area. They utilized a combination of the coupling coordination degree model, spatial gravity model, and panel Tobit model to analyze the temporal and spatial variations in the degree of coupling coordination between the tourism industry and the process of urbanization from 2010 to 2019. Additionally, they explored the factors that influence this coordination. Similarly, Yu [10] employed a GIS-based spatial network analysis method and population distribution data to assess the arrangement of primary healthcare facilities in the built-up area of Lhasa. By implementing the location-configuration model, they were able to optimize the arrangement of these facilities. Existing research on urban environments in underdeveloped ethnic areas primarily focuses on topics such as urban resilience, urbanization, and healthcare. However, there is a lack of research on street perception and the evaluation of urban street environment quality using deep learning methods.

Urban streets play a crucial role in connecting individuals within a city and facilitating daily social interactions among its residents. These streets comprise various elements, including roads, surrounding buildings, and roadside vegetation, all intricately intertwined [11]. The quality of urban streets significantly contributes to enhancing the livability of cities and positively affects the well-being of inhabitants [12]. Previous academic investigations on street quality have primarily focused on comprehending individuals' visual or sensory perceptions of the built environment, employing techniques such as surveys, face-to-face interviews, and onsite observations [13]. Surveys and interviews commonly serve to collect individuals' subjective evaluations of the constructed environment, often employing open-ended queries and numerical rating scales. Nonetheless, concerns have been raised regarding potential biases in participants' responses associated with these methods [14,15]. Personal audits serve as the primary approach to capturing human perceptions and emotional experiences related to urban design across various scales [12]. However, personal audits are characterized by high costs, time constraints, and practical limitations, thereby restricting urban researchers from conducting small-scale empirical studies [14,16].

In recent years, notable advancements have been witnessed in the realm of machine technology. These progressions can be credited to the emergence of mapping services that rely on contributions from the crowd and the wide availability of images with geographic tags. Notably, Baidu Street View images (BSVIs), Google Street View images, and Tencent Street View images are among the convenient services that provide a diverse array of visual data. These data represent the viewpoint of local inhabitants and offer significant insights into the conditions of streets. The influx of such data has made a substantial contribution to achieving a comprehensive understanding and visualization of urban landscapes [17]. The merging of machine technology and research on urban streets has captured the interest of urban planners and researchers. Consequently, there has been a surge in studies that employ advanced learning methods to investigate the perception of street quality and extract valuable information from street view images [18,19]. However, most studies in this area have focused on developed regions, leaving a gap in research on underdeveloped ethnic regions. These regions have unique characteristics and face more pronounced issues of unbalanced urban development. Therefore, utilizing deep learning to assess the quality of streets in underdeveloped ethnic regions is crucial for promoting balanced development in these areas.

The effectiveness of deep learning in the classification and mapping of urban streets has been well documented. This advanced technology allows for the automatic extraction of semantic data, including pedestrian counts, as well as the categorization of visual elements and representation of scene features in images [20,21]. By understanding how individuals perceive street space, deep learning can provide valuable insights for efficient and thorough urban planning. To demonstrate the capabilities of deep learning in this field, Zhang [22] introduced a data-driven approach that utilizes deep learning to predict human perception of street images. In a similar vein, Zhou [15] employed deep learning techniques to segment and extract physical characteristics from Baidu Maps Street View (BMSV) images in Shenzhen, China, with the goal of evaluating street visuals and promoting healthy urban development. In a groundbreaking study, Zhao [23] developed an innovative deep learning methodology that considers both the street view and subjective factors of human perception to estimate and analyze land value. Moreover, Rui [24] conducted a study in central Düsseldorf, utilizing Google Street View (GSV) and pedestrian views to extract pixel-level semantic information using the DeepLabV3+ segmentation technique. The objective of this research was to identify variations in quantitative measurements of perceptual quality in the street view from different perspectives, thereby facilitating a comprehensive analysis of perceived quality and walking potential to determine priorities for street renewal efforts.

Scientists have expressed a significant interest in investigating the way people perceive the overall quality of urban streets. In a recent study, the researchers [25] delved into this topic by employing both spatial syntax and deep learning techniques. Their focus was specifically on the Binjiang District, which is situated in Hangzhou City, Zhejiang Province, China. In order to assess the quality of the streets, the scientists utilized semantic image segmentation, making use of the SegNet code structure. To facilitate a comparison between the perceptions of humans and machine analysis, a random forest model was implemented. Additionally, the researchers employed a multivariate linear regression approach to understand the relationship between the spatial characteristics of a street and its constituent elements. This study presents a framework that utilizes deep learning, street view images, and iterative feedback mechanisms to assess street perception on an urban scale. The Cityscapes Dataset, considered the most up-to-date, accurate, and comprehensive street view dataset, was utilized for training, validation, and testing using the DeeplabV3+ semantic segmentation method for road images. Visual elements in 53,820 street images from 14,002 segmented sampling points in Lhasa's city center were categorized. This study employed a random forest model within a human–computer confrontational framework to score six dimensions of street perception (beautiful, wealthy, safe, lively, boring, and depressing). Streets with the highest percentage of visual elements and top street perception scores in each dimension were identified and overlaid in GIS. Correlation and spatial distribution analyses were conducted to explore the relationship between visual elements and street view perception in Lhasa's city center. This research sheds light on the spatial heterogeneity of street view perception in underdeveloped ethnic areas, offering valuable insights for urban planning departments in such regions. The findings are particularly relevant for the urban development of underdeveloped ethnic areas, addressing the existing gap in urban street perception studies within these regions.

2. Literature Review

2.1. A Study of City Development and Streets in Underdeveloped Ethnic Areas

At present, many scholars are actively exploring the urban development of underdeveloped ethnic areas. Among them, monitoring changes in land use in coastal regions is crucial for promoting sustainable development worldwide. However, previous studies have primarily focused on modifications in developed areas, creating a knowledge gap regarding economically undeveloped regions experiencing rapid progress. To bridge this gap, Liang [26] conducted a study in the coastal vicinity of Vietnam. Using the random forest model and the land use alteration index, they assessed spatial variations in land use changes and identified key factors. Their findings offer valuable insights into land use research in economically undeveloped regions undergoing rapid progress. In a separate investigation, Liang [27] examined land use alterations along Highway No. 4 in Cambodia. They analyzed land use data from the GlobeLand30 2020 database for the years 2000, 2010, and 2020. The study focused on the characteristics and drivers of these changes and proposed that leveraging tourism for urbanization is a beneficial strategy in underdeveloped ethnic areas. Expanding upon this research, Zhang [9] studied the mountainous region along the western border of Yunnan, which serves as an exemplary underdeveloped ethnic area. They utilized the coupling coordination degree model, spatial gravity model, and panel Tobit model to investigate the spatio-temporal characteristics of the coupling coordination degree between tourism and urbanization from 2010 to 2019 while identifying influential factors. The primary objective of this study is to investigate the relationship between tourism and urbanization, specifically focusing on understanding the interconnected evolutionary process that connects these two phenomena. Moreover, it examines the advancement of street vending in economically disadvantaged ethnic streets, an area that has attracted considerable academic attention. As described in the research results by Hossam [28], the presence of merchants from various ethnic backgrounds in places like Malmi significantly contributes to the liveliness of public streets and plays a crucial role in enhancing the dynamism of the streets. Additionally, informal settlements in the Greater Cairo region of Egypt exhibit a street arrangement that promotes social interaction among residents. By conducting a comprehensive analysis on an urban scale, it becomes possible to identify specific intersections within the street system of these settlements that may form potential connections with the broader transportation network, thereby influencing regional integration [29].

2.2. Geospatial Big Data and Street Environment Perception

In recent years, there has been a rapid increase in multi-source big geospatial data, especially the rise of multiple image datasets with geographic tags [30]. These images, when combined with location data, contain a plethora of visual information [31] that accurately portrays the visual aspects of daily life [32]. Since landscapes play a crucial role in shaping how city dwellers perceive their surroundings [33], these geotagged image collections offer fresh possibilities for comprehending urban perspectives on a large scale. Street View Imagery (SVI), providing comprehensive views of various streets, has emerged as a valuable asset for deducing urban perceptions. Street view photos are geotagged pictures gathered, processed, and managed by popular mapping services such as Google Maps and Tencent Maps through standard techniques. Captured using specialized gear at multiple locations and landmarks, primarily along city roads [34], these images showcase the physical attributes of urban environments [35,36].

Salesses [37] suggested using street view images to collect human opinions by comparing them in pairs to analyze the impact of urban environments on social and economic results. Expanding on this idea, Dubey [38] utilized online crowdsourcing and machine learning in computer vision to broaden the survey scope to major global cities, generating a detailed dataset of urban perceptions. This method overcomes the challenges of small sample sizes and limitations associated with traditional interview and questionnaire approaches. Prior studies have evaluated city perceptions using data from the MIT Place Pulse program (Place Pulse 1.0 and 2.0) [39–42].

2.3. Deep Learning and Street Quality Evaluation Research

Acquiring a comprehensive understanding of the perception of city streets by users is of immense importance to urban planners and managers. This understanding is crucial for optimizing the overall living environment and effectively attracting talented individuals, investors, and businesses [43]. The level of perceived quality plays a significant role in the emotional connection residents have with city streets, as indicated by research conducted by Low [44]. Specifically, place attachment refers to the emotional connection individuals establish with a particular street, resulting in increased comfort, security, and a stronger inclination to remain in that area [45]. As a result, in recent years, multiple fields such as geoinformatics, sociology, and urban and rural planning have actively contributed to establishing the relationship between street attributes and residents' perceptions [46].

After AlphaGo, developed by Google, defeated the Go world champion, there has been a remarkable increase in the fascination with deep learning. As a result, scientists have extensively utilized deep learning techniques to extract visual semantic features from images for their research purposes [47]. In the previous era, the extraction of semantic characteristics from images mainly depended on methods centered around object recognition. Nevertheless, these techniques faced certain constraints, including difficulties in distinguishing objects with similar colors or detecting elusive elements within the image [48].

Deep convolutional neural networks have been widely used in recent times for processing visual information found in images. The advancements in this field have been driven by the implementation of deep learning algorithms like FCN, ResNet, and SegNet. These algorithms have shown great effectiveness in identifying various visual elements in images, including roads, buildings, the sky, pathways, trees, and vegetation. This has significantly contributed to enhancing the reliability of research involving urban roads and human perception [49–51]. In recent research conducted by Ma [17], they focused on extracting semantic information at the pixel level and organizing visual components using the SegNet approach for image segmentation. They utilized a dataset consisting of one million panoramic street view images from Shenzhen, China. Additionally, they investigated the spatial characteristics of five perceptual elements and studied how perceptual outcomes differ across different functional streets. This allowed them to determine the overall impact of Urban Renewal Projects (URPs) on street landscape transformations. Donghwan [52] conducted a study that emphasized the computation of the Green View Index (GVI) using FCN semantic segmentation. They specifically focused on pedestrians' viewpoints using Google Street View (GSV) and deep learning methodologies. Another research team led by Wang [53] investigated the spatial quality of streets in Xiamen City. To assess the correlation between greenery variables and the Green View Index (GVI), the researchers' main aim was to utilize conventional correlation analysis and a multiple regression model. Additionally, they conducted a study to understand the relationship between walking duration and green space variables. Various resources, such as Street View Imagery (SVI), Points of Interest (POIs), and data collected from social media comments, were employed to support their analysis. In a separate investigation conducted by William [54], they employed images sourced from Google Street View (GSV) and utilized the SegNet technique for image segmentation. The primary focus of their research was to provide valuable insights for enhancing urban streets. To achieve this objective, they developed a concatenated convolutional neural network.

The field of machine science has increasingly focused on street perception modeling, capitalizing on the well-known DeeplabV3+ learning framework, renowned for its ability to extract extensive and precise information about street scenes. Yu [55] proposed a lightweight technique that applies the DeepLabV3+ with complex-valued computation to conduct semantic segmentation on Polarimetric Synthetic Aperture Radar (PolSAR) images. To tackle the issue of overfitting and improve segmentation accuracy in the analysis of polarized SAR data with limited samples, a specific technique was developed. In order to conduct a comprehensive exploration of the variables influencing the perceived excellence of street views and establish priorities for the revitalization of streets, Rui [24] amassed a total of 5300 images of pedestrian landscapes in the downtown vicinity of Düsseldorf. These images were obtained through a combination of Google Street View

(GSV) and self-captured photographs. To attain comprehensive insights, the advanced segmentation technique known as DeepLabV3+ was utilized to extract highly detailed semantic information at the pixel level. The primary aim of this inquiry centered on assessing diverse perspectives regarding perceived street quality and pedestrian potential, ultimately facilitating an exhaustive assessment that informs decision-making in street refurbishment projects.

3. Data and Method

3.1. Study Areas

Lhasa, situated in southwest China on the Tibetan Plateau, serves as the capital of the Tibet Autonomous Region (TAR). Nestled in the valley plain of the Lhasa River, a tributary of the Yarlung Zangbo River, it stands as the political, economic, cultural, and educational hub of Tibet. As a prime example of a city in an underdeveloped ethnic region, Lhasa is the focus of this study (Figure 1).

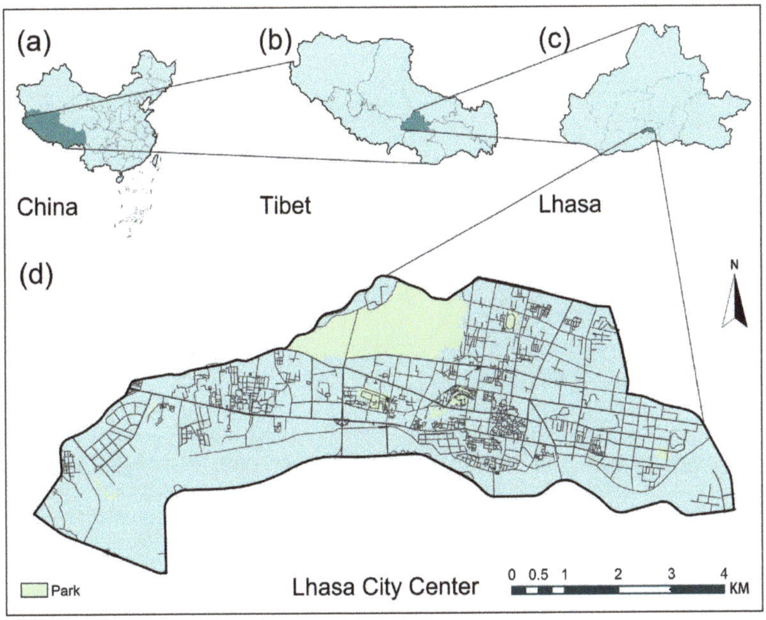

Figure 1. Overview of the study area for the present study: (**a**) location of Tibet in China, (**b**) location of Lhasa within Tibet, (**c**) the central areas of Lhasa City, and (**d**) the road network in the central areas of Lhasa City.

Located in Lhasa City at an elevation of 3650 m above sea level and spanning 29,640,000 square kilometers, the city had a population of 581,200 as of the conclusion of 2022. With a diverse population of 31 ethnic groups, including Tibetans, Han Chinese, and Hui, recent efforts by the Lhasa Municipal Government have focused on improving urban infrastructure, environmental quality, cityscapes, and resident well-being and safety. Streets are vital components of urban environments, influencing human spatial perception greatly [56]. Assessing the street quality in the central city is crucial for understanding visual transformations linked to street enhancements. The wealth of Baidu Street View images (BSVI) data in Lhasa City increases the precision of research outcomes, making it an ideal setting for the study detailed in this paper.

3.2. Research Framework

The analytical framework and methodology used in this paper are shown in Figure 2.

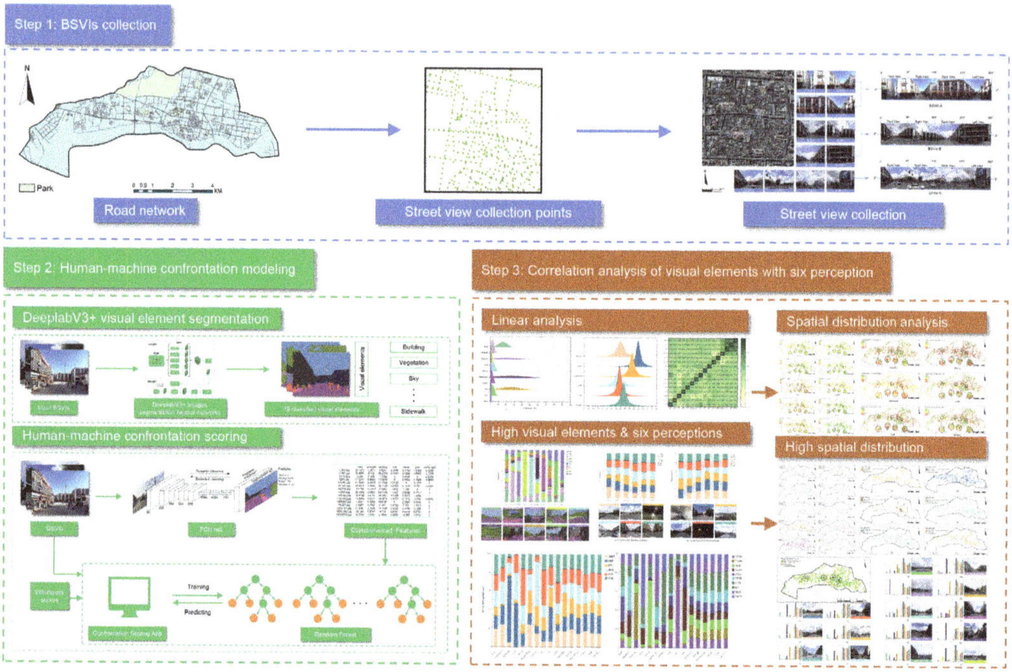

Figure 2. Research framework for this study.

In the first stage, in order to initiate the project, essential data about the research area were gathered using Open Street Map (OSM). Following this, the collected data were fed into ArcGIS to create data points for street view, maintaining a uniform gap of 50 m between each point. Subsequently, the Application Programming Interface (API) of Baidu Street View mapping technology was employed to retrieve Baidu Street View images (BSVIs). Consequently, tailored software was developed to establish a connection with the Baidu Street Map Application Programming Interface (API) and obtain the BSVIs.

In the second stage, our study utilized a Python Torch framework to develop a neural network model for segmenting BSVI data with full-element representation in the second phase. This particular model was trained on the BSVI dataset to establish a confrontational human–computer model for scoring based on perception. Individuals with expertise in the historical and cultural context of the research area were recruited to evaluate the perception of city streets across six dimensions. Utilizing the confrontational model, predictions regarding perception scores were made for the remaining images, aligning with the assessors' scores. Calculation of final spatial perception scores across all six dimensions was achieved through a process of weighting and averaging individual assessors' scores, considering variations from model predictions. To examine the relationships between the perceptual data dimensions and the top visual elements, a Pearson correlation analysis was performed. Following this, the top 20% of images with the highest score across each dimension and the top ten visual elements were selected for a detailed analysis of street view elements.

In the third stage, the following research introduces an innovative method for evaluating the perception of streets in underdeveloped ethnic regions by combining the spatial perception of urban streets with a confrontational analysis framework involving human–computer interaction. Through the integration of visually striking streets with those exhibiting high scores on six spatial perception metrics, this study seeks to discern spatial connections. The discovery of dimensions associated with the spatial layout and quality of streets in Lhasa's central city, known for its significant visual component, may provide valuable guidance for urban street design and development in underdeveloped ethnic regions.

3.3. BSVI Data Collection

The analysis of city environments and urban streets using street view images has gained significant attention in the realm of urban science [36]. This visual representation primarily emphasizes the human encounters within urban surroundings, enabling a distinctive viewpoint [57]. Numerous platforms dedicated to Street View Imagery provide accessible data and services to delve into this particular kind of visual content.

In order to comprehensively evaluate how residents perceive the central streets, we implemented a data collection system for streets. This system allowed us to gather data every 50 m during the investigation. To generate a comprehensive visualization of the city's perception of the streets, we utilized a Python script to merge the collected images from all four cardinal directions. Specifically focusing on the central district, we obtained a total of 14,002 data collection points from the street network provided by Open Street Map (OSM). By utilizing Python programming and the Baidu Street Map API, we successfully acquired 53,820 images of the street view for subsequent analysis.

To ensure the authentic representation of residents' perspectives, we formulated specific criteria for the Baidu Street View images (BSVIs) derived from the Application Programming Interface (API). These criteria comprise the subsequent elements: the longitude coordinates, which have been converted to the Baidu Street View map BD09; a fixed vertical angle (pitch) of 0°; and a field of view (fov) width of 90°. In the case of each sampling point, we amalgamate perspectives from four distinctive directions to obtain a comprehensive 360° panoramic depiction of the street. It is noteworthy to mention that the image is confined to a maximum resolution of 400 × 300 pixels. Figure 3 illustrates the procured street view image. To assure the dependability of the overall evaluation of street quality, we implemented a filtration procedure to eliminate any accessible street view images that were flagged as invalid.

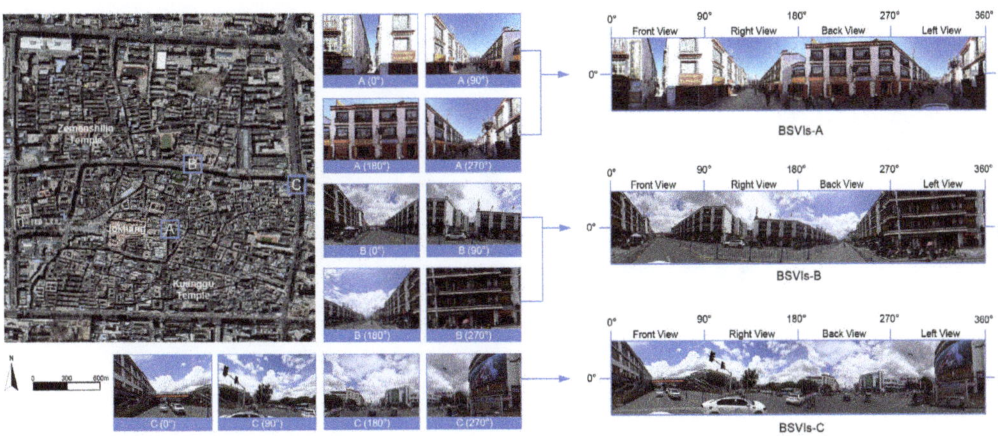

Figure 3. Example of the BSVI collection process in the central areas of Lhasa City.

3.4. Deep Learning-Based Semantic Segmentation and Visual Element Classification for Street View Images

The utilization of the street scene perception model with the DeeplabV3+ learning framework enables the extraction of more comprehensive and accurate street scene information. In this study, a DeeplabV3+-based semantic segmentation method for road images was implemented using the experimental environment of Windows 11. The training process utilized an NVIDIA GeForce RTX 2080Ti GPU (NVIDIA Corporation, Santa Clara, CA, USA) with 11 GB of graphics memory, and Python Torch was employed as the learning framework. To ensure a robust foundation for the experiments, the Python Torch framework, known for its proficiency in deep learning, was selected. Regarding the dataset employed in this investigation, the Cityscapes Dataset, developed by Mercedes-Benz, was utilized. This dataset serves as a standard to assess the performance of vision algorithms in scene semantic comprehension. It consists of street views from 50 different cities, each with distinctive backgrounds, seasons, and contextual elements, which allows for comprehensive analysis. To effectively categorize various elements, the dataset utilized 19 semantic labels, ranging from 0 to 18. These labels aid in the classification of entities such as roads, buildings, street trees, sky, people, and cars, as illustrated in Table 1 and Figure 4. It is important to note that the dataset was divided into three subsets for accurate evaluation: training, validation, and testing. The training subset consisted of 2975 images, while the validation subset comprised 500 images, and the testing subset encompassed 1525 images. The model xception71_dpc_cityscapes_trainfine, trained by DeepLab-V3+ on the Cityscapes Dataset, was selected for street map recognition in this study due to its algorithm and dataset compatibility.

Table 1. Cityscapes Dataset semantic labeling.

Group	Classes
Flat	Road (0), sidewalk (1)
Buildings	Building (2), wall (3), fence (4)
Object	Pole (5), pole group, traffic light (6), traffic sign (7)
Trees	Vegetation (8), terrain (9)
Sky	Sky (10)
Human	Person (11), rider (12)
Vehicle	Car (13), trunk (14), bus (15), train (16), motorcycle (17), bicycle (18)

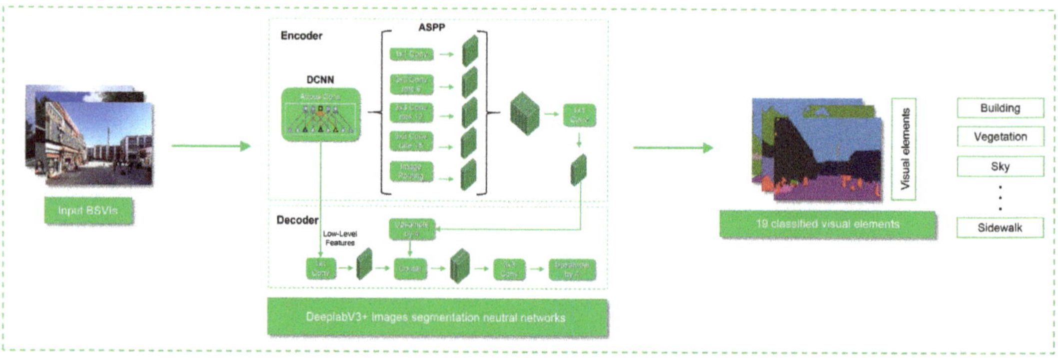

Figure 4. Street view segmentation in central Lhasa City based on DeepLabV3+.

In our scientific investigation, we employed the Cityscapes Dataset for our research investigation. Initially, the training images and labeled files were provided to us in PNG format for the purpose of semantic segmentation. However, to enhance clarity and streamline the training procedure, we purposely decided to convert the format of the training images to JPG. Despite this alteration in format, the critical correspondence between the information conveyed in the training images and the labeled files remains unaltered. Taking into account the specific requirements of the algorithmic criteria and dataset under scrutiny, we intentionally selected the xception71_dpc_cityscapes_trainfine model, which underwent training utilizing DeepLabV3+ on the Cityscapes Dataset, as our designated model for recognizing street maps.

3.5. Scoring Street Perception Using a Human–Machine Confrontational Scoring Framework

The innate ability of the human brain to recognize scenes gives it a natural advantage in identifying global attributes within images. This advantage serves as the basis for our strategy in the 'human–machine confrontation' paradigm, where we leverage this innate ability to assist in human evaluations of street scene perception. The program development and deployment involved predicting perceptions using the human–computer confrontational model algorithm proposed by Yao [58] and deploying it on Tencent cloud servers to collect perceptual data. Training urban environment perception using the MIT Place Pulse dataset [35] aimed to uncover the visual elements that influence how a place is perceived in terms of safety, vibrancy, despondency, and others. The dataset categorizes perceptions into six classes: safe, lively, wealthy, beautiful, boring, and depressing perceptions. Studies have explored the relationship between these perception categories and urban residents' well-being [59], urban security [22], and the recognition of urban film locations [60]. These studies have shown that these six indicators effectively capture human perception irrespective of cultural backgrounds, income levels, or racial differences among residents [22].

The approach of human–computer confrontational scoring integrates a random forest component to examine the visual attributes of images of urban environments in relation to user assessments. This well-suited model is automatically created once the user rates 50 images, offering suggested ratings for future assessments. The mechanism adapts these recommendations based on user actions. In instances where the suggested ratings notably deviate from user ratings by more than 10 points for over 5 images, the random forest component is retrained to adjust the model settings. The user evaluation procedure concludes when the out-of-bag validation error of the model is under 10 points, leading to the generation of the human–machine confrontational scoring dataset (Figure 5).

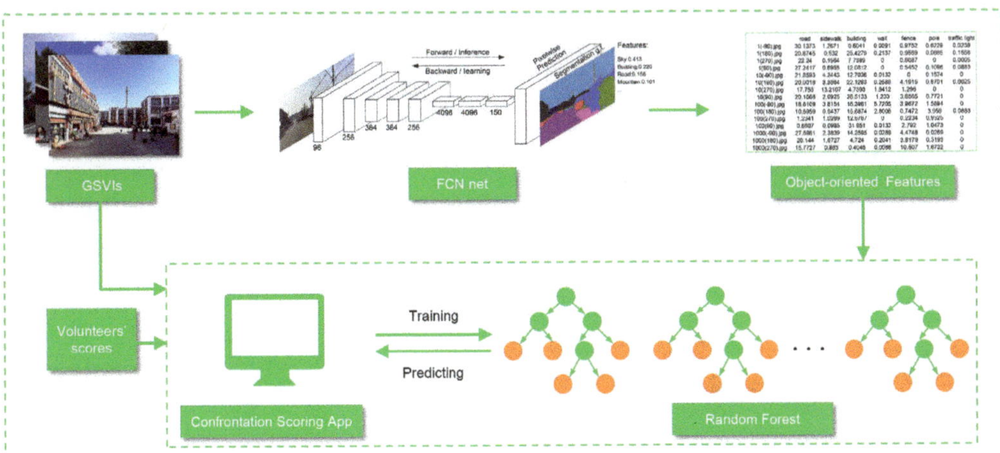

Figure 5. Construction of human–machine confrontation model for street view scoring in central Lhasa.

In the process of training with random forest, the dataset is split randomly into two sets: the training dataset (in-bag) and the testing dataset (out-of-bag). The out-of-bag dataset is used specifically to evaluate the accuracy of the model at each iteration while training. The average validation error from the out-of-bag dataset is a key metric for assessing the effectiveness of the random forest classification model in achieving the highest accuracy possible. Studies have indicated that using out-of-bag estimation yields better results compared to cross-validation [61].

This research employs various statistical metrics, including the Pearson correlation coefficient (Pearson R), the standardized R-squared (R^2), Root Mean Square Error (RMSE), and Mean Absolute Error (MAE), to assess the precision of fitting the human–robot confrontational score to random forests and the RF fitting of city perceptions relative to poi-based city functionalities. The mathematical formulas for Pearson R, standardized R^2, RMSE, and MAE are delineated in Equations (1) through (4), respectively.

$$Pearson\ R = \frac{\sum_{i=1}^{n}(y_i - \overline{y_i})(\hat{y}_i - \overline{\hat{y}_i})}{\sqrt{\sum_{i=1}^{n}(y_i - \overline{y_i})^2}\sqrt{\sum_{i=1}^{n}(\hat{y}_i - \overline{\hat{y}_i})^2}} \quad (1)$$

$$R^2 = 1 - \frac{\sum_{i=1}^{n}(y_i - \hat{y}_i)^2}{\sum_{i=1}^{n}(y_i - \overline{y})^2} \quad (2)$$

$$RMSE = \sqrt{\frac{\sum_{i=1}^{n}(y_i - \hat{y}_i)^2}{n}} \quad (3)$$

$$MAE = \frac{1}{n}\sum_{i=1}^{n}|y_i - \hat{y}_i| \quad (4)$$

where y_i is the base truth value, $\overline{y_i}$ is equal to $\frac{1}{n}\sum_{i=1}^{n} y_i$, and $\hat{y}_i$ is the prediction result of the fitted model.

The research methodology employed in this investigation is grounded on the volunteer scoring approach introduced by Wang and colleagues [25], which encompasses six different perception categories for evaluating the quality of urban streets in underdeveloped ethnic regions, specifically concentrating on the central district of Lhasa. To guarantee a thorough assessment of street quality perception, a group of 50 volunteers from various backgrounds including urban planners, architects, sociologists, tourists, college students, and community members well versed in the historical and cultural milieu of the region were recruited. The demographic characteristics of the participants are outlined in Table 2, indicating an average age of 34.60 years, with a majority being male (56.67%). In terms of educational attainment, 26.67% had primary school education or lower, 46.66% had secondary school education, and 26.67% had tertiary education or beyond. The ethnic makeup was predominantly Tibetan (76.44%), with local residents constituting 86.67% of the sample.

Table 2. Statistical table of volunteer information.

Variables	Proportion/Mean (SD)
Gender (%)	
Male	56.67
Female	43.33
Age	34.60 (32.11)
Education (%)	
Primary school or below	26.67
College and above	46.66
High school	26.67
Nation (%)	
Tibetan	76.44
Chinese Han	23.56
Residents (%)	
Local resident	86.67
Non-local resident	13.33

A schematic representation depicting the assessment of urban streets via the human–robot confrontational framework is illustrated in Figure 5. Initially, volunteers evaluated six criteria: aesthetics, monotony, somberness, vibrancy, wealth, and security. Using the human–computer confrontational framework, the perceived ratings for these criteria were scored on a scale from 1 to 50 to assess the streets' perceived quality, where a higher score indicated a closer alignment with the specific criterion being evaluated. To aid the scoring process, each volunteer was responsible for subjectively rating the first 2000 street views for each criterion. Following this, the model generated a random forest dataset and offered score recommendations for each criterion based on the relationship between the volunteer's prior ratings and the visual elements detected in the street views.

4. Results

4.1. Street Quality Analysis Based on BSVIs

By employing cutting-edge DeepLabV3+ technology for image segmentation, this research investigates the spatial arrangement of the top ten visual components found in streets located in the central area of Lhasa (Figures 6 and 7). The central region of Lhasa showcases a greater overall proportion of visual components associated with buildings and vegetation, with buildings constituting 88.23% and vegetation at 89.52%. In contrast, poles make up the tiniest portion at 3.14%. Roads (31.10%), sky (56.27%), terrain (16.27%), and walls (19.49%) display comparable spatial traits in the dispersion of visual components, primarily clustered along the primary thoroughfares. Figure 8 emphasizes the highest percentage of each visual element among the top ten rated streets within Lhasa's central district. Visual elements like walls, fences, cars, terrain, and poles exhibit a higher ratio of the maximum value, whereas buildings and vegetation closely trail behind. Conversely, roads and sky display a lesser percentage of the maximum value in comparison to the other visual components.

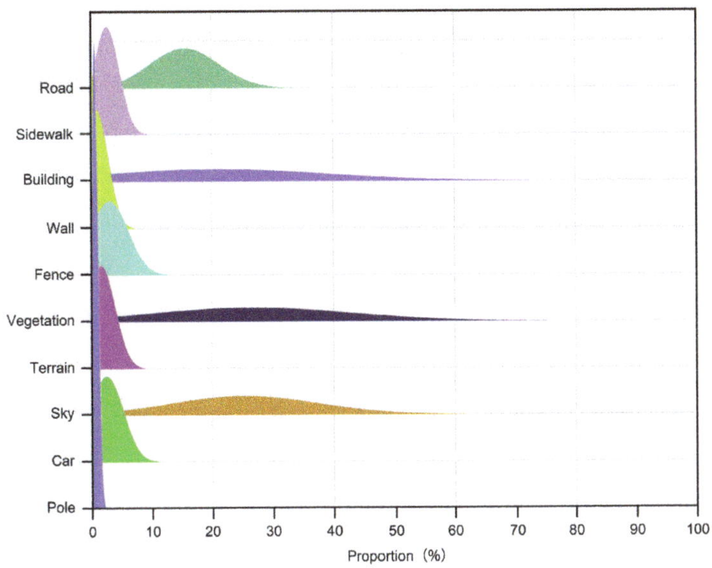

Figure 6. The proportion of the top ten visual elements in Lhasa's central urban area. The colors in the figure, from top to bottom, represent road, sidewalk, building, wall, fence, vegetation, terrain, sky, car and pole.

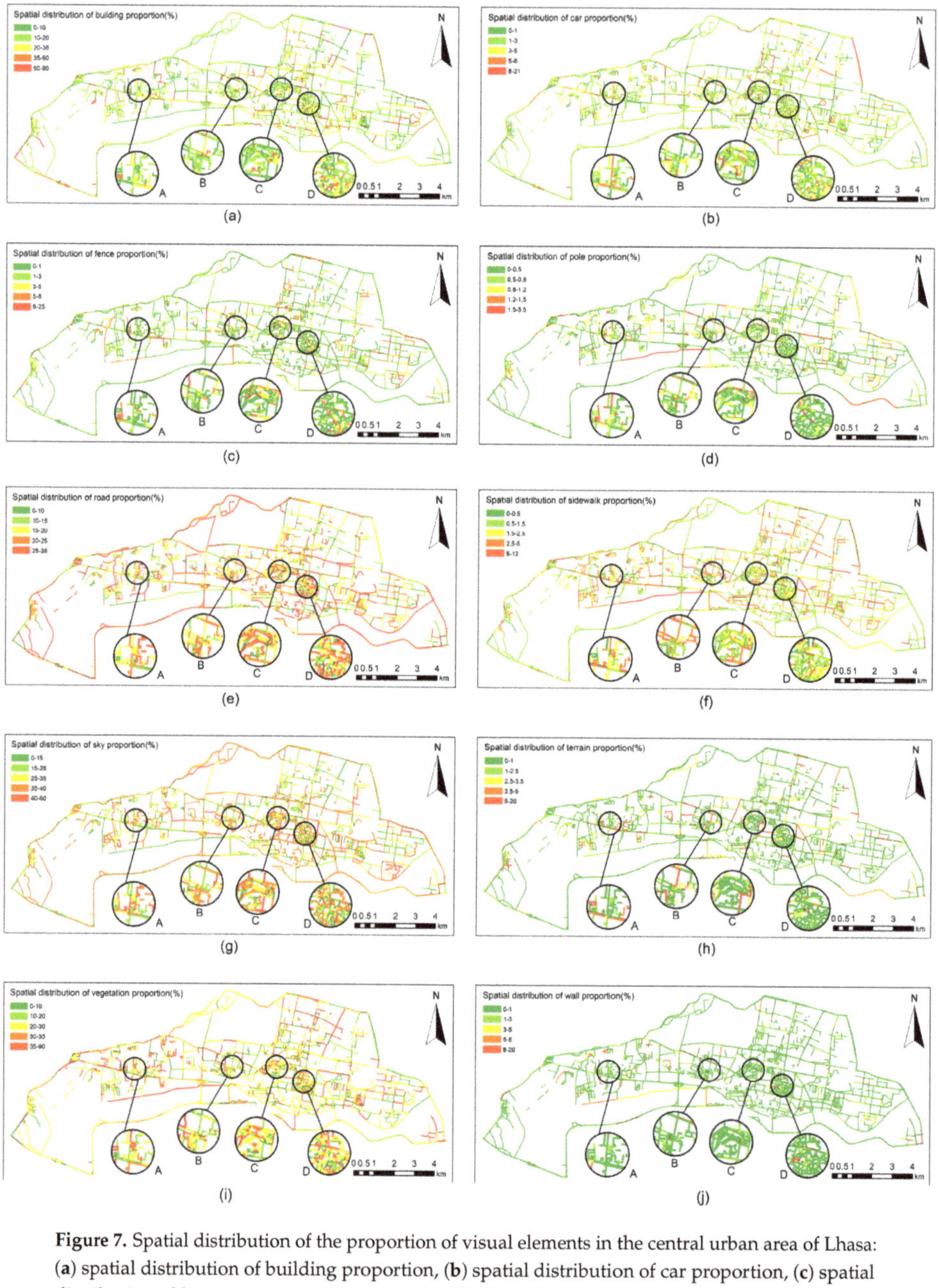

Figure 7. Spatial distribution of the proportion of visual elements in the central urban area of Lhasa: (**a**) spatial distribution of building proportion, (**b**) spatial distribution of car proportion, (**c**) spatial distribution of fence proportion, (**d**) spatial distribution of pole proportion, (**e**) spatial distribution of road proportion, (**f**) spatial distribution of sidewalk proportion, (**g**) spatial distribution of sky proportion, (**h**) spatial distribution of terrain proportion, (**i**) spatial distribution of vegetation proportion, (**j**) spatial distribution of wall proportion. In the selected area, A represents new residential developments and industrial zones, B represents the central business area, C represents the Potala Palace Scenic Area, and D represents the historic city center, showcases Dazhao Monastery and Barkhor Street.

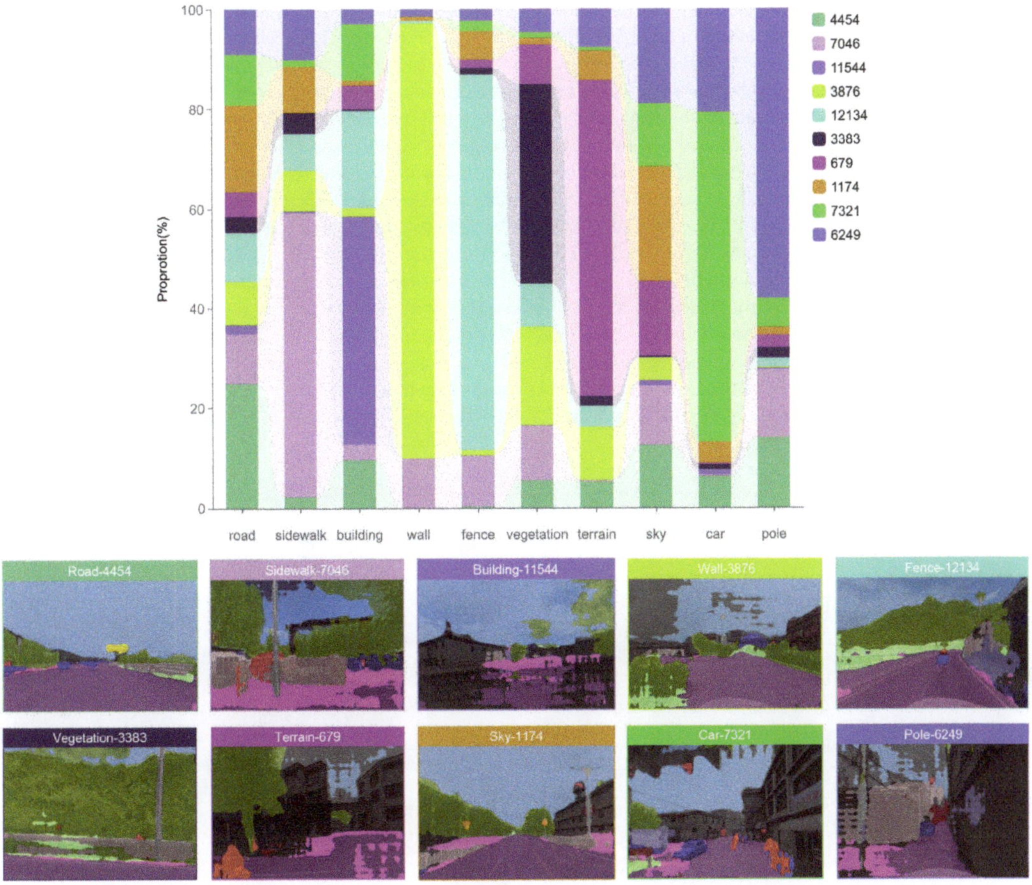

Figure 8. The maximum proportion of each visual element in the top ten urban areas in central Lhasa.

To gain a deeper insight into the spatial distribution characteristics of street visual elements in the central area of Lhasa, this study evaluates four distinct zones. The visual elements of building and vegetation incrementally increase from Zone A to Zone D. Zone A, located near the city boundaries, comprises primarily new residential developments and industrial zones. Zone B serves as the central business area, featuring traditional commercial streets and shopping centers. Zone C is characterized by the presence of the Potala Palace, a significant tourist attraction. Zone D, the historic city center, showcases Dazhao Monastery and Barkhor Street, which exhibit traditional Tibetan street features. Due to the emphasis on Dazhao Monastery and Barkhor Street, Zone D has the largest proportion of road and sky, reflecting Tibetan architectural designs. In contrast, Zone C has a reduced number of visual elements because city planners have restricted building heights to complement the Potala Palace. Zones A and B, being nearer to the suburban areas, are less densely populated with buildings and visual elements. Zones B and C exhibit more topographical variations, whereas Zone D is dominated by walls and roads.

4.2. Street Quality Analysis Based on Six Perceptions

This study examined the perceptual score from 50 volunteers across various dimensions, as illustrated in Figures 9–11, to assess perceptions of the Lhasa city center. The negative dimensions, 'boring' (69.70) and 'depressing' perceptions (67.76), received the highest score, whereas the positive dimensions, 'beautiful' (60.66) and 'wealthy' perceptions (59.91), followed closely. The dimensions of 'lively' (56.68) and 'safe' perceptions (50.64) were rated the lowest. The central streets of Lhasa City are predominantly characterized by tourist attractions and traditional Tibetan residences, inhabited by a mix of Tibetan locals, tourists, and commercial activities. The lively and safe perceptions were more favorable in the eastern sections and certain central streets, especially those adjacent to Dzongjiao Lukang Park. These streets are enveloped by iconic landmarks such as the Potala Palace and Dazhao Monastery, contributing to a vibrant atmosphere. Additionally, the scores for boring and depressing perceptions were significantly low in Lhasa's city center. High boring scores in Lhasa are predominantly observed along the city's principal roads, suggesting that the planning and construction of these pathways overlooked the incorporation of regional characteristics intended to create landscapes with distinct Tibetan features. Rather, there was an emphasis on a homogeneous and standardized approach, which culminated in increased boring perception. On the other hand, the lower score regarding depressing perception might be linked to the relatively older infrastructure of some primary roads in the city center and their expansive design, which does not significantly affect residents in a negative manner.

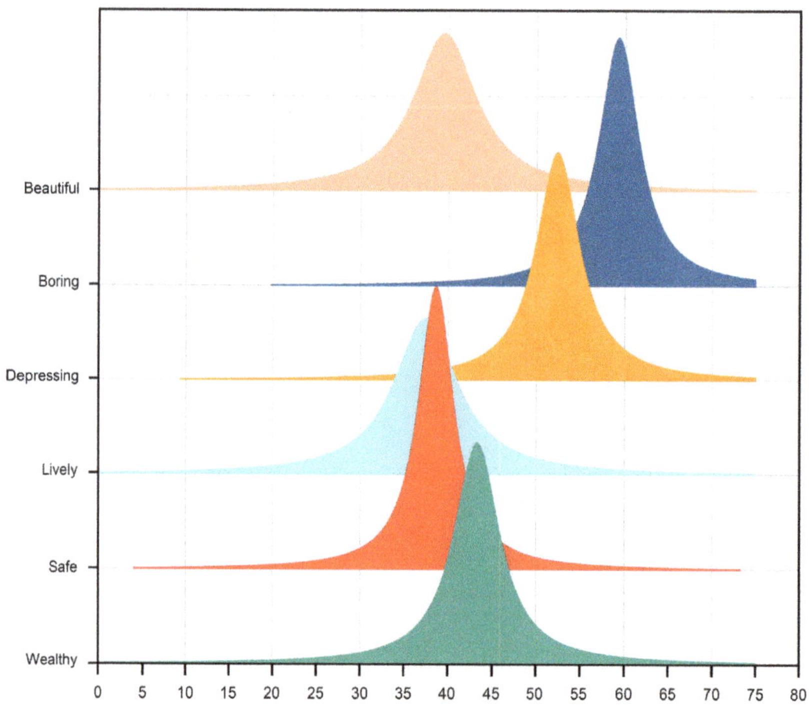

Figure 9. Six types of perception scores in Lhasa's central urban area.

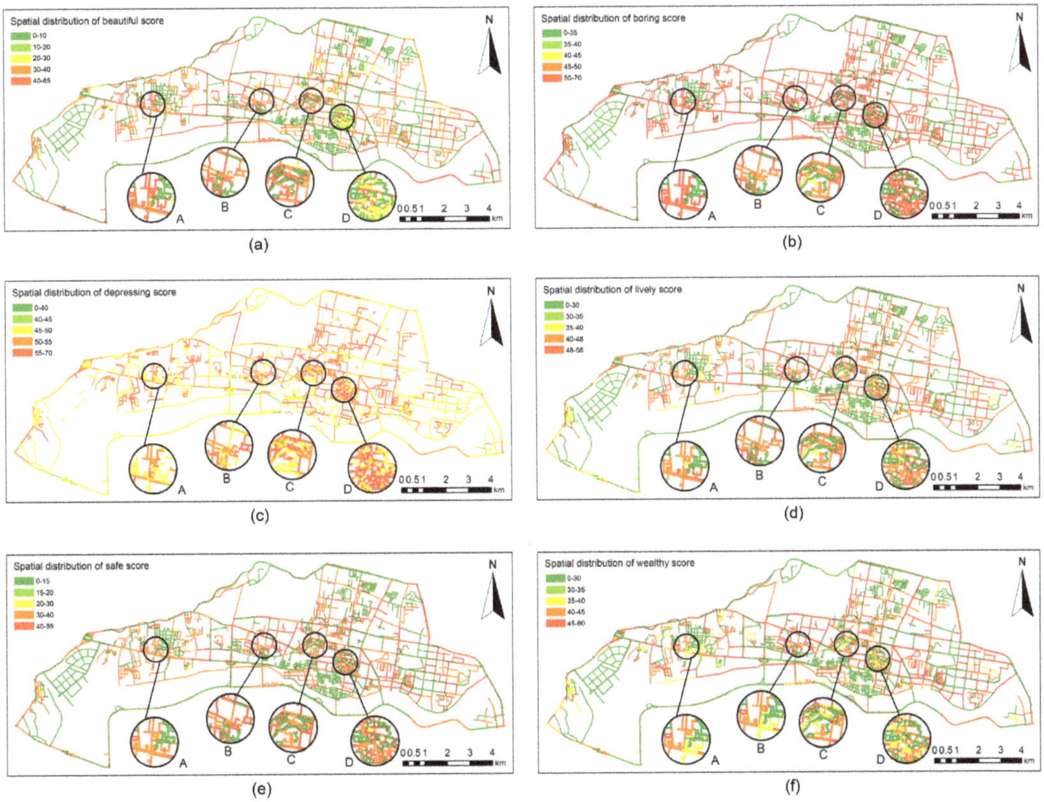

Figure 10. Spatial distribution of six perception scores in Lhasa's central urban area: (**a**) spatial distribution of beautiful perception, (**b**) spatial distribution of boring perception, (**c**) spatial distribution of depressing perception, (**d**) spatial distribution of lively perception, (**e**) spatial distribution of safe perception, (**f**) spatial distribution of wealthy perception. In the selected area, A represents new residential developments and industrial zones, B represents the central business area, C represents the Potala Palace Scenic Area, and D represents the historic city center, showcases Dazhao Monastery and Barkhor Street.

To analyze the spatial distribution of the six perceptions of streets in Lhasa's city center more clearly, this study thoroughly examines the four sub-districts with unique features. Zone A, located on the city outskirts with mainly residential neighborhoods, ranks the highest in terms of beauty and wealth. Zone B, serving as the city's commercial center, features a mix of traditional and modern streets. Zone C is home to the Potala Palace scenic area, with minimal surrounding streets to preserve the landscape. As a result, Zone B and Zone C scored lower compared to Zone A. Zone D, housing the Da Zhao Monastery and Barkhor Street, displays distinct Tibetan street characteristics but scored the lowest for the beautiful and wealthy perceptions due to narrow streets surrounding the old city center. In terms of lively and safe perceptions, Zone B scored the highest as it is situated in the commercial district, showcasing how a suitable scale of commercial streets can enhance street vitality and people's sense of security. On the other hand, Zone D scored highest for boring and depressing perceptions, attributed to its location in an old urban area with narrow, environmentally unfavorable streets. While Zones A, B, and C share similar spatial characteristics, the highest scores are primarily concentrated on the main city streets.

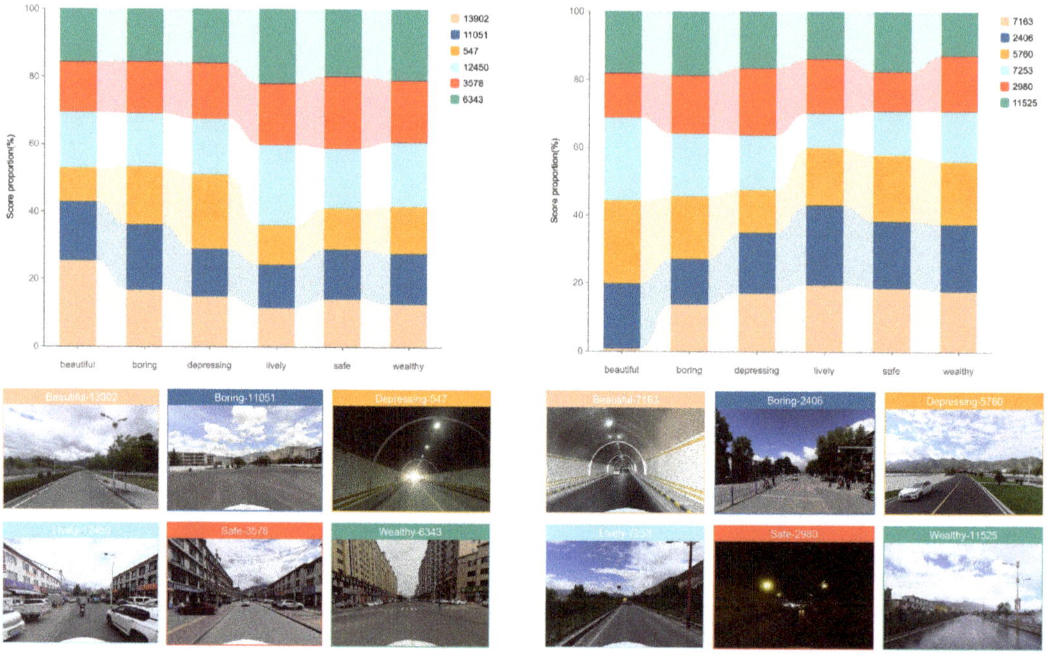

(a) (b)

Figure 11. Maximum and minimum value of each perception score in the top ten urban areas in central Lhasa: (**a**) six-dimensional perception maximum, (**b**) six-dimensional perception minimum.

4.3. Linear Regression Analysis of Visual Elements and Six Perceptions in Street View Images

In this analysis, multiple linear regression was employed to explore the relationship between the ten primary visual components found in urban landscapes and the six facets of perception within the central area of Lhasa. The presence of a positive correlation signifies a beneficial impact, while a negative correlation indicates a detrimental influence. The data depicted in Figure 12 demonstrate that building aspects exhibit a positive effect (0.030) on the aesthetic aspect but a negative impact (−0.030) on the diversity dimension. This contrast can be traced back to the presence of traditional Tibetan design in the aesthetic aspect, in contrast to the absence of Tibetan attributes in the modern residential areas, which leads to the negative effect in the diversity dimension, subsequently impacting the feelings of safety and liveliness. The subpar quality of features like walls, vegetation, terrain, and cars contributes to decreased perceptions of both safety and diversity among inhabitants, resulting in undesirable outcomes. Concerning adverse spatial reception, roads (−0.094), sidewalks (−0.031), fences (−0.036), terrain (−0.020), sky (−0.098), cars (−0.016), and poles (−0.075) significantly impact the feeling of monotony in a negative manner, while the other visual elements positively influence boredom. Conversely, roads (0.066), sidewalks (0.008), sky (0.078), cars (0.049), and poles (0.028) have a considerable positive impact on boring perception, whereas buildings (−0.009), walls (−0.072), fences (−0.002), vegetation (−0.067), and terrain (−0.018) exhibit a notable negative effect on depressing perception.

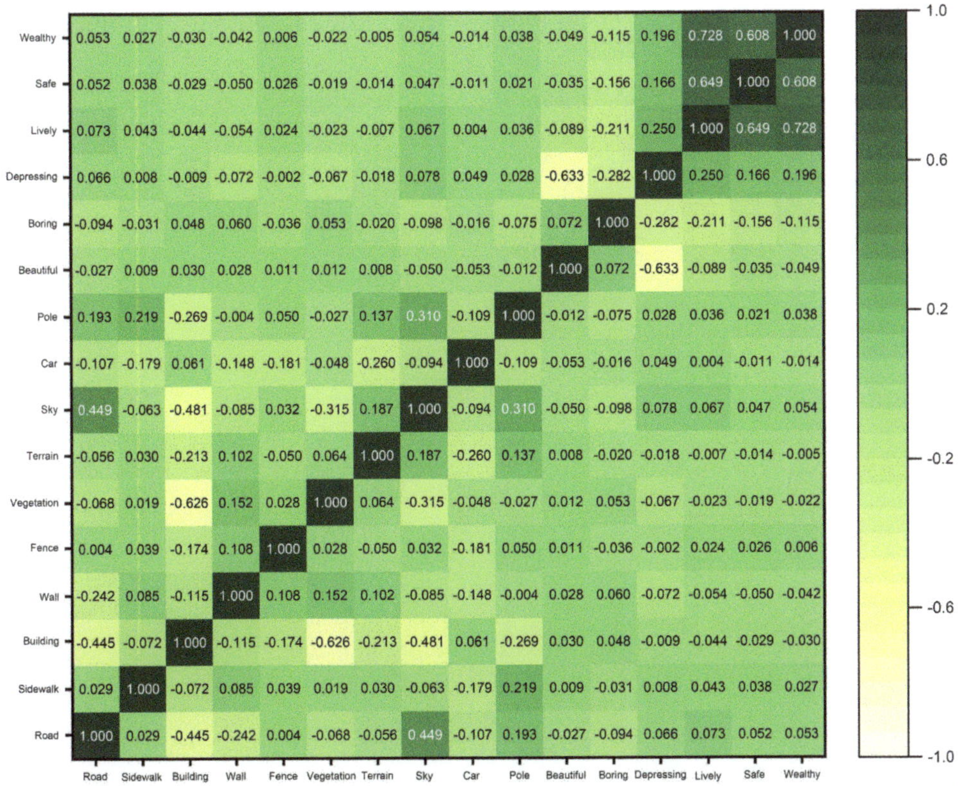

Figure 12. Multiple linear regression analysis results of visual element proportion and perception score.

4.4. Combination Analysis of Street Visual Elements and Six Perceptions

This research analyzes how individuals perceive street space in Lhasa's central city from different perspectives. The goal is to offer valuable insights to urban planners for designing future urban streets in underdeveloped ethnic regions. Streets with the greatest visual elements were categorized as high-visual-element-percentage streets, and those with the top perception score for each aspect were chosen as representatives. High-visual-element-percentage streets were then paired with high-scoring streets throughout all research areas. Areas that scored highly on the lively perception showcase landscape features that enhance the space's energy, promoting a feeling of liveliness and motion. Regions with a strong sense of opulence exhibit elaborate structures, street components, and high-end landscape attributes. On the other hand, regions with a heightened sense of beauty display stunning landscapes, while those with a diminished sense of beauty present dull landscapes and unvarying street designs.

Drawing from previous studies on the six perceptions of streets in the city center of Lhasa, this research revealed that the dimension of boredom received the highest score of 69.70. Figure 13 illustrates the top 20% of streets with high boring scores overlaid with the top 20% of shares of the top ten visual elements for further examination. The overlay highlights specific areas of interest in contrasting colors to denote streets with elevated perception scores and visual element shares. Notably, elements such as vegetation, sidewalks, and poles, which contribute to the boring perception, were primarily concentrated on main streets. Conversely, buildings, fences, and cars led to a higher boring perception score due to their prevalence in the old town and side streets, which generally have a less favorable urban environment. Terrain and walls displayed similar spatial distribution characteristics with a consistent scatter layout. In contrast, the negative perception of the sky was predominantly observed on secondary streets and small alleys, potentially influenced by the density and height of buildings in these areas.

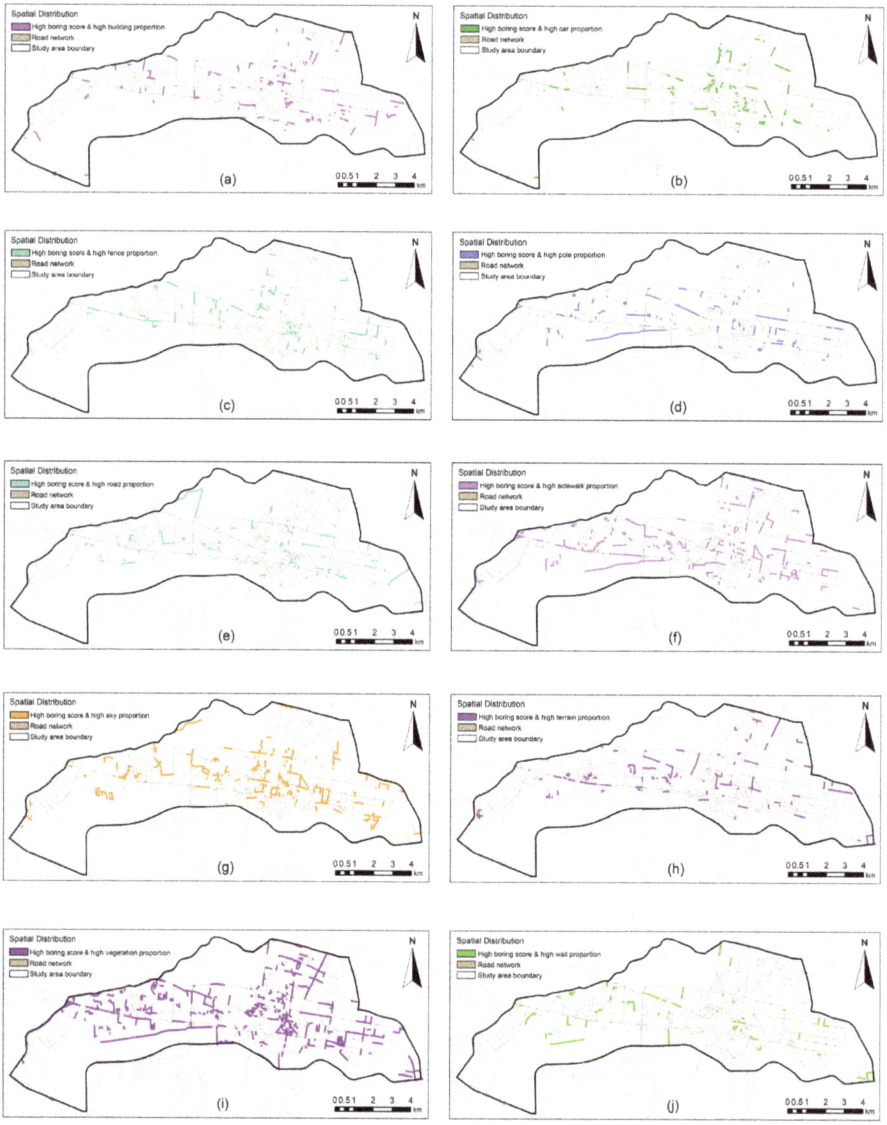

Figure 13. Spatial distribution of high boredom perception (top 20%) and high visual elements (top 20%): (**a**) high boring score and high building proportion, (**b**) high boring score and high car proportion, (**c**) high boring score and high fence proportion, (**d**) high boring score and high pole proportion, (**e**) high boring score and high road proportion, (**f**) high boring score and high sidewalk proportion, (**g**) high boring score and high sky proportion, (**h**) high boring score and high terrain proportion, (**i**) high boring score and high vegetation proportion, (**j**) high boring score and high wall proportion.

Figure 14 demonstrates the overlay of the highest 20% of vegetation scores with the top 20% scores for each of the six perceptions. The spatial distribution of vegetation displays similarities in beautiful, wealthy, safe, and lively perceptions, showcasing a consistent layout pattern. Some primary roads also display unique attributes. Nevertheless, concerning the boring perception, the dense concentration of vegetation is situated centrally within the historic city, along with other primary roads. This phenomenon can be ascribed to the compact buildings in the old city and the singular alignment of vegetation along primary roads. Additionally, the feeling of melancholy derived from vegetation is predominantly observed in the central historic town and the northern region, where streets are tightly packed and densely distributed. This indicates that a well-designed street layout and vegetation configuration can have a positive impact on the emotional perceptions of the street.

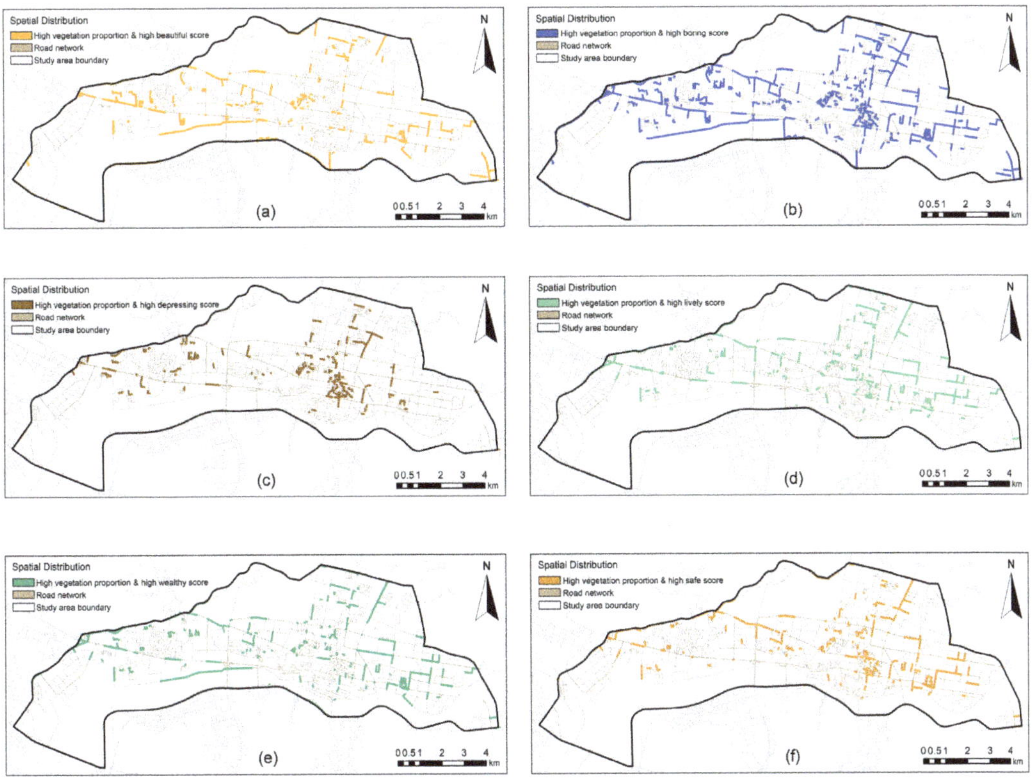

Figure 14. Spatial distribution of high vegetation percentage (top 20%) and high perception score (top 20%): (**a**) high vegetation proportion and high beautiful score, (**b**) high vegetation proportion and high boring score, (**c**) high vegetation proportion and high depressing score, (**d**) high vegetation proportion and high lively score, (**e**) high vegetation proportion and high wealthy score, (**f**) high vegetation proportion and high safe score.

Figures 15 and 16 show various street views comparing levels of boring perception and visual element presence. The results suggest that boring perception is heavily influenced by the overall street environment quality. For example, the perception of visual elements like cars (20.74%), sidewalks (11.18%), sky (57.02%), and fences (24.04%) is closely tied to their placement on major streets. Conversely, roads (30.50%), walls (19.38%), and poles (3.19%) show higher boredom levels in suburban settings. Vegetation (89.51%) is linked to the historic city center, while the terrain (16.29%) is more impacted by overpasses. In Figures 17 and 18, scenes with abundant greenery and respective perception scores are depicted. The findings reveal that while greenery is generally seen positively, it can also adversely affect street perception. Scores such as boring perception (59.91) are noticeable in tall building neighborhoods, lively perception (58.46) in urban commercial areas, beautiful perception (61.13) in city parks, safe perception (55.55) in older urban neighborhood side streets, and boring perception (69.68) and depressing perception (67.26) on main thoroughfares in the city center.

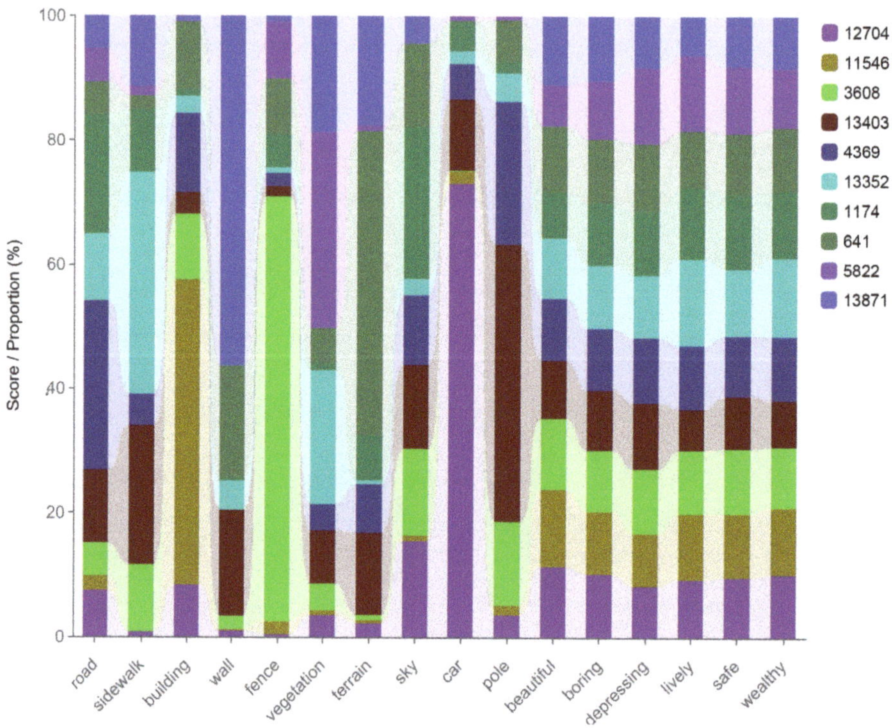

Figure 15. Representative street view relationship distribution of high boring perception (top 20%) and high visual elements (top 20%).

To enhance the analysis of street distribution, this research focuses on the top 20% of regions with high visual element share and perception scores for spatial overlay analysis. Detailed examination is conducted on four distinct regions (See Figure 19). Zone A, situated on the city outskirts, and Zone D display similar distribution patterns, with an equal spread of high visual element share and perception scores. Zone D is bordered by the Da Zhao Monastery and the traditional Tibetan commercial street (Barkhor Street), consisting of numerous alleys, urban villages, a blend of tourists and locals, and intricate traffic and pedestrian movement. In contrast, Zones B and C showcase a distribution highlighted by high perception scores. Zone B is a residential district in Lhasa with consistent building styles and a pleasant street ambiance. Zone C encompasses attractions like the Potala Palace and parks like Dzongkok Lukang Park, underlining the significance of humanistic landscapes in enriching the overall experience for visitors and residents.

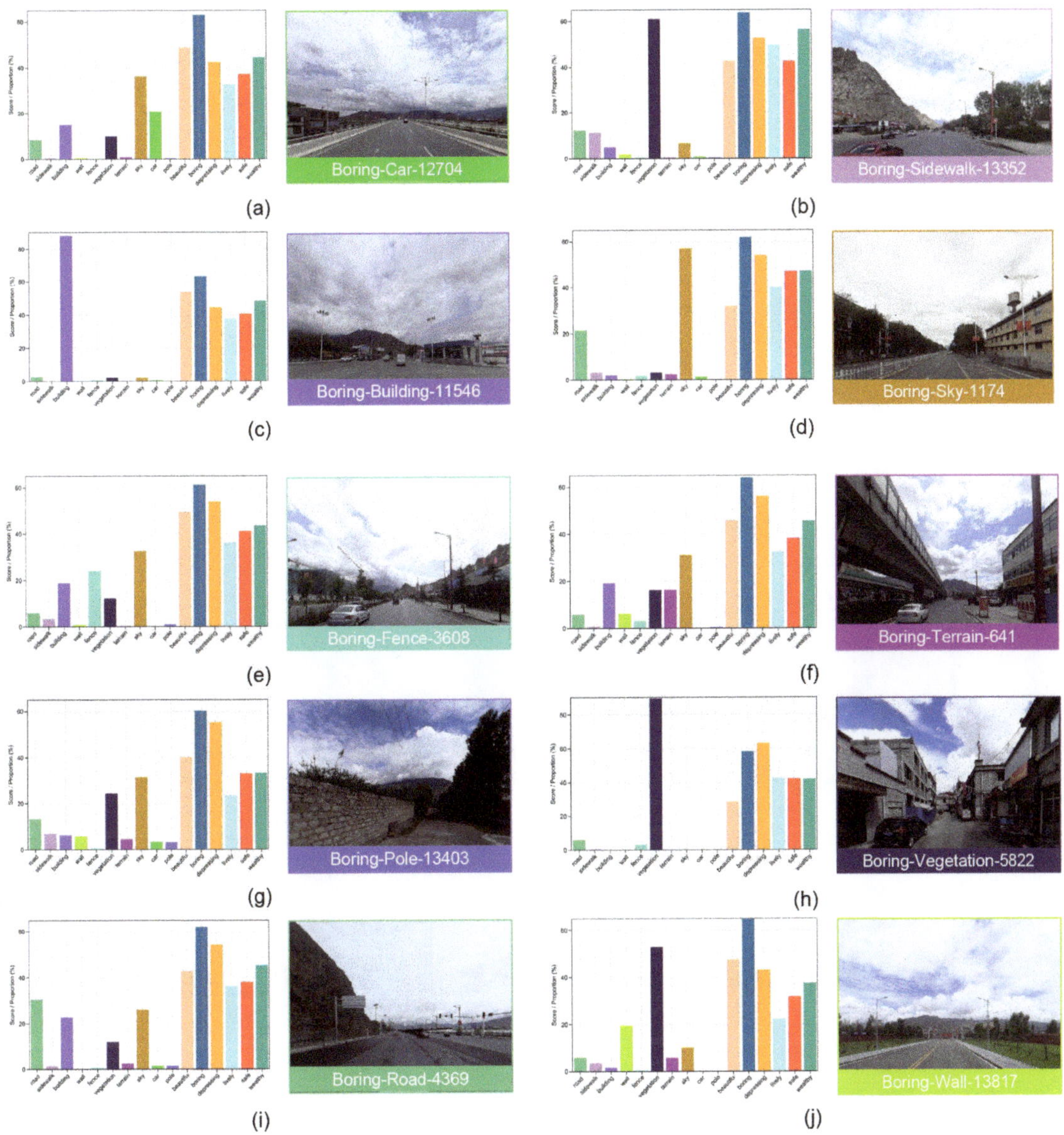

Figure 16. Representative street views with high boring perception (top 20%) and high visual elements (top 20%): (**a**) high boring score and high car proportion, (**b**) high boring score and high sidewalk proportion, (**c**) high boring score and high building proportion, (**d**) high boring score and high sky proportion, (**e**) high boring score and high fence proportion, (**f**) high boring score and high terrain proportion, (**g**) high boring score and high pole proportion, (**h**) high boring score and high vegetation proportion, (**i**) high boring score and high road proportion, (**j**) high boring score and high wall proportion.

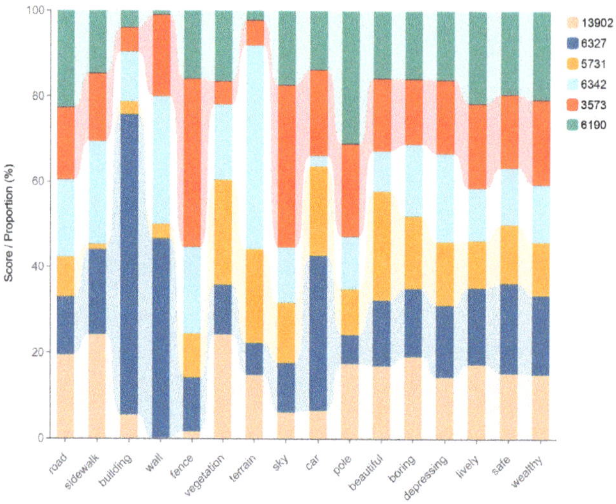

Figure 17. Representative street view relationship distribution between high vegetation proportion (top 20%) and high perception score (top 20%).

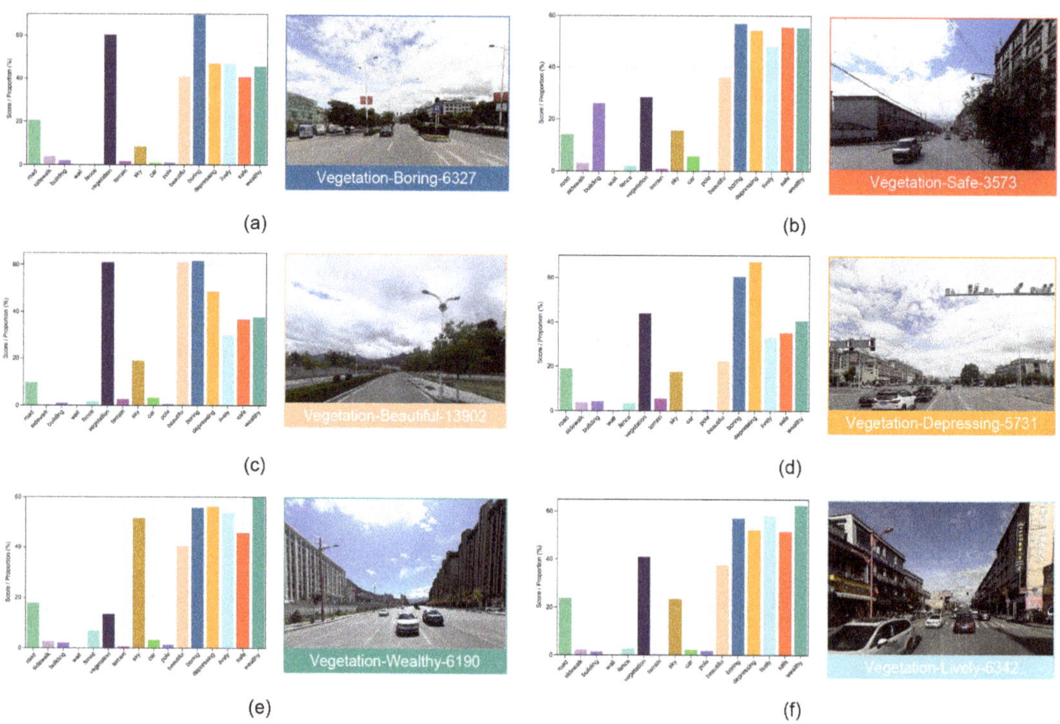

Figure 18. Representative street views with high vegetation proportion (top 20%) and high perception score (top 20%): (**a**) high vegetation proportion and high boring score, (**b**) high vegetation proportion and high safe score, (**c**) high vegetation proportion and high beautiful score, (**d**) high vegetation proportion and high depressing score, (**e**) high vegetation proportion and high wealthy score, (**f**) high vegetation proportion and high lively score.

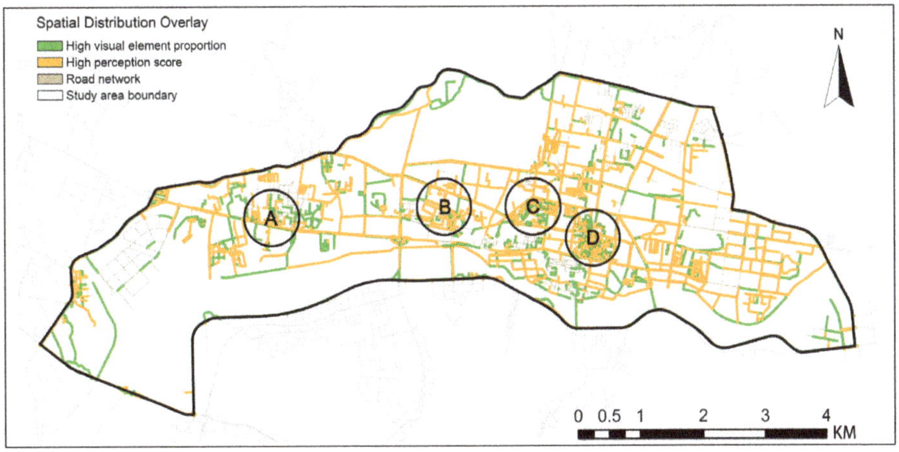

Figure 19. Spatial distribution of high visual element proportion and high perception score. In the selected area, A represents new residential developments and industrial zones, B represents the central business area, C represents the Potala Palace Scenic Area, and D represents the historic city center, showcases Dazhao Monastery and Barkhor Street.

5. Discussion

5.1. The Influence of Visual Elements on the Perception of Street Quality

This research aims to evaluate the perceptual assessment of streets within Lhasa's city center to identify the overlap between areas with a high share of visual elements and elevated spatial perception scores. The results provide vital insights for planning urban street renewals in underdeveloped ethnic regions. Various perceptual dimensions emerge within Lhasa's city center: high-rise neighborhoods are wealthy in perception, business districts exude vitality, city parks are perceived as beautiful, and secondary roads in the old city are felt to be safe, while city center main roads convey boring and depression perceptions. Elements like roads, sidewalks, fences, terrain, sky, cars, and poles significantly impact feelings of boring perception in the old city area and suburbs. Essential visual components such as roads, vegetation, buildings, sky, and sidewalks are critical for enhancing street quality in Lhasa's city center. Future initiatives should focus on these elements to improve street conditions. Moreover, the combination of buildings, vegetation, and roads largely contributes to the monotonous and dreary street atmosphere along Lhasa's central main roads. To improve existing street conditions, the Lhasa municipal government can diversify vegetation, incorporate unique Tibetan architectural features, and enhance the quality of roads, sidewalks, building height, and regional traits for future planning. These findings offer a valuable reference point for improving street quality in other underdeveloped ethnic areas.

5.2. Implications for Urban Development Policy Practices in Underdeveloped Ethnic Areas

The current research also examines the implications of urban policy and design practices, improving the street space quality and assisting city managers in gaining a more comprehensive insight into urban streets. Through the combination of deep learning techniques and overlay analysis of visual elements and perceptions, a more efficient and accurate understanding of the urban landscape in underdeveloped ethnic regions can be attained. This method can support the construction and advancement of cities in these areas, providing scientific backing for city planners. Evaluating the spatial quality of streets in the central city of Lhasa by analyzing a high percentage of visual elements and spatial perception score not only reinforces street perception in underdeveloped ethnic regions but also lays the groundwork for future city development and enhancement of street user experiences. Additionally, the incorporation of deep learning methods in this study offers a way to assess urban street quality perceptions in underdeveloped ethnic areas, aligning with the real experiences of urban users in these regions. This promotes the improvement of street quality in underdeveloped ethnic regions while preserving the unique cultural characteristics of the Tibetan people.

5.3. Scientific Contribution of Research Methods

In recent years, there has been a surge in interest in analyzing street space perception using street view data within the urban domain. While many studies have employed various methods to gather perception data and assess urban street space quality, the focus has mainly been on developed cities, overlooking underdeveloped ethnic neighborhoods. This research presents a significant scientific advancement by introducing a comprehensive technique to evaluate the perception of urban street space quality in these marginalized ethnic areas through the utilization of deep learning. By integrating street visual components with six perception variables, this approach pinpoints crucial aspects of urban street perception in these regions, offering important benchmarks for evaluating spatial quality and pushing forward research in this area. The developed analytical framework enables the measurement of urban street usage frequency and high street spatial perception in underdeveloped ethnic neighborhoods while also investigating the influence of visual elements on urban street perception. The results of this investigation hold promise for stakeholders in underdeveloped ethnic regions, providing valuable insights for sustainable street planning and urban infrastructure development and introducing a new angle for spatial analysis of urban streets in these locations.

5.4. Research Limitations

This study delves into the perception of urban streets in underdeveloped ethnic areas, highlighting certain issues and gaps that warrant further exploration in future research. By overlaying visual elements and six perceptions, the study aims to pinpoint specific areas within urban street space in underdeveloped ethnic areas that require attention. However, the reliance on a singular type of research data poses challenges in gaining comprehensive insights into the characteristics of the built environment and the evolution of urban street quality in underdeveloped ethnic areas. This limitation hinders the provision of a multi-level explanation for the origins of urban perception outcomes in such areas. To address these limitations, it is crucial to encompass a broader range of underdeveloped ethnic regions, such as Shannan and Linzhi City, in order to evaluate the primary features of visual elements influencing urban street perception in these areas. Furthermore, transcending the constraints of the time dimension by utilizing historical images of BSVIs and Google Street View images can offer valuable insights into the historical progression of urban street quality in underdeveloped ethnic areas. This approach can aid researchers in urban science and anthropology in gaining a deeper understanding of how the built environment influences human perception across different temporal contexts.

6. Conclusions

This research utilizes deep learning methods to evaluate the perception of urban street quality in economically disadvantaged ethnic regions, with a particular emphasis on Lhasa's city center. In Lhasa's city center, the urban area demonstrates a high percentage of visual elements, especially in terms of buildings and vegetation. Specifically, buildings accounted for 88.23% and vegetation for 89.52%, whereas poles represented the lowest percentage at 3.14%. Other visual elements, including roads (31.10%), sky (56.27%), terrain (16.27%), and walls (19.49%), showed similar spatial characteristics, largely concentrated along the main road. Out of the six perceptions evaluated, boring (69.70) and depressing (67.76) perceptions registered the highest scores, followed by beautiful (60.66) and wealthy perceptions (59.91). Lively (56.68) and safe perceptions (50.64) received the lowest scores.

In the analysis of the correlation between visual components and perceptions in architectural design, the results indicated that architecture positively impacted the beautiful perception (0.030) while negatively affecting the wealthy perception (−0.030). In terms of unfavorable spatial impressions, negative effects on boring perception were noted for roads (−0.094), sidewalks (−0.031), fences (−0.036), terrain (−0.020), sky (−0.098), cars (−0.016), and poles (−0.075). Conversely, other visual elements demonstrated a favorable impact on the boring perception. Moreover, roads (0.066), sidewalks (0.008), sky (0.078), cars (0.049), and poles (0.028) were associated with a positive effect on depressing perception. Conversely, buildings (−0.009), walls (−0.072), fences (−0.002), vegetation (−0.067), and terrain (−0.018) exhibited a negative impact on the depressing perception.

The high levels of boring perception are chiefly influenced by the quality of the street environment. Visual aspects such as cars (20.74%), sidewalks (11.18%), sky (57.02%), and fences (24.04%) are closely associated with main thoroughfares. Suburban areas significantly impact the boring perception of roads (30.50%), walls (19.38%), and poles (3.19%). Vegetation (89.51%) is linked with historic towns, whereas terrain (16.29%) is more subject to viaducts. Interestingly, although vegetation is generally a positive visual element, it can also negatively affect street perception. Wealthy perception (59.91%) is predominantly observed in high-rise neighborhoods, while lively perception (58.46%) is more concentrated in urban commercial districts. Beautiful perception (61.13%) is prevalent in urban park areas, safe perception (55.55%) is focused on secondary streets in older urban districts, and boring perception (69.68%) and depressing perception (67.26%) are mainly encountered on main roads in central city areas.

Author Contributions: C.L. and Y.Y. participated in all phases; X.Y. helped with data collection and paper organization. All authors have read and agreed to the published version of the manuscript.

Funding: This study was supported by the National Natural Science Foundation of China (Grant No. 52178059) for the project titled 'Research on the evolution mechanism and planning methods of the spatial structure of Tibetan towns along the Sichuan-Tibet Railway'.

Institutional Review Board Statement: Not applicable.

Informed Consent Statement: Informed consent was obtained from all subjects involved in the study.

Data Availability Statement: The data presented in this study are available on request from the corresponding author on reasonable request. The data are not publicly available due to privacy policies.

Conflicts of Interest: The authors declare no potential conflicts of interest with respect to the authorship and/or publication of this article.

Nomenclature

FCN
ResNet
SegNet
AlphaGo
Cityscapes Dataset
Green View Index (GVI)
MIT Place Pulse program
Google Street View (GSV)
Street View Imagery (SVI)
Urban Renewal Projects (URPs)
Baidu Street View images (BSVIs)
DeeplabV3+ semantic segmentation
Polarimetric Synthetic Aperture Radar (PolSAR)

References

1. Cao, S.; Lv, Y.; Zheng, H.; Wang, X. Challenges facing China's unbalanced urbanization strategy. *Land Use Policy* **2014**, *39*, 412–415. [CrossRef]
2. Drobnjaković, M.; Panić, M.; Đorđević, J. Traditional undeveloped municipalities in Serbia as a result of regional inequality. *Eur. Plan. Stud.* **2016**, *24*, 926–949. [CrossRef]
3. Hutárová, D.; Kozelová, I.; Špulerová, J. Tourism development options in marginal and less-favored regions: A case study of Slovakia's Gemer Region. *Land* **2021**, *10*, 229. [CrossRef]
4. Fang, C. A theoretical analysis of the mechanism and evolutionary law of urban-rural integration development. *Acta Geogr. Sin.* **2022**, *77*, 759–776.
5. Huang, Z.; Lu, L.; Su, Q.; Zhang, J.; Sun, J.; Wan, X.; Jin, C. Research and development of rural tourism under the background of new urbanization: Theoretical reflection and breakthrough of predicament. *Geogr. Res.* **2015**, *34*, 1409–1421.
6. Pang, X.; Wang, R.; Wang, W. Research on coupling of tourism and urbanization in underdevelopment regions: A case study of Fusong county in Jilin Province. *Geogr. Geo-Inf. Sci.* **2014**, *30*, 130–134.
7. Matlovičová, K.; Kolesárová, J.; Demková, M.; Kostilníková, K.; Mocák, P.; Pachura, P.; Payne, M. Stimulating poverty alleviation by developing tourism in marginalised Roma communities: A case study of the Central Spis Region (Slovakia). *Land* **2022**, *11*, 1689. [CrossRef]
8. Zhou, R.; Yu, Y.; Wu, B.; Luo, X. Quantitative evaluation of urban resilience in underdeveloped regions: A study of six cities in Sichuan & Tibet, China. *Front. Environ. Sci.* **2022**, *11*, 1266487.
9. Zhang, P.; Zhang, L.; Han, D.; Wang, T.; Zhu, H.; Chen, Y. Coupled and Coordinated Development of the Tourism Industry and Urbanization in Marginal andunderdeveloped Regions—Taking the Mountainous Border Areas of Western Yunnan as a Case Study. *Land* **2023**, *12*, 640. [CrossRef]
10. Yu, Y.; Zhou, R.; Qian, L.; Yang, X.; Dong, L.; Zhang, G. Supply-demand balance and spatial distribution optimization of primary care facilities in highland cities from a resilience perspective: A study of Lhasa, China. *Front. Public Health* **2023**, *11*, 1131895. [CrossRef]
11. Abusaada, H.; Elshater, A. Effect of people on placemaking and affective atmospheres in city streets. *Ain Shams Eng. J.* **2021**, *12*, 3389–3403. [CrossRef]
12. Tang, J.; Long, Y. Measuring visual quality of street space and its temporal variation: Methodology and its application in the Hutong area in Beijing. *Landsc. Urban Plan.* **2019**, *191*, 103436. [CrossRef]
13. Mcginn, A.; Evenson, K.; Herring, A.; Huston, S.; Rodriguez, D. Exploring associations between physical activity and perceived and objective measures of the built environment. *J. Urban Health* **2007**, *84*, 162–184. [CrossRef] [PubMed]
14. Li, X.; Zhang, C.; Li, W.; Ricard, R.; Meng, Q.; Zhang, W. Assessing street-level urban vegetation using Google Street View and a modified green view index. *Urban For. Urban Green.* **2015**, *14*, 675–685. [CrossRef]
15. Zhou, H.; He, S.; Cai, Y.; Wang, M.; Su, S. Social inequalities in street visual walkability: Using street view imagery and deep learning technologies to facilitate healthy city planning. *Sustain. Cities Soc.* **2019**, *50*, 101605. [CrossRef]
16. Wang, R.; Lu, Y.; Zhang, J.; Liu, P.; Yao, Y.; Liu, Y. The relationship between visual enclosure for neighbourhood street walkability and elders' mental health in China: Using street view images. *J. Transp. Health* **2019**, *13*, 90–102. [CrossRef]
17. Ma, X.; Ma, C.; Wu, C.; Xi, Y.; Yang, R.; Peng, N.; Zhang, C.; Ren, F. Measuring human perceptions of street views to better inform urban renewal: A perspective of scene semantic parsing. *Cities* **2021**, *110*, 103086. [CrossRef]
18. Gonzalez, D.; Rueda-Plata, D.; Acevedo, A.; Duque, J.; Ramos-Poll'an, R.; Betancourt, A.; García, S. Automatic detection of building typology using deep learning methods on street level images. *Build. Environ.* **2020**, *177*, 106805. [CrossRef]
19. Nagata, S.; Nakaya, T.; Hanibuchi, T.; Amagasa, S.; Kikuchi, H.; Inoue, S. Objective scoring ofstreet view walkability related to leisure walking: Statistical modeling approach with semantic segmentation of Google Street View images. *Health Place* **2020**, *66*, 102428. [CrossRef]
20. Hu, C.; Zhang, F.; Gong, F.; Ratti, C.; Li, X. Classification and mapping of urban canyon geometry using Google Street View images and deep multitask learning. *Build. Environ.* **2020**, *167*, 106424. [CrossRef]
21. Li, H.; Páez, A.; Liu, D. Built environment and violent crime: An environmental audit approach using Google Street View. *Comput. Environ. Urban Syst.* **2017**, *66*, 83–95.
22. Zhang, F.; Zhou, B.L.; Liu, L.; Liu, Y.; Fung, H.; Lin, H.; Ratti, C. Measuring human perceptions of a large-scale urban region using machine learning. *Landsc. Urban Plan.* **2018**, *180*, 148–160. [CrossRef]
23. Zhao, C.; Ogawa, Y.; Chen, S.L.; Oki, T.; Sekimoto, Y. Quantitative land price analysis via computer vision from street view images. *Eng. Appl. Artif. Intell.* **2023**, *123*, 106294. [CrossRef]
24. Rui, J. Measuringstreet view perceptions from driveways and sidewalks to inform pedestrian-oriented street renewal in Dusseldorf. *Cities* **2023**, *141*, 104472. [CrossRef]
25. Wang, L.; Han, X.; He, J.; Taeyeol, J. Measuring residents' perceptions of city streets to inform better street planning through deep learning and space syntax. *ISPRS J. Photogramm. Remote. Sens.* **2022**, *190*, 215–230. [CrossRef]
26. Liang, Y.; Zeng, J.; Li, S. Examining the Spatial Variations of Land Use Change and Its Impact Factors in a Coastal Area in Vietnam. *Land* **2022**, *11*, 1751. [CrossRef]
27. Liang, Y.; Zeng, J.; Sun, W.; Zhou, K.; Zhou, Z. Expansion of construction land along the motorway in rapidly developing areas in Cambodia. *Land Use Policy* **2021**, *109*, 105691. [CrossRef]

28. Hossam, H.; Johanna, L. The death and life of Malmi neighbourhood shopping street: Is ethnic retail a catalyst for public life recovery in Helsinki? *Eur. Plan. Stud.* **2021**, *30*, 336–358.
29. Abozied, E.; Vialard, A. Reintegrating informal settlements into the Greater Cairo Region of Egypt through the regional highway network. *Reg. Stud. Reg. Sci.* **2020**, *7*, 333–345. [CrossRef]
30. Zhou, B.; Liu, L.; Oliva, A.; Torralba, A.; Fleet, D.; Pajdla, T.; Schiele, B.; Tuytelaars, T. Recognizing city identity via attribute analysis of geo-tagged images. In Proceedings of the 13th European Conference on Computer Vision (ECCV), Zurich, Switzerland, 6–12 September 2014; pp. 519–534.
31. Xu, G.; Zhu, X.; Fu, D.J.; Dong, J.W.; Xiao, X.M. Automatic land cover classification of geo-tagged field photos by deep learning. *Environ. Model. Softw.* **2017**, *91*, 127–134. [CrossRef]
32. Hu, Y.J.; Gao, S.; Janowicz, K.; Yu, B.L.; Li, W.W.; Prasad, S. Extracting and understanding urban areas of interest using geotagged photos. *Comput. Environ. Urban Syst.* **2015**, *54*, 240–254. [CrossRef]
33. Ulrich, R.S. Visual landscapes and psychological well-being. *Landsc. Res.* **1979**, *4*, 17–23. [CrossRef]
34. Anguelov, D.; Dulong, C.; Filip, D.; Frueh, C.; Lafon, S.; Lyon, R.; Ogale, A.; Vincent, L.; Weaver, J. Google street view: Capturing the world at street level. *Computer* **2010**, *43*, 32–38. [CrossRef]
35. Cheng, L.; Chu, S.S.; Zong, W.W.; Li, S.Y.; Li, M.C. Use of tencent street view imagery for visual perception of streets. *ISPRS Int. J. Geo-Inf.* **2017**, *6*, 265. [CrossRef]
36. Gebru, T.; Krause, J.; Wang, Y.L.; Chen, D.Y.; Deng, J.; Aiden, E.L.; Li, F.F. Using deep learning and Google Street View to estimate the demographic makeup of neighborhoods across the United States. *Proc. Natl. Acad. Sci. USA* **2017**, *114*, 13108–13113. [CrossRef]
37. Zhang, F.; Hu, M.Y.; Lin, H.; Fang, C.Y. Framework for virtual cognitive experiment in virtual geographic environments. *SPRS Int. J. Geo-Inf.* **2018**, *7*, 36. [CrossRef]
38. Salesses, P.; Schechtner, K.; Hidalgo, C.A. The collaborative image of the city: Mapping the inequality of urban perception. *PLoS ONE* **2013**, *8*, e68400. [CrossRef]
39. Dubey, A.; Naik, N.; Parikh, D.; Raskar, R.; Hidalgo, C.A. Deep learning the city: Quantifying urban perception at a global scale. In Proceedings of the 14th European Conference on Computer Vision (ECCV), Amsterdam, The Netherlands, 11–14 October 2016; pp. 196–212.
40. Ordonez, V.; Berg, T.L. Learning high-level judgments of urban perception. In Proceedings of the 13th European Conference on Computer Vision (ECCV), Zurich, Switzerland, 6–12 September 2014; pp. 494–510.
41. Porzi, L.; Bulò, S.R.; Lepri, B.; Ricci, E. Predicting and understanding urban perception with convolutional neural networks. In Proceedings of the ACM International Conference on Multimedia (ACM Multimedia), Brisbane, Australia, 26–30 October 2015; pp. 139–148.
42. Naik, N.; Kominers, S.D.; Raskar, R.; Glaeser, E.L.; Hidalgo, C.A. Computer vision uncovers predictors of physical urban change. *Proc. Natl. Acad. Sci. USA* **2017**, *114*, 7571–7576. [CrossRef]
43. Yin, L. Street level urban design qualities for walkability: Combining 2D and 3D GIS measures. *Comput. Environ. Urban Syst.* **2017**, *64*, 288–296. [CrossRef]
44. Low, S.; Altman, I. (Eds.) Place attachment: A conceptual inquiry. In *Human Behavior & Environment: Advances in Theory & Research*; Plenum Press: New York, NY, USA, 1992; Volume 12, pp. 1–12.
45. Hidalgo, M.; Hernandez, B. Place attachment: Conceptual and empirical questions. *J. Environ. Psychol.* **2001**, *21*, 273–281. [CrossRef]
46. Andrew, L.; Gu, X.; Chen, L.; Perry, H. Predicting perceptions of the built environment using GIS, satellite and street view image approaches. *Landsc. Urban Plan.* **2021**, *216*, 104257.
47. Liu, J.; Li, T.; Xie, P.; Du, S.; Teng, F.; Yang, X. Urban big data fusion based on deep learning: An overview. *Inf. Fusion* **2020**, *53*, 123–133. [CrossRef]
48. He, T.; Li, X. Image quality recognition technology based on deep learning. *J. Vis. Commun. Image Represent.* **2019**, *65*, 102654. [CrossRef]
49. He, N.; Li, G. Urban neighbourhood environment assessment based on street view image processing: A review of research trends. *Environ. Chall.* **2021**, *4*, 50090. [CrossRef]
50. Wang, R.; Liu, Y.; Lu, Y.; Zhang, J.; Liu, P.; Yao, Y.; Grekousis, G. Perceptions of built environment and health outcomes for older Chinese in Beijing: A big data approach with street view images and deep learning technique. *Comput. Environ. Urban Syst.* **2019**, *78*, 101386. [CrossRef]
51. Wu, J.; Feng, Z.; Peng, Y.; Liu, Q.; He, Q. Neglected green street landscapes: A re-evaluation method of green justice. *Urban For. Urban Green.* **2019**, *41*, 344–353. [CrossRef]
52. Donghwan, K.; Sugie, L. Analyzing the effects of Green View Index of street streets on walking time using Google Street View and deep learning. *Landsc. Urban Plan.* **2021**, *205*, 103920.
53. Wang, M.; He, Y.; Meng, H.; Zhang, Y.; Zhu, B.; Joseph, M.; Li, X. Assessing Street Space Quality Using Street View Imagery and Function-Driven Method: The Case of Xiamen, China. *ISPRS Int. J. Geo-Inf.* **2022**, *11*, 282. [CrossRef]
54. William, T.; Matthew, N.; Chyi-Lin, L.; Christopher, P. Implementing a deep-learning model using Google street view to combine social and physical indicators of gentrification. *Comput. Environ. Urban Syst.* **2023**, *102*, 101970.

55. Yu, L.; Zeng, Z.; Liu, A.; Xie, X.; Wang, H.; Xu, F.; Hong, W. A Lightweight Complex-Valued DeepLabv3+ for Semantic Segmentation of PolSAR Image. *IEEE J. Sel. Top. Appl. Earth Obs. Remote Sens.* **2022**, *15*, 930–943. [CrossRef]
56. Kim, J.; Lee, S.; Hipp, J.; Ki, D. Decoding urban landscapes: Google street view and measurement sensitivity. *Comput. Environ. Urban Syst.* **2021**, *88*, 101626. [CrossRef]
57. Rzotkiewicz, A.; Pearson, A.L.; Dougherty, B.V.; Shortridge, A.; Wilson, N. Systematic review of the use of Google Street View in health research: Major themes, strengths, weaknesses and possibilities for future research. *Health Place* **2018**, *52*, 240–246. [CrossRef]
58. Yao, Y.; Liang, Z.; Yuan, Z.; Liu, P.; Bie, Y.; Zhang, J.; Wang, R.; Wang, J.; Guan, Q. A human-machine confrontational scoring framework for urban perception assessment using street-view images. *Int. J. Geogr. Inf. Sci.* **2019**, *33*, 2363–2384. [CrossRef]
59. Li, Y.; Miller, H.J.; Root, E.D.; Hyder, A.; Liu, D.S. Understanding the Role of Urban Social and Physical Environment inOpioid Overdose Events Using Found Geospatial Data. *Health Place* **2022**, *75*, 102792. [CrossRef]
60. Zhang, G.; Zu, J.; Hu, M.; Zhu, D.; Kang, Y.; Gao, S.; Zhang, Y.; Huang, Z. Uncovering Inconspicuous Places Using Social Media Check-Ins and Street View Images. *Comput. Environ. Urban Syst.* **2020**, *81*, 101478. [CrossRef]
61. Fernández-Delgado, M.; Cernadas, E.; Barro, S.; Amorim, D. Do we need hundreds of classifiers to solve real world classification problems. *J. Mach. Learn. Res.* **2014**, *15*, 3133–3181.

MDPI AG
Grosspeteranlage 5
4052 Basel
Switzerland
Tel.: +41 61 683 77 34

Buildings Editorial Office
E-mail: buildings@mdpi.com
www.mdpi.com/journal/buildings

www.ingramcontent.com/pod-product-compliance
Lightning Source LLC
Chambersburg PA
CBHW061546220926
51850CB00005B/263
9783725854837